ESTATES GAZETTE LAW REPORTS

Estates Gazette Law Reports

1982

Volume 1

Series edited by
Barry Denyer-Green LLM PhD FRICS
of the Middle Temple, barrister

Cases edited by
J Muir Watt OBE MA
of the Inner Temple, barrister

Audrey Boyle BA
Publisher's editor

Estates Gazette

Routledge
Taylor & Francis Group
LONDON AND NEW YORK

EG Books is an imprint of Routledge
2 Park Square, Milton Park, Abingdon, Oxon OX14 4RN
711 Third Avenue, New York, NY 10017, USA
Routledge is an imprint of the Taylor & Francis Group, an informa business

ISBN for complete set of 2 volumes: 0 7282 0236 0
ISBN for this volume 0 7282 0237 9
ISSN 0951-9289

For information on all EG Books publications
visit our website www.routledge.com

CONTENTS

TABLE OF CASES

Titles of cases shown in bold type
Names of parties reversed shown in ordinary type

INDEX OF SUBJECT-MATTER

LANDLORD AND TENANT — *continued*
Lease or licence — *continued*

Business tenancy
Rent review clause requiring improvements to be taken into account if tenant at time of carrying them out was licensee — Whether, at any time, occupation under agreement to grant a lease that of licensee or tenant — Pre-condition that works were required to be carried out by tenant if plans were certified by landlord — Held, no equitable tenancy — While pre-condition remained unperformed principle in *Walsh* v *Lonsdale* inapplicable — On true construction agreement created licence.
***Euston Centre Properties Ltd* v *H&J Wilson Ltd (QB)*, 57**

Leasehold enfranchisement
Price for freehold
House with higher rateable value under Leasehold Reform Act 1967, s9(1A) — Certificate for fair rent of £4,250 pa — Agreed unencumbered freehold value £187,500 — Held, marriage value would be shared — Support from settlements — Lessor's valuation of £113,000 accepted.
***Lloyd-Jones* v *Church Commissioners for England (LT)*, 209**

Possession
Evidence
Action for possession for breach of covenant not to, *inter alia*, sublet or part with possession or occupation or share occupation — Landlord's case depended on inference of occupation by third party drawn from letter from electoral registration officer — No evidence called to rebut this — Held, even if entry in register evidence of occupation, that evidence on its own was insufficient — No notice to admit served on tenant nor was relevant Form A produced and proved.
***Metropolitan Properties Co Ltd* v *Griffiths (CA)*, 101**

Rent — See also **Landlord and tenant** (Landlord and Tenant Act Part II — New tenancy application) (Landlord and Tenant Act Part II — New tenancy grant)
Payment in advance
Application to set aside judgment for possession, rent arrears and mesne profits — Tenants claim that credit ought to have been given for payment on account of arrears that had been overlooked in the compromise of previous action — Held, judgment not irregular — No implied condition that, when money is paid in advance of date when rent is due, the landlord will apply it to discharge any future rent that might fall due, and not just the rent in respect of which it was paid.
***Morris Gore Ltd* v *Westbourne Hill Properties Ltd (CA)*, 55**

Rent Acts
Matrimonial home
Tenancy vested in husband — Leaves matrimonial home, no intention of returning — Renews lease after departure — Wife and son remain in occupation — Decree absolute — Whether ex-wife or son could assert statutory tenancy through ex-husband — Whether ex-wife "real" tenant and husband nominal — Held, wife's vicarious protection ended with decree absolute — Ex-husband actual tenant as tenancy agreement not sham — No "abandoned child" exception to rule in *Skinner* v *Geary*.
***Metropolitan Properties Co Ltd* v *Cronan (CA)*, 104**

Possession
Case 9 in Sched 15 to Rent Act 1977 — Claim by administrators of estate, with no beneficial interest in property, to provide home to deceased's children whom they had adopted — Whether landlords for purposes of Case 9 — Held, no rule that personal representatives could not be a landlord, irrespective of whether had beneficial interest — Normally cannot seek possession as would be in breach of trust — Order for possession.
***Patel* v *Patel (CA)*, 98**

Security of tenure — One of two extensions occupied by tenant — Whether tenant's dwelling-house formed "part only of a building" under s12(1)(*a*) Rent Act 1977 — Held, question of fact — Borderline case but whole property could reasonably be regarded as one building, even though no internal connections linking main body to extensions — No protected tenancy.
***Griffiths* v *English (CA)*, 96**

Rent assessment
Main complaint of landlord that he did not receive notice of hearing — In applying for rent assessment, which he did in respect of six flats, landlord had used two different addresses — That chosen for posting notifications of referral had resulted in landlord acknowledging notifications — At previous oral hearing landlord had given committee another address — Held, by sending notice of hearing to same address as that use for notifications, reg 11 of Rent Assessment Committees (England and Wales) Regulations 1971 had been complied with.
***Dundon* v *Kenna (QB)*, 100**

Rent assessment committee
Refusal to grant long adjournment so that tenants could be represented by counsel of their choice — Tenants believed application would be formality but warned they should come fully prepared — Whether breach of natural justice — Held, correct approach was whether tenants had been denied, through no fault of their own, reasonable opportunity to present their case and tribunal knew refusal would cause such injustice — Committee entitled to opinion it was tenants' own fault they came unprepared.
***R* v *Rent Assessment Committee for London, ex parte Ellis-Rees (QB)*, 107**
Revision of fair rent approximately four years after previous registration — Complaint by tenant, by reference to s67(3) Rent Act 1977, that rent assessment committee ought only to have fixed rent by applying cumulative rate of inflation since date of application to register — Held, s67(3) irrelevant — Applicable only to landlord's application for redetermination less than three years after earlier determination — Rate of inflation an improper multiplier.
***Kovats* v *Corporation of Trinity House (QB)*, 103**

Rent review clause
Arbitration — See also **Agricultural holdings** (Arbitration)
Application for leave to appeal against award under Arbitration Act 1979 — Allegations that in calculating award arbitrator had regard to letting prices and trading results occurring after review date — Held, on facts only second allegation made out — This was marginal point — Applying s1(4) of 1979 Act application dismissed.
***Duvan Estates Ltd* v *Rossette Sunshine Savouries Ltd (QB)*, 20**

Fixtures
Lease surrendered by operation of law immediately followed by new lease with rent review clause — Rent to be determined at "open market value" — Whether value of fixtures to be taken into account when determining increased rent — Dependent upon whether tenants' right to remove their fixtures existed under new lease — Held, in absence of contrary agreement, right to remove tenants' fixtures continued provided tenant remained in possession — No such agreement — Premises to be valued as if without fixtures.
***New Zealand Government Property Corporation* v *HM&S Ltd (CA)*, 52**

Notice and counternotice
Underlease providing for two-year period in which landlords' notice to be served and that in absence of such notice being served, landlords were entitled, at any time thereafter, to a rent review, on one months' notice — Negotiations outside machinery of rent review clause result in agreement that no rent increase justified — Agreement not formally endorsed — Whether landlords entitled to serve type of notice — Held, negotiations constituted effective rent review, the formal requirements of which had been waived — Estoppel *in pais* argument rejected.
***Esso Petroleum Co Ltd* v *Anthony Gibbs Financial Services Ltd (Ch)*, 124**

Time-limits
Clause providing strict time-limits within which landlord and tenant were to agree rent or, failing that, surveyor to make determination — Thereafter, surveyor to be appointed by president of RICS, but no time-limit specified — Whether third stage could be initiated without any attempts to agree rent or surveyor within times specified — Held, no conditions precedent to be implied for invocation of rent review machinery that notice to be served and to be served before date specified for agreement — Landlords could properly seek determination of rent by arbitration.
***Laing Investment Co Ltd* v *GA Dunn & Co (Ch)*, 129**

Clause specifying notice could be served to set in motion rent review at any time not earlier than specified date — Notice served some two and a quarter years after start of review period — Whether unreasonable delay — Whether implied term that notice to be served by beginning of review period — Held, such term inconsistent with express terms of clause and could not be justified on basis of necessity — *Scalter* v *Horton* distinguished — Not meaningful to refer to interval of time as "delay", given that notice could be served "at any time".
***London & Manchester Assurance Co Ltd* v *GA Dunn & Co (Ch)*, 117**

Declaration in previous proceedings that time not of the essence — Landlords seek second declaration that entitled to increased rent from date specified in lease — Whether trigger notice invalid because of unreasonable delay or delay to be treated as evidence of abandonment — Held, as no evidence of prejudice or hardship notice to be treated as valid in equity — No abandonment — Part of delay could be discounted because of uncertainty of law and remainder needed for advice and consultations.
***Amherst* v *James Walker (Goldsmith & Silversmith) Ltd (Ch)*, 121**

Landlords' notice served 18 months after time specified in clause — Arbitrator appointed and award made — Whether time of the essence — Held, time not of the essence — Immaterial that initiative to review rent solely landlords — Disadvantage in tenant being required to produce large amount of arrears illusory — Delay not unreasonable and doubted whether delay *per se* can prevent landlord from relying on clause.
***H West & Son Ltd* v *Brech (QB)*, 113**

Rent review clause linked to lessees' option to determine — Landlords serve notice over three years after time specified in clause — Whether time of the essence — Held, it was — Presumption that time of the essence when rent review clause linked with break clause not successfully rebutted — Treatment of *Samuel Properties (Developments) Ltd* v

AGRICULTURE

A **COURT OF APPEAL**

November 5 1981

(Before Lord Justice CUMMING-BRUCE, Lord Justice TEMPLEMAN and Dame Elizabeth LANE)

E D & A D COOKE-BOURNE (FARMS) LTD v MELLOWS AND ANOTHER

Estates Gazette April 17 1982
(1981) 262 EG 229

Agricultural Holdings Act 1948 — Sixth Schedule, para 6 —
B **Whether arbitrator should have given consent to amendment or addition to statement of case — Following a "Tomlin Order" in the Chancery Division, the tenancy of an agricultural holding was surrendered and the tenants served notice of intention to claim on the landlords — An arbitrator was appointed and statements of case delivered to him — Tenants' particulars did not make clear whether claims were under section 13 in respect of fixtures or under section 47 in respect of improvements — Landlords' statement contended that a claim under section 13 was not competent as the conditions required by section 13(2) were not satisfied and that any alternative claim for improvements was defeated by lack of evidence of landlords' consent — Tenants then submitted to arbitrator a "supple-**
C **mental statement of case", accepting that a claim in respect of fixtures was not competent, but claiming in respect of improvements — Question then raised as to whether arbitrator should give consent to the addition — Landlords opposed consent and arbitrator refused it — County court judge on case stated upheld arbitrator's refusal — Different view taken by Court of Appeal — Interesting contrast drawn between section 16(2) of Agricultural Holdings Act 1923, with its "Draconian" one-stage procedure containing no provision for amendments or additions if claim did not give full particulars, and the 1948 Act two-stage procedure with preliminary notice of claim followed by statement of case and particulars, with provision for amendments or additions if arbitrator consented — Analysis of the well-known 1923 Act decisions in *Re O'Connor and Brewin's Arbitra-*
D ***tion* and *Spreckley v Leicestershire County Council* — These old cases still provide guidance if it is applied not merely to sufficiency of original statement of case but to the case as amended or as subject to additions — It is not fatal if a party does not get everything right first time — Advice to arbitrators as to considerations which should influence them in deciding whether to consent to amendments or additions — *Church Commissioners for England v Matthews* misunderstood — No doctrine preventing fundamental change in amendments or additions — "As much relevance to this case as the principle of 'No Popery' " — Consent should have been given to supplemental statement of case — Appeal allowed**

E This was an appeal from the decision of Judge Kellock in Spalding County Court on a case stated by an arbitrator raising the question whether he should give consent to the delivery by E D & A D Cooke-Bourne (Farms) Ltd, tenants of an agricultural holding, High Park Farm and Low Park Farm, Aslackby, in the county of Lincoln, of an amendment or addition to their statement of case, taking the form of a 'supplemental statement of case", under paragraph 6(a) of Schedule 6 to the Act of 1948. The landlords, A R Mellows and H C O'Neill, executors of the late E D Cooke, objected. The judge upheld the view taken by the arbitrator that consent to the supplemental statement of case should be refused. F
The tenants appealed.

J Knox QC and V Chapman (instructed by Andrews, Stanton & Ringrose, of Bourne, Lincolnshire) appeared on behalf of the appellants; Anthony Cripps QC and P M Darby (instructed by Turner, Peacock) represented the respondents.

Giving judgment, CUMMING-BRUCE LJ said: On October 8 1975 the landlord, described as the executors of the late Mr E D Cooke, who had died in 1960, entered into a tenancy agreement with the tenant, described as E D and A D Cooke-Bourne (Farms) Ltd, whereby Bourne Farms Ltd became, under the agreement, tenants of an agricultural holding, described as High Park and Low Park, to hold from October 11 1975 on a yearly tenancy determinable on G
October 11 in any given year for a rent of £6,000. There were elaborate but usual covenants by the landlord and the tenant and the obligations of the landlords included obligations for repair, and I refer in particular to clause 3 of the Second Schedule to the agreement and clauses 4 and 5 of those agreements. By clause 6 of the agreement the landlord was under no obligation to "execute repairs or replacements or to insure buildings or fixtures which are the property of the tenant". There was a special provision inserted which read as follows:

> It is hereby agreed between the parties that the following buildings erected on High Park Farm shall be regarded as tenants fixtures: The ten bay Dutch H
> barn, workshops and glasshouses and the grain and potato store with two fan tunnel equipment. Also brick and asbestos open shed and workshop at Brickpits off West Road, Bourne.

Whatever effects that special provision may or may not have had, it clearly had a relevance to the obligations of the landlord to repair, since these buildings are to be regarded as tenant's fixtures. It may well be that, in arriving at the rent which the tenant was to pay during the period that he enjoyed the occupation of the land with its buildings and fixtures, that rent may have been agreed between the parties by reference to the provision in the special provision.

By the same agreement, when one comes to the rights of the parties upon the determination of the tenancy, the landlord J
accepted the obligation in clause 2.

> On termination of the tenancy to pay the tenant compensation:
> (a) for the unexhausted value of the improvements referred to in the Second and Third Schedules to the Act, subject to and in accordance with the provisions of the Act,
> (b) for fixtures paid for on entry at a fair valuation and in respect of other fixtures subject to and in accordance with the provisions of the Act.

The parties did address their minds to the question of compensation on the termination of the tenancy, and agreed that the landlord should be under an obligation to pay for the value of improvements referred to in the Second and Third Schedules to the Act in accordance with the provisions of the Act. K

This agreement succeeded an earlier oral tenancy. The background of the whole matter was that Mr Cooke senior, the late E D Cooke who died in 1960, by his will constituted trusts of the will and it emerged that there were disputes between the beneficiaries which led to an action the title of which appears in the title to the Tomlin Order at p 1 of the bundle. The action was compromised upon the terms of the Tomlin Order which appears exhibited in the bundle.

It is not necessary for me to refer at this juncture to other provisions of the agreement which took its final form in the Tomlin

Order, save to remark that under those provisions that come under the heading of "The Final Distribution" and provide for the final distribution from the trust fund as soon as practicable after the death of Mrs Cooke there are careful provisions providing for the change in the voting rights in the company "in order to procure Bourne Farms Ltd would surrender its tenancy of the land comprised in the trust fund without compensation being payable in respect thereof, save as provided in paragraph 12.2 hereof." Paragraph 12.2 of the Tomlin Order provided:

> Tenant right and compensation for tenants improvements and tenants fixtures shall be payable to Bourne Farms in accordance with the provisions of the Agricultural Holdings Act 1948 (or any re-enactment thereof) . . .

and further provisions for final distribution provided that

> Andrew shall have the option of purchasing the land comprised in the Trust Fund at the value determined in accordance with paragraph 2 hereof.

In paragraph 2 the provisions for valuation had been set out, and the subsequent provisions dealt with the exercise of the option. I think it is unnecessary to refer further to the Tomlin Order. That was made on November 18 1976, that is to say just over a year after the commencement of the tenancy under the subsisting tenancy agreement.

In September 1978 Mrs Cooke died, which brought into practical operation the provisions for the final distribution following the death of Mrs Cooke, and the tenancy in accordance with those provisions was duly terminated by surrender on March 28 1980.

On April 3 1980 agents for the tenants served a notice under the Agricultural Holdings Act upon the trustees of E D Cooke deceased and gave notice as follows:

> I HEREBY GIVE YOU NOTICE pursuant to sect 70 of the above Act of my intention to make against you certain claims arising out of the termination of the tenancy of the above holding or part thereof, the nature of which claims is set out in the Schedule hereto.
>
> THE SCHEDULE
>
> Nature of claim — Statutory provisions, Customs or term of Agreement under which the claim is made.

Those are the headings. Under the nature of claim the endorsement was "Tenantright Fixtures of the above Farms £60,000. Tenantright of the above farms £15,000". The statutory provisions: "Agricultural Holdings Act 1948 (or any re-enactment there of) As provided under clause 12.2 of the statement in the High Court of Justice Chancery Division", namely the Tomlin Order.

That notice was addressed to the trustees of the deceased. It was followed by a second notice dated May 22. That notice recited again that it was a notice pursuant to section 70 of the Agricultural Holdings Act 1948, and set out a different schedule. Under "Nature of claim", the following words: "Compensation for new Improvements, viz: the erection of a grain and potato store in the field Ordnance Survey Number 1 and the erection of a Dutch barn in the field Ordnance Survey Number 42 the sum of £75,000.00". Further, under "Nature of Claim": "Compensation for Tenant-Right matters, viz: seeds sown and cultivations fallows and acts of husbandry performed on the holding at the expense of the Tenant."

When one looks at what was endorsed under the heading "Statutory Provisions . . ." etc, it reads:

> Section 47 of the Above Act. Third Schedule or part I of the Fourth Schedule and Minute 12.2 of the Minute of Order of the High Court of Justice Chancery Division in the matter of the Estate of Everitt Delanoix Cooke deceased and in the Matter of the Trusts of the Will of Everitt Delanoix Cooke deceased dated November 29 1960 and in the Matter of the Judicial Trustees Act 1897 and in the Matter of the Trustee Act 1925 between Joy Delanoix Richardson and Robert Kevin Andrew Feltham (formerly Kevin Andrew Richardson) as Plaintiffs and Kate Elizabeth Cooke and Andrew Delanoix Cooke and Paul Delanoix Tointon as Defendants

Then:

> Section 47 of the above Act and Part II of the Fourth Schedule and the above mentioned Minute of Order

On July 2 the arbitrator was appointed, and we are told that the practice was followed of serving documents thereafter upon the arbitrator. Thus it came about that on July 14 in the same year a tenant's statement of case was served upon the arbitrator. That statement of case reads as follows:

> *Statement of Case for the Arbitrator*
> with particulars on behalf of the Tenant
>
> 1. The Tenant, a Limited Company, was formed on May 10 1951 and from that date assumed the Tenancy of the Holding on a Verbal agreement only.
> 2. On October 8 1975 a written Tenancy Agreement was drawn up and remained in force until it was terminated by the provisions of an Order of the High Court of Justice, Chancery Division (Group A) given by Mr Justice Templeman on November 18 1976. The date of termination was March 28 1980.
> 3. With one exception every provision of the aforementioned Order of the High Court has been discharged by mutual agreement between Landlord and Tenant.
> 4. Under Clause 12(2) of the High Court Order, the Tenant was to be paid for Tenantright and Compensation for Tenants Improvements and Tenants Fixtures in accordance with the provisions of the Agricultural Holdings Act 1948 (or any re-enactment thereof).
> 5. The Tenant claims Compensation in respect of Tenants Fixtures situate at High Park Farm, Aslackby as Scheduled in the Tenancy Agreement.
> 6. The Tenant claims the sum of £75,000 (SEVENTY FIVE THOUSAND POUNDS) being the value of the two buildings referred to in paragraph 5 above.
> 7. The Tenant will produce evidence in support of the sum claimed as compensation which fairly represents the value of the two buildings to an incoming tenant.
> 8. The Tenant reserves the right to be legally represented at the Arbitration Hearing if such representation should be considered necessary to support his case.
>
> *Particulars of Claim referred to above*
> Compensation for Tenants Fixtures viz:
> The 120-ft × 105-ft steel framed and asbestos-clad corn and Potato Store Shed, together with the concrete apron surrounds £60,000
> The 180-ft × 30-ft steel framed and corrugated-iron clad Dutch barn .... £15,000.

On July 28 the landlords served upon the arbitrator their statement of case. It was submitted on behalf of the landlords by their experienced agents, and I summarise it by saying that it began by reciting that the tenants had claimed under section 70 of the Agricultural Holdings Act under a formal claim dated April 3, which was attached to the landlords' statement of case. The landlords then pleaded their case in respect of tenants' fixtures, which would be a claim under section 13. Messrs Strutt & Parker recited the provisions of section 13 and accurately pointed out that the provisions of section 13(2) had not been followed, and therefore the section 13 claim could not get off the ground. They appreciated, and drew the attention of the arbitrator to, the fact that the section 13 claim, based as it had to be on a notice by the landlords to purchase the buildings, could not arise on the determination of this tenancy in the circumstances that had happened. The landlords' case continued by referring to an alternative method of considering claims for tenantright fixtures, namely to regard them as tenants' improvements. They did not in that passage of the landlords' statement of case refer to the notice of May 22. I know not the explanation and I do not think it matters. They did not refer to it, but they set out that in so far as there was an alternative claim for tenants' improvements the landlords claimed to defeat such claim because the tenants had not produced to them, the landlords' agents, "any evidence that a written landlords' consent exists in connection with the improvements for which compensation is claimed." They go on: "We thus submit that any claim made by the Tenants in this matter, either for Tenantright Fixtures or alternatively as Tenants' Improvements is invalid: and that under the legal requirements of the Agricultural Holdings Act 1948 the Landlords are entitled to dismiss any claim for compensation."

Then they went on in their statement of case to "Assume an unconditional landlords' consent, compensation for New Improvements must be assessed in accordance with section 48 of the Act". They recite part of that section, and then they plead a case on the merits, leading to the conclusion that if there were a legal enforceable claim under section 47 the amount of compensation would not exceed £35,000 as compared to the claim for £75,000

A which had been particularised in the tenants' statement of case. At that stage those were the statements of case which the respective parties served upon the arbitrator.

On October 14 the tenants' agents submitted to the arbitrator the document entitled "Supplemental Statement of Case" which read as follows:

> 9. The tenant has read the Landlord's statement of case dated July 28 1980.
> 10. The tenant accepts that no compensation is due under s 13 of the Agricultural Holdings Act 1948.
> 11. The tenant's case is that compensation is due under s 47 of the Agricultural Holdings Act 1948.
> 12. The landlord's written consent to the erection of the Corn and Potato
> B Store Shed was given on March 2 1965.
> 13. The tenant will ask the arbitrator to infer that the landlord's written consent to the erection of the Dutch barn was also given in that the said barn was erected with the assistance of a Government grant, a precondition of which grant is production of evidence of landlord's written consent to erection.

On November 23, we are told, the tenants wrote to the landlords giving them notice of their application for leave to amend. The arbitrator fixed a preliminary meeting to deal with the tenants' application for leave to amend, and fixed that meeting for February 9 1981. When that hearing came on before the arbitrator the tenants sought leave, and, if they were given leave to file their
C supplemental statement of case, leave to file a further amendment pleading that the tenants had quitted the holding, the subject-matter of the claim for compensation. There was an amendment for which the landlords were seeking leave on the same basis, and it was agreed before the arbitrator and accepted by the arbitrator that if he gave leave to the tenants to file a supplemental statement of case the further amendment would follow and the landlords' amendment to their statement of case should also be allowed.

Those further amendments, therefore, gave rise to no issue, and the only issue before the arbitrator was whether he should give leave to file and serve the supplemental statement of case.

The arbitrator heard argument upon that issue. He understood
D counsel for the landlords to make the following submissions:

> 19. The Landlord appeared by counsel who submitted:
> 1. The Tenant's statement of case of July 14 1980 is the Statement of Claim and in paragraph 5 is restricted by reference to "Tenants fixtures" that is to say s 13 of the 1948 Act.
> 2. Tenant's counsel has already admitted s 13 is not a sustainable basis of claim in this case.
> 3. The claim for cultivations and acts of husbandry referred to in the Notice of Intention to Claim dated May 22 1980 was dropped before the claim was made and the claim, so far as the buildings are concerned, relates to tenants fixtures and hence a s 13 claim.
> 4. Once the buildings had been accepted by both parties as Tenant's
> E Fixtures, no claim for compensation could be sustainable in respect of them under any other category than as Tenant's Fixtures.
> 5. To consent to the Tenant's request would give rise to injustice to the Landlords as they would then have to come again for a two to three day and then a one to two day hearing on two further separate matters.
> 6. Whilst the sixth paragraph of the Sixth Schedule to the 1948 Act gives an arbitrator power to consent to an amendment it follows that the arbitrator also has the power not to consent, there is nothing automatic about it, the arbitrator should decide what is just.

The decision of the arbitrator, with his reasons, is set out in paragraphs 20, 21 and 22 of his statement of case, which read as follows:

> F
> 20. Although not referred to by Landlord's or Tenant's counsel my subsequent researches brought to light the case of *Church Commissioners for England v Matthews* (1979) 251 EG 1074 which seems to have followed previous precedents established in County Court judgments which prevents fundamental changes to Statements of Case under the 1948 Act.
> 21. Having considered in detail the arguments put forward at the preliminary hearing and having read the legal research referred to and having done my own research I conclude that following precedent I could not accept the fundamental amendment to the Tenant's Statement of Case and accordingly my Decision in writing dated February 20 1981 indicated that I would refuse my consent to the admission of the Tenant's supplemental statement of case.
> 22. At the preliminary hearing counsel for both Landlords and Tenant requested me to state in the form of a special case for the opinion of a G
> County Court the following question of law arising in the course of the arbitration.

On that reasoning the arbitrator refused leave to amend by filing the supplementary statement of case and the further amendment, therefore, fell with it.

The tenants asked for a case stated, for an opinion of the court. That case was stated, from which I have already quoted, and the matter came before the learned county court judge, who gave judgment and made an order on September 23 ordering that there be judgment in accordance with the copies delivered to the advocates, which does not reveal very much. That refers back, I suppose, to the note of judgment which the learned county court judge H
signed on September 18 1981. In effect, the learned judge upheld the decision of the arbitrator, on different grounds, and gave leave to appeal.

With regard to the decision of the learned judge, he took the view:

> In this case, subject to the fundamental point as to there being a valid Statement of Case to be amended, I have no doubt that the amendments or additions proposed could be effected without manifest or grave injustice. The Landlord has had it well in mind from the beginning that these claims might be being made.

But he decided the case had upheld the decision of the arbitrator on J
the ground that there was no valid statement of case. So there was nothing to amend. That is to say, on a scrutiny of the statement of case filed by the tenant, on the proper understanding of that statement of case there was no valid statement of case within the relevant legislation.

I now, therefore, turn, before examining the reasoning of the learned judge, to the scheme of the legislation, and I begin by referring to the 1923 Act, since it was in relation to the statutory scheme under that Act that two important decisions of this court were made. By the 1923 Act provision was made for compensation to the tenant for various kinds of loss, compensation for improvements on holdings, compensation for damage by gale, compensa- K
tion for disturbance. Section 16 provided the procedure for settling any question or difference arising out of any claim by the tenant of a holding against a landlord, or for any sums claimed to be due to the tenant from the landlord for breach of contract. The provision which is relevant for consideration is section 16(2), which reads:

> Any such claim as is mentioned in this section shall cease to be enforceable after the expiration of two months from the termination of the tenancy unless particulars thereof have been given by the landlord to the tenant or by the tenant to the landlord, as the case may be, before the expiration of that period.

There is a provision for extension in the proviso under a certain L
contingency.

When one considers section 16(2), it is apparent that there is a statutory and mandatory period within which the claimant is under an obligation to give particulars of the claim, and there is no provision in subsection (2) or elsewhere in that Act for any amendment or correction or addition to the particulars required by subsection (2). Therefore, if the tenant failed to give particulars such as were in contemplation by the subsection he was for ever precluded from proceeding with his claim. That is the subsection which was under consideration in this court in 1933 and 1934. In *Re O'Connor and Brewin's Arbitration* [1933] 1 KB 20 the matter for consideration was whether sufficient particulars had been delivered M
pursuant to section 16(2), and the court decided that on the facts in that case no sufficient particulars had been delivered and, therefore, the claim was barred. Giving his judgment in that case, Slessor LJ referred to a passage in the judgment of Warrington LJ in the case of *Jones v Evans* [1923] 1 KB 12, where the lord justice said:

> It is in general sufficient if the document or the conversation which contains the particulars gives an indication to the landlord or the tenant, as the case may be, of the particular kind of claim which is going to be made in order that he may have an opportunity of himself examining the subject matter and seeing what evidence he will have to adduce, or what information he will have to give the arbitrator.

A Slessor LJ approving that passage in the judgment considered the facts of the case before him, and decided that on those facts the landlord had not by the particulars given by the tenant been given the particulars to which he was entitled. Slessor LJ went on:

> Indeed, I find great difficulty in seeing that he has really been given any particulars at all. It is required under the Act that the landlord shall receive notice of a claim, and that notice of a claim is to be made before the tenant had ended the tenancy. What is called "particulars" in this case appears to be no more than a notice of a claim given after the tenancy has come to an end.

In the following year, in the case of *Spreckley v Leicestershire County Council* [1934] 1 KB 366, the court decided, reversing the
B decision of the learned county court judge

> that no sufficient particulars had been delivered and, therefore, that the claim was barred on the grounds: (1) that the notice of November 23 1931 was merely a notice that a claim would be made and not a rendering of particulars, . . .

So there again, the question was against the background of the statute which imposed upon the claimant the obligation to particularise his claim sufficiently to enable the opposite party to appreciate the nature of the claim and to determine what evidence was required in order to meet it. The court decided on the facts that there were no particulars which would sufficiently arm the opposite
C party with the opportunity for understanding the nature of the claim and for appreciating the evidence that it would be necessary to give in order to meet it.

Those were the cases which gave guidance as to the kind of particularity which was appropriate to valid particulars complying with the statutory requirement laid down in section 16(2) of the 1923 Act.

I have referred to those cases because they evidently played a part, not only in the subsequent decision of *Church Commissioners v Matthews*, but also in the reasoning of the learned judge when he was considering the instant case.

By contrast, I come to the scheme of legislation in the Agricul-
D tural Holdings Act 1948, which incorporates provisions that first appeared in the Agriculture Act 1947, to which it is unnecessary to make separate reference. In the 1948 Act, I refer first to two provisions for compensation.

By section 13(1) it was provided

> (1) Subject to the provisions of this section —
> (a) . . . (b) any building (other than one in respect of which the tenant is entitled to compensation under this Act or otherwise) erected by him on the holding; . . . shall be removable by the tenant at any time during the continuance of the tenancy or before the expiration of two months from the termination of the tenancy, and shall remain his property so long as he may remove it by virtue of this subsection.
> E (2) The right conferred by the foregoing subsection shall not be exercisable in relation to a fixture or building unless the tenant —
> (a) has paid all rent . . . (b) has, at least one month before both the exercise of the right and the termination of the tenancy, given to the landlord notice in writing of his intention to remove the fixture or building.
> (3) If, before the expiration of the notice aforesaid, the landlord gives to the tenant a counter-notice in writing electing to purchase a fixture or building comprised in the notice, subsection (1) of this section shall cease to apply to that fixture or building, but the landlord shall be liable to pay to the tenant the fair value thereof to an incoming tenant of the holding.

It is to be observed that by virtue of that section it is provided that the tenant may give notice of the intention to remove fixtures or buildings, but if after such notice the landlord gives a counternotice
F electing to purchase the fixture or building, then the landlord may purchase, but on such purchase shall be liable to pay a fair value thereof to an incoming tenant of the holding.

I turn next to section 47. By that section and the succeeding sections there is provision for compensation for new improvements and for other matters.

> 47(1) The tenant shall, subject to the provisions of this Act, be entitled on the termination of the tenancy, on quitting the holding, to obtain from his landlord compensation for a new improvement carried out by the tenant and for any such matter as is specified in Part II of the Fourth Schedule to this Act: . . .

subject to certain provisos which do not arise in the present case.

By section 48 the measure of compensation for long-term new improvements was enacted: G

> The amount of any compensation under this Act for new improvement specified in the Third Schedule thereto shall be an amount equal to the increase attributable to the improvement in the value of the agricultural holding as a holding, having regard to the character and situation of the holding and the average requirements of tenants reasonably skilled in husbandry.

That measure of compensation may be contrasted with the different measure of purchase price imposed upon the landlord under section 13 if he elects to purchase.

I refer next to section 50, which provides that where the landlord has refused consent there is an alternative provision that the tenant H may apply to the minister to override the landlord and to grant a consent which binds the landlord.

By section 70 the Act sets out the scheme for the settlement of claims between landlord and tenant on termination of tenancy, and specifies that disputes should be settled by arbitration; it sets out the provisions relating to the appointment of an arbitrator and has an important and quite new provision in subsection (2) of section 70:

> No such claim as aforesaid shall be enforceable unless before the expiration of two months from the termination of the tenancy the claimant has served notice in writing on his landlord or tenant, as the case may be, of his J intention to make the claim.
> A notice under this subsection shall specify the nature of the claim, and it shall be a sufficient specification thereof if the notice refers to the statutory provision, custom or term of an agreement under which the claim is made.

It is to be observed that that imposes upon the claimant the duty in pursuit of a claim which has not been settled to give a notice in writing of an intention to make the claim, and that notice is sufficient if it specifies the nature of the claim by specifying the statutory provision, custom or term of agreement under which the claim was made.

So, to use the language which has appeared in some of the judgments given in this court in relation to the 1923 Act Parlia- K ment, by subsection (2), has provided that the first stage in pursuing a disputed claim shall be to give a notice which specifies the class of claim as compared to the particulars of claim.

After such notice, the procedure is enacted in the Sixth Schedule. After appointment of the arbitrator, by paragraph 6 of the Sixth Schedule:

> The parties to the arbitration shall, within

what is now by a subsequent amendment to the Act a period of 28 days

> from the appointment of the arbitrator, deliver to him a statement of their L respective cases with all necessary particulars and —
> (a) no amendment or addition to the statement or particulars delivered shall be allowed after the expiration of the said 28 days except with the consent of the arbitrator;
> (b) a party to the arbitration shall be confined at the hearing to the matters alleged in the statement and particulars delivered by him and any amendment thereof or addition thereto duly made.

Thus, in contrast to the statutory procedure prescribed under section 16 of the 1923 Act, the 1948 Act prescribes a procedure materially different in two respects. It became a two-stage procedure instead of a one-stage procedure. The two stages were to be: a notice specifying the nature of the claim, followed by a statement M of case pursuant to paragraph 6 of the Sixth Schedule, which instead of stating particulars has to state all necessary particulars. Secondly, whereas under the 1923 Act the particulars served were final and did not provide for amendment or addition of any description, by paragraph 6 of the Sixth Schedule to the 1948 Act, whereas the statement of case had to state all necessary particulars as compared to all particulars, it provided that those particulars could be amended or added to provided that the arbitrator gave consent. By subparagraph (b) to paragraph 6 the party to the arbitration is confined at the hearing to matters alleged in the statement and particulars delivered by him and amendment thereof or addition thereto duly made.

A It seems to me clear that the explanation of this important relaxation of the procedure enacted in 1923 was that Parliament had appreciated by 1947 that an arbitration under the Agricultural Holdings Act had characteristics arising from the subject-matter which made the arbitration something importantly different from arbitrations under the Arbitration Acts or from proceedings at law. In particular, it was likely that a significant number of claimants would be small people, landlords or tenants of smallholdings who, competent, no doubt, as farmers, were likely to be ill-versed in the delicacy of pleading for the purposes of a statutory arbitration. The Draconian character or the Draconian effect of section 16(2) of the 1923 Act was recognised by Parliament as being inappropriate to
B the kind of arbitration which was provided for in the Agricultural Holdings Acts. Against that background it is easy to understand how it was that Parliament, while imposing upon the claimant a duty to file not only particulars, but more elaborately all necessary particulars within a stated time, decided that there should be room for amendment or addition to those particulars, provided that the arbitrator gave his consent to the amendment or addition. The scheme was completed by the provision under subparagraph (b) of paragraph 6 that at the arbitration the parties shall be confined to the statement and particulars delivered and any amendment or addition thereto duly made, that is to say with the consent of the arbitrator.

C The relevance of the observations in the judgments in the cases to which I have referred, being cases under the 1923 Act, has to be very carefully considered in the light of the novelty of the statutory scheme under the 1948 Act, because those cases under the 1923 Act were concerned with section 16(2) of the 1923 Act, which made it mandatory upon the claimant to complete his particulars without any opportunity for addition or amendment. The courts decided that in that legislation those particulars which alone were the particulars which fell for consideration at the arbitration must be such as would enable the opposite party to appreciate the nature of the claim, and to appreciate the evidence that he should produce in order to meet that claim.

D For the purposes of the scheme in the Sixth Schedule, the guidance given in the judgments in the cases to which I refer remains, in my view, an entirely appropriate guidance, provided that the arbitrator focuses his attention on the sufficiency of the particulars given by the claimant, not only in his original statement of case but also in the statement of case which has been amended or been the subject of addition on supplemental statements, leave to file which has been the subject of consent by the arbitrator.

It seems to me that it would be entirely misleading and quite wrong to apply the reasoning of this court in the two cases to which I have referred to the statement of case filed under the Sixth Schedule, rather than applying the reasoning of those judgments to
E a statement of case, together with such amendment or addition as an arbitrator thinks it right to consent to. The schemes of legislation are significantly different in the ways to which I have already referred. It is now a two-stage instead of a one-stage procedure; the second stage of that procedure does not impose upon the claimant an obligation to get everything right the first time, but an obligation to do his best to get everything right, provided that he can get it right by amendment and addition to which the arbitrator gives his consent.

So I come in the light of the present statutory scheme to consider the judgment in this case. First I refer to the decision of the arbitrator. It is quite clear to me that the arbitrator took the view,
F expressed in paragraphs 20, 21, and 22 of his judgment, that he was precluded by authority from exercising his discretion because he took the view, in my view quite wrongly, that he had no power, in the light of the guidance given in the county court in the case of *Church Commissioners v Matthews*, to consent to the particulars. That view, which I think he clearly expressed, is based on a complete misunderstanding of the case to which he refers.

Church Commissioners v Matthews is reported in (1979) 251 EG 1074. What the case was concerned with was the sufficiency of a landlord's statement of case, for the purposes of paragraph 6 of the Sixth Schedule, and the point of the case was that the landlords had delivered a statement of case which they never sought to amend, and the learned county court judge decided that the purported statement of case hopelessly failed to comply with the G
statutory requirement of a statement of case with all necessary particulars. The statement of case there under consideration read as follows:

> If the arbitrator finds that the above notice was valid, he will be asked to give an award on the validity and/or quantum of 30 times of dilapidations to house, buildings, cottage and land still outstanding between valuers.

The learned county court judge, and I would respectfully submit correctly, took the view that there was no particularity at all about the landlord's statement of case, and it therefore failed to comply with the provisions of paragraph 6. That case is *nihil ad rem* to the issues in the instant case, and the arbitrator was wrong in thinking H
that it would assist him, because the point in the case was that the statement of case, without any application for amendment, was hopelessly inadequate, and there was no conclusion that the learned county court judge could reach other than that which he did reach. The problem before the arbitrator was not the problem that faced the learned county court judge in *Church Commissioners v Matthews*. It was a quite different problem. Faced with an imperfect statement of case, as the arbitrator undoubtedly was in the instant case, he had to ask himself "Ought I to permit it to be made clear by amendment or addition as drafted by the landlords' agent in the application to file a supplemental statement of case?" I find it impossible to see how in answering that question there is any reason J
for thinking that any assistance would be derived from the learned county court judge's judgment in the case of *Church Commissioners v Matthews*. But it is clear that the arbitrator thought that he was bound, as he said: "Following precedent I could not accept the fundamental amendment to the tenant's statement of claim." That was a misconception, because he was not bound at all by anything that appeared in the case that he was relying upon.

So the matter came before the learned county court judge. It is to be observed that in his note of judgment the learned county court judge began by expressing his conclusion that he had no doubt that the amendments or additions proposed could be effected without manifest or grave injustice, although there had been apparently an K
argument to the contrary. On the facts, that finding of the learned judge was clearly right.

The judge then went on to consider another question, and he put it like this:

> The landlords allege that the amendments produce such a fundamental change of case that they should not be allowed even if the statement of case is valid in itself. I find this a difficult argument to accept as the tenant's statement of case is so vague in itself that a claim under section 13 or section 47 can be justified under it.
>
> This is the real difficulty in this case. The rules in the Sixth Schedule to the Act lay down that "the parties . . . shall, within 28 days from the appointment of the arbitrator, deliver to him a statement of their respective cases with all necessary particulars." The time-limit is mandatory and inflexible. L
> If this is not done the party is unable to produce any evidence in support of his case.

Therein the learned judge was wrong. The time-limit is not mandatory and inflexible, because the time-limit in the sixth paragraph is a time-limit of 28 days for the delivery of the statement with all necessary particulars, but there is provision for amendment or addition to a statement or particulars with the consent of the arbitrator. There is nothing mandatory or inflexible about the scheme of the Sixth Schedule. Indeed, as I understand it, the reason for the legislative change between the scheme in section 16(2) of the 1923 Act and the scheme in the sixth paragraph of the Sixth M
Schedule to the 1948 Act was to introduce a flexibility that was absent from the 1923 Act. The object of the change was to enable the claimant to put his house in order. I quite understand that there could be, hypothetically, a filing of a purported statement of case which was so hopelessly at variance with the intention of a statement of case within paragraph 6 that it could be said that there was nothing to amend. I think it is unnecessary to elaborate the kind of hypothetical situations that I have in mind. If the statement of case were merely frivolous or something of that sort it might well be held that there was nothing to amend. But that is not this case at all.

The notice dated May 22 gave notice of a claim for compensation for new improvements under section 47. As the judge found, it

A superseded the first notice. The trouble arose, of course, from the extraordinary muddled statement of case which the tenant's agents filed in purported compliance with the duty of the claimant to give all necessary particulars. It was an odd document. It used language that on its face would leave a respondent in profound obscurity as to what kind of case the claimant was really presenting. Paragraph 4 recited a provision of the Tomlin Order. Paragraph 5 read:

> The tenant claims compensation in respect of tenants fixtures . . . as scheduled in the Tenancy Agreement.

Paragraph 6 claimed £75,000, "being the value of the two buildings B referred to." If the draftsman thought he was presenting a claim under section 47 he would only have pleaded that measure of loss if he had completely failed to appreciate the measure of compensation provided for the purposes of a claim under section 47, because the measure of compensation under section 48 is not the value of the buildings.

There was a sufficient particularity of the physical subject-matter of the claim because two buildings were sufficiently described and a figure was put against each of them in respect of their value. I do understand that as long as the opposite party was faced only with that document he could have been in great difficulty in deciding what was the case that he had to meet.

C The statement of the landlords' case does indicate that the landlords' agents realised there was confusion, because they pleaded reasons for negativing the section 13 claim and then proceeded to plead an affirmative case in the event of the claim being eventually regarded as a section 47 claim, as had been foreshadowed in the notice of May 22. But all those difficulties were swept away as soon as the supplementary statement of case was considered by the arbitrator or by the landlords' agents, because in the supplemental statement of case the tenants, by paragraph 10, accepted that no compensation was due under section 13, pleaded that the case for compensation was under section 47, and pleaded that the landlords had given written consent for the erection of the corn and potato D shed, and pleaded the facts relied on as leading to an inference of consent by the landlords to the erection of the Dutch barn. That was still imperfect and defective, because they still failed to plead that the tenants had quitted the holding. That was the subject-matter of the further supplement, which everybody agreed should be allowed if the supplemental statement of case was consented to as an amendment or addition.

For my part, once the supplemental statement of case is scrutinised it clears up all the difficulties arising from the muddle of language and confused thinking in the statement of case originally filed; and faced with the statement of case, plus the supplemental statement of case, a respondent would have no difficulty in E appreciating the nature of the tenants' case, the particulars that describe that case, and in mobilising his resources by way of evidence in order to meet it. When the two are read together, all the difficulties are swept away. If the amendment is allowed, the statement of case thus amended and added to, with the further addition to which I have referred, will comply completely with the prescribed requisite of pleading all the necessary particulars.

In this court I think it is unnecessary to dwell longer on the question of whether those amendments could have been consented to without injustice, because I find it quite impossible to apprehend any injustice or any difficulty confronting the respondents that could not adequately be dealt with by an adjournment, if any adjournment were necessary, and by costs if costs were thrown F away. I say adjournment if any adjournment were necessary because the appointment for the arbitrator in February was only an appointment for the purpose of deciding a preliminary point, namely consent to the amendment, and as far as the information before this court goes no date for the arbitration on the merits was ever fixed.

So I come to the reasons relied upon by the learned judge as the basis of his decision in upholding the determination of the arbitrator to refuse his consent. In my view, the error into which the learned judge fell is manifest on reading the last two pages of his note of judgment. He was misled by the cases of *O'Connor v Brewin* and *Jones v Evans* into thinking that faced with an application for leave to amend it was appropriate to look at the defective statement G of case to decide whether it by itself afforded all necessary particulars, and on reaching that decision the judge evidently thought that he was precluded from giving leave to amend or that it would be wrong to give leave to amend. But that reasoning was based on a misapprehension of the scheme set out in the 1948 Act. This two-stage scheme, culminating in a statement of case which may be amended or added to with the consent of the arbitrator, has this effect: on an application to amend ordinary considerations of justice and fairness make it appropriate to amend, if after amendment the respondent will know sufficiently the particulars of the case that he has to meet. In such a situation, leave to amend should ordinarily be granted. The consent should ordinarily be given. It is manifest H that when the Sixth Schedule expressly enacted the power to amend or add, Parliament contemplated that the statement of case first filed would be defective in some respect. Otherwise there would be no point in seeking leave to add necessary particulars. The mere fact of the absence of all necessary particulars in the original statement of case cannot itself be a ground for refusing consent to an attempt to remedy the deficiency.

As I understand it, if the notice given on May 22 is read with the statement of case, together with the amendments and additions sought to be made, all the requirements of the Sixth Schedule to the 1948 Act would be amply satisfied. For those reasons I would, respectfully, reverse the learned judge's decision and state that the J arbitrator was wrong in refusing his consent and that the judge was wrong in upholding his decision.

Two other points arise for mention. Mr Cripps drew the court's attention to the familiar case of *Gray v Ashburton (Lord)* [1917] AC 26 which brought home the point that an arbitration under the Agricultural Holdings Acts has characteristics which differentiate and distinguish it from an arbitration under the Arbitration Acts or an action at law. So that, as was held in that case, in their Lordships' House, the way in which the arbitrator under the Agricultural Holdings Act exercised his discretion in relation to costs was not bad because he did not follow the practice of the High Court in relation to the judicial exercise of discretion as to costs in the High Court. But, having said that, I am unable to discover in the guidance K in the speeches in their Lordships' House any assistance for the purpose of deciding the question before this court, which is whether the arbitrator and the learned county court judge were wrong in law in the decisions which they respectively reached.

Another point was raised by Mr Cripps which turned on the effect of a special provision which was a provision of the tenancy agreement of October 8 1975 in relation to the obligations of the landlord. That special provision stated:

> It is hereby agreed between the parties that the following buildings erected on High Park Farm shall be regarded as tenant's fixtures: The ten bay Dutch barn, workshops and glasshouses and the grain and potato store with two for tunnel equipment. L Also brick and asbestos open shed and workshop at Brickpits off West Road, Barnes.

Mr Cripps' argument was that, because those buildings had been the subject of that agreement, they should be regarded as tenant's fixtures, so that that agreement in some way precluded the tenant from exercising his right to claim for new improvements. This is a very puzzling submission in the light of clause 2(1) of the tenancy agreement, which provided that:

> On the termination of the tenancy to pay to the Tenant compensation: (a) for the unexhausted value of the improvements referred to in the Second and Third Schedules to the Act, subject to and in accordance with the M provisions of the Act, (b) for fixtures paid for an entry at a fair valuation and in respect of other fixtures subject to and in accordance with the provisions of the Act, . . .

In spite of Mr Cripps' argument I remain unable to follow how he could submit successfully that the special provision which I have recited could possibly have the effect of negativing or modifying or reducing any claim of the tenant for unexhausted value of the improvements without meeting the difficulty provided for by section 65 of the 1948 Act, which expressly provides that contracting out is not permitted. The grounds that he submitted suggesting a special agreement to the contrary seem to me difficult to sustain.

For those reasons, I would reverse the order of the learned county court judge, and I would move that consent to this amendment should have been given by the arbitrator because nothing in the facts recited in the case stated by the arbitrator could be consistent with any just determination of the application for leave to amend, save consent. There is, therefore, to my mind, no room for sending the case back to the arbitrator, because the facts stated by the arbitrator point only to one correct legal conclusion, which is that consent should be given to the amendments sought. I would allow the appeal.

Agreeing, TEMPLEMAN LJ said: I agree that there are no grounds upon which consent to the additions and amendments to the statement of case sought by the tenants could properly have been withheld. But as the arbitrator and the judge in their different ways have gone sadly astray in construing and applying the relevant provisions of the Agricultural Holdings Act 1948, I will as shortly as possible endeavour to give my reasons.

On May 22 1980 the tenants served on the landlords a notice that they were seeking under section 47 of the Act compensation for new improvements identified as the erection of a grain and potato store in OS No 1 and the erection of a dutch barn in OS 42 on the holding known as High Park Farm and Low Park Farm, Aslackby, in the County of Lincoln, in the sum of £75,000. Nothing could be plainer or clearer. However, the tenants' statement of case which followed was defective. First, the statement of case dated July 14 1980 did not refer either to section 13 or to section 47 in express terms. So far as the landlords were concerned section 13 was out, because section 13 can come into play only if the landlords had elected to purchase the fixtures in question, and the landlords above all knew that they had not made any such election. Secondly, the statement of case gave the wrong measure of damages. But that did not affect the landlords, who could prove and give evidence of the right measure of damages. Thirdly, the statement of case did not specify the date of quitting. But it did give the date of termination of March 28 1980, which was the same as the date of quitting, and it does not seem to me that the landlords were in any difficulty, as they presumably knew as much as the tenants; and the arbitrator waiting in the wings could always be enlightened by both of them. Fourthly, and more importantly, the statement did not allege or give particulars of the landlords' consent, which is a prerequisite to an entitlement to compensation under section 47. Therefore, if the landlords chose to challenge the tenants on the question of consent — and at that stage nobody knew whether they would or would not — and if, that consent being challenged, the tenants' statement of case was not amended, then clearly the tenants would be bound to fail in their application.

That this was all clear to the landlords appears from the landlords' own statement of case, dated July 28 1980. First of all, they dealt with a possible claim under section 13, which in my view was an unnecessary elaboration, but understandable in view of the difficulties in the tenants' statement. Secondly, they dealt with the tenants' failure to allege and prove that the landlords' consent had been given for the purpose of section 47. They took the point, and the point being taken was good so long as the tenants' statement of case remained unamended. Then, no doubt feeling the possibility that the tenants might wish to amend, no doubt feeling in their bones that there might be some possibility that amendment might be allowed, the landlords' statement of case very efficiently and in great detail dealt with the merits of the application for compensation under section 47, ending with the conclusion that the compensation payable to the tenants could not exceed £35,000.

When the landlords' statement of case was received by the tenants, or when for some other reason the tenants woke up, they realised the defects in their original statement of case, and so they applied for additions and amendments. On November 17 1980 the landlords received the tenants' supplemental case in which the tenants proposed to make additions and amendments to their statement of case which would repair all the defects which I have mentioned in their original statement of case. In particular, the tenants supplied the necessary particulars with regard to what they said was the landlords' consent given for the purposes of section 47. At that stage, the landlords had a choice. They could either go along to the arbitrator and say "We consent to your making the order which the tenants seek", or they could choose to oppose. They chose to oppose and that is the reason we are here today. I say that, not to blame them: it was a choice perfectly open to them and they took it.

The arbitrator on February 21 1981 gave two reasons for refusing to consent to the tenants' application to repair the defects in their original statement of case. The first reason he gave was:

> After consideration of legal argument and accepting that the Tenancy Agreement clearly scheduled both buildings in question as Tenants Fixtures, I concluded that no claim for these items under section 47 of the Agricultural Holdings Act 1948 could possibly be sustained.

In the first place, that conclusion as a matter of law on the construction of the tenancy agreement was quite wrong. In the second place the arbitrator was not being asked to construe the tenancy agreement or the claim; he was being asked to allow an amendment. Then the arbitrator states in paragraph 21 that his researches had established "County Court judgments which prevent fundamental changes to Statements of Case under the 1948 Act." There is no trace in the Act of any reference to fundamental changes, and one reason why the arbitrator may have gone wrong is that he entirely failed to set out and consider the perfectly simple words of paragraph 6 of the Sixth Schedule, which told him what he ought to be doing. The principle of "no fundamental change" has as much relevance to this case as the principle of "No Popery". However, those were the reasons given by the arbitrator, and so the parties, not unnaturally, went along to the county court judge.

The county court judge was not impressed with the arbitrator's reasons, and he considered what attitude the arbitrator should have taken. In the first place, and quite correctly, as my lord has pointed out, he said: "I have no doubt that the amendments or additions proposed could be effected without manifest or grave injustice. The landlord has had it well in mind from the beginning that these claims might be being made." But then the learned judge proceeded, so far as I can see, without applying his mind to the exact words of paragraph 6 of the Sixth Schedule with which he ought to have been concerned. He began to point out defects in the original statement of claim. He said: "I see that there is no reference to either section 47 or section 13 in it." That is precisely why the tenants sought an amendment, and it is no answer to a request for an amendment to say that the original document was wrong and to knock out the amendment because of the defect in the earlier document. He said that the tenants' original statement of case would leave any landlord in a state of confusion; but the object of the tenants' amendment was to remove the state of confusion and make everything clear to the landlords to whom, as I have said, it was already pretty clear in the first place. It cannot be right to refuse an amendment because an amendment is necessary.

In my judgment, one simply has to look at the words of paragraph 6 of the Sixth Schedule as amended. Parliament has provided that a claimant shall deliver a statement of case with all necessary particulars within 28 days from the appointment of the arbitrator. Parliament has recognised that the draftsman of the statement of case may be fallible, whether he be farmer, bailiff, chartered surveyor, solicitor or counsel. Parliament in its wisdom has given the draftsman an opportunity to repent. He can repent at will within the 28 days, because the paragraph goes on: "No amendment or addition to the statement of particulars delivered shall be allowed after the expiration of the said 28 days, except with the consent of the arbitrator." So that if within that 28 days the draftsman has a rush of commonsense to the head and realises that he has left out a necessary particular, he can rapidly fill in the gap. Or, if he has another rush of commonsense and realises that he has got it wrong and that the statement needs amending he can do so. Parliament in its wisdom has realised that people do not necessarily come to their senses within a fixed limit of 28 days. So Parliament has given the draftsman another chance of repenting: after the 28 days he can amend with the consent of the arbitrator, and he can go along and say "I have made a mistake. I have left this out. I have put this wrongly. I have not given the necessary particulars, and now I want to do so." If the arbitrator and the learned county court judge are

A right, it is open to them to say: "Oh, how wicked you have been! Look what a mess you have made of it. We are not going to help you out of your difficulty." That is not the object of the power given by that paragraph; that is not the object of requiring the consent of the arbitrator. The object is to enable the arbitrator to allow an amendment or an addition to be made, and the reason that his consent is required and can be refused is simply this. That he has to say to himself: "What about the poor landlord? Is he going to be any worse off than he would have been if the amendment had been incorporated in the original statement of case?" It is no good the landlord saying: "Oh, but if you allow this amendment the tenants might be able to get £35,000 out of me." The £35,000 does not B become payable because of the amendment. It becomes payable because the necessary conditions are satisfied, if they are satisfied, which Parliament has laid down for the payment of £35,000. But, of course, if the landlord is going to be in any worse position as a result of the amendment than he would have been if it had originally been in the statement of case, then the arbitrator may pause. If he finds that since the original statement of case by lapse of time or because the landlord has altered his position or for some other reason it would be unfair on the landlord to impose the amendment upon him, then the arbitrator has power to refuse consent. But if the answer is, as it is in the present case, that the landlord is no worse off, if all these amendments had been put in the original statement C of case he would not have been any better or worse off, then consent ought to be given to the amendment. Although there may be some difficulty in knocking sense into the Agricultural Holdings Act, we have not yet got to the stage where arbitrators and county court judges can foist on the Act the kind of machinery that is more appropriate to the ages of Lord Eldon and the Circumlocution Office than it is to the twentieth century. The object and proper application of paragraph 6 of the Sixth Schedule are quite clear and I hope that there will be no difficulty in the future. Amendments ought to be allowed as of course, unless the landlords have a strong case for saying that because the tenants got it wrong in the first place and because of actions taken thereunder or for various other D reasons it would be unfair to allow the amendment; and the test as I have adumbrated it is whether the landlords are any worse off than they would have been if the statement of case had been properly drafted in the first instance.

That only leaves Mr Cripps' point that all this is beating the air anyway, because the tenants cannot get their claim under section 47 off the ground, because of the addition of the special provisions of the Second Schedule to the tenancy agreement.

That addition makes it quite clear that the Dutch barn and the grain and potato store are not to be taken into account when fixing the rent, because they are tenants' fixtures. They are not to pin on the landlords' liability for repairing them because they are tenants' E fixtures. It makes it quite clear that notwithstanding that this appears to be a new tenancy agreement, and notwithstanding that The Dutch barn and the grain and potato store have been there from the beginning, they were nevertheless in the tenants' part of the world instead of the landlords' part of the world. That means to my mind that there was recorded not only the facts as regards these two buildings but also maintenance of any claim which the tenants might have in respect of them either under section 13 or under section 47 or under any other section of the Agricultural Holdings Act. The insertion of that special provision had advantages from the landlords' point of view and the tenants' point of view. It may be that it was unnecessary, because the Agricultural Holdings Act, F section 53, provides that the rights of a tenant with regard to improvements and compensation can be preserved from one tenancy through a succession of tenancies, but no doubt it was put in *ex abundanti cautela*. That this particular provision has to be wrenched by construction into an abandonment by the tenants of any rights they may have under section 47 seems to me quite unarguable; and, as my lord has pointed out, it is not only unarguable, it is also probably illegal.

For those reasons I agree with my Lord.

Also agreeing, Dame Elizabeth Lane said: I would make only these comments. In the second paragraph of section 70(2) of the 1948 Act there is a description, if not a definition, of what amounts G to a sufficient specification, for the purposes of a notice of claim. But when it comes to paragraph 6 of the Sixth Schedule and the mandatory requirement of a statement of case with all necessary particulars, no guidance whatsoever is given as to what particulars are necessary, and there are no statutory prescribed forms such as are to be found, for example, in the Rent Acts Rules and Regulations or in the Matrimonial Causes Rules.

So, a layman, seeking as he may under this Act to draft a statement of case, who is unlikely to have studied such authorities as those to which we have been referred or a report of this case or any helpful textbook, is left to guess what particulars are necessary.

Unless and until such forms are prescribed, in my view whoever H has drafted the statement of case, an arbitrator, exercising his discretion under this Act should expect to be, and be, liberal in the exercise of the discretion to consent to amendments so long as such amendments will not cause injustice to the opposite party, as was the case here; and if there are any consequential costs, those can be ordered in the opposite party's favour.

I agree that this appeal should be allowed, and the order suggested made.

The appeal was allowed with costs in the Court of Appeal and in and about the appeal to the county court, including costs of the case stated. Leave to appeal to the House of Lords was refused.

J

COURT OF APPEAL

December 21 1981

(Before Lord Justice EVELEIGH, Lord Justice GRIFFITHS and Lord Justice KERR)

UNIVERSITY COLLEGE, OXFORD v DURDY

Estates Gazette April 24 1982

(1981) 262 EG 338 K

Agricultural Holdings Act 1948, section 8 — Important Court of Appeal decision — Date of appointment of arbitrator by minister reinterpreted — Tenant challenged validity of appointment because he did not receive notice of the appointment until after "the next ensuing day", ie after the term date on which any adjustment of rent would take effect — In this case the minister made his appointment on January 29 1981, the next ensuing day was February 2 and the tenant received notice of the appointment on February 3 — Held by the Court of Appeal, contrary to a fairly general belief, that the receipt of notification by the parties is not necessary to complete the arbitrator's appointment — *Tradax Export SA v Volkswagenwerk AG* and old case of *Tew v Harris* distinguished — Arbitrator's L **appointment complete when the instrument of appointment is signed and sealed by minister — But although receipt of notice by the parties is not necessary to complete the appointment, there is an implication that a party must have received notice before the 28 days begin to run against him for the purpose of paragraph 6 of Schedule 6 to the 1948 Act — As to arbitrator's acceptance of appointment, this is clearly necessary, but, as the court pointed out, in practice the minister will ascertain the arbitrator's willingness before making the appointment — Appeal from decision of county court judge (who decided that the arbitrator had no jurisdiction) allowed — The judgment of Kerr LJ, who expressed unorthodox views on the absence of any precise correlation between the "date of the reference" and "the next ensuing day", should be noted — As a result of** M **the Court of Appeal's decision the rules formulated by Judge Meurig-Evans in *Richards v Allinson* (1979) 249 EG 59 as to the requirements for perfecting the arbitrator's appointment need revision**

This was an appeal from the decision of Judge J D Walker at Doncaster County Court in favour of the tenant, respondent to the appeal, Michael John Durdy, who had challenged the appointment of the arbitrator, J G Cleverly FRICS, by the Minister of Agriculture in connection with a rent arbitration. The appellants, the

A landlords, were the Master and Fellows of University College, Oxford. The two agricultural holdings, the rents of which were the subject of the arbitration, were Cockwood and Kilham Farms, Cantley, South Yorkshire.

Derek Wood QC and Miss Joanne Moss (instructed by Hepworth & Chadwick, of Leeds) appeared on behalf of the appellants; E C Evans Lombe QC and A De Freitas (instructed by Oxley & Coward, of Rotherham) represented the respondent.

Giving judgment, EVELEIGH LJ said: By a lease dated March 16 1965 the appellants let to the respondent two agricultural holdings known as Cockwood and Kilham Farms, Cantley, in South Yorkshire, for a term of 10 years from February 2 1965 at a yearly rent of
B £6,000 payable by equal half-yearly instalments on February 2 and August 1 in every year, "the first payment to be made on the 1st day of August next". The respondent's tenancies subsequently became converted into a yearly tenancy by virtue of the Agricultural Holdings Act 1948. On November 29 1979 the landlord served notice upon the tenant demanding a reference to arbitration under section 8 of the Act. Application was made to the minister for the appointment of an arbitrator, and by notice dated January 5 1981 the minister informed the landlord and the tenant that he was about to appoint an arbitrator. On January 29 there was signed and sealed, on the authority of the minister, a document which stated that in exercise of his powers conferred by the Agricultural Holdings Act
C 1948 the minister "does hereby appoint JG Cleverly Esq, FRICS, to be the arbitrator for the purpose of settling the claims, questions or differences between the landlord and the tenant of the holding mentioned in the First Schedule hereto which are specified in the Second Schedule hereto". The First Schedule specified Cockwood and Kilham Farms, and the Second Schedule read: "The rent to be paid for the said holding as from the next ensuing date on which the tenancy could have been determined by notice to quit given at the 29th day of November 1979 being the date of the demand for arbitration made by the landlord to the tenant". A copy of the document was sent to the landlord and to the tenant. The landlord received his on February 2. The tenant received his on February 3.
D The tenant challenged the jurisdiction of the arbitrator, and the learned county court judge decided that the arbitrator had no jurisdiction.

The learned judge held that the tenancy commenced on February 2 and that the next ensuing date (after November 29 1979) upon which the tenancy could have been determined by notice to quit was February 1 1981 and that the decision in *Sclater v Horton* [1954] 2 QB 1 required the arbitrator to be appointed before that date, the appointment did not take effect until February 3 (the date when the tenant received the notice) and consequently the arbitrator had no jurisdiction.

Mr Wood for the appellants contended that as the term was
E expressed to run from February 2 it began on February 3. Further notice to quit can be given to expire on the last day of the current term or the day after that. Consequently it could have been given to expire on February 3. Therefore, as notice of the appointment was received by the tenant on February 3, the appointment of the arbitrator was in time because, he says, seeing that February 3 ended at midnight, the appointment was made before that date. This ingenious rescue operation was not fully argued by the respondent, as the court indicated that it wished to hear other points first. I mention it lest anyone reading the judgments in this case is puzzled by some apparent discrepancies in the dates. I do not find it necessary to deal with the argument.

F It was also argued that if *Sclater v Horton* contained the principle which the learned judge applied and which in effect could be said to make time the essence, then it was no longer to be followed, in view of the decision of the House of Lords in *United Scientific Holdings Ltd v Burnley Borough Council* [1978] AC 904. Again I do not find it necessary to go into the argument in detail. That case was dealing with the approach to an arbitration clause as machinery for settling disputes. In section 8 " the date of the reference" is not a time-table for a procedural role. It is a fact relevant to the determination of the arbitrator and it lies at the basis of his decision, for he has to determine the appropriate rent at that date. The machinery for arbitration laid down in an arbitration clause must depend upon the construction of each particular clause, and this was clearly recog-
nised in *United Scientific Holdings Ltd v Burnley Borough Council.* G
I am therefore unable to regard the decision in that case as an authority which overrules the decision of the court in *Sclater v Horton*, which was dealing with the interpretation of a statute.

In my opinion, this case admits of a simple solution. I have come to the conclusion that the arbitrator was appointed at the moment when the minister performed the official act of appointment by signing and sealing the document. It accords with the natural meaning of the word "appointment". It is true that paragraph 6 of the Sixth Schedule to the Act might cause injustice if literally construed, but that is a reason for the court to avoid a strict construction of the paragraph and not a reason for departing from the natural meaning of the word "appoint". Paragraph 6 reads: H

> The parties to the arbitration shall, within fourteen days from the appointment of the arbitrator, deliver to him a statement of their respective cases with all necessary particulars and — (a) no amendment or addition to the statement or particulars delivered shall be allowed after the expiration of the said fourteen days except with the consent of the arbitrator; (b) a party to the arbitration shall be confined at the hearing to the matters alleged in the statement and particulars delivered by him and any amendment thereof or addition thereto duly made.

A benevolent construction in accordance with natural justice is readily achieved by reading into the section the requirement that a party must have notice before time begins to run against him. J

In *Tradax Export SA v Volkswagenwerk AG* [1970] 1 QB 537 Lord Denning and Salmon LJ both stated that notice was necessary to complete the appointment of an arbitrator. They were dealing with a commercial arbitration agreement. Each party had the right to appoint. What one did could be relevant to the action taken by the other. Moreover, the court was concerned with the failure to notify the arbitrator himself, and the question of notice to a party was not relevant to the decision. Edmund Davies LJ did not deal with that point. Furthermore, Salmon LJ clearly regarded it as a question of construction of the arbitration clause. At p 545 at "F" he said:

> No doubt the word "appointed" has different meanings in different contexts. We, however, are only concerned to construe it as used in the Centrocon arbitration clause which my Lord has read and which I need not read again. K

The case of *Tew v Harris* (1847) 11 QB 7, upon which Denning LJ and Salmon LJ relied, was itself a case where the procedure set out in the clause indicated that notice was to be given to the other party. For these brief reasons and for those more fully developed in the judgments of Griffiths LJ and Kerr LJ, I cannot accept the argument that the appointment is not complete until notice thereof is received by each party. That being so, the appointment was in time. I would allow this appeal.

I do not find it necessary to decide the further arguments submitted on behalf of the appellants, namely, that it is possible to read the decision in *Sclater v Horton* as not requiring strict observance of the need to appoint the arbitrator before the next ensuing day. However, I have great difficulty in accepting this argument. On the other hand, I must confess that upon reading section 8 I treated the words "at the date of the reference" as indicating a period rather than a particular day. A reference to arbitration is a process rather than a specific event. Nonetheless I regard the decision of *Sclater v Horton* as deciding that for the purpose of section 8 the date of the reference is the date of the appointment. Romer LJ at p 14 said: L

> Both the landlord and the tenant are in agreement that the words "the date of the reference" in section 8(1) of the Act mean the date of the effective appointment of the arbitrator, either by agreement or, in default of agreement, by the Minister, and I think that they are right in that view. M

Denning LJ, at p 13, said:

> On the true construction of the section I am in full agreement with the Master of the Rolls that the reference must take place within that year, by which I mean that the arbitrator must be appointed and must accept his appointment within that time. If the arbitrator is not appointed within the year, then the landlord loses his right to increase the rent; but if the arbitrator is appointed within the year, then the arbitrator has to fix the new rent. He must fix it as at the date of the reference, and that means, I think, as at the date of his appointment.

A I further see great difficulty in saying that the principle enunciated in *Sclater v Horton* does not apply if the arbitrator is appointed so that he can give his decision within the year for which the new rent is to be determined. I appreciate that the facts in *Sclater v Horton* were different from those in the present case. However, the court approached the problem by first determining the true construction of section 8. This was done by considering the section in isolation and not by reference to the facts of the case. The Master of the Rolls, at p 12, scrutinised the wording of the section and the tenses used in it. Denning LJ was in full agreement with the Master of the Rolls in his construction of the statute: p 13. I find it difficult to discover a *ratio decidendi* of the case other than that based upon B the construction of the statute so positively stated as it was.

Agreeing that the appeal should be allowed, GRIFFITHS LJ said: The contractual term for which the farm was let was expressed to run for a term of 10 years from February 2 1965. Thereafter, by virtue of the provisions of the Agricultural Holdings Act 1948, the tenant remained in occupation as a tenant from year to year with the security of tenure provided by the Act.

By a notice dated November 29 1979 the landlord demanded of the tenant an arbitration as to the rent that the tenant should pay for the three years commencing "from the next ensuing day on which the tenancy thereof could be determined by Notice to Quit given at the date of this demand". That is from February 2 1981.

C The landlord and tenant could or did not agree upon an arbitration; so on December 10 1980 the landlord called upon the minister to appoint an arbitrator. By an instrument dated and sealed on January 29 1981 the minister appointed JG Cleverly Esq as the arbitrator, and by letters of the same day notified the landlord and tenant of the appointment. The landlord received the letter notifying him of the appointment on February 2 but the tenant did not receive his letter until February 3.

The tenant, relying upon the decisions of this court in *Sclater v Horton* [1954] 2 QB 1 and *Tradax Export SA v Volkswagenwerk AG* [1970] 1 QB 537, has taken the technical point that because he did not receive his notification until February 3 the reference to D arbitration was a day too late and, if this is right, the landlord will have to wait another two years before a fair rent can be fixed for the farm which will reflect current inflation.

The county court judge decided the issue in the tenant's favour, holding that upon the authority of *Sclater v Horton* the arbitrator had to be appointed before February 2 1981 and that the appointment of the arbitrator was not complete until notice of the arbitration had been received by the tenant, for which proposition he relied upon the *Tradax* case.

Mr Wood on behalf of the landlord has attacked the judgment upon a number of grounds. First he says that, in view of the decision of the House of Lords in *United Scientific Holdings Ltd v Burnley* E *Borough Council* [1978] AC 904, *Sclater v Horton* is no longer binding on this court; it has, he says, been implicitly overruled because, in the language of Lord Greene in *Young v Bristol Aeroplane Company* [1944] 1 KB 718, it can no longer stand with the decision in *United Scientific v Burnley*. That case in the House of Lords decided that there was no presumption that time was of the essence in a rent review clause in a lease. Part of the reasoning that led their Lordships to that conclusion was that there was no impediment to the landlord recovering retrospectively an increase in rent awarded pursuant to a rent review.

In *Sclater v Horton* this court construed section 8 of the Agricultural Holdings Act of 1948 as incorporating a timetable into appli- F cations for a rent review. There are arguments both for and against such a construction and it is clear from the report that the matter was fully argued before the court. All the members of the court agreed that section 8 must be construed as requiring that the date of the reference to arbitration must precede the date upon which the tenancy could next be determined by a notice to quit. All three members of the court gave as one of their reasons that, as the arbitrator had to fix the new rent "at the date of the reference", it must have been intended that that date should precede the commencement of the new three-year term for which it would apply. That appears to me to be a very powerful reason for adopting the construction they did. To allow a rent to be fixed towards the end of a three-year period based upon values existing at that date and G require it to be paid retrospectively would in an inflationary age produce a most unfair result for the tenant. This reason is wholly untouched by the decision in the House of Lords.

The other principal reason given by the Master of the Rolls and with which Denning LJ agreed depended upon the actual language of the subsection and in particular the use of the future tense. I am unable to see how a statement of general principle about the proper approach to the construction of a commercial lease can be said to overrule a decision of this court based upon the close analysis of the wording of a subsection in a statute.

The only reason for the decision in *Sclater v Horton* that can no longer stand in the light of the decision in *United Scientific Holdings* H *Ltd v Burnley Borough Council* is the doubts expressed by the Master of the Rolls about the ability to recover retrospectively over- or under-paid rent. But that was not one of the principal reasons for the decision and it is a wholly insufficient basis upon which to hold that *Sclater v Horton* is not binding upon this court. I cannot therefore accept Mr Wood's first attack upon the county court judge's decision.

Sclater v Horton also decided that "the date of the reference" was the date by which the arbitrator had been appointed and accepted his appointment. Denning LJ said at p 13:

> On the true construction of the statute I am in full agreement with the Master of the Rolls that the reference must take place within that year, by J which I mean that the arbitrator must be appointed and must accept his appointment within that time.

It was not suggested by any member of the court that the appointment was not complete until the parties had received notification of the appointment. In the present case the appointment was dated January 29 and there is no suggestion that the appointment had not been accepted by February 2. The county court judge, however, felt impelled to hold that the appointment was not complete because of the decision in *Tradax Export SA v Volkswagenwerk AG*. In that case the Court of Appeal was dealing with the appointment of a commercial arbitrator under the Centrocon arbitration clause. The point that the court was called upon to decide was K whether an appointment of an arbitrator could be complete before the arbitrator had been asked if he was willing to act. Not surprisingly, the court said it could not. But in the course of their judgments Lord Denning and Salmon LJ also said that the appointment of an arbitrator required notification to the parties to the arbitration before it was complete. For this they both relied upon the old case of *Tew v Harris* (1847) 11 QB 7. That case turned upon whether a party required to nominate a valuer could be said to have done so if he had not given his valuer's name to the other side.

The agreement between the plaintiff and defendant in that case provided that the defendant would purchase the plaintiff's crops at a valuation to be made by two persons, one named by each party, L and, if they disagreed, by a third person to be named by them before entering on the valuation. But the agreement also provided that if one party failed to appoint a valuer, the valuer of the other could make the final decision. Such an agreement is clearly to be construed as requiring notice of the appointment of the valuer to be given to the other party. Suppose the plaintiff appoints a well-respected valuer in whose judgment the defendant has trust, the defendant may well be content to leave the valuation in his hands and not appoint another valuer. On the other hand, if the defendant has no confidence in the plaintiff's valuer, he will want to appoint his own. In either case the defendant must know the name of the plaintiff's valuer. But, for myself, I should have had considerable M hesitation in using that decision as authority for the general proposition that in all commercial arbitrations the appointment of an arbitrator was not complete until the parties to the arbitration had been told of his appointment. Whereas I can readily see that the arbitrator cannot be said to be appointed until he had been asked to act and agreed to do so, which is the point that had to be decided in *Tradax*, it is not clear to me why in all cases it necessarily follows that his appointment is not complete until both parties to the arbitration have been notified of the appointment. Much would depend, I should have thought, upon the particular wording of the arbitration agreement.

A But, whatever may be the position in commercial arbitrations which require co-operation between the parties in appointing an arbitrator, I can see no reason why, in the case of the appointment of an arbitrator by the minister under this Act, the appointment should not be complete once the appointment has been made in writing and the arbitrator has agreed to act. As a matter of practice, I would expect the minister to communicate with the arbitrator and ask whether he is willing to act before he makes the appointment. On this assumption, there is a certain date to which the parties can look to see if the reference was made within time, namely the date of the written instrument of appointment, in this case January 29 1981. This is a convenient procedure, for it allows the appointment B to be made very close to the date upon which the current rental period expires, which is the appropriate date at which a new rent should be assessed.

Neither tenant nor landlord is given any right to object to the arbitrator appointed by the minister; having failed to agree upon an arbitration themselves, they are bound to accept the minister's appointment. The position is quite different from commercial arbitrations in which the parties have to co-operate with one another in the appointment of the arbitrator.

In *Palgrave Gold Mining Co v McMillan* [1892] AC 460 the question before the Privy Council was whether it was necessary to give notice to the parties before exercising a statutory power to C appoint an arbitrator. The trenchant language in which Lord Hobhouse dismissed the requirement of notice lends some support to the view that notice to the parties, whether before or after the event, is not a necessary ingredient to perfect the appointment of an arbitrator under a statutory power.

I must now consider the wording of paragraph 6 of the Sixth Schedule, which reads:

> The parties to the arbitration shall, within fourteen days from the appointment of the arbitrator, deliver to him a statement of their respective cases with all necessary particulars and — (a) no amendment or addition to the statement or particulars delivered shall be allowed after the expiration of the said fourteen days except with the consent of the arbitrator; (b) a party D to the arbitration shall be confined at the hearing to the matters alleged in the statement and particulars delivered by him and any amendment thereof or addition thereto duly made.

The arbitrator has no power to extend the short period of 28 days in which the statement of the case must be delivered and, as it would be absurd that time should start running before the party knew of the appointment of the arbitrator, Mr Evans-Lombe submitted that the appointment cannot be complete until both parties have notice of it. This, it is said, must have been in the mind of the draftsman; otherwise he could not have produced a procedure so unfair as to have the result of debarring a party from submitting his case if his notice of the appointment of the arbitrator had gone astray in the E post. This Act, however, was drafted over 20 years before the observations in *Tradax* and, much as I admire and hold in awe the powers of parliamentary draftsmen, I am unwilling to assume that the author of this Act had at the forefront of his mind when drafting the Sixth Schedule the case of *Tew v Harris* to be found in the English Reports and decided over a century earlier.

The Sixth Schedule is dealing with procedural matters, not the question of substantive rights. The Court of Appeal has been compelled to read a statutory time-table into section 8 to give effect to its manifest intention, and I have no inhibition in reading paragraph 6 as implicitly requiring that the landlord or tenant has received notice of the appointment before time starts to run against him. F This solution is, I think, far preferable to the uncertainty that would arise if the question whether the reference to arbitration under section 8 had been made in due time depended upon when notice of the appointment had been received by both the landlord and the tenant.

Accordingly, I would hold the appointment is complete once the arbitrator has agreed to act and the minister has made the appointment in writing. I would therefore allow this appeal.

I had written this judgment before I had the opportunity to read the draft of the judgment about to be delivered by Kerr LJ. The Court of Appeal in *Sclater v Horton* did not have the advantage of the powerful argument deployed by Lord Justice Kerr based in particular upon the elasticity of the meaning to be attached to "the G reference" in arbitration proceedings. If the Court of Appeal had received that argument, I readily accept that they might have construed section 8 differently. But that the court regarded the true construction of section 8 as vital to their decision is, I think, with all respect, clear beyond doubt. Indeed, it was because the court appreciated that the construction they placed upon section 8 might affect the way in which the minister exercised his powers that they adjourned the first hearing of the appeal so that the minister could be represented by counsel as *amicus curiae*. The minister in argument pointed out that he had no power to impose a time-limit for the arbitration. I will content myself by quoting two short passages from the judgments of Sir Raymond Evershed MR and Denning H LJ. At p 9 Sir Raymond Evershed says:

> In order to answer the problem presented to the court, it is necessary to construe section 8. Three dates are mentioned in subsection (1), as regards each of which it is important to come to a clear conclusion. The first date is the date of the demand of a reference to arbitration. The second date is the next ensuing day on which the tenancy could have been determined by notice given at the date of the demand. The third is the date of reference. In stating the dates in that order, I am not indicating their chronology, but stating the order in which the dates appear in the subsection;

And later he says:

> Mr Binney says, therefore, that, in the absence of express language, the J court cannot hold that there is any specific limit of time applicable. I do not agree with that argument. As I construe this subsection, though it is true that it makes no express reference to time (as is made in other sections) yet I think that the necessary and inevitable implication of the language which Parliment has used is that the termination date, that is the date which is here called "the next ensuing day", must be after the date of the reference.

Denning LJ, on p 13, says:

> On the true construction of the statute, I am in full agreement with the Master of the Rolls that the reference must take place within that year, by which I mean that the arbitrator must be appointed and must accept his appointment within that time.

I regard *Sclater v Horton* as binding authority for the proposition K that the arbitrator must be appointed before the next ensuing day, and that the decision must be followed unless and until some different construction is placed upon section 8 by the House of Lords.

Also agreeing that the appeal should be allowed, KERR LJ said: Section 8 of the Agricultural Holdings Act 1948 provides the statutory basis for the review of rents for agricultural holdings by means of arbitration. If either the landlord or the tenant serves a notice demanding arbitration for this purpose, there is to be a reference to a single arbitrator by virtue of section 77 and the Sixth Schedule to the Act. Under paragraph 1(1) of this schedule the L arbitrator is to be appointed by agreement between the parties or, in default of agreement between the parties, by the minister upon application by either of the parties. Subject to the option as to whether the arbitrator is to be appointed by the parties or by the minister, or whether any appointment made by the minister is revoked by the consent of both parties (see paragraph 3 of the schedule), the effect of section 77 is that any rent review in relation to agricultural holdings can only take place by means of the arbitral machinery laid down in the Act.

Upon a demand for a reference to arbitration, the question to be decided by the arbitrator is "what rent should be payable in respect of the holding as from the next ensuing day on which the tenancy M could have been determined by notice to quit given at the date of demanding the reference?" In this connection three general matters concerning the scheme of the Act are relevant. First, once the contractual term of the tenancy has expired (as was the case here), there is a continuing tenancy from year to year. Secondly, such a tenancy can only be determined by at least one year's notice to quit. Thirdly, once there has been a rent review by arbitration, it operates as a determination of the rent for the next three years irrespective of whether the rent is increased, reduced, or left unchanged by the arbitrator's award.

Having regard to the second and third of these factors, it follows that if a demand for arbitration is made in (rental) year A, the

A determination of the rent by arbitration cannot take effect before the beginning of year C, and it then applies to the triennium of years C, D and E. Further, under section 8 the arbitrator must determine the appropriate rent "at the date of the reference", and the effect of the concluding paragraph of this subsection is that this is to be based, in effect, on the then prevailing rents in the open market, subject to certain qualifications in subsection (2) which are immaterial for present purposes.

Subject to the foregoing matters, section 8 does not contain anything in the way of a chronology, let alone a prescribed timetable. Thus, the demand for a reference can be made at any time during year A, and no time is stipulated for the appointment of an
B arbitrator thereafter. There is no definition of "reference", and the expression "the date of the reference" is also left undefined. In the context of arbitrations generally, the word "reference" is a loose term capable of covering a number of matters, beginning (at the latest) with the appointment of the arbitral tribunal and covering the period up to and including the final hearing, and perhaps up to the award.

One has only to read through the Arbitration Act 1950, consolidating the Acts of 1889 and 1934 which were then in force, to see that the word "reference" is used in a variety of senses in the context of the process of procuring the resolution of a dispute by means of arbitration. It denotes events over a period of time and not
C a particular moment in time. Moreover, in the general context of section 8 there is also no need to assign any precise chronological meaning to the words "the date of the reference". These words govern only the then prevailing market rents, and since paragraph 13 of the schedule provides that, subject to any enlargement of time by the minister, the arbitrator is to make his award within 56 days from his appointment, a period during which market rates are unlikely to fluctuate violently, the scheme of section 8 does not necessitate any precise determination of any particular day as "the date of the reference". However, since the arbitrator's determination will fix the rent as from the beginning of year C in the way which I have described, the sensible way of operating the section is
D clearly that there should be a close degree of proximity in time between the date of his appointment and the beginning of year C. Provided that the arbitrator's appointment is made close to that date, I cannot see that it makes any legal or practical difference from the point of view of the operation of the section whether it is made before, or on, or shortly after this date, though the most sensible course is no doubt to make it shortly before. Even in that event, however, the final award is unlikely to be published until one or two months after the beginning of year C, and it will then date back to this extent.

The foregoing approach is illustrated by the way in which the section was operated in the present case. The anniversary date of
E the tenancy was February 2 and the rent was payable in arrears on that date and on August 1 of each year. The demand for a reference had been made by the landlord in November 1979, and we were told that he applied to the minister to appoint an arbitrator in December 1980. The rent review could accordingly not take effect earlier than February 2 1981, and on January 5 1981 the ministry informed both parties that the minister would make an appointment. Mr J G Cleverly was evidently one of the persons who were on the panel of arbitrators referred to in paragraph 1 of the Sixth Schedule, and on January 29 1981 his appointment on behalf of the minister was sealed to determine the rent "as from the next ensuing day on which the tenancy could have been determined by notice to
F quit given at November 29 1979", which was February 2 1981. Notice of his appointment was duly sent to both parties; it was received by the landlord on February 2 but by the tenant only on February 3. It is this latter fact alone which is now said to have invalidated the whole process under section 8; and indeed the learned county court judge has so held.

To my mind this is a wholly astonishing proposition in the face of section 8. What is said on behalf of the tenant, however, is that this court is bound by authority to accept it. The main authority relied upon for this purpose was the decision of this court in *Sclater v Horton* [1954] 2 QB 1, to which I turn in a moment, and then — superimposed upon this decision — we were referred to a number of decisions in other contexts to which I turn briefly thereafter. The G
upshot of the whole argument was that there had to be an effective and complete appointment of the arbitrator on or before February 2 and that this also required that both parties should by then have had notice of his appointment. For the reasons explained below, I cannot accept any part of this argument. However, before dealing with it, I must mention two matters in passing, though in my view neither affect the result. First, it was common ground that Mr Cleverly had in fact accepted the appointment by February 2. Secondly, it was argued on behalf of the landlord that, since the tenancy was held "from" February 2 and the rent was payable in arrears, notification to the tenant on February 3 was still in time. I express no view on this question, since in my view it has no bearing H
on the decision.

I then turn first to *Sclater v Horton*. In my judgment what was said by this court in that case, albeit in relation to section 8, was said in a context which was so far removed from the present that it would be an abuse of the doctrine of *stare decisis* to rely upon the language of the judgments as governing the interpretation of section 8 for present purposes. I would therefore not accept that the way in which the judgments were expressed in that case governs the present case to any extent, ie quite apart from the further edifice of other authorities unconnected with section 8 which it was then sought to superimpose upon that decision. The landlord in that case gave a notice demanding arbitration shortly before the end of year A, and if this had been followed by a reference to arbitration J
the award would have operated as from the beginning of year C in the way explained above. However, the landlord did nothing to procure a reference, perhaps deliberately because rents were rising: see the judgment of Sir Raymond Evershed MR at p 9. Year B and nearly the whole of year C then went by. The landlord then served a further notice demanding arbitration shortly before the end of year C. Since any notice to quit given at that time could not operate until the end of year D, it followed that any reference to arbitration pursuant to *this* notice could not result in an award to review the rent earlier than from the beginning of year E. However, in relation to *this* notice the landlord proceeded to demand a reference to arbitration by applying to the minister to appoint an arbitrator. He K
in fact did so on the basis of the demand which he had made in year A (see *ibid*), and the minister duly appointed an arbitrator shortly after the beginning of year D, but (no doubt deliberately) in terms which did not identify which demand for arbitration was regarded as relevant for this purpose. The landlord then contended that the relevant demand for arbitration was the one which he had made in year A, so that, on the basis of market values in year D, the arbitrator was (according to the landlord) required to determine the appropriate rent retrospectively for the triennium which had begun with year C. The tenant, on the other hand, contended that, since that demand for arbitration had not been followed by any reference, the arbitrator could only determine the rent for the L
triennium beginning with year E.

That was the only issue before this court. All that fell to be decided was whether a request to the minister to appoint an arbitrator at the end of year C, which was followed by an appointment in year D, could be based on a demand for arbitration which had been made in year A, so that the arbitrator would have to determine, retrospectively in year D, the triennial rent as from the beginning of year C; or whether, as the tenant contended, the arbitrator could only determine the rent for the triennium as from the beginning of year E. This issue, as to which was the relevant triennium, was clearly stated in the judgment of the Master of the Rolls (see *ibid*). M

Although every word of the judgments in that case was directed to this issue alone, many passages in the judgments were subjected to minute verbal analysis in the arguments before us for the purpose of deciding the present case, in which the issues are wholly different. It is quite clear that none of the members of the court then had in mind any precise chronological correlation between the date of the arbitrator's appointment and the beginning of the period from which his award was to be determinative. This can be seen, for instance, from the following passages. First, the Master of the Rolls at p 11:

If a rent is to be fixed, the normal and fair date at which it should be fixed is

A the date when it begins to operate. I should have thought it plain that the section is contemplating that the rent is fixed by reference to a fair commencing date

And per Romer LJ:

If, on the other hand, the date of the reference (and, therefore, the date of fixing the rent) was at or just before the beginning of the new period, the parties would at least start off with a rent which, presumably, was a fair and proper rent at that time. That seems to me to be the contemplation of the legislature.

It therefore seems to me to be clear that *in substance* this court then approached section 8 in much the same way as I have
B indicated above. However, in formulating the judgments on the issue as to which of the two demands for arbitration, and which of the two triennia, were the relevant ones, the language which expressed the reasoning of the court incorporated two steps, which have been used as the sheet-anchor of the tenant's argument in the present case. This is clear from the passages which have already been quoted. First, the judgments proceeded on the basis that "the reference" must precede "the next ensuing day", ie the date as from which the rent review would be operative. Secondly, as the result of a concession in which both counsel acquiesced, and which was accepted by the court, the judgments also proceeded on the basis that the words "the date of the reference" were synonymous with "the effective appointment of the arbitrator". However, on the
C issues before the court in that case, no question arose as to any chronological precision in relation to either of these steps in the reasoning, or as to any precise correlation between "the date of the reference", let alone "the effective appointment of the arbitrator", and "the next ensuing day".

In these circumstances I cannot accept that these steps in the reasoning of the judgments for the purposes of that case govern the present one. In relation to what was there held to be "the next ensuing day", viz the beginning of year E and not of year C, the arbitrator having been appointed shortly after the beginning of year D, the precise chronological meaning of "the date of the reference", let alone of the "effective appointment of the arbitrator",
D was wholly irrelevant. The substance of the decision was that, in order to fix a fair rent on the basis of then prevailing market rentals, the date of the reference to arbitration and the beginning of the period from which the rent review would operate should be reasonably and fairly correlative.

Thus, suppose that in the present case the minister had, due to some administrative oversight, failed to seal the appointment of Mr Cleverly by February 2 or even 3, that the landlord had then pointed this out as a matter of urgency, and that the minister had then made the necessary appointment a few days later. Having regard to the scheme of section 8 and the substance of the decision in *Sclater v Horton*, is it then to be said that the landlord had lost his
E right to a rent review as from February 1981, and that he had to give a further notice demanding arbitration which could not have taken effect until February 1983? Similarly, suppose that it could not be shown that Mr Cleverly had accepted the appointment by February 2 or 3, but only a few days later, because he was away or because the letter of appointment had miscarried. There could be a number of slips of this nature covering the first few days of February. None of them would have any real significance in relation to the arbitration or its subject-matter. However, on the tenant's argument – so far as I can see — the landlord would have no remedy against anyone for the loss of the chance of obtaining an increase in rent over the next two years. There is nothing in the wording or purpose of section 8
F which compels this extreme conclusion, and in my view it would be an abuse of the doctrine of precedent to extract this conclusion from the judgments in *Sclater v Horton*. In any of these events the scheme of section 8 and the substance of the decision in *Sclater v Horton* would both have been met.

Accordingly, I would allow the present appeal on this ground alone. However, I also consider, in agreement with the judgments which have already been given, that the tenant must fail on the next step of his argument in any event. This was that there was no "effective appointment" of Mr Cleverly until the appointment had been made *and* both parties had had notice of it; or alternatively — which comes to the same thing — that the appointment of an arbitrator has not been effectively completed until it has been G
notified to the parties. This argument was based on two 19th-century cases and the acceptance of one of these, *Tew v Harris* (1847) 11 QB 7, in the decision of this court in *Tradax Export SA v Volkswagenwerk AG* [1970] 1 QB 534, although the point was *obiter* for the purposes of that case. However, all these cases were concerned with the appointment of an arbitral tribunal *inter partes*, and indeed with the appointment of an umpire or third arbitrator by two arbitrators appointed respectively by the parties. The constitution of an arbitral tribunal by such methods clearly requires co-operation and communication between the parties. In my view these cases cannot govern the question of what constitutes an "effective appointment" when the appointment is to be made by some outside H
agency. All that is necessary for that purpose, in my view, is the making of the appointment and its acceptance by the arbitrator or other tribunal. Once this has happened, the appointment as such is complete. However, there may then be a further factor, which was also present in all these cases. This is that the date of the appointment may have consequences in terms of time for the subsequent rights and obligations of the parties and for the conduct of the arbitration. For this purpose it is accordingly necessary to add a further implication to any provisions concerning the appointment of the tribunal in order to give business efficacy to the consequences which flow from such appointment. This implication is that for such purposes the appointment only takes effect as from the time when J
the parties have had notice of it, or at any rate that party — which is normally the claimant — on whom some obligation is imposed as from the date of the appointment. This is also the position in relation to paragraph 6 of the Sixth Schedule to the Act of 1948. I would therefore construe the obligation concerning the delivery of the statements of cases within 28 days from the appointment of the arbitrator in this provision as being subject to the implication that this requirement only begins to run as from the time when the parties have had notice of the appointment.

For all these reasons I agree that this appeal should be allowed.

The appeal was allowed with costs of the appeal and of the hearing before the county court judge. The question of law in K
paragraph 10 of the stated case, ie as to whether the arbitrator had jurisdiction, was answered in the affirmative.

CHANCERY DIVISION

November 4 1981

(Before Mr Peter BAKER QC, sitting as a deputy judge of the division)

SECRETARY OF STATE FOR SOCIAL SERVICES v BEAVINGTON L

Estates Gazette May 8 1982

(1981) 262 EG 551

**Agricultural Holdings Act 1948 — Minister of Agriculture's approval under section 2(1) of a licence to use the land as agricultural land for a period of 364 days — Land not suitable for grazing or mowing but intended to be used for arable farming — Object of obtaining minister's approval to avoid the creation of a tenancy from year to year with security of tenure — Period of 364 days expired without any steps having been taken to renew the licence, it being assumed by all concerned in the subsequent transactions that the minister's approval continued in force — Tenders were invited M
for the grant of the licence, it being stated that the approval of the minister had been obtained — The defendant submitted the highest tender, which was accepted by the authority which managed the land in question on behalf of the nominal plaintiff, the Secretary of State — The result was the creation of a contractual licence for 364 days which was not approved by the Minister of Agriculture and was not for grazing or mowing — Unless any of the points put forward on behalf of the plaintiff were valid, the transaction resulted in the conversion of the licence into a tenancy from year to year by virtue of the main part of section 2(1) of the 1948 Act — None of the plaintiff's objections held to be valid — The licence was not gratuitous; the fact that no date was specified for the commence-**

ment of the licence was in the circumstances immaterial; and the licence gave the defendant exclusive occupation — Plaintiff's action for possession dismissed.

This was an action for possession of two parcels of agricultural land, together amounting to some 45½ acres, which were part of the grounds of Leybourne Grange Hospital, near West Malling, Kent. As the hospital was in the national health system the title was vested in the Secretary of State for Social Services, the nominal plaintiff in the action. The defendant was Dr Frank Beavington, his doctorate being in agricultural subjects. The main question concerned the application of section 2 of the Agricultural Holdings Act 1948 to a licence granted to Dr Beavington.

David Neuberger (instructed by Bird & Bird) appeared on behalf of the plaintiff; Rodger Bell (instructed by Bracher, Son & Miskin, of Maidstone, Kent) represented the defendant.

Giving judgment, MR PETER BAKER QC said: This is an action for possession of land which is part of Leybourne Grange Hospital near West Malling in Kent. The land comprises some 45½ acres. The hospital is a psychiatric hospital which stands in extensive grounds. It is a National Health Service hospital and as such it is vested in the Secretary of State for Social Services, who is the nominal plaintiff in this case. But in fact he is of course assisted in the discharge of his responsibilities by a considerable administrative structure, and the particular administrative organ which has the general supervision of this hospital is the South-East Thames Regional Health Authority of which a Mr G F Goodall is the officer principally concerned with the matters which arise in this case. He is the senior assistant secretary for planning. In addition the hospital itself has an administrative staff, and the principal officer there is Mr Malcolm George Hodgson, who is the assistant sector administrator, and he is responsible for the farming and gardening activities. As I have said, there are extensive grounds, and these grounds in so far as they are not occupied by the hospital itself are used for farming or gardening. For that purpose Mr Hodgson is assisted by a farm manager and other staff who actually do the physical work, the farm manager being a Mr Osbourne.

The land in question is two parcels of land, one on each side of the main drive which leads to the hospital. One parcel is of some 29.5 acres and the other some 16 acres. Up until the events with which I am concerned this land had been farmed by Mr Osbourne under the supervision of Mr Hodgson. In April 1979 the decision was reached to give up direct farming or gardening in regard to those lands. Mr Goodall arrived at that decision and wrote about it to the Ministry of Agriculture. In his letter he said that these areas were not immediately needed for hospital purposes but were likely to be required in a year or two under plans which are currently being developed for improvement of mentally handicapped services. He went on:

> We wish to ensure the land is fully used until its future is determined and as we are unable to farm it ourselves any longer, we would be obliged if you would agree to a 364 day licence to avoid establishing an agricultural tenancy for the short period involved, which would then prevent us from selling the land.

That was the opening and the general background to it.

They were looking for a temporary letting which would avoid creation of an agricultural tenancy which would bring into train the protection of the Agricultural Holdings Act 1948. That was the plan, and it is based on section 2 of the Agricultural Holdings Act which, when the conditions are satisfied, provides for the creation of licences and contracts of tenancy which, though otherwise protected, would escape that provision. The section reads as follows:

> Subject to the provisions of this section, where under an agreement made on or after the first day of March, nineteen hundred and forty-eight, any land is let to a person for use as agricultural land for an interest less than a tenancy from year to year,

that is one case

> or a person is granted a licence to occupy land for use as agricultural land, and the circumstances are such that if his interest were a tenancy from year to year he would in respect of that land be the tenant of an agricultural holding,

that is another case

> then, unless the letter or grant was approved by the Minister before the agreement was entered into, the agreement shall take effect, with the necessary modifications, as if it were an agreement for the letting of the land for a tenancy from year to year.

That brings into protection lettings for less than a tenancy from year to year and licences to occupy land for use as agricultural land, but with this exception, that if the letting or grant was approved by the minister, that is the Minister of Agriculture, Fisheries and Food, before the agreement was entered into, then the agreement takes effect. Then there is a proviso which allows another exception and which is not dependent upon the minister's consent. It reads as follows:

> Provided that this subsection shall not have effect in relation to an agreement for the letting of land, or the granting of a licence to occupy land, made (whether or not the agreement expressly so provides) in contemplation of the use of the land only for grazing or mowing during some specified period of the year, or to an agreement for the letting of land, or the granting of a licence to occupy land, by a person whose interest in the land is less than a tenancy from year to year and has not by virtue of this section taken effect as such a tenancy.

That exception is limited to the particular agricultural purpose contemplated, for grazing or mowing. The land in question here was not fit for grazing or mowing because the fences were not in a good enough condition for them to contain animals, so this matter proceeded on the basis that it would be arable farming going on there, and therefore it was essential, if protection was to be avoided, that the letting or grant should be approved by the minister before the agreement was entered into. For that reason the minister was approached, and the letter, part of which I have already read out, is an application for such a licence. On June 12 1979 the minister gave his consent in these terms:

> In exercise of the powers conferred upon him by section 2 of the Agricultural Holdings Act 1948, the Minister of Agriculture, Fisheries and Food hereby approves but only for the purposes of the said section the grant of a licence to occupy by an agreement to be entered into after the date hereof the land at Leybourne Grange Hospital Farm, West Malling, comprising 36.52 acres and 17.12 acres, and shown hatched black on the plan annexed hereto signed by the undersigned for use as agricultural land for a period of 364 days commencing after, but not more than three months after, the date hereof.

That licence in terms expired by September 12 1979. In fact it was allowed to run out before any further steps were taken to find a licensee for this land. It is not part of my function to allocate any blame for overlooking that or indeed to consider whether any blame attached to anybody over it. But what is clear is that everybody concerned in the subsequent transactions in this case thought that the consent continued to run and continued in force. It was not until July 11 1980 at the earliest when people began to wake up to the fact that the licence given by the minister had expired before any further steps were taken pursuant to it. On July 11 1980 the solicitor then acting in the matter for the regional health authority alerted the authority's official there to the matter. Thereafter it quickly came to light that there were difficulties concerning this licence.

The case was pleaded on the basis that despite its apparent expiry the licence did continue to govern the situation and take the subsequent licensing out of the section, but that point was abandoned earlier in the hearing and in my judgment quite rightly abandoned as it seemed hopeless to run that particular point.

As I have said, there was a licence granted, and I must look into what did in fact happen subsequent to September 12, which is the final date on which the consent was available for the grant of the licence. In order to find a licensee or person to take this land the method adopted was to engage a local firm of auctioneers and estate agents, Page & Wells. The assistant who actually dealt with it was a Mr Murray James Wills, and the method adopted was to invite tenders. Particulars were prepared. Some of the particulars I must refer to. I do not want to state all the particulars, but they took the form of inviting tenders. After setting out the description of the land, which I have already mentioned, they stated: "The land is easily defined although in part it is unfenced and therefore unsuitable for keeping stock." So quite clearly the whole thing proceeded right from the outset on the basis that this was not for grazing or mowing or for any sort of use of animals but was for arable farming. There were no buildings included in the licence. "The licence shall be subject to the approval of the Ministry of Agriculture, Fisheries and Food and full vacant possession will be required when the licence expires." Then they go into the means of access, which I need not go into, at all events at this stage. "Rent shall be paid in advance by two equal instalments, the first on completion of the licence and the remainder six months later." Then there is a point about sporting rights which are reserved. Then they go on: "For the benefit of prospective tenderers we attach a list of terms which may form the basis of the licence, although the licensor reserves the right to vary the form of licence appropriate to the letting of these parcels of land. Applicants wishing to submit an offer for the licence of this land should complete the attached form of tender and return to the licensor's agents, Messrs Page & Wells, not later than 12 noon on Friday November 2 1979." I think that is all I need to refer to in the particulars.

In the attached form of licence there were the usual sort of terms for an agricultural licence, and there are two paragraphs which I should refer to: "8. A right of access and use without prior notice for all purposes other than the subject of this licence is to be reserved at all times for the landlords and their agents." Then, "10. The approval of the Ministry of Agriculture, Fisheries and Food has been obtained for the creation of this licence and full vacant possession will be required at the termination of the agreement." That was not put forward as the terms on which it would be allowed but it was the sort of framework of the terms that the licensor had in mind while reserving the right to vary them.

A number of farmers were interested in taking this land and went to see it and looked round it and submitted their tenders. One gets to the point that the tenders are in, and the highest tender was that of the defendant, Dr Frank Beavington. He is a doctor of philosophy in agricultural subjects. He was already farming in the area at Church Farm, Ryarsh, which is I understand quite near, so that he was obviously looking at this as an addition to his farm for the period of the licence and working it in with his farm. I should read the tender which he submitted. It was on a printed form supplied by the estate agents and he filled it up without alterations. He merely filled in the blanks as he was required and did not attempt to vary the terms in any way. It starts off: "Form of Tender Land at Leybourne Grange Hospital, near West Malling, Kent. I/We Dr Frank Beavington of Church Farm, Ryarsh, Maidstone, Kent hereby offer the following sums for the grant of a 364 day licence for the land described in the letting particulars. £652.00 for the parcel of land described as Block 1." That was the smaller parcel. "£1,344.00 for the parcel of land described as Block 2. I/We agree that in the event of an offer being accepted I/We will complete without undue delay the form of licence provided by the licensor and will pay six months' rent on completion of the document." Then he signs it and "Dated November 2 1979", the last day for tenders. Then it says: "The solicitors/land agents (if appropriate) instructed to act for me/us are Messrs Bracher Son & Miskin", and then, "If required, the agents who would act on my behalf are Messrs Geering & Colyer." He was asked to set out details of his farming experience and he sets that out, but I do not think I need read that. Then, "Please state if you are not prepared to take a licence of one block of land only." That does not arise because he was willing to take the two and he was offered the two.

That tender went in on November 2 1979. On November 5 Mr Wills reported to his principal, Mr Goodall of the South East Thames Regional Health Authority, on the result of the tender operation. He sets out a list of some nine farmers who had offered and Dr Beavington's was the highest in an aggregate value of £1,996 for the two parcels. He says: "I am quite pleased with the response and hope that you are satisfied with the offers which have been submitted. All the applicants are anxious to have access to the land in the hope that winter cereal can be planted and if possible it would be preferable if a licence could be completed before the end of this month." That was sent on November 5. After consultations with the district valuer among other people it was decided to accept Dr Beavington's tender. Mr Wills, and there is no question that he had authority to write the letter, wrote to Dr Beavington in these terms:

> Further to my letter dated November 6 I am now pleased to advise you that your offer of £1,996 has been accepted by the South East Thames Regional Health Authority and the letting has been approved by the Ministry of Agriculture. I have asked my clients to submit the licence document to Messrs Bracher, Son & Miskin and unless I hear from you to the contrary I trust this is in order. It has now been decided that the proposed route of access to the smaller parcel of land south of the main entrance drive would be inadequate for large farm vehicles and I enclose a plan on which I have marked in red the proposed alternative route which I hope you will agree is better for all concerned.

On the face of it, that exchange of documents, the tender submitted by Dr Beavington and accepted by that letter, would appear to be a contract. The point of having tenders is to bring about a contract in which the purchaser or taker, the lessee, licensee or whoever he is, makes an offer which is capable of being accepted by the owner. Once it is accepted, in the ordinary way a contract results. Two reasons have been put forward as to why that result should not ensue. The first rests on the fact that it was to be followed up by a subsequent licence. When the plaintiff's solicitors came into the matter to deal with the conveyancing, they wrote a letter to Dr Beavington, it having emerged in the meantime that Dr Beavington was going to deal with the matter himself and was not at any rate at that stage going to engage solicitors. They wrote: "We thank you for your letter of January 18 which crossed our letter to you of the same date. We have copied your letter to our clients with a request that they let us have their instructions on the proposal you make. For the present however we are proceeding on the basis of your original tender which, subject to licence, has been accepted." That was a letter by the solicitor, Mr Zonena, who was then an assistant solicitor with Messrs Bird & Bird, the solicitors acting in this matter then and now for the plaintiff. What he told me about it was that "the last sentence reflected my understanding. I added the words 'subject to licence' to indicate that the whole matter, the whole range would be subject to a formal licence being actually expedited." I am quite unable to accept that view of the matter.

As I see it this was intended to be a binding contract. There was nothing in there that it would be subject to contract or subject to licence or anything of that sort when it was being handled by Mr Wills. And the normal intention is that a tenderer should be bound, once his tender is accepted, and it is not intended that he should be able to back off from it. Otherwise why go through the system of tendering? What it is, as I see it, is that the tenderer makes an offer to take a grant of the licence in terms of the particulars, and when that is accepted then the landlord or owner becomes under an obligation to make the grant corresponding to it. As I say, I cannot accept that as a reason. It was introduced subsequently, and in my judgment ineffectually, by the solicitor.

Then a more formidable point was taken by Mr Neuberger for the plaintiff. The grant was said to be defective because there was no date in the documents which I have read out for the commencement of the term. Therefore, even assuming that it was intended to be effective, the whole matter was too uncertain to be enforceable. That is a point which is not new in relation to leases. There is quite old authority but it was recently considered in the case of *Harvey v Pratt* [1965] 1 WLR 1025. This concerned an agreement to lease a property known as a service station "including offices therein, at an inclusive annual rent of £2,125 per annum exclusive of rates for a period of 21 years with option to renew or purchase at the end of that period." There was nothing in that agreement to say when the 21 years was to begin. Nothing further was done about it in the sense that the proposed lessee did not go into possession in any way or get anything started or pay any rent or anything of that sort. The landlord brought an action claiming that the document was not binding or an enforceable contract. It seems what had happened was that the lessee or potential lessee had registered an estate contract and these proceedings were to get rid of that on the basis that there was no enforceable agreement. I think I can best take from Davies LJ's judgment on p 1027 the principle on which this failed.

> The analogy which [counsel for the proposed lessee] seeks to draw between an agreement or contract for sale of freehold and an agreement for a lease

A is, in my opinion, quite an unsound one. In the case of a contract for the sale of freehold, the subject-matter is ascertained, namely, the land. In the case of an agreement for a lease, if the length of the term and the commencement of the term are not defined, then the subject of the agreement or contract is uncertain. Therefore, there is no agreement.

Then Russell LJ on the same page:

> Never has it before been suggested that in the case of an alleged contract to grant a lease such as this, where nothing whatever is said to indicate to what date the term is to commence, the law will imply that it will commence at the expiration of a reasonable time from the contract, although opportunity to make such a suggestion in reported cases has by no means been lacking.

B Then he deals with some authority, and the learned lord justice goes on:

> The truth is that the parties must themselves define the subject-matter of their bargain, and a term of years can only be defined by indicating the commencement and the termination.

One notices that in that case there was a term of years, that that was the subject-matter of the contract, the grant of a term of years, and there was no circumstance to indicate when it would begin. The situation here is, as I see it, very different. What was offered here was a licence. There was certainly no question of any interest in the land under a licence. The 364-days period is not the subject-matter. What was being offered here was a contractual right for the use of the land for the next growing season. In the context as I see it
C the important matter was that the land was being offered in the autumn for the next growing season, and what was being offered was the chance to cultivate and take the crops off this land. The actual date in the sense of one day or the other on either side of a few days was secondary. There is also this matter, that the contract, which is one of the contractual documents as I have found, did provide that in the event of the offer being accepted, "I", that is to say Dr Beavington "will complete without undue delay the form of licence provided by the licensor and will pay six months' rent on completion of the document." So there was a further step to be taken which would fill in terms and make the matter more certain. As I see this type of contract, having regard to what I have said as
D to its timing and being agricultural, the circumstances as to the background of this, that as soon as the tender was accepted, the landlords then became under an obligation as soon as reasonably possible to let Dr Beavington in. That was the basis on which the contract had been made. It could vary one day to another. That might be dictated by the time needed for the preparation of the licence and also for the time needed for the previous occupants, the hospital administration itself, to vacate the ground and get it ready. In that contractual situation as I see it, within those limits the obligation was to grant the licence to Dr Beavington as soon as reasonably possible, and certainly in time for him to sow and reap for the next season. But within those limits the actual date could be
E varied. That is on the basis that nothing further had been done at all. The licensor could not have turned round and said, having accepted the tender, we have had second thoughts and we are not going ahead at all. But in this case something else did occur and took it further than *Harvey v Pratt* in that occupation did ensue. Part of the defence is put in this way in paragraph 3: "The contractual licence was made by: (i) the defendant's written tender and offer", and (ii) the letter of acceptance, and (iii) "the oral agreement of Mr Hodgson acting for and on behalf of the plaintiffs, on an occasion between about November 21 and 30 1979, that the defendant could take possession of the premises and start the cultivation thereof as soon as he wished." I have heard some evidence about this. It is
F common ground that towards the end of November Dr Beavington went to the hospital and saw both Mr Hodgson and Mr Osbourne. And I think it is common ground that Mr Hodgson said, "it is okay to go in". This was a meeting towards the end of November, I think about the 26th. But there is a curious conflict about the whereabouts of the meeting, in which particular office. I do not propose to resolve that. There is also a conflict as to the true background of the meeting. Mr Hodgson said he got to understand from Dr Beavington that he, Dr Beavington, was the highest bidder, that the defendant knew that he had been successful, "and I was not aware that any tender had been accepted at that time." He said in relation to the documents that he had seen lists, a copy of the report of tenders was sent to him but he had not seen a par- G
ticular letter in which Mr Wills reported further to Mr Goodall that he was proposing to accept one tenderer and write to the others advising the other tenderers that they were unsuccessful. So when Dr Beavington arrived that was the first he knew that Dr Beavington had been the successful tenderer, but he, Mr Hodgson, accepted it from Dr Beavington and said, "well if that is so you can go on the land."

Dr Beavington puts it differently, in this way. After he had spoken to Mr Wills and he learned that he was the successful tenderer, then he spoke to Mr Wills about when he could get on to the land. Mr Wills told him he would have to talk to the staff at Leybourne Grange. So he telephoned the farm manager and said H
he would like to come and talk to him, and the next day he went there and saw the farm manager in his office. "He appreciated what I had come for and said you had better go and talk to Mr Hodgson. He said, I think you will want to get on the land as soon as possible. I said I do because it is getting late to farm wheat and it is already too late for barley. He took me to another office where Mr Hodgson was and introduced me to Mr Hodgson." He had not been to that office before. And according to Dr Beavington Mr Hodgson knew what it was all about, and it appeared from his manner that Mr Hodgson knew that Dr Beavington was the successful tenderer. In so far as it is necessary to resolve that conflict I much prefer the evidence of Dr Beavington. It seems to me in- J
conceivable, having regard to the fact that Mr Hodgson was furnished with a copy of Mr Wills' report of the outcome of the tender, that he was not subsequently advised that Dr Beavington was the successful applicant. After all, it came directly within the range of his responsibilities, the handover of the farming land to Dr Beavington. And looking at the letter of November 20 1979, a letter to Mr Goodall from Mr Wills, Mr Wills says: "I note that the district valuer has approved the letting to Dr Beavington", and he would be writing to the other applicants who tendered for this land advising them they were unsuccessful. "I have also received a telephone call from Mr Hodgson of the hospital regarding the right of access to the land and he informs me that you intend to use the original access to the northern larger area." Then he goes on about K
that. So at the same time that he was in communication with Mr Goodall he was in communication with Mr Hodgson over the details of the handover and the access that the successful tenderer was to use. I do not find specifically that Mr Hodgson had this letter or a copy of it, I accept his evidence on that point. But I do find it inconceivable that Mr Hodgson should not have been informed that Dr Beavington was the successful tenderer. I think his recollection has faded. I accept Dr Beavington's evidence on that episode.

The outcome of that was that Dr Beavington got the green light, as it were, from Mr Hodgson and within a day or two, on December 3 as I find, he was starting work on the land. This was L
evidenced by a diary which is kept by Mrs Beavington and which has an entry which I accept as the day he did start. That as it seems to me was an admission, the letting into occupation of Dr Beavington on to the land. It was in anticipation of the licence. At that point no licence had been sent to Dr Beavington, so that there is no question of his failing to sign it and return it at that stage. In my judgment they were already under an obligation to grant the licence to Dr Beavington. This seems to me to be just a matter of the performance of their obligation to let him in as soon as reasonably possible. I am quite satisfied that Mr Hodgson had authority to do that, and I accept the submissions of Mr Bell who appears for Dr Beavington on this point. The plaintiff had put Mr Hodgson in the M
position where he had authority. Dr Beavington was sent by the estate agent, Mr Wills, to the hospital staff itself and not to any higher authority to ascertain the details of taking possession of the land and getting on with the work, when it would be convenient to do so, and it was Mr Hodgson's direct responsibility to see to the proper management of the land at Leybourne Grange Hospital which I take it included the access of tenants and licensees to it.

Accepting, as I do, that there was a binding contract on the grant of this licence, there is no question that it was gratuitous, the alternative way it was put. The licence was not in any way gratuitous. It was, as I see it, enforced by the obligation that was created by the

A tender and acceptance. By having his tender accepted Dr Beavington came under liability to pay what he had tendered in the way of rent, and the terms for that in the tender were that on the offer being accepted he would complete the form of licence and pay the first six months' rent on completion of the documents. As I see it he came under an obligation to pay that amount, but the actual time for payment depended upon the licence being produced to him and being completed. That was a matter wholly within the control of the licensors. Dr Beavington's obligations in the matter were to complete it without undue delay as soon as it had been proffered to him. Of course the reason for that is obvious, that he was not to delay the completion of the licence and thereby delay the actual B date when the first instalment of rent was to be made. But that was wholly in the control of the licensors to proffer the licence and then the liability would crystallise to make immediate payment.

I thus find that there was a contract. I now have to consider a further point taken by the plaintiff on the terms of it, an alternative ground for saying that what is claimed in this case by the defendant does not arise, even assuming a contract to grant a licence exists. I must explain this a bit, reverting to section 2 of the Agricultural Holdings Act. The method by which this works in relation to licences is that if there is granted a licence to occupy land for use as agricultural land and the circumstances are such that if the interest were a tenancy from year to year he would in respect of that land C be a tenant of an agricultural holding. Then "the agreement shall take effect, with the necessary modifications, as if it were an agreement for the letting of the land for a tenancy from year to year." The effect is that where you have a licence to occupy land for use as agricultural land, it gets converted into a tenancy from year to year and that of course brings into train the protective provisions of the Agricultural Holdings Act, protective of agricultural tenants that is. The courts have had certain difficulties with making the modifications necessary to convert some of the licences which have come before them into tenancies from year to year. And one of the difficulties which has arisen is: if it is not a licence which gives exclusive control or occupation, it is impossible, whatever modifica- D tions you make to it, to convert that into a tenancy from year to year because it is of the essence of a tenancy that there shall be exclusive possession. That is the setting of that, and I was referred to some illustrative cases where what are undoubtedly licences to use land for farming purposes have nevertheless escaped from the provisions of section 2 owing to the impossibility of converting them into tenancies from year to year bringing them within the ambit of the Act.

One case was *Harrison-Broadley v Smith* [1964] 1 WLR 456. The essence of it was there was a partnership agreement. One party was the joint administratrix of a deceased husband and she remained in occupation of the farm of 176 acres. She had no power to grant a E lease. She entered into a partnership agreement with the defendant, a working farmer. Each was to contribute £1,000. The defendant was to work the land, had the powers of management, and was to take nine-tenths of the profits and the other party was to take the remaining one-tenth. The partnership was terminable. That was the essence of the agreement, but the point there was that the defendant, while he was to do all the work on the land, was in partnership with somebody else. He was not exclusively the farmer, so that it was the joint parties on the ground. There was only one farming business that was going on, but it was shared between two people, the landlord as it were, and the licensee. Although he had a licence to go on it was not exclusive in the sense that he could ex- F clude the licensor from the land, because she also had a share in that business and indeed took a tenth of the profits of it. It is not difficult to see that that partnership arrangement could not be converted into a tenancy from year to year which would have had the effect of excluding totally the joint administratrix.

Another case is *Bahamas International Trust Co Ltd v Threadgold* [1974] 1 WLR, 1514. In this case there was a separate farming business being carried on. The defendant's mother had been the tenant, but on her death the tenancy determined. The arrangement with the defendant, who continued to carry on the dairy business, was that he was allowed to remain in the farmhouse rent and rates free for a limited period. By clause 2 of the licence document they paid him as caretaker of the farm. "By clause 3 they G allowed him to retain his cows and other livestock and to sell milk until October 31 1969." This was a farm of 11 acres of pasture, quite a small place, but it was larger than was necessary for the cows that he was allowed to retain. He was not allowed to introduce other cows, but just maintain the cows that were already there. And one can see that that arrangement could be given effect to without at the same time excluding the licensor from the land to make what use of it he could which did not in some way impinge on the rights granted to the defendant. This case somewhat surprisingly went to the House of Lords, and it turned on the construction of the document I have referred to. Lord Diplock delivered the only reasoned speech. I think the essence of it is this, that he would H dismiss the appeal "upon the ground that the agreement of March 27 1969, the terms of which must be unique, did not upon its true construction grant to the appellant the right to use the land for agricultural purposes to the exclusion of its use for agricultural purposes by the respondents." That seems to be the *ratio* of that case, that it was not excluding the right to use the land for agricultural purposes by the respondents. It is pressed on me that one would have to show that not only was the lessor excluded from the land for agricultural purposes but also for any purposes. If he could use it for any purpose, although not agricultural, that would take it out of the ambit of section 2, but I do not see that as the *ratio* of the decision in the *Bahamas International Trust* case. J

Those are the authorities on agricultural holdings which bear on exclusivity. Another case was cited to me on this point. It is not an agricultural case. I propose at this stage to return to the facts before dealing with that case.

We must now go forward in time from the point when Dr Beavington first went on the land in December 1979. The licence which was envisaged in the tender document was not sent out until June 13 1980. The reason was not explained to me; but it was explained this far, and this I accept, that the solicitor concerned prepared it in January and sent it to the regional authority but it did not come back from there until shortly before June 13 1980. We find a letter from Mr Zonena, the solicitor, sending the draft to Dr Beavington. K In that draft there is this clause, no 9: "The licensee shall not impede at any time the entry of the owner's servants or agents on to the land for any purpose whatsoever." And I mention that in clause 12 there is stated that the approval of the Ministry of Agriculture, Fisheries and Food had been obtained for the creation of this licence and full vacant possession will be required on or before the end of the period. And I see in clause 6 that the land is to be used solely for the purposes of agriculture. Then in clause 1: "The term of occupancy to run from 1.11.79 to 30.10.80."

As to clause 9, much reliance is placed on this clause as indicating that this is a licence which does not fall within section 2 of the Agricultural Holdings Act. Its provenance is as follows: it is L not to be found in the specimen document which was sent out by the estate agents which I have already mentioned. It was introduced by Mr Zonena into clause 9 because he thought that was consistent with what had been agreed. Dr Beavington was not being given an agricultural tenancy. So that was a form which he thought was appropriate to create a licence as opposed to a tenancy, without specially differentiating between exclusive and non-exclusive licences. This document in general was acceptable to Dr Beavington. He had no real quarrel with any of its terms except the first one, that the period of occupancy was to run from 1.11.79 to 30.10.80. It did not accord with the facts as he understood them. He had not been there until December, not been allowed on, or had M not got on, until December, so, as he saw it, it was unfair that the year should start to run from a month earlier which in any event was the day before the tenders had to be submitted. That is a reasonable objection. Dr Beavington amended the draft and sent it back and sent his cheque. He altered the provision to begin on January 1 1980, and run for 364 days from then. So it went back with an amendment that he put in, January 1 1980 to December 30 1980. That again did not quite accord with the facts. That was a bit too generous to Dr Beavington if one accepts that he went in on December 3. But be that as it may, that is what he did, he amended it in that sense. He signed it. He had no quarrel with the rest of it,

A and he sent it back dated July 2 as one can see from the completed document. However, that alteration was not acceptable to those acting for the plaintiff, or not immediately acceptable. They were insisting that it had to be as originally drawn. It was sent back to Dr Beavington to be altered back again, but while it was in Dr Beavington's hands the plaintiff resiled from the position he had taken up in regard to the date and was prepared to accept Dr Beavington's proposals. That emerges from the letter of July 21 1980. So one has the position that Dr Beavington accepted the terms in general but the period was unacceptable and he signed the document in that sense. That was returned to Dr Beavington but subsequently the date that he had first proposed was accepted, but of B course by that time the document was in Dr Beavington's hands and, as I have said, that particular point was resolved by July 21, but in the meantime much more serious matters had arisen over the minister's consent.

On July 11 1980 there was a series of telephone calls by the solicitor and a note in the bundle of his of what occurred on that occasion. Apparently what woke the parties up to the fact that there was a serious difficulty about consent was this, that the solicitor, as the date might be going to be changed, thought that it might be necessary to check that the licence covered the proposed new date. Up to that date, as I find, no one had the slightest suspicion that the licence that was proposed was not within the original C consent. Of course a very unpleasant surprise was in store, because when the licence was inspected it was apparent that not only did it not cover the revised dates, it did not cover any date that was under consideration at the time. That unpleasant fact having been unearthed, the solicitor's note goes on: "In the afternoon I received a call from Dr Beavington who said that he had received our latest letter and was just wondering where I had got the dates specified for the licence, since he could not recall ever having agreed them." Then having gone into that with the doctor Mr Zonena said: "I then mentioned the Minister of Agriculture point and that I was waiting to hear from the clients on this. If that consent did not restrict the client it might be that they would agree to the dates pro- D posed by Dr Beavington." And it is a fact that Dr Beavington on the same day himself was getting in touch with the Minister of Agriculture to check the position as to the licence because we see that the appropriate official of the Ministry of Agriculture is writing to Mr Goodall that he had received a call from Dr Beavington on July 11 1980. It was Dr Beavington's evidence that that is purely coincidental. He denied that he was prompted to do that by the solicitor but he did it independently. But on that point I cannot accept the evidence of Dr Beavington, and I accept the evidence of Mr Zonena supported as it is by the contemporaneous note.

I have gone into that because I thought I should give my findings on the matter in case the matter goes further. For my part I am E prepared to accept and deal with the case on the basis that Dr Beavington did agree to the terms of the licence, and the effect of what was done was that he went into occupation on the basis of the terms of the agreement which he signed on July 2 1980, and those were the terms on which he was and still is occupying the premises. It is true it was not completed and was not accepted by the plaintiff, but I am prepared to accept that those are the terms on which this land is held or was held by Dr Beavington.

So the question then is: is clause 9, which as I find is under the terms on which he is holding, exclusive? It was very much pressed on me that that was sufficient, because it reserved rights to the licensor which were inconsistent with an exclusive licence and was F so inconsistent with it that it was not capable of a conversion into a tenancy from year to year and fell on the *Bahamas International Trust Company* and the *Harrison-Broadley* side of the line. There was prayed in aid of that and pressed on me the case of *Shell-Mex & BP Ltd v Manchester Garages Ltd* [1971] 1 WLR 612. This was a case which had nothing to do with agricultural holdings, but was a case where the question was whether the transaction fell within the ambit of the Landlord and Tenant Act 1954, Part II, the business premises protection. In that case the issue was whether there is a licence or a lease. Any licence, whether it is exclusive or non-exclusive, will escape the ambit of that 1954 Act because that Act only applies to leases. There is no question in that Act of converting licences into leases in the way that there is in the Agricultural G Holdings Act whereby certain licences are converted into annual tenancies. So it would not be sufficient for the plaintiff's purposes to show that this was a licence. To be of assistance in this case he would have to show that it was a licence which was non-exclusive and covered the present situation. The subject-matter was a filling station with land behind it in Manchester. It was owned by the plaintiff oil company, and they allowed the defendant garage company to go into occupation. The back land which was a storage and servicing area was covered by a genuine underlease for 21 years. The filling station was covered by a document called a licence which expired on July 31 1970 and by the terms of the licence the defendants agreed to do all they could to foster the sale of the H plaintiff's products. Under clause 19 the defendants agreed not to impede the plaintiff's rights of possession and control of the premises. It is suggested that that clause, which I have summarised, is parallel to the clause 9 we have in this case. The plaintiff petrol company attempted to put an end to it, and then the defendants claimed that they had exclusive possession of the filling station and the licence constituted a tenancy to which Part II of the Landlord and Tenant Act 1954 applied. "Held — that clause 19 showed that the transaction concerning the filling station constituted a licence and not a tenancy. Accordingly since the licence had ended the judge was right to give summary judgment for an injunction." J This went to the Court of Appeal and the essence of the judgment of the Master of the Rolls, I think, is this. I am reading from p616D so far as it concerns this matter. The learned Master of the Rolls sets out the conditions of the licence. He sets out at C the full terms of the clause in question, the clause not to impede: ". . . not to impede in any way the officers, servants or agents of the company in exercise by them of the company's rights of possession and control of the premises, and in particular to give all reasonable assistance and facilities to such officers and servants or agents for the alteration at any time of the layout, decorations or equipment of the premises." Then he goes on: "That shows that Shell-Mex Company's men can go and visit the premises whenever they like. K The Manchester Garages are not to impede them in any way, they are to give them assistance. Those provisions point to a licence and not to a tenancy." Then the Master of the Rolls goes on: "Mr Dillon" (he was for the defendants) "says that Manchester Garages Ltd have exclusive possession and that that carries with it a tenancy. That is old law which is now gone. As I have said many times, exclusive possession is no longer decisive. We have to look at the nature of the transaction to see whether it is a personal privilege or not." As I read it, the Master of the Rolls was saying, even given that it is exclusive, nevertheless it may still be a licence and that is sufficient in that case to exclude it from the Landlord and Tenant Act 1954, Part II. If that is the true basis of the decision then obviously this is of no assistance in this case, because an exclusive L licence would be fatal to any point that the plaintiff takes here. Sachs LJ, who is looking more at what went on, starts his judgment in these terms:

> Upon looking at the substance of the matter as a whole it becomes apparent that the dominant objective of the contractual relationship between the parties was to further the promotion of the sale of the Shell-Mex Co's products on the site which they had selected with the aid of the structures and equipment which they had provided and over which they were to exercise a right of control including the right to deal with the layout and equipment as they desired from time to time. The general tenor of the relevant document of August 1 1969 points strongly in this direction when taken as a whole.

M

Then he goes on to say that the tenor of the document is consistent with that assessment. The essence of that judgment as I see it is that you look at the factual situation and see that this really was a joint enterprise between Shell-Mex on the one hand and the garage company on the other for the exploitation of this land. It would not be sufficient simply to find a clause which perhaps permitted it, taken in isolation, but one has got to look at the circumstances to see what the parties intended. The clause assists, but one cannot by putting in a clause convert what is one arrangement to another. If it is an arrangement for exclusive possession on the true facts of it, one cannot convert that by reserving a power to intervene if that would be totally inconsistent with what the parties had intended to do. One can well understand that on the basis described by Sachs LJ it was a non-exclusive licence, or could be argued to be a non-exclusive licence, because looking at the matter as a whole that is what the parties intended. Shell-Mex were to go in, not actually to run the business, but were to have an important part in it. It was their goods that were being sold, they provided the selling area and structures, they could go in and change them as required and direct the conduct of the business.

But when one comes to the facts of this case the matter is very different. This was an arable piece of land that was intended to be and was in fact used as arable land by Dr Beavington and no one else during 1980. As Mr Wills put it, right at the beginning I think of his cross-examination: "I understood that only the licensee would farm the land. There was no question of anyone from the hospital farming the land." That was the basis on which they went ahead, and there was no evidence that anyone else farmed it. The activities that did go on from the hospital, I think the evidence showed, were these. There was a piece of woodland at the far side of the land for which the hospital authorities remained responsible and to which they required access across a portion of the land. I was shown the route on which they went in order to get access to that. The land is used to carry drains and similar services to the hospital. There was a tree in the middle of the land which from time to time required attention. Tree surgeons had to go to attend to it. The land was used for cross-country running at times. Cross-country events were organised and went round it. I think the only other matter was occasionally the inmates of the hospital were found in the land or might wander into it and had to be gathered or shepherded back by the hospital staff.

None of that in my judgment detracts from the exclusive nature of the grant to Dr Beavington. They are all intermittent. Certainly the drains and the access to the woodland come under the heading of ordinary easements.

Nobody has ever suggested that an easement was inconsistent with a tenancy. The cross-country running, in so far as Dr Beavington had to submit to it, was not inconsistent with his exclusive occupation of the arable land. But in fact Dr Beavington's consent was sought for that, and I think it is extremely doubtful whether it comes within clause 9 at all. Whether the hospital could insist there should be cross-country running around the field I very much doubt having regard to the context of this agreement. And as to the inmates, it was not proved to me that it was necessary to have this land as a sort of cordoning off of the hospital from the outside world. Some patient might wander off and go anywhere, but that does not mean to say that there were rights of property involved in that. So I find that this is an exclusive licence, and on that basis, which is caught by section 2, converted into a tenancy from year to year. As it is conceded that on that basis the action must fail it is quite obvious that there is nothing for me to do but to dismiss this action.

The action was dismissed with costs.

ARBITRATION

A **QUEEN'S BENCH DIVISION**

March 4 1981

(Before Mr Justice Robert GOFF)

DUVAN ESTATES LTD v ROSSETTE SUNSHINE SAVOURIES LTD

Estates Gazette January 30 1982

(1981) 261 EG 364

**Arbitration Act 1979 — Application for leave to appeal on a question of law arising out of an arbitrator's award — Application made
B by tenants of business premises alleging errors of law in award — Arbitration arose out of rent review clause in lease — Applicants criticised award on the ground that arbitrator had had regard to matters, facts or events subsequent to the review date — Criticism rejected in so far as it related to figures per sq ft in respect of lettings taking place after review date which merely confirmed his findings — Criticism accepted in so far as it related to consideration by the arbitrator of the trading results for a period subsequent to review date — Held, however, applying section 1(4) of the Act of 1979, that this particular question of law (concerning the taking into account of the subsequent trading results) did not "substantially affect the rights of one or more of the parties to the
C arbitration agreement" — It was a marginal point — Application for leave to appeal dismissed**

By this application Rossette Sunshine Savouries Ltd, tenants of premises at Pontypridd, Mid Glamorgan, sought leave under section 1(3)(b) of the Arbitration Act 1979 to appeal from an arbitrator's award in respect of the rent review clause in the lease. The respondents to the present application were the landlords, Duvan Estates Ltd. The lease, dated April 30 1974, was for a term of 21 years and contained provision for rent reviews after seven and 14 years.

D E Neuberger (instructed by Kanter, Jules & Co, agents for Cyril Goldstone & Co, of Swansea) appeared on behalf of
D the applicants for leave; L F R Cohen (instructed by Kennedys, agents for Peter Hughes & Co, of Swansea) represented the respondents.

Giving judgment, Robert Goff J said: There is before the court an application for leave to appeal under the Arbitration Act 1979 against an award made by Mr J R Probyn in respect of a rent review under a lease between the parties. The award in question was made on October 15 1980.

The lease which is before the court, is of premises in Pontypridd, in the county of Glamorgan, of which the applicants, Rossette Sunshine Savouries Ltd, are the tenants and the respondents, Duvan Estates Ltd, are the landlords. The lease is dated April 30
E 1974 and is for a term of 21 years. It provides for reviews of rent. The first review is to take place after a period of seven years and the second review after 14 years. The yearly rent for the first seven years of the lease was £2,100. The lease then provides:

> (A) (ii) For the next seven years of the said term a yearly rent (not being less than that first above reserved) to be agreed by the parties or in default of agreement to be determined in accordance with the provisions in that behalf hereinafter contained;

There is a similar provision in respect of the last seven years.

Clause 2 of the lease provides for what is to happen if the parties F
do not agree on the rent for the second, or the third, seven-year period. Clause 2 provides as follows:

> 2. The Lessee shall in default of agreement between the parties pay during the second and third period of seven years of the total period of 21 years hereinbefore granted a yearly rent which shall be determined in accordance with the following formula that is to say such rent shall be the rent (but not less than the rent reserved in the period of seven years immediately before expired) at which the demised premises might reasonably be expected to be let in the open market by a willing landlord by a lease for a term equal to the residue of the term unexpired at the end of the period of seven years immediately before expired with vacant possession on the same terms and subject to the same incidents in all other respects as this present demise . . . G

and there are various other assumptions which I need not mention.

There is also the proviso:

> . . . that if no agreement is reached between the parties by the date six calendar months prior to the commencement of the second or third period of seven years respectively as to the rent at which the demised premises might reasonably be expected to be let in the open market on the basis hereinbefore described then the question shall be referred to the decision of a single arbitrator to be appointed . . .

in accordance with the provisions of the clause. Mr Probyn was the arbitrator appointed pursuant to the provisions of the clause to decide that particular question. H

The only other provision to which I must refer is clause 3(15) under which the lessee covenants:

> (15) Not to carry on or permit to be carried on upon the demised premises or any part thereof any offensive noisome noxious noisy or dangerous trade business or manufacture and without prejudice to the foregoing not to use suffer or permit the demised premises or any part thereof to be used for any purpose other than for making meat pies and pasties.

which indeed, as their name indicates, was the nature of the tenants' business.

Now the matter proceeded as follows. The parties failed to agree
on the rental for the second seven years. I was told that the rival J
figures advanced, at least in the arbitration, for an appropriate annual rent were £7,500 by the landlord and £4,100 by the tenants. At all events, Mr Probyn must have been appointed in due course by the president of the [Royal] Institution [of Chartered Surveyors]. Written submissions were placed before him in September 1980, some time after the relevant review date which was July 17 1978; then it appears that oral submissions were addressed to him, and he made an award in October 1980, which must have been very shortly after he heard these oral submissions. The rent which he awarded was a figure of £6,750 per annum.

It is plain from the reasons which he produced in his award how
he reached that rental. He came to the conclusion that, as at the K
review date, July 17 1978, an appropriate figure for that type of premises was £1.25 per sq ft. On that basis, he took the view that £7,500 was the appropriate annual rental; but that was ignoring the restrictive user clause, and it appears that, on account of the restrictive user clause, he made a deduction of £750, thereby reaching his figure of £6,750.

Today, Mr Neuberger, on behalf of the tenants, asks for leave to appeal from this award. He does so on the basis that the arbitrator erred in two respects in his award. At one time he advanced three

A arguments, but one of those arguments he, in my judgment quite rightly, did not pursue. The point, in broad terms, Mr Neuberger says, is that in considering a valuation of this kind it is proper to have regard to the relevant facts existing as at the review date, which is July 17 1978, and not to have regard to facts and events existing after that date. In support of that proposition he relied upon a passage at the end of the judgment of Whitford J in *Ponsford v HMS Aerosols Ltd*, a case which went on appeal [1979] AC 63; (1978) 247 EG 1171; but I was provided with a transcript of Whitford J's judgment because on this point apparently the appeal was not pursued. As a general principle, I entirely accept that. Indeed, I do not understand that Mr Cohen, on behalf of the
B landlords, contested the principle as such. Now Mr Neuberger said, first of all, that the arbitrator erred in reaching a figure of £1.25 per sq ft. He referred me to a passage in paragraph 2 of the arbitrator's reasons where, at the end of that paragraph, he said:

> It is evident to me that a rate of £1.25 per square foot has been established for the standard 6,000 square foot unit and generally applied for the rent reviews from September 1978-February 1979, and which confirms the valuation basis adopted at the July 1978 review date. On the evidence presented the rental value, as at July 17 1978 and in accordance with the terms of the Lease but without a very restrictive user clause, would be £1.25 per square foot, ie £7,500.

C

There, said Mr Neuberger, was the arbitrator doing the very thing which he ought not to be doing; because what he was doing was looking at figures resulting from rent reviews after the review date. Mr Cohen was able to rebut Mr Neuberger's assertion to some extent on this point; but it turned out that the evidence placed by both parties before the arbitrator did in part infringe this particular principle. However, I put that on one side. I have to ask myself the simple question: did the arbitrator err in law having regard to these figures in respect of reviews made between September 1978 and February 1979? In considering that question, I have to consider what the arbitrator was doing in the light of the evidence before him and also consider the way he expressed his reasons in his
D award.

It appears that what happened was this. The landlords called evidence relating to premises let on this particular estate. They called evidence of a particular arbitrator's award in respect of a review as at a review date in July 1978. The figure which that particular award produced was indeed the figure of £1.25 per sq ft. They also produced a schedule showing other figures (I think largely settlements) which had been fixed in respect of reviews of rent in the succeeding nine months or so after that date. This showed a consistent picture of £1.25 per sq ft except in the case of one set of premises which was considerably smaller in size and, not surprisingly, the figure was somewhat larger, namely, £1.50 per
E sq ft.

The tenants, on the other hand, called evidence of premises not on this particular estate, but I assume not far away, where the premises were put on the market in July 1978 but as to which agreement was not reached on the rent until December 1978. In the light of that evidence the arbitrator was plainly entitled to hold, as indeed he did hold, that the evidence as at July 1978 pointed to a rental of £1.25. But he went on to say that that was, in effect, the evidence before him of the particular letting as at July 1978, but so as to put beyond doubt any question as to his conclusion he recorded the fact that he was comforted to find that subsequent lettings in the succeeding few months confirmed that figure. I do not
F think that the arbitrator was relying upon this latter evidence to reach his conclusion. He was simply saying: "I am thankful to see this evidence does not cause me to depart from the conclusion which I have formed on the basis of the relevant evidence, as at the review date." In those circumstances I do not think the arbitrator, although one quite understands why he referred to this particular schedule which the landlord placed before him, was basing his conclusion upon evidence other than evidence as to the position as at the review date. In those circumstances, I do not think, on a fair reading of the reasons of the award, that the arbitrator was infringing the principle which both parties accept should be applicable. Therefore, I reject Mr Neuberger's first submission.

Then I come to his next submission, which is based on paragraph G
3 of the award. It appears that before the arbitrator reliance was placed upon *Plinth Property Investments Ltd v Mott, Hay & Anderson* (1978) 249 EG 1167 for the proposition that there should be a rent reduction where there is a restrictive user clause, and it is plain that that authority does indeed support that proposition — the point being, quite simply, that where there is a restrictive user clause then the market would be reduced and that could have a depressing effect on the rent. But, as Mr Neuberger was constrained to admit, and I am sure he was right in this, what the arbitrator did was to accept that principle and to apply it, because when he refers to the case he says: "I accept this principle". Then it becomes plain from the figure which he chose, in paragraph 4, that H
he did indeed accept it and made a reduction in respect of it. Of course, the reduction which he made was a matter purely for his own judgment and one with which this court could not possibly interfere. So there is no error of law by the arbitrator in so far as he applied that principle.

But then another point was made by Mr Neuberger, which he did pursue before me, and that was as follows. In paragraph 3 the arbitrator went on to say:

> However, I do not agree with the suggestion of Mr Staddon, the lessee's consultant, that the manufacture of meat pies and pasties and the like was not a viable operation. J

and then came the sentence of which Mr Neuberger complains:

> The 1978 results were not supported by those for 1979, which I am entitled to take into account to verify whether or not the misgivings expressed at the review date were justified.

Mr Neuberger said: "There we have yet another example of the arbitrator taking into account events and facts after the review date and that he ought not to have done. Therefore he erred in law and the court should give leave to appeal and remit the matter to the arbitrator to reconsider his valuation." K

Now first of all, Mr Cohen said that, in fact, the only basis upon which trading results could be relevant was simply and solely for the purpose of showing that the site or location was not a good place for the business in question. I have no doubt that there are cases, indeed one was cited to me, where this may be so, and evidence can be called to this effect: "Now this is the way my business has gone over the last few years and that really shows that this is not a good spot for the business which is the subject of the restrictive user clause and, for that reason, the rent ought to be reduced." A case concerning hotels was cited to me which showed that this is so. But I am bound to say that I do not accept Mr Cohen's submission that that is the only purpose for which you can look at trading results. It could be that, quite apart from the location of the business, the L
particular business which is the subject of the restrictive user clause could be so depressed that applicants for leases, under the lease containing that clause, would be less ready or willing to pay a high rent; and therefore a general depression in that business, or a general depression in that business in that area, might be relevant in a particular case as having a bearing upon the rent which is payable as at the review date.

Now it seems to me that this was basically the argument which was being advanced in this case. If one looks at the written submissions put in on behalf of the plaintiffs I find the sentence:

M

> We propose calling a director of our client company to give further evidence to the arbitration as to the nature of the user and his personal knowledge of competitors and the state of the industry as at July 1978.

So this does seem to be the point being made on behalf of the tenants here. They are saying that the state of the industry as at the review date was very bad and that that must be taken into account in considering the rent as at the review date. It seems to me to be quite a legitimate argument to advance but, of course, it would have to be supported by evidence.

If one looks at the arbitrator's award and the sentence of which

Mr Neuberger complains, what one finds is this. The arbitrator was saying:

> In the context of this argument being advanced by the tenant, the 1978 results (which I read to be the trading results for the tenant's business for the calendar year 1978 which are obviously being relied upon in support of this argument) were not supported by those for 1979, which I am entitled to take into account to verify whether or not the misgivings expressed at the review date were justified.

As I read it, the tenants were saying to the arbitrator: "Look at our 1978 results. They were bad. Therefore, as at the review date, we had misgivings as to the state of the pie and pasty manufacturing industry in this area." Now it seems to me that in relation to that submission and in relation to the weight, if any, to be attached to the 1978 accounts, it was not right for the arbitrator to look at the 1979 results, because there he was doing what Mr Neuberger complains of, namely, he was looking at subsequent events to discover not whether it was justifiable to have misgivings as at 1978 but whether those misgivings were proved in the result to have had a sound basis, which, of course, was not the point that he had to consider. Therefore it can be said that the arbitrator did, on the basis of these reasons, err in law in having regard to 1979 accounts for that purpose.

But the question arises whether or not, in those circumstances, I should give leave to appeal on this particular point. I must have regard to the provisions of the 1979 Arbitration Act and it is, I think, clear under the Act that the applicant for leave has to point to, first, a question of law arising out of the award, and I think they have done that. But that is not enough. They have to go further, because section 1(4) provides:

> The High Court shall not grant leave . . . unless it considers that, having regard to all the circumstances, the determination of the question of law . . . could substantially affect the rights of one or more of the parties to the arbitration agreement.

and the question arises whether this particular point of law is one which could substantially affect the rights of the parties.

Mr Neuberger says that I cannot tell, looking at this award, how far the 1979 results had an influence upon the arbitrator's conclusion as to the sum he should knock off the £7,500 by virtue of the restrictive user clause. Now I look at the arbitrator's reasons and I see that the arbitrator rejected the submission being advanced by a consultant, Mr Staddon, on behalf of the lessee, that the manufacture of meat pies and pasties was not a viable operation. I look at it this way. If the 1978 trading results of one company were the only hard evidence placed before the arbitrator as to manufacturing meat pies generally, it seems to me that that was the most fragile basis upon which to advance a submission as to the general viability of manufacturing meat pies and pasties. By itself it could carry very little weight indeed. If, on the other hand, there was other evidence upon which the arbitrator could proceed, then he must look at the 1978 results in the context of the other evidence; but, once again, the 1978 results of one company must be of marginal importance to the wider question whether or not there was a general depression in the manufacturing industry in meat pies and pasties. It seems to me that given the nature of the submission being advanced and the particular item, namely the 1978 results of one company, in respect of which the arbitrator took a matter wrongly into account, it is impossible for me to say that taking into account an illegitimate matter as undermining this one particular factor — the 1978 results — is something that could substantially affect the rights of the parties. It seems to me that the 1978 results, taken by themselves, must be something of very minor importance indeed. The only way in which this arbitrator could have been persuaded that the whole manufacture of meat pies and pasties, even limiting it to the particular area in question, was not viable would be to look at a far wider range of evidence before he could reach any such conclusion.

For that reason I am not prepared to hold that this particular question of law, namely, a question of law relating to the taking into account of the 1979 accounts, is a question of law which could substantially affect the rights of one or more of the parties to this arbitration agreement. It seems to me, in truth, that the present case is precisely the sort of case which the Arbitration Act 1979 is designed to exclude from reconsideration by the courts. This is a marginal point in fact, and only a marginal point, and because it is marginal it is just the sort of question for which the court ought not to invoke the process of review. I therefore conclude that this is a case in which I should not grant leave to appeal for the reason I have given, namely, that I do not consider that, having regard to all the circumstances, the determination of the question of law could substantially affect the rights of one or more of the parties to the arbitration agreement. It follows that the application for leave to appeal is dismissed.

The application was dismissed with costs.

COMPULSORY PURCHASE AND COMPENSATION

A **COURT OF APPEAL**

November 6 1981

(Before Lord Justice WALLER, Lord Justice TEMPLEMAN and Sir David CAIRNS)

WILKINSON AND OTHERS v MIDDLESBROUGH BOROUGH COUNCIL

Estates Gazette February 20 1982

(1981) 261 EG 673

Compulsory acquisition — Compensation — Whether compensation should be on equivalent reinstatement basis under rule (5) in
B **section 5 of Land Compensation Act 1961 — Appeal from Lands Tribunal's decision that basis of compensation was rule (2) — Compulsory purchase of premises occupied by veterinary surgeons' multi-principal practice — Normal methods by which interests in veterinary surgeons' practices are dealt with — Distinction between single-principal and multi-principal practices — Not usual for veterinary premises and practices to be offered for sale in the open market — Whether nevertheless there was a "general demand or market for land for that purpose" — Held by majority of Court of Appeal (Waller LJ dissenting), upholding decision of Lands Tribunal, that the claimants had not discharged the burden of proof of establishing that there was no general demand or market for prop-**
C **erty for the purpose of carrying on the practice of veterinary surgeons — Rule (5) accordingly not applicable — Appeal dismissed**

This was an appeal from a decision of the Lands Tribunal (E C Strathon FRICS). The appellants (claimants) contended that the basis of compensation for the compulsory acquisition of premises at 126 Newport Road, Middlesbrough, which were devoted to the carrying on of the appellants' practice as veterinary surgeons, should be in accordance with rule (5) in section 5 of the Land Compensation Act 1961. The respondent council contended that rule (2) applied and the Lands Tribunal decided in the respondents' favour. The decision of the Lands Tribunal was reported at (1979)
D 250 EG 867.

K R Bagnall QC and S Lightwing (instructed by Newby, Robson & Cadle, of Middlesbrough) appeared on behalf of the appellants; G E Moriarty QC and R Fookes (instructed by the Solicitor, Middlesbrough Borough Council) represented the respondent council.

Giving judgment, WALLER LJ said: This is an appeal from a decision of the Lands Tribunal on a preliminary point of law concerning the basis of compensation payable on the compulsory purchase of premises at 126 Newport Road, Middlesbrough. The member found that the basis of assessment of compensation was rule (2) of the rules for assessing compensation under section 5 of
E the Land Compensation Act 1961, and not rule (5). The appellants claim that the assessment should be on the basis of rule (5).

The appellants own the freehold and carry on the practice of veterinary surgeons at 126 Newport Road. The property is about 90 years old and was formerly an end-terrace dwellinghouse. It now comprises on the ground floor a waiting room, inquiry office, small accounts office, a consulting room, x-ray/consulting room, laboratory/operating room, preparation room, a range of four kennels and food-preparation room, an exercising yard and additional kennels and stores. According to a letter written in 1976, that part of the premises was largely custom built in the 1920s. There is also in F
the original building on the first floor a caretaker's flat of three rooms and bathroom. The property lies about 550 m to the south west of the town centre in a locality which has been the subject of compulsory purchase orders made under Part III of the Housing Act 1957. A blight notice dated July 8 1977 in respect of the subject property was served under the provisions of the Town and Country Planning Act 1971 and the Land Compensation Act 1973.

The only question arising on this appeal is whether rule (5) applies or not. Rule (5) provides:

> Where land is, and but for the compulsory acquisition would continue to be, devoted to a purpose of such a nature that there is no general demand or G
> market for land for that purpose, the compensation may, if the Lands Tribunal is satisfied that reinstatement in some other place is *bona fide* intended, be assessed on the basis of the reasonable cost of equivalent reinstatement.

A veterinary practice was established at 126 Newport Road, Middlesbrough in 1900, and the appellants are the owners of the practice and the building. Mr Wilkinson joined the practice in 1948, Mr Gale in 1963 and Mr Hall in 1968; no 126 is the headquarters with branches at Hartlepool, Stockton and Billingham. The whole of the ground floor of 126 is used for the practice and the caretaker deals with telephone calls at night and also helps in emergencies. In addition to the three partners there are five veterinary surgeons in
the practice; at no 126 the total workforce comprises a partner, four H
veterinary surgeons and 10 lay staff. Work at no 126 and at the branches is closely related, with communication by telephone, and each surgeon has radio telephone equipment in his car. The practice provides a 24-hour service. The headquarters is responsible for the purchase and storage of drugs and equipment for the practice and has expertise on call for the branches; operations are mainly carried out at no 126 but some at Hartlepool.

In 1977 the "large animal" part of the practice was sold and there were mutual restrictive covenants covering an area of a radius of 25 miles from 126 Newport Road. The appellants estimate that their practice covers 50 per cent of the work in the area. There is a plan
with the papers showing the layout of the premises and this plan J
shows that two-thirds of the ground-floor area is occupied by kennels, consulting rooms etc and presumably was the custom-built part referred to in the letter. The plan shows that, with the exception of the caretaker's flat on the first floor, the whole of the ground floor is devoted to the practice, the inquiry office and waiting room being underneath the caretaker's flat.

No evidence was called by the acquiring authority. On behalf of the appellants, two partners, two chartered surveyors and the hon secretary of the RSPCA clinic at Halifax were called.

In the decision of the tribunal some of the evidence called on
behalf of the claimants is set out and the decision summarises it as K
follows [(1979) 250 EG 867 at 870]:

> In the instant case the evidence about general demand or market comes from Mr Gale and Mr Catterall. Mr Gale says that veterinary practices in the Middlesbrough area are not bought or sold in the open market; it is customary to take partners into the firm. First the single-principal practice which is sold by the owner to the incoming principal after a period of working in dual harness. I think that as such a transaction is the normal method of disposal it points to the existence of a market.
>
> I had no evidence as to the extent of such a market; on the one hand the witnesses said they had no knowledge of sales of practices and, on the other

hand, that practices are normally disposed of in the manner described. Secondly, incoming partners to the multi-principal practice purchase shares in the practice, which continues to operate as a partnership. The instant case provides an example by the purchase of a partnership by Mr Hall in 1968 on terms which Mr Catterall said he would expect from an incoming partner to his own multi-partnership practice of surveyors. There is also the example of the disposal of the "large animal" part of the claimants' practice to Mr Russ to carry on that part of the business elsewhere. Thus in a multi-partner practice, so long as partners are taken into the business to add to the number of working partners or to take the place of retiring partners the practice as a whole is not offered for sale. In the instant case Mr Gale puts the capital value of the whole practice at about £100,000, and Mr Catterall's valuation of the goodwill is of the order of £139,000. Mr Gale accepts that a number of veterinary surgeons might band together into one firm and that the subject premises might prove an attraction to such a partnership, but he did not know of any source from which finance could be secured for such a purchase. In the instant case the practice is flourishing; it is commercially viable with every prospect of improving in performance and Mr Catterall considers that having regard to the practice being so well established with income secure, monopoly value attaches to the subject property.

The decision concludes:

It seems to me that there is a demand and market for premises for the purposes of the single-principal practice of a veterinary surgeon. In respect, however, of multi-principal veterinary practices these never do reach the market, for the simple reason that the aim of such a partnership is to continue to carry on the practice as veterinary surgeons rather than to dispose of the partnership assets. I see no reason to doubt, nevertheless, that for such a practice there *would be* a market just as there is for the single-principal practice.

Mr Bagnall submits: 1. That there is an over-emphasis on the sale of practices as opposed to the sale of land in the decision of the tribunal. 2. There was no evidence of a general demand or market for veterinary surgeons' practices. 3. Alternatively there was no evidence of a general demand or market for multi-principal veterinary practices. Put another way, although the onus was on the claimant to prove that there was no evidence, having regard to the fact that no evidence was called by the acquiring authority and the member, although a chartered surveyor, was sitting judicially, the only conclusion from the evidence called was that there was no general demand or market.

Mr Moriarty, for the acquiring authority, relied on the word "demand" in the rule and submitted that there was some evidence of a market in single practices and, the onus of proof being on the claimant, that the member was entitled to find that it was not proved that there was no general demand for land occupied for multi-principal veterinary purposes.

We were referred to a number of cases where rule (5) had been applied and for the most part they related to charitable or non-profit-making purposes; see, for example, *Vaughan (Viscount) v Cardiganshire Water Board* (1963) 14 P & CR 193, 185 EG 949; *Nonentities Society Trustees v Kidderminster Borough Council* (1970) 22 P & CR 224, 215 EG 385, 455; *Manchester Homeopathic Clinic (Trustees) v Manchester Corporation* (1970) 22 P & CR 241, 215 EG 1033; and *Sparks v Leeds City Council* (1977) 34 P & CR 234, 244 EG 56 (LT). In *Festiniog Railway Society Ltd v Central Electricity Generating Board* (1962) 13 P & CR 248 it was held that the rule governed the case of a commercial concern, although as a matter of discretion the tribunal did not apply the rule. Nowhere is there reported a case similar to that with which we are concerned, although the *Manchester Homeopathic* case is perhaps the nearest. In that case a homeopathic clinic purpose-built in 1939 was compulsorily acquired, and the trustees sought compensation on the basis of rule (5). The tribunal refused to restrict the purpose to that of a homeopathic clinic and regarded the purpose as that of a clinic but came to the conclusion that the evidence was that sales of land used for the purpose of clinics were so few that it was not possible to infer that there was a general demand or market for such purpose.

In giving his decision in the present case the member, while dealing with single-principal veterinary practices, came to the conclusion that there was a market but that there would be a period of working in dual harness as the normal method of disposal.

The period of dual harness of course would be the safeguard for the personal goodwill. In the present case the tribunal was concerned with land on which are premises largely custom built and occupied by a multi-principal practice. A clinic for domestic animals, submitted Mr Bagnall, is not unlike the medical clinic in the *Manchester* case. This clinic is ancillary to the multi-partner practice, which depends to a large extent on goodwill. The goodwill of this practice must be largely personal and not local. The personal part of goodwill was aptly described by Sir George Jessel MR in *May v Thompson* (1882) 20 Ch D 705, at 718, when he said:

What is the meaning of selling a medical practice? It is the selling of the introduction of the patients of the doctor who sells to the doctor who buys, he has nothing else to sell except the introduction. He can persuade his patients probably who have confidence in him to employ the gentleman he introduces as being a qualified man and fit to undertake the cure of their maladies, that is all he can do.

This would apply very much to a veterinary practice. Indeed the sale of the large animal part of the practice may be an indication of personal goodwill because much the greater part of that work will be done on the farm and not at the premises of the veterinary surgeon (the plan shows one horse box as opposed to five to six kennels).

No doubt there is a local element as well, what was known in the days of the Landlord and Tenant Act 1927 as the cat element as opposed to the dog element, ie the customers who stay with the premises as opposed to those who follow the practitioner (see *Whiteman Smith Motor Co Ltd v Chaplin* [1934] 2 KB 35, per Scrutton LJ at p 42). But it is because of the personal element that no sale takes place, because the only way to transfer the goodwill is by personal introduction. The result is that, as the evidence shows, multi-principal practices are not sold in the open market. The transfer of the goodwill is done gradually by introducing a new partner who buys his share of the partnership over a period of years.

In this case the land consists for the greater part of custom-built kennels, x-ray room, etc, and Mr Moriarty realistically accepts that the land would not be sold without the goodwill. The purpose is that of a multi-principal veterinary practice and in this respect is to be contrasted with professional offices of surveyors, solicitors or other similar occupations. Although there was no evidence about it, there are obviously other multi-principal veterinary practices in various parts of the country. In the Cleveland district there is this one, and it has roughly 50 per cent of the domestic animal practice within 25 miles of 126 Newport Road, which probably covers a substantial part of Cleveland.

In considering rule (5) it is necessary first to consider what is the purpose to which the land is devoted. There is no dispute that it is devoted to the purpose of a multi-principal veterinary practice, and in my opinion it is a fair inference from the letter of June 10 1976, which the acquiring authority quoted, that the premises are such as to require several veterinary surgeons for running them. The land therefore is devoted to the purpose of a multi-principal veterinary practice and but for the compulsory acquisition would have continued to be devoted to that purpose.

In the context of rule (5) I must consider whether the purpose is of such a nature that there is no general demand or market for land for that purpose. As I have already said, there was no evidence of the existence of other multi-principal practices save that they do exist but they are not bought and sold as a whole. Is there a demand for such land and, if there is, is it a general demand? If the practice is separated from the land the evidence shows that there is a demand for a share in a pre-existing practice, but there is no evidence of the sale of a multi-principal practice and the fact of the demand for a share indicates that it is the personal goodwill which is being bought and accordingly the sale of the whole practice would be unlikely.

However, it is not the practice but the land with which this case is concerned. Rule (5) is concerned with the situation where there is to be reinstatement in some other place. If that fact has to be taken into account there would certainly be no demand for land for this purpose, or for this land. But if the assumption is that all the partners are retiring and so some of the personal goodwill will remain, there would be a demand for this land. But the rule pro-

A vides that there must be no general demand or market for *land* for that purpose. That indicates to me that it is not sufficient that there should be a demand for the land in question but that there should be a general demand, ie a demand not only for that land but for other land elsewhere for the same purpose. Furthermore the use of the word "market" connotes in my opinion something more than the fact that there are potential users of the land for the named purpose but that there is a buying and selling for that purpose.

In the *Manchester* case the tribunal did not pay attention to evidence of lettings. The member thought that only sales were relevant to market and not lettings. Whether he was correct in his conclusion or not, I am quite satisfied that buying a share in a B practice is not evidence of a general demand for the purchase or sale of that practice. In my opinion the evidence is against the buying and selling of whole multi-principal practices. In other words, in my opinion the evidence indicates that there is no buying and selling of land for multi-principal practices.

In his decision the member said: "It seems to me that there is a demand and market for premises for the purposes of the single-principal practice of veterinary surgeon". In my opinion he was entitled so to find on the evidence. However, he went on:

> In respect however of multi-principal veterinary practices these never do reach the market for the simple reason that the aim of such a partnership is to continue to carry on the practice as veterinary surgeons rather than C dispose of the partnership assets. I see no reason to doubt nevertheless for such a practice there would be a market just as there is for the single principal practice.

In my opinion the member was in error in finding there would be a market. For the reasons which I have given I am of opinion that in relation to multi-principal veterinary practices the evidence was all one way, namely that there is no general demand or market for that purpose. I am of opinion that the member was in error in his final conclusion. The question is not whether there would be a market, it is whether there is a general demand or market. In my judgment the only conclusion on the facts found by the member is that there was D no general demand or market for the sale of land for a multi-principal veterinary practice.

Taking the contrary view, namely, that the appeal should be dismissed, TEMPLEMAN LJ said: In 1979 the three claimants who are the present appellants were carrying on in partnership a practice of veterinary surgeons at the premises 126 Newport Road, Middlesbrough, held by the claimants in fee simple. The premises were on September 10 1977 compulsorily acquired by the respondents, the Middlesbrough Borough Council, and the claimants became entitled to compensation to be agreed or determined by the Lands Tribunal.

The premises were originally an end-terrace house. The ground-E floor rooms were used by the claimants as offices, waiting rooms and consulting and operating rooms with kennels. The former garden of the house was used as an exercise yard and outbuildings were installed to provide additional kennels, storerooms and ancillary rooms in connection with the practice. The first-floor rooms of the premises were used as a caretaker's flat comprising three rooms and a bathroom.

The partnership practice was carried on from the premises 126 Newport Road, Middlesbrough, with subordinate branches at Hartlepool, Stockton and Billingham.

The profits of the partnership practice derived by the three F claimants, who worked full-time in the practice, increased from £36,000 in 1973 to £56,000 in 1977 and £71,000 in 1978.

By section 5 of the Land Compensation Act 1961 compensation shall be the amount which the land if sold in the open market by a willing seller might be expected to realise (rule (2)) and there shall be added compensation for disturbance or any other matter not directly based on the value of the land (rule (6)). If rules (2) and (6) apply, the claimants will be entitled to the value of the premises and to the value of any goodwill of the practice which they lose.

But section 5 also provides that:

> Where land is, and but for the compulsory acquisition would continue to be, devoted to a purpose of such a nature that there is no general demand or G market for land for that purpose, the compensation may, if the Lands Tribunal is satisfied that the reinstatement in some other place is *bona fide* intended, be assessed on the basis of the reasonable cost of equivalent reinstatement (rule (5)).

If rule (5) applies the claimants will be installed in or entitled to the costs of acquiring equivalent premises.

The respondent council contend that rules (2) and (6) apply. The claimants contend that rule (5) applies. The Lands Tribunal decided in favour of rules (2) and (6). The claimants appeal to this court.

The premises 126 Newport Road constitute land devoted to the purpose of carrying on the practice of veterinary surgeons in part-H nership. But for the compulsory acquisition the land would continue to be devoted to that purpose. On behalf of the claimants it is submitted that there was no or no sufficient evidence to support the finding of the Lands Tribunal that the claimants had not satisfied the tribunal that there is no general demand or market for land for the purpose of carrying on the practice of veterinary surgeons in partnership.

It is clear that the class of purchasers interested in land to be used for the purpose of carrying on the practice of veterinary surgeons in partnership must be limited to veterinary surgeons. The claimants did not suggest that there was no demand for veterinary surgeons in Middlesbrough or elsewhere or that veterinary J surgeons were not being trained or were leaving the profession. In these circumstances the Lands Tribunal found it impossible to believe that there was not a general demand by veterinary surgeons for premises devoted to veterinary surgery or that there was no general demand by veterinary surgeons for 126 Newport Road for the purpose of veterinary surgery in view of the fact that a practice was carried on at those premises and was producing £71,000 per annum for three full-time partners. I share the incredulity of the Lands Tribunal. It was submitted in this court that evidence and authority precluded the Lands Tribunal and this court from giving effect to such incredulity.

The evidence as set out in the case stated by the Lands Tribunal K was as follows:

(1) Until 1977 the practice comprised a small animal practice concerned with pets and a large animal practice concerned with horses and farm animals. By an agreement dated March 2 1977 the goodwill of the large animal practice was sold to a vet for £14,500. The claimants covenanted not to carry on a "large animal" practice for five years within 25 miles of 126 Newport Road, Middlesbrough; the purchaser of the large animal practice covenanted not to carry on a small animal practice for five years within 25 miles of his address at Maltby, Cleveland. It appears from this that it was possible to carry on a "large animal" practice from 126 Newport Road, that there was a purchaser for a "large animal" L practice formerly carried on at 126 Newport Road and the inference is that there would have been a purchaser of 126 Newport Road for the purpose of carrying on an animal practice from those premises if the premises had been offered for sale by a willing seller.

(2) It appears from the case stated that "the claimants say that the goodwill of the practice attached to the subject site where there has been a veterinary practice since 1900; they would not voluntarily leave the site". The inference is that 126 Newport Road would be attractive to veterinary surgeons as potential purchasers because of the goodwill attached. The inference is also that the evidence of the claimants with regard to the difficulties of selling, though no doubt frank and truthful, must be regarded with some caution in the M present context because they are clearly not the "willing sellers" postulated by rule (2).

(3) Mr Gale, one of the claimants, said that "in the Cleveland area veterinary practices are not bought or sold; in the open market the normal course is to take a partner. A single partner practice would advertise for a partner; the newcomer would work with the principal for a period and would then take over the practice and pay out the principal." From this it appears that there are purchasers for single-partner practices. The fact that the interests of both vendor and purchaser lead to a transitional takeover period does not disguise the fact that practices are bought and sold and it would be little use in most cases in selling the practice without the premises

upon which the practice was being carried on. Similarly, interests in partnership practices change hands at negotiated prices. It appears that partners in veterinary practices do not choose to offer for sale either the premises or the practice because it suits them not to sell as a whole. But there may be a general demand for such practices and premises nevertheless. The demand is evinced by the fact that interests in partnerships are sold to new partners. In the present case if the three claimants had been killed in 1976 before any threat of compulsory acquisition, I find it impossible to believe that the executor of the last survivor would not have found veterinary surgeons willing and anxious to acquire 126 Newport Road for the purpose of carrying on a veterinary practice. Similarly if partners quarrel or move to another part of the country so that a veterinary practice or premises occupied for the purposes of a veterinary practice come on the market, I can see no reason why purchasers should not appear from the ranks of veterinary surgeons.

(4) Mr Catterall, an experienced chartered surveyor, said that in his firm, which practised within an area of 1,000 square miles, there was no experience of veterinary practices coming on to the market. I do not find this surprising in view of the evidence given by Mr Gale that it is not common to find veterinary practices and premises being offered for sale and even less surprising that they are not offered for sale through surveyors and estate agents. But it is significant that Mr Catterall was able to value the premises 126 Newport Road and the goodwill of the practice for the purposes of rules (2) and (6). He "considered that three years' purchase would be the measure to apply where the practice is so well established. It is the figure that Mr Catterall would apply to his own practice when seeking an incoming partner. *It was supported by the three years' purchase paid by Mr Hunt one of the claimants when he entered the practice in 1968*". In my judgment this evidence demonstrates a demand for veterinary practices and for premises adapted for the purposes of that practice. The fact that demand can usually only be met by buying into the partnership which owns the premises does not mean that there is no demand.

(5) Mr Gale "accepts that a number of veterinary surgeons might band together into one firm and that the subject premises might prove an attraction to such a partnership, but he did not know of any source from which finance could be secured for such a purpose". But there was no evidence that veterinary surgeons were a bad risk or that the practice of veterinary surgeons was so peculiar that money could not be provided or borrowed to purchase a practice which was operating at a profit.

In my judgment there was evidence from which the Lands Tribunal could properly reach the conclusion that the claimants had failed to show that there was no general demand for premises for the purpose of carrying on the practice of veterinary surgeons in partnership or that there would be no general demand if a willing seller offered to sell 126 Newport Road for the purpose of carrying on such a practice. The relevant evidence consisted of the profitability of the partnership carried on upon the premises, the assertion by the claimants themselves that goodwill attached to the premises, the sale of the large animal practice, the admitted sales of veterinary practices by sole proprietors to other sole proprietors, the admitted sales of an interest in the veterinary practice carried on at 126 Newport Road to new partners on the basis of three years' profits, the admission that interests in other veterinary partnerships were sold to new partners and the admission that the premises if offered for sale might prove an attraction to a veterinary practice. There was an entire absence of any evidence that premises occupied by veterinary surgeons would be any more difficult to sell than premises occupied by chartered accountants or solicitors or members of any other profession. I cannot think that rule (5) was intended to apply for the purposes of reinstating professional occupiers. The fact is that the claimants understandably resist rules (2) and (6), not because 126 Newport Road could not be sold by a willing seller for the purpose of a veterinary practice to be carried out in partnership but because the claimants fear, rightly or wrongly, that the value of 126 Newport Road plus any further compensation under rules (2) and (6) would not enable them to construct a new veterinary surgery in the neighbourhood.

The relevant authorities to which we were referred are as follows:

(1) In *Festiniog Railway Society Ltd v Central Electricity Generating Board* (1962) 13 P & CR 248, 176 EG 271 the Lands Tribunal exercised its discretion not to assess compensation under rule (5), although all the requisites of that rule were satisfied, because the cost of reinstatement was wholly disproportionate to the total value of the assets of the claimants. In the present case the Lands Tribunal have postponed consideration of the question whether, if rule (5) is satisfied, the Lands Tribunal should exercise its discretion in assessing compensation under that rule. In the *Festiniog* case this court [(1962) 181 EG 787] declined to interfere with the decision of the Lands Tribunal because by section 3(4) of the Lands Tribunal Act 1949 a decision of the Lands Tribunal should be final provided that any person aggrieved by the decision's being erroneous in point of law may require the tribunal to state a case. In the present case no point of law arises unless there was no evidence from which a reasonable Lands Tribunal could reach the same conclusion.

(2) In *Vaughan (Viscount) v Cardiganshire Water Board* (1962) 14 P & CR 193, 185 EG 949 the Lands Tribunal refused to apply rule (5) to a cottage which was used for the purposes of estate management. It was held that there was a general demand or market for premises devoted to estate management, though no general demand for premises devoted to the particular estate in question. I do not find this decision helpful to the submissions made on behalf of the claimants.

(3) In *Nonentities Society Trustees v Kidderminster Borough Council* (1970) 22 P & CR 224, 215 EG 385, 455 there was evidence that there was no general market or demand either for a commercial theatre or a non-commercial theatre in Kidderminster or in any similar town. There was evidence that many theatres had been closed. Rule (5) was applied. In the present case there is no evidence that premises for use by veterinary surgeons in practice are not profitable or that there are no veterinary surgeons willing and anxious to acquire premises for the purpose of carrying on the practice of veterinary surgeons from 126 Newport Road or from similar premises in Middlesbrough and elsewhere.

(4) *In Manchester Homeopathic Clinic Trustees v Manchester Corporation* (1970) P & CR 241, 215 EG 1033 it was held that there was no general demand or market for premises to be used for the purposes of a clinic, group medical practices, for example, being limited by a licensing system to one per area and other possible purchasers of such premises being very few and far between. That case was complicated by the licensing system and by the fact that medical practices under the National Health Service cannot be bought and sold and by the fact that the numbers of persons interested in clinics, particularly a clinic which had formerly been carried on as a homeopathic clinic, were necessarily limited. There was no evidence, as there is in the present case, of sole practitioners and partners being ready and willing to buy into the practice of a veterinary surgeon. The test submitted by counsel and accepted by the Lands Tribunal in the *Manchester* case at p 252 was as follows: "Assume that the premises were on the market for use for a purpose of the kind already discussed" (in that case a clinic and in the present case a practice in partnership of the profession of veterinary surgeons) "and assume that that purpose was to be continued, then the real question was would the vendor say to himself that he could reasonably expect to find a purchaser of the premises for that purpose at a fair price". In my judgment if that test is applied to the present case there was ample evidence before the Lands Tribunal which justified the Lands Tribunal in deducing that a willing vendor could reasonably expect to find a purchaser of 126 Newport Road for the purpose of carrying on a veterinary practice in partnership at a fair price.

In the result I can see no reason for interfering with the decision of the Lands Tribunal in the present case.

Sir David Cairns said: I agree with Templeman LJ that this appeal should be dismissed.

In order to bring into play rule (5) of section 5 of the Land Compensation Act 1961 it is necessary for him who asserts that that rule is applicable to establish (a) that but for the compulsory purchase the land would continue to be devoted to a certain purpose;

A (b) that that purpose is of such a nature that there is no general demand for land for that purpose; (c) that there is no market for land for that purpose; (d) that reinstatement in some other place is *bona fide* intended.

If and only if the Lands Tribunal is satisfied of *all* these matters it *may* award compensation on the basis of the reasonable cost of reinstatement.

I make the following comments on some points of construction of rule (5) as I have divided it up: In (a) it is obviously the land which is the subject of the compulsory purchase that is referred to. It seems to me equally obvious that in (b) and (c) it is not that very land which has to be considered. For (b) I think it has to be established B that practically nobody in the area wants to obtain land for the specified purpose and under (c) that practically no sales for the purpose take place in the area. The effect of the word "general" is no doubt to indicate that the fact that one particular person wants to obtain a piece of land for the specified purpose would not make the rule inapplicable. I am not sure that the word "general" is intended to qualify "market" as well as "demand" but, however that might be, I should not consider that the fact that an isolated transaction had taken place would constitute a "market".

Now in the present case the appellants established points (a) and (d) but they failed to satisfy the Lands Tribunal of either (b) or (c) and in my judgment the member was right in his conclusion that C rule (5) was not applicable.

As to (b) it was clear from the evidence that people did from time to time take up the practice of veterinary surgery in partnership with others. I do not consider that the fact that no evidence was given of a firm of veterinary surgeons acquiring premises in which to practise showed that there was no demand for land for such a purpose. Clearly veterinary surgeons can find work in Cleveland. Clearly it is not unusual for them to do such work in partnership. There must therefore be a demand for land for that purpose. Whether the demand takes the form of several partners seeking premises to move into together or of one veterinary surgeon seeking premises where others will later join him is irrelevant.

D I agree with Waller LJ that it is a demand for land for the practice rather than a demand for a share in a practice that is relevant, but I can see no reason to suppose that a person who buys a share in a practice does not wish to buy a share in the premises where the practice is carried on. So far as financing such a purchase is concerned, I agree with Templeman LJ that there was no evidence that veterinary surgeons would have any greater difficulty in obtaining the money for such a purchase then would the members of any other profession.

As to (c) I feel more doubt whether the transactions described in the evidence could properly be said to constitute a "market", but on the whole I am of opinion that they do.

E Of the authorities that were cited to us, I do not find in any of them any support for the view that the decision of the Lands Tribunal was wrong in law and I agree with Templeman LJ that the test applied in *Manchester Homeopathic Clinic Trustees v Manchester Corporation* is a suitable test and that by that test the member was fully justified in reaching his decision in favour of the corporation.

Therefore, with great respect for the differing view of Waller LJ, I agree with Templeman LJ that the appeal should be dismissed.

The appeal was dismissed with costs. Leave to appeal to the House of Lords was refused.

F

COURT OF APPEAL

July 31 1981

(Before Lord Justice STEPHENSON, Lord Justice OLIVER and Lord Justice FOX)

DUTTONS BREWERY LTD v LEEDS CITY COUNCIL

Estates Gazette March 6 1982 and March 13 1982

(1981) 261 EG 885 & 989

Compensation for compulsory acquisition of public house — Dispute as to date at which compensation should be assessed — Appeal by acquiring authority against judge's declaration that compensation should be assessed as at date of entry — Court of Appeal G **divided, Stephenson LJ dissenting — Whether compensation should be assessed as at May 1968 when an alleged agreement for compensation was made or as at June 1976 when entry by acquiring authority took place — Value of licensed premises based on "barrelage" had increased substantially between these two dates from an "agreed" figure of £15,000 to £78,000 claimed by brewers — Dispute as to whether the phrase "subject to a Contract to be approved by me" in a letter from town clerk to brewers' solicitor bore the usual conveyancing meaning, so that there was no binding agreement, or whether in the context the agreement was binding — Stephenson LJ, expressing a minority view, considered** H **that the agreement was binding, so that the figure for compensation should be £15,000 — Oliver and Fox LJJ held that the words "subject to a Contract to be approved by me" should be given their normally understood meaning, so that there was no binding agreement and the compensation should be assessed as at date of entry — Decision of trial judge upheld, but on different grounds, and appeal dismissed**

This was an appeal from a decision of Nourse J on an originating summons issued by the plaintiffs, Duttons Brewery Ltd, respondents to the present appeal, for a declaration that the defendants, the present appellants, Leeds City Council, were bound to compensate the respondents for the acquisition of licensed premises on the J basis of the value at the date of entry by the appellants. The premises in question were The New Inn, Upper Town Street, Bramley, Leeds. Nourse J's judgment is reported at (1980) 256 EG 919.

P Ground QC and J Behrens (instructed by Sharpe, Pritchard & Co, agents for James Rawnsley, Leeds City Council) appeared on behalf of the appellants; Peter Millett QC, B J Knight QC and R J C Wilmot-Smith (instructed by Farley, Parker & Pickles, of Blackburn) represented the respondents.

Giving judgment, STEPHENSON LJ said: This appeal concerns the compulsory purchase by the Leeds City Council ("the council") of The New Inn, Upper Town Street, Bramley in that city, a tied public K house owned by Duttons Brewery Ltd ("the brewers"). The brewers let it to a licensee named Johnson in 1960. It was one of over 400 properties included in the City of Leeds (Bramley) Part V Compulsory Purchase Order 1964 made by the council on March 4 1964 and confirmed by the minister on March 19 1965.

Thereafter, things moved slowly as such things often do. The council gave the brewers and all others interested in the property notice to treat on April 3 1967. In May 1968 the figure for compensation to be paid was "agreed" at £15,000. I use that word in its ordinary sense, but put it between inverted commas to preserve the brewers' contention that the agreement was not legally binding. But the council was in no hurry to take possession and the brewers were L in no hurry to give it. The sales at The New Inn, which had been going down during the years 1962 to 1967, were going up. The council did not give notice of intention to enter and take possession until May 4 1976 and possession was not given and taken until June 30 1976. The council did not execute a deed vesting in themselves the fee simple in the property until November 10 1978.

By that time the value of these licensed premises, a value based on "barrelage", had risen above £15,000 to a figure which the brewers claim to be £78,000. So on January 24 1979 the brewers issued an originating summons for a declaration that the council was bound to compensate them for their freehold interest in the premises known as The New Inn, "such compensation to be assess- M ed in accordance with the statutory rules, the said premises to be valued *rebus sic stantibus* as at the date of entry by the council, namely June 30 1976".

That declaration Nourse J granted on June 27 1980 and from his order the council appeals to this court.

Before the judge the council took a preliminary point that he had no jurisdiction to determine the issue raised by the summons because it was a question of disputed compensation which had to be referred to the Lands Tribunal by virtue of section 6 of the Compulsory Purchase Act 1965. The learned judge had no hesitation in rejecting the point and the council's appeal against that part of his

A decision is not pursued. I need therefore say no more about it except that it has led the brewers, at the suggestion of Oliver LJ, to amend their summons to raise the real questions argued below by seeking a declaration that on June 30 1976 "there was no subsisting binding and enforceable agreement (1) for the sale by (the brewers) and the purchase by (the council) of the said premises; or (2) for the amount of compensation payable (by the council to the brewers) for the compulsory acquisition thereof".

The learned judge decided those questions partly in the council's favour and held that there was a binding agreement as to price or compensation in 1968. But he decided in favour of the brewers that the agreement as to compensation no longer stood because it was B subject to an implied term or condition that completion by entry should take place within a reasonable period after June 1968, and that reasonable period expired in 1971 or shortly thereafter. The judge found it unnecessary to decide whether the agreement had been "abandoned" by the council, and whether it was still enforceable by specific performance. The implication upon which the judge rested his judgment in favour of the brewers is challenged by the council in their notice of appeal, Mr Ground for the council pointing out that it accepted what was and still remains a subsidiary argument not expressed in the summons or any affidavit. Mr Millett for the brewers asks us to affirm the judge's decision both on that ground and on the other grounds which the judge rejected or did C not decide.

The correspondence between the parties, their solicitors and their valuers shows that the figure of £15,000 was negotiated on the trade (barrelage) figures of The New Inn and included a figure for depreciation to fixtures and fittings. On April 19 1968 the brewers wrote to their solicitors and to their valuers. To the solicitors they wrote as follows:

> The above property

that is, The New Inn at Bramley

> has been the subject of a compulsory purchase order by the Leeds Corporation and our agents, Messrs Spencer, Son & Gilpin, have now agreed D compensation with the licensed property valuer at £14,500; in addition to this figure the licensed property valuer has agreed to pay this company £500 in respect of depreciation of fixtures and fittings. The corporation have also agreed to pay your charges and those of Messrs Spencer, Son & Gilpin.
>
> The sale is, of course, with vacant possession but as yet we have had no intimation from the town clerk as to when the corporation might require possession of these premises and we propose to carry on trading until such time as the corporation can give a firm date as to when completion of the purchase of this property can be expected.
>
> We should be pleased if you would bear in mind that under the terms of the tenancy agreement the tenant would normally require three months' notice to terminate the tenancy and we should be obliged if you would take this point up with the town clerk as and when his department contact you in E connection with the sale of this property, as the licensed property valuer's report will not yet have been forwarded to the town clerk.

To the valuers they wrote:

> We have considered at some length what you say as regards this property and in the circumstances we are prepared to accept the licensed property valuer's offer of £15,000 in respect of these premises, though we feel in this case that the licensed property valuer's offer is not over generous.

Their valuers wrote to them and to their solicitors on April 23 and 26 respectively the letters which I shall read. In the first letter they wrote:

> We note that you are prepared to accept the licensed property valuer's offer F of £15,000 in connection with this property and when we receive the formal offer from him we will inform your solicitors accordingly.
>
> We note what you say with regard to the licensed property valuer's offer. This is an unfortunate case in as much as the trade at this house has shown a steady decline over the last few years and it cannot be attributed to slum clearance where we could have argued that this had a detrimental effect. We feel that in all the circumstances we have got the best possible figure for this house.

In the other letter they wrote:

> On behalf of our mutual clients we have negotiated the compensation in connection with the above property. The details of the agreement reached with the licensed property valuer are as follows:
>
> 1. The premises to be acquired are freehold and comprise the New Inn G

and they refer to the plan which identifies it.

> 2. The sale is to be with vacant possession.
> 3. The licence authorising the sale of beer, wines and spirits for consumption on the premises is excluded from the sale and remains the property of the vendors.
> 4. There are no fixed charges or onerous restrictions or liabilities affecting the property.
> 5. Subject to the above conditions, compensation in settlement of all heads of claim to be the sum of £15,000 (fifteen thousand pounds).
> 6. Included in the compensation at 5 above is an item representing depreciation of fixtures, fittings and trade effects. These remain the property of the vendors, subject to their removal before delivery up of possession. H
> 7. In addition to the compensation set out at 5 above, the acquiring authority to pay the proper legal costs of the vendors and to make a contribution of £184 16s (one hundred and seventy-six guineas) towards surveyor's fees.
>
> No doubt you will be hearing from the corporation in the matter in due course.

Thereupon the brewers' solicitors wrote to the town clerk of the council on April 29; and that letter reads:

> We understand that you will shortly be hearing from the licensed property valuer indicating that compensation has been agreed in the above matter at £14,500 and in addition you have been recommended to pay £500 in respect of the depreciation of fixtures and fittings and the legal charges and surveyor's fees. If your corporation are to accept these terms we shall be glad if you will let us know when you will require vacant possession and in this J respect perhaps you will bear in mind that under the terms of our clients' tenancy agreement with the present tenant the tenant would normally require three months' notice.

On May 6 they sent the town clerk a copy of the valuers' letter to them of April 26; their letter reads:

> We refer to our letter of the 29th ultimo and enclose a copy of Messrs Spencer Son & Gilpin's letter setting out the terms. We propose to dispose of Contracts and enclose therefore full abstract of title to enable you to proceed.

To that the town clerk replied on May 29:

> K I refer to your letter of May 6 and have to inform you that the city council have now approved the purchase of the above property from your clients at the price of £15,000 subject to a Contract to be approved by me. The corporation will also pay your legal costs in accordance with the extra statutory scale, amounting to £100 and surveyors' fees of £184 16s on completion.
>
> In view of the fact that notice to treat has been served in respect of the property, I agree that contracts may be dispensed with and will let you have a draft conveyance for approval in due course.
>
> I have requested the city architect to let me know when vacant possession will be required and I will write to you again on this point as soon as possible.

There followed a further letter from the town clerk to the brewers' solicitors, dated June 7: L

> Further to my letter of May 29, I am now informed by the city architect that it is unlikely that this property will be required for redevelopment before 1971.

On June 12 the town clerk sent the brewers' solicitors the draft conveyance for approval and on June 27 they returned the draft conveyance "duly approved". This the town clerk acknowledged on July 2 and on July 23 he wrote:

> I now enclose engrossment of the Conveyance for execution by your client. Please let me know when you are ready to complete.

M That letter was followed by the following exchange of letters on July 29 and August 5. The brewers' solicitors wrote to the town clerk as follows:

> We have heard from our clients in this matter and they are surprised at receiving the document for sealing so soon. When they instructed us in the sale of this property on April 19 last, they notified us that they had no intimation as to when your corporation would require the above premises and that they intended to carry on trading until such times as your corporation could give a firm date. They inform us that you will not require this public house until 1971 though you will require the adjoining unlicensed properties prior to this date. Perhaps you will kindly let us know whether or not there has been any change in your requirements.

A The reply from the town clerk was:

I refer to your letter of July 29 and confirm that the corporation's requirements regarding possession of this property are unchanged. The engrossment of the conveyance was forwarded to you in the routine way and I overlooked the fact that completion will be deferred for some three years.

On August 8 the brewers' solicitors informed the brewers accordingly: "We have now heard from the Leeds Corporation that they will not require the public house until 1971".

1971 came and went, and so did 1972. Then, on February 28 1973, the brewers' valuers took up the question of compensation with the council's valuer in the following letter:

B You will recall that in 1968 we negotiated compensation with you in respect of the above property. These premises were never taken over by the local authority and in fact are still open and trading. We have now received a letter from our clients and they request that we reopen negotiations with you for revised compensation. The trading has increased considerably over the last four years and for your information we set out details below

and they there set out the years 1969–70, 1970–71, and the nine months to December 31 1972, with the figures for draught beer (barrels), bottled beer (barrels) and wines and spirits (gallons).

When you have had an opportunity of considering this we shall be pleased to hear further from you in the matter and arrange to discuss revised terms.

C The brewers' valuers made the same request direct to the council on March 13:

We negotiated compensation with the licensed property valuer in respect of the above property in April 1968. The sale of the property was not completed and in fact it has continued to be occupied and trading since that date. The premises are still as of to date trading and there are no indications as to when the corporation will want the property. In the circumstances, our clients requested that we write to you with a view to reopening negotiations for a revised figure of compensation. We should be pleased if you would confirm that you are prepared to reopen negotiations and that you will instruct the licensed property valuer accordingly.

To that the council's chief legal officer replied on March 21 as
D follows:

As you are no doubt aware notice to treat was served on April 3 1967 and the compensation settled on April 29 1968

that should be May 29; it is an obvious mistake

thus creating a binding contract. In the circumstances the corporation are not prepared to reopen negotiations for a revised figure of compensation.

In addition, I would remind you that the accepted date for assessment of compensation is either the date when possession is taken or the date when the compensation is agreed, whichever is the earlier.

On April 11 1973 the brewers' solicitors wrote to the council expressing surprise at the council's attitude and said "we feel that
E our clients do have an arguable case in that compensation was settled on the basis of a fairly early completion, and it is, of course, some five years before our clients were called upon to complete". The council's chief legal officer rejected this request; he saw no reason to refer the matter to the council and was equally unmoved by a subsequent appeal to fair play over a moral issue. He took his stand on the principles laid down by statute and case law. "The relevant date," he wrote, "for assessment of compensation where notice to treat applies was considered by the House of Lords in the case of *Birmingham Corporation v West Midland Baptist (Trust) Association Inc* [1970] AC 874 when it was decided that in their view the right date for assessment of compensation would be either
F the date on which the acquiring authority take possession of the property or the date when the value is agreed, whichever is the earlier".

There the matter rested until the notice of entry on August 22 1975, followed by possession on June 30 1976 after a further unsuccessful attempt to reopen negotiations on the amount of compensation payable. By July the brewers' solicitors and the council were discussing in correspondence the issue whether there was or was not a binding contract to complete, based on the compensation assessed in 1968 on an average maintainable trade basis.

Was there? I agree with the judge that there was, and I find nothing which happened after 1968 to put an end to it or vary it or prevent the council from relying on it. I shall deal with this question G first, together with all the additional grounds raised in the respondents' notice on the brewers' behalf, before I consider the finding of an implied term or conditional agreement on which the judge based his decision in favour of the brewers.

I put the cart before the horse because it was put there — wisely — by counsel for the brewers both below and before this court. Those additional grounds are:

(1) No binding and enforceable contract was ever made by the plaintiff and the defendant because (a) It was "subject to contract", and/or (b) the notice to treat upon which the acquisition proceeded was in respect of the plaintiff's freehold interest subject to a tenancy whereas the price of £15,000 was negotiated in respect of the plaintiff's freehold interest with vacant possession; alternatively, (2) On the assumption that the terms of purchase H were agreed between the plaintiff and the defendant in May 1968 for the purchase of the plaintiff's freehold interest at £15,000, by May 1976 the defendant (a) had abandoned their contractual entitlement to purchase the plaintiff's said interest at the said price or (b) by the date the defendant entered and took possession of the plaintiff's premises in June 1976 (pursuant to notice of entry served pursuant to section 11 of the Compulsory Purchase Act 1965) the defendant has lost any right to specific performance of the said agreement.

Ground (1)(b) seems to have been argued first below, and the judge disposed of it as follows:

First, he J

that is, Mr Millett

says that the parties were not negotiating the amount of compensation which would be payable by the council for the acquisition of the interest to which the notice to treat applied, that is to say Duttons' freehold interest subject to Mr Johnson's tenancy, but that they were negotiating for the acquisition of the whole freehold interest with vacant possession. Mr Millett says that the agreement by Duttons was to give vacant possession, and accordingly that was the interest in respect of which compensation was being or must be taken to have been negotiated. I must say that I find this argument a little unrealistic. It seems to me that the parties were in practice negotiating the amount of the compensation which would be payable in respect of the freehold subject to the tenancy. That is borne out both by the K fact that the council knew about the tenancy before the price was finally agreed, and also by the fact that the council later paid Mr Johnson a substantial amount of money

and I interpolate that it was in fact £10,000

for his own interest. And even if Mr Millett's view of the negotiations was correct, it does not, in my judgment, follow that the agreement as to the £15,000 was not binding so far as Duttons' actual interest was concerned. In the circumstances, I reject Mr Millett's argument on that point.

I agree with all that and have only to add that, while it is true that vacant possession was referred to in the letter of April 26 1968, the council knew that Mr Johnson was the tenant of these licensed L premises, the valuation was the valuation of the council's licensed property valuer and the council met Mr Johnson's claim for disturbance compensation with a separate payment of £10,000. The council were clearly valuing and purchasing the brewers' freehold interest subject to Mr Johnson's tenancy, which was itself in terms determinable immediately upon the giving of notice to treat and notice to cease trading; so that it would not have reduced the purchase price appreciably, if at all.

Ground (1)(a) has more substance, so much more that it has been accepted by both my brothers and will defeat this appeal. What the judge said about that was this:

The second argument on this part of the case is based on the terms of the M town clerk's letter of May 29 1968, to which I have already referred. Mr Millett says that that letter makes it clear that the council only entered into the agreement as to the price subject to a contract to be approved by the town clerk, that is to say, subject to contract in the sense in which that term is understood by conveyancers. He has referred me for this purpose to the *Munton* case

I shall refer to that later

from which it is clear that if that was a correct view of the relationship between the parties, then there certainly was not a binding agreement as to price. However, that is not, in my view, a correct reading of the town clerk's letter of May 29. That letter, like many which come to be scrutinised in the

A course of proceedings in this court, is not perhaps with hindsight as happily worded as it might have been. But, in my judgment, it would be putting altogether too strained an interpretation upon it to say that the words "subject to a Contract to be approved by me" meant that the council's approval of the purchase price of £15,000 had only been on terms that there would actually be a contract in a form approved by the town clerk and that that contract would be signed and exchanged on both sides in accordance with normal conveyancing practice. Clearly the town clerk did not understand it in that way because he went on to say that he agreed that contracts might be dispensed with. It seems to me that the natural and normal meaning of the words "subject to a Contract to be approved by me" is that the council were leaving it to the town clerk to deal with the contract, and if he decided that this was a case where it was not necessary to have contracts, B that would be all right by them. So I reject Mr Millett's argument on that point as well.

With that also I agree. The true construction of the town clerk's letter of May 29 1968 has to be ascertained against the background of the notice to treat and the relevant facts and law.

I take the law from what Lord Denning MR said in *Munton v GLC* [1976] I WLR 649 at pp 652 and 653 and 655, and from the statement by the editors of vol 8 of ed 4 of *Halsbury's Laws of England*, para 136, there mentioned by Lord Denning, and quoted by the judge in his judgment. The statement in that paragraph reads:

> C The effect of serving a notice to treat is to establish a relation analogous in some respects to that of vendor and purchaser, a relation which binds the owner of the land to give up the land subject to his being paid compensation, and which binds the undertakers or acquiring authority to take the land, but there is no contract of sale until the price is ascertained and the land remains the property of the landowner. Both parties have the right to have the price ascertained and the purchase completed in manner provided by the Lands Clauses Acts or the Compulsory Purchase Act 1965 or any Acts modifying those Acts.

The Master of the Rolls' statement, with which both Scarman LJ and Goff LJ agreed, is this:

> Two propositions are settled. First, when a notice to treat is given, it binds D the acquiring authority to purchase and the owner to sell at a price to be ascertained: see *Mercer v Liverpool, St Helen's and South Lancashire Railway Co* [1903] 1 KB 652, 664, approved by the House of Lords [1904] AC 461, 463. Second, when there is an unconditional agreement fixing the price — whether in writing or by word of mouth — it is the equivalent of a binding contract between the parties: see *Harding v Metropolitan Railway Co* (1872) 7 Ch App 154. Lord Hatherley LC said at p 158: ". . . when the price is ascertained . . . you have then all the elements of a complete agreement, and, in truth, it becomes a bargain made under legislative enactment between the railway company and those over whom they were authorised to exercise their power".
>
> In *Dart on Vendors and Purchasers*, 8th ed (1929), vol 1, p 256, it is said: "Where, however, the price is ascertained . . . the contract is complete, and E may be specifically enforced by or against the company". Similar statements will be found in *In re Cary-Elwes' Contract* [1906] 2 Ch 143 and *Halsbury's Laws of England, 4-ed, vol 8 (1974) para 136.*

The paragraph I have just read.

> Once therefore a notice to treat has been served and there is an agreement on the price, a binding obligation is created which is equivalent to a contract between the parties.

Thereafter the acquiring authority can call for a conveyance and completion, as indeed can the vendor whose property is acquired. Nothing further remains to be agreed unless the agreement is made F conditional or subject to contract. If the vendor will not execute the conveyance, the authority has only to pay the ascertained price to the vendor or into court with interest: Lands Clauses Consolidation Act 1845, section 76, Compulsory Purchase Act 1965, section 11, replacing Acquisition of Land (Authorisation Procedure) Act 1946, Sched 2, para 3(1). The council, we were told, have paid the £15,000 with interest into an account pending the resolution of this dispute.

But a compulsory purchase may be subject to contract and if the compensation is agreed subject to contract there is no concluded agreement to buy and sell. That was plainly the position in *Munton's* case [1976] 1 WLR 649 and that would be the position here if the first paragraph of the letter of May 29 1968 stood alone. But, as the judge pointed out, it has to be read with the second paragraph, G and read against the legal position created by a notice to treat and the preceding letters from the brewers' side, agreeing the £15,000 and suggesting that contracts should be "disposed of", which was rightly understood by the town clerk as meaning "dispensed with". We have not got the resolution of the council which gave the town clerk authority to approve the contract in this case or generally, but I think we must take it to have been correctly reproduced in the first sentence of the letter and to have required the agreement between the council and the brewers to be indeed subject to contract, but to a contract approved by him. That must, however, mean that if he approved the contract the ordinary exchange of contracts would follow without any need to refer their terms to the council. If then H the council were content to leave the approval of the contract to their town clerk, why should he not dispense with the exchange of contracts altogether if the brewery did not insist on them but indeed suggested doing without them, and go straight to a conveyance reciting, as the draft conveyance did, the agreement come to and leaving undecided only the date for completion? The town clerk understood his authority in this way, so did the brewers and so did the council, as is shown by the course of subsequent events and correspondence. So did the judge, and so, with the greatest respect to the opinion of those who think differently, do I.

The court cannot, of course, look at subsequent correspondence to aid the interpretation of a written contract or the letters in which J it was concluded; but it is a striking fact that no one questioned the existence of a binding contract until the letter of July 7 1976 from the brewers' solicitors, and though there was in that letter a reference to the case of *Munton* just decided, the point that the agreement was not binding because it was subject to contract was not made even then. Now it is taken on behalf of the party which invited the council to dispense with contracts and kicked, not against the compulsory purchase but the price paid for it; and the absence of that which was treated by both parties as an unnecessary formality is to be declared fatal to a purchase disliked and disputed only because it has turned out to be underpriced.

I would accordingly hold, like the judge, that the agreement to K sell at £15,000 was not subject to contract.

Grounds (2)(a) and (b) I must now consider because I feel bound to differ from the ground of the judge's decision which made it unnecessary for him to consider them. It is not easy to infer abandonment of contractual rights from inactivity or of an interest in land from delay. But the courts have long ago come to the aid of unwilling vendors of property compulsorily acquired by purchasers who delay completion to the prejudice of those helpless vendors and have held that a notice to treat can be abandoned by lapse of time. What Lord Romilly MR said in 1868 in *Richmond v North London Railway Co*, LR 5 Eq 352 at pp 358-9 has been endorsed L by Lord Upjohn both when sitting at first instance in *Grice v Dudley Corporation* [1958] Ch 329 at p 339; and when giving the judgment of this court in *Simpsons Motor Sales (London) Ltd v Hendon Corporation* [1963] Ch 57 at p 82, which was affirmed in the House of Lords [1964] AC 1088. There Lord Evershed, with the agreement of all their Lordships, pointed out (at pp 1117 amd 1118) that "as regards delay the court will and should be somewhat less strict in the case of a local authority exercising the powers and duties relating to housing and like matters than it was in the past in cases of railway corporations, which were regarded as private undertakers seeking to make profits for themselves". In *Ystalyfera Iron Co v Neath and Brecon Railway Co* (1873) LR 17 Eq 142, at p 149 Sir M George Jessel MR was less sympathetic to the landowner, but Lord Upjohn preferred (in *Grice's* case) the view of Lord Romilly MR in *Richmond's* case that "a notice to treat cannot operate for an indefinite time: it must be acted upon within a reasonable time or it must be deemed to be abandoned", for it would be "obviously inequitable to permit the railway company for an indefinite time to keep this notice over the head of the landowner, fettering his enjoyment of his property".

In the present case it is common ground that, in the absence of any agreement to the contrary, the council would be under a duty to complete the compulsory purchase in a reasonable time and that the brewers could have executed the conveyance and, in the words of Sir George Jessel MR in the *Ystalyfera Iron Co* case, forced the council on by giving notice to complete at any time and so making time of the essence; but they naturally chose to continue trading. And although it is not true to say that they requested the council to delay completion, as their trade turned out they welcomed the long postponement. Told in 1968 that the property would not be required before 1971 and completion would be deferred for some three years, and told in 1970 that it would not take place until some time in 1971 they made no protest and never suggested that the delay would have any effect on what their own valuers described in a letter to their solicitors of December 18 1968 as the compensation agreed with the district valuer of the council, or that the compensation was provisionally agreed, until their valuers requested a reopening of negotiations for revised compensation in their letter of February 28 1973. Thereafter the council were affirming the contract, including the agreement to pay £15,000 compensation, and in my judgment the delay and inactivity before and after 1973, surprising, annoying and even regrettable though it may have been, was not such as to amount to laches or misconduct or to cause such injustice to the brewers as to constitute abandonment or repudiation of the contract or the council's right under it to purchase the brewers' freehold interest in the premises at £15,000.

If that be right, I do not see that there would be any justification for the court's refusing to decree specific performance of the contract of sale and purchase to either party. There are incidents of ordinary contracts of sale and purchase of land which apply to what have been called parliamentary or statutory contracts after notice to treat, and there may be cases where delay in completing a sale and purchase of land may be such as to entitle the prompt and eager party to accept it as a repudiation, or the court to refuse specific performance. But where, as here, there is initially mutual consent to long delay, there must in my opinion be a clear intimation by that party that his consent is exhausted and he wants his contract performed or terminated. The brewers gave no such intimation but claimed revision of a term of the contract. Furthermore, the council needed no decree of the courts to enforce their contract; they had their statutory rights and I agree with Mr Ground's submission that it is irrelevant to consider the question whether, if they had required a decree, a court of equity would have granted it.

My rejection of both these grounds is, I think, supported by the affidavit evidence of Mr Duxbury for the council that the council did not need the land until 1976 and by the further affidavit evidence of Mr Booth Lee, the council's compulsory purchase officer, which we admitted to deal with the implied term point, as to the practice of the council, resulting from the unpredictable delays to which compulsory purchase orders are subject, to allow owners to continue trading from the premises and to complete without delay if the owner wishes to cease trading and to vacate the premises.

I come finally to the ground on which the judge felt able to decide the case in the brewers' favour. The judge dealt with Mr Millett's argument that the agreement as to price was conditional on completion taking place by 1971 or shortly thereafter, and in any event long before June 30 1976 in this way. The learned judge said:

> ... the first question here is: Was there an agreement that the compensation should still be £15,000, even if completion did not take place during the period which was within the contemplation of the parties in June 1968? It seems to me that if you take into account the method of assessing compensation in this case and the fact that the date of entry was very much at the dictation of the council, the answer to that question must be "No". I find it impossible in the circumstances to construe the agreement between the parties as being that the price should be £15,000, irrespective of the date on which entry was actually made. Entry might not have been made until 1980, 1985 or even later. In my judgment, Duttons' agreement to accept the £15,000 necessarily assumed that entry would be made and the sum paid within a reasonable period after June 1968. That reasonable period expired in 1971 or shortly thereafter. It is not necessary for me to be more precise than this because it is on any footing clear that June 30 1976, or even (so there should be no doubt about it) January 1 1976, was well outside any period which was reasonable, or which could have been within the contemplation of the parties in June 1968. On that ground, therefore, I hold that the agreement as to price no longer stands and that it had ceased to stand by January 1 1976 at the latest. In the circumstances, it must follow that compensation falls to be assessed as at the date of entry on June 30 1976 in accordance with the decision of the House of Lords in the *Birmingham Corporation* case, to which the council's chief legal officer had referred in the correspondence which I have read.

I have considered the further evidence, not only of Mr Booth Lee for the council but of Mr Sampson for the brewers on this point, and I am not surprised to find Mr Booth Lee saying that if during the negotiations in 1968 the brewers had asked when completion would take place the council's answer would have been "when we need the premises or at any time earlier should you wish to (complete)", and Mr Sampson saying that the brewers would never have agreed to the sum of £15,000 if they had been told in 1968 that entry would be delayed until 1976. Whether he would have sworn to the same answer if the barrelage of The New Inn had continued to decline during those years, I feel some doubt. But then there would have been no originating summons.

Much as I sympathise with the judge's implication of this term or condition, I see no commercial necessity for it and no room for it in the circumstances of this case. The judge found it "very strange" if the council, having put off entry until 1976, were still able to assert a right to pay compensation agreed on the basis of the most recent barrelage figures in 1968. So at first did I, and I am not unhappy to find that my brothers are able to uphold the judge's decision, but on a different ground and one for which there is more to be said, and much more has been said, by Mr Millett. If, however, as I would hold, the brewers made a binding agreement to sell for £15,000, with every necessary term expressed except the date of completion left uncertain by agreement, they cannot get out of it by implying a term which could so easily have been expressed but if expressed might not have been agreed to by the council. The compensation could have been agreed provisionally at £15,000 and made subject to revision or renegotiation after the expiration of a particular number of years or if the value of the property or its trade increased. But it was not. It is therefore not open to the brewers to take the compensation to the Lands Tribunal for assessment.

I am accordingly of the opinion that the brewers should be held to their agreement as to price which still stands, the appeal allowed and the declaration they obtained from the judge set aside. But as my brothers are of the contrary opinion, this appeal will be dismissed.

Giving judgment in favour of dismissing the appeal, OLIVER LJ said: I entirely agree that, for the reasons given by my Lord, the existence of an implied term to the effect stated by the learned judge cannot be supported, for I do not see how a term can be implied into a contract except upon the supposition that, if it had been brought to the attention of the parties, they would both have agreed to its inclusion. It is, in my judgment, perfectly plain that the defendants would not have agreed to anything of the sort, and if that were the only matter in issue, I too would be in favour of allowing the appeal.

There are, however, a number of points which are raised by the respondent's notice — points which were either decided by the learned judge against the plaintiffs or which, because of the view which he took of the case, it became unnecessary to decide; and on the principal and critical point of the initial agreement I have felt compelled to a conclusion different from that reached by the learned judge and by my Lord.

Logic and convenience suggest that I should deal first with that point, which is the first ground advanced in the respondent's notice, for one thing that is common ground is that if the defendants are to succeed on this appeal they must demonstrate that the amount of compensation payable was contractually and unconditionally agreed. That is not disputed and it appears clearly from the decision of this court in *Munton v Greater London Council* and, in particular, from the passage from the judgment of the Master of the Rolls at p 652 of the report in [1976] 1 WLR 649, the passage already read by my Lord. In that case a negotiation of the amount of compensation had taken place in correspondence which as expressly made "subject to contract" and it was held that the amount so agreed could not be treated, there never having been any completed contract, as the agreed compensation payable under a statutory contract constituted by a notice to treat followed by agreement of the amount of compensation. The landowner's right on a compulsory purchase is to have the compensation assessed in accordance with statutory rules and he is not to be deprived of that right save by a binding contract to relinquish it.

Mr Millett's submission is that the instant case is substantially indistinguishable from *Munton's* case and his submission rests, as regards this point, on the true construction to be put upon the letter from the town clerk to the plaintiffs' solicitors of May 29 1968 read in conjunction with the surrounding circumstances, and the earlier and immediately subsequent correspondence.

Mr Millett's argument proceeds in two connected, but to some extent independent, steps. First he says that what was under negotiation between the parties was not merely the agreement of the sum of compensation payable under the notice to treat but the terms of a privately negotiated contract of sale — negotiated no doubt under the threat of the notice to treat, but otherwise independently of it. That was what was offered to the defendants and it was that which they accepted subject to contract; and the town clerk's acceptance of a suggestion (of which I surmise the defendants themselves were in fact unaware) that contracts should be dispensed with does not bring into being a contract of wholly different nature, that is to say a statutory contract.

This submission as to the nature of the agreement under negotiation is not critical to Mr Millett's argument but it assists him (if correct) in his approach to the true construction to be placed on the correspondence. In substance it bases itself on the fact that the property was subject to Mr Johnson's tenancy and that all that the notice to treat purported to affect was the plaintiff's interest in the land subject to that tenancy. What was offered in their valuers' letter was the property with vacant possession. Thus, the argument proceeds, although the notice to treat no doubt triggered the negotiations, the contract being negotiated was a different contract from the statutory contract which would have resulted from the mere agreement of the compensation.

The learned judge rejected this as artificial and I think he was right to do so. The correspondence prior to the plaintiffs' valuers' offer of April 26 was entirely upon the footing that what was being agreed was compensation only and although it is true that, technically, the only interest which could be compulsorily acquired under the notice to treat was the freehold subject to the tenancy, Mr Ground has pointed out that the tenancy itself contained a term terminating the tenant's interest upon service of a notice to treat and reducing it to a mere licence which would determine immediately the landlords chose to give up possession to the acquiring authority. From a valuation point of view, therefore, the tenancy was of minimal significance.

This, however, does not dispose of Mr Ground's difficulty, because the question remains, was the amount of compensation, assuming that to be the sole subject-matter of the negotiation, ever unconditionally agreed? This brings me to the second step in Mr Millett's argument, and it is necessary, at pain of repeating to some extent ground already covered by my Lord, to refer again to the correspondence. For relevant purposes it starts with the defendants' licensed property valuer's letter of August 18 1967 from which it is quite clear that what is being negotiated is not a contract but an amount which he can recommend for the approval of the defendants. It is unnecessary to refer in any detail to the course of the negotiations prior to April 1968, when the plaintiffs, having taken their own valuers' advice, wrote to their solicitors approving the figure of £15,000 suggested and asking them to take up with the town clerk the question of when completion would be required — a matter of some little importance to them because they were intending to carry on trading until completion and they wanted to give three months' notice before terminating Mr Johnson's occupation (although in fact they were not bound to do this). On the same day, April 19, they wrote to their valuers saying that they were prepared to accept the licensed property valuer's figure.

On April 26 1968 the plaintiffs' valuers wrote the important letter to the plaintiffs' solicitors to which my Lord has already referred. Thereupon the plaintiffs' solicitors wrote to the town clerk the letter dated April 29 1968.

Two things are, I think, clear from this letter. First, it is clear that it was not in the contemplation of the writer that any agreement between the valuers bound anybody. The terms suggested were being put forward as an offer which was open for acceptance, not as a record of a completed bargain already made. Secondly, the letter stresses the importance of some clear information as to a completion date.

No reply to this letter was received at this stage and on May 6 the offer which it contained was superseded by a new offer, because the solicitors sent to the town clerk a copy of the valuers' letter to them of April 26, to which I have already referred. They added "We propose to dispose of" (and I think that is a malapropism for "dispense with") "Contracts" (with a capital C) "and enclose therefore full abstract of title to enable you to proceed". So here was an unconditional offer, accompanied by an intimation that the vendors' solicitors were prepared to dispense with formal contract — that is to say to let the matter rest on the correspondence alone and proceed straight to conveyance.

It is, I think, unarguable that anybody was bound at this stage. I come then to the crucial reply of the town clerk, which begins by referring to the letter of May 6 — that is, the letter of offer. My Lord has already read this letter, but I think it will be convenient for me to read it again. It is in these terms:

> Dear Sirs, I refer to your letter of May 6 and have to inform you that the city council have now approved the purchase of the above property from your clients at the price of £15,000 subject to a Contract to be approved by me. The corporation will also pay your legal costs in accordance with the extra statutory scale, amounting to £100 and surveyors' fees of £184.16s. on completion".

Now if this letter had terminated there, it would in my judgment be equally unarguable that any binding agreement had been concluded. Indeed, I do not think that Mr Ground was disposed seriously to quarrel with that proposition. What is said, however — and this was the argument which appealed to the learned judge — is that the *prima facie* significance of that first paragraph is not only neutralised but nullified by the next paragraph. So I will read the remainder of the letter. It continues:

> In view of the fact that notice to treat has been served in respect of the property, I agree that contracts may be dispensed with and will let you have a draft conveyance for approval in due course. I have requested the city architect to let me know when vacant possession will be required and I will write to you again on this point as soon as possible.

Nothing further occurred until June 7, when the town clerk wrote again saying that the property was unlikely to be required before 1971. Five days later he sent a draft conveyance for approval and that was returned approved on June 27. This document recites an agreement for the payment of the sum of £15,000 by way of compensation and Mr Ground prays this in aid as evincing the intention of the parties. It does not appear to me, however, that this document can be of any relevance in determining the question of the contractual or non-contractual effect of the correspondence, much less as in itself constituting a contract. This was merely the town clerk's draft of a suitable form of conveyance to give effect to a transaction which both parties then clearly intended to effect and it cannot, in my judgment, have any bearing upon the question of whether either party was contractually bound by the correspondence to effect it. Equally the mere approval of a form of conveyance by the plaintiffs' solicitors cannot in itself constitute a contract.

Following approval of the form of conveyance, the town clerk had an engrossment prepared and sent it to the plaintiffs' solicitors for execution by their clients, on July 23 1968. They replied expressing some surprise in the light of the information which had been given regarding the date on which the premises were likely to be required, and asking whether the defendants had changed their requirements. The response was that the conveyance had been sent as a matter of routine and that completion would be deferred for some three years. Accordingly the conveyance never was executed and had not been executed in 1973, when the plaintiffs sought to reopen the question of compensation.

That suggestion was resisted by the defendants on the footing that there was and remained a binding agreement between the parties fixing the compensation once and for all at £15,000 no matter when the transaction came to be completed and that contention has been maintained both before the learned judge and in this court. It rests, and can only rest, upon the construction of the town clerk's letter of May 29 1968, for there is nothing before or after that which could arguably give rise to any binding obligation. Mr Ground submits that the case is a simple one of a coalescence of three ingredients — a compulsory purchase order, a notice to treat issued under it, and an agreement of the amount of compensation to be paid.

Those three ingredients in conjunction constitute a statutory contract. The second paragraph of the letter of May 29 1968, he submits, simply reflects the town clerk's appreciation that the only element missing at that stage was the agreement of the purchase price, an element which he himself supplied by agreeing to dispense with the formal contract which had been suggested in the first paragraph. The letter has, he submits, to be read against the background that the parties were perfectly familiar with the mechanics of a compulsory purchase order and, in particular, that formal exchange of documents was unnecessary to produce a binding agreement. Against that background the announcement in the first paragraph that the defendants had approved the purchase at £15,000 "subject to a Contract to be approved by me" does not, it is argued, carry the connotation that the lawyer would expect in a normal conveyancing transaction, but merely signifies an authority to the town clerk to deal with the matter in the way which he thinks best. By agreeing to dispense with contracts he signified the defendants' final and unconditional acceptance of the plaintiffs' offer and a binding agreement resulted as soon as the letter was dispatched.

That was an argument which impressed the learned judge. He said:

> That letter, like many which come to be scrutinised in the course of proceedings in this court, is not perhaps with hindsight as happily worded as it might have been. But, in my judgment, it would be putting altogether too strained an interpretation upon it to say that the words "subject to a Contract to be approved by me" meant that the council's approval of the purchase price of £15,000 had only been on terms that there would actually be a contract in a form approved by the town clerk and that contracts would be signed and exchanged on both sides in accordance with normal conveyancing practice. Clearly the town clerk did not understand it that way because he went on to say that he agreed that contracts might be dispensed with. It seems to me that the natural and normal meaning of the words "subject to a Contract to be approved by me" is that the council were leaving it to the town clerk to deal with the contract, and if he decided that this was a case where it was not necessary to have contracts, that would be all right by them.

Despite Mr Ground's persuasive argument, I find myself unable to take that view of the letter. Accepting that it was written in the context of a compulsory purchase, with the procedure under which the town clerk was doubtless familiar, it has also to be borne in mind that it is a letter written by someone whose position makes him, almost by definition, a skilled and experienced conveyancer, with the most complete familiarity with the invariable — or almost invariable — and certainly the natural and *prima facie* meaning and effect of the expression "subject to contract" in any conveyancing correspondence. It seems to me quite inconceivable that the significance of this expression in the defendants' resolution, which must be assumed to be correctly set out in the letter, could possibly have escaped him. Equally it seems to me to be impossible to resist the conclusion that the resolution had, and was intended to have, the effect classically attributed to the formula used, that is to say, that the defendants were to be free to withdraw unless and until a formal agreement in terms approved by their legal adviser had been executed and exchanged. That such a formal agreement was contemplated seems to me entirely clear from the use of the capital C in the word "Contract". It was not merely to be subject to arriving at a binding agreement but subject to "a Contract" (ie a formal written contract). In *Chillingworth v Esche* [1924] 1 Ch 97 Sargant LJ at p 114 observed: "I do not say that the phrase makes the contract containing it necessarily and whatever the context a conditional contract. But they are words appropriate for introducing a condition, and it would require a very strong and exceptional case for this clear *prima facie* meaning to be displaced".

This was quoted by Goff LJ in the *Munton* case at p 656 where, explaining his own decision in the exceptional case of *Michael Richards Properties Ltd v St Saviours, Southwark* [1975] 3 All ER 416, he went on to say: "There may be other cases — I do not know — in which the facts might again be so strong and exceptional as to require the court to treat the words 'subject to contract' as meaningless. If there be any such other cases, they must indeed be rare . . .".

For myself I find no strong or exceptional circumstances in the instant case which would lead me to the conclusion that the defendants' resolution meant anything other than exactly what it said. If that is right, then I find myself unable to read the first paragraph of the town clerk's letter as signifying anything other than this: "The city council has authorised me to write to you accepting your offer subject to a formal agreement to be prepared by me and exchanged in the usual way".

Then comes the question, what is the effect of the second paragraph? I cannot see that this carries the matter any further. What this appears to me to be signifying with tolerable clarity is that, although the offer is accepted "subject to contract", so that, at this stage, there is no binding agreement, the fact that a notice to treat has been served means that the plaintiffs are bound to convey with or without a binding agreement, so that a formal contract can safely be dispensed with and the matter can proceed to completion without any binding agreement ever coming into existence. In other words, they can take matters on trust until conveyance. But to dispense with a contract cannot *make* a contract, and I find myself unable to accept Mr Ground's submission to the contrary. What he suggests, if I understand his argument correctly, is that what this paragraph means is in effect this: "The council have agreed, subject to my preparing a formal agreement; but we both know that a formal agreement is unnecessary when a notice to treat has been given. All that is needed is that the figure of compensation should be unconditionally agreed, and on behalf of the council, I now unconditionally agree it".

Now apart from the difficulty that this is not what the letter says — it says merely that "I agree that contracts may be dispensed with" — it is flatly contradictory of the first paragraph, where the town clerk sets out the limits imposed by his principal on his own authority. He is the defendants' solicitor for the purpose of this transaction and would not ordinarily have any ostensible authority to make a contract on their behalf. *A fortiori* he would and could not be thought to have any such authority when he has, in this very letter, set out the extent of the terms which the defendants have authorised him to accept.

Furthermore there still remained an outstanding matter which, whether crucial to a statutory contract or not, was undetermined, namely the date for completion, a gap which can hardly be said to have been satisfactorily filled by an intimation (on June 7) that "it is unlikely that this property will be required for redevelopment before 1971".

As to the argument that the town clerk himself did not attribute to the words used the well-established and sanctified meaning which they have to every conveyancer, this, assuming it to be true, seems to me to be irrelevant, for whether the offer made met with an unqualifed and unconditional acceptance has to be judged not by what the writer may have intended in his private thoughts, but by an objective appraisal of the words which he has used in the context in which he has used them. Equally irrelevant, as it seems to me, is the fact that no express suggestion that there had not been any binding agreement was made prior to July 1976, after the report of the *Munton* case had appeared. It may well be that up to that time, the parties did in fact think that they were bound by a statutory contract, since it was not, prior to that case, clear that an agreement of the amount of compensation had, in order to constitute a statutory contract, to be a final and binding agreement.

I find myself unable, therefore, to accept the learned judge's analysis of the position, and I am unpersuaded by Mr Ground's arguments that the parties ever reached the stage of a binding agreement on the amount of the compensation. Accordingly, although for different reasons from those given by the learned judge. I take the view that his conclusion was correct and I would dismiss the appeal.

A The view which I take of the case makes it unnecessary to deal with the further grounds argued by Mr Millett, but I should perhaps add this: I am quite satisfied that, on any analysis, the argument based upon abandonment must fail. Abandonment of a contract must depend upon either the expression or an inference of an intention to abandon and if there is one thing clearer than another in this case it is that the defendants, if they ever achieved the contractual position which they claim, never intended to abandon it.

The argument advanced by Mr Millett and based on the inability of the defendants, after this lapse of time, successfully to claim specific performance is one that I find more difficult. I am not sure, B speaking for myself, that it is a complete answer to assert, as Mr Ground claims, that the defendants had no need to seek the assistance of equity to enforce the contract because they could obtain possession and convey the legal estate without any assistance from the court by reliance on their statutory powers. This is no doubt the case, but the contract that is sought to be performed is not simply a contract to convey the land and give possession. It is a contract to forgo the right to have the compensation assessed in accordance with the statutory rules. Now, as Mr Millett points out, there is ample authority for the propositions (a) that in a case of a compulsory purchase where no compensation has been agreed or assessed, the right to rely on a notice to treat may be lost by failure of the C acquiring authority to proceed within a reasonable time (*Grice v Dudley Corporation* and *Simpsons Motor Sales (London) Ltd v Hendon Corporation*); and (b) that under an ordinary contract for sale a party may lose his right to enforce by long delay even though there has been no notice to complete (see *Mills v Haywood* (1877) 6 ChD 196 and *Accuba Ltd v Allied Shoe Repairs Ltd* [1975] 1 WLR 1559).

I am not wholly convinced, therefore, that it necessarily follows that, simply because the amount of compensation has been agreed, the owner of land compulsorily acquired is for ever precluded from demanding that compensation be assessed in accordance with the statutory rules if there has been a long and unexplained delay on the D part of the acquiring authority in implementing the statutory contract thus arrived at. Having regard, however, to what I believe to be the correct analysis of the contractual position between the parties, I prefer not to express a concluded view on the point.

Agreeing that the appeal should be dismissed, Fox LJ said: I agree with my Lords, and for the reasons which they have given, that the judge's conclusion as to the implication of a term cannot be supported.

The next question is whether any binding contractual relationship for the disposition of the land ever came into existence at all. If it did not, then the council must fail.

E In considering that question, I disregard Mr Johnson's tenancy, which seems to me to be of no practical significance; it came to an end upon the service of the notice to treat and became a mere licence.

A notice to treat having been served, there were two modes by which a contract could have come into existence. The first is by an ordinary private contract for sale at a price. The second is by what is sometimes called a parliamentary contract. The position was stated by the Court of Appeal in *Munton v GLC* [1976] 1 WLR 649 at p 625 in the judgement of Lord Denning MR as follows:

> First, when a notice to treat is given, it binds the acquiring authority to F purchase and the owner to sell at a price to be ascertained: see *Mercer v Liverpool, St Helen's and South Lancashire Railway Co* [1903] 1 KB 652, 664, approved by the House of Lords [1904] AC 461, 463. Second, when there is an unconditional agreement fixing the price — whether in writing or by word of mouth — it is the equivalent of a binding contract between the parties: see *Harding v Metropolitan Railway Co* (1872) 7 Ch App 154. Lord Hatherley LC said at p 158: "... when the price is ascertained ... you have, then all the elements of a complete agreement, and, in truth, it becomes a bargain made under legislative enactment between the railway company and those over whom they were authorised to exercise their power".

What is asserted in the present case is that there was a parliamentary contract by reason of the unconditional agreement of the purchase price between the parties. Negotiations as to the price commenced in August 1967. It is evident from the valuation G officer's letter of August 18 1967 to the plaintiffs' valuers that the valuation officer did not regard it as his function to conclude any agreement as to the amount. He says: "I am instructed to negotiate and if possible agree with you terms of compensation which ... I can recommend for the approval of the Corporation".

Negotiations proceeded. On April 19 1968 the plaintiffs wrote to their own solicitors approving the figure of £15,000 which was suggested.

On April 26 the plaintiffs' valuers wrote the letter to the plaintiffs' solicitors setting out the details of the agreement which they had reached with the valuation officer. On April 29 the plaintiffs' solicitors wrote to the town clerk, saying: H

> We understand you will shortly be hearing from the licensed property valuer indicating that compensation has been agreed in the above matter at £14,500 and in addition you have been recommended to pay £500 in respect of the depreciation of fixtures and fittings and the legal charges and surveyors' fees. If your corporation are to accept these terms we shall be glad if you will let us know when you will require vacant possession....

It is clear that that letter does not regard the valuers as having committed their clients to anything. The letter was not recording an agreement; it was asking if certain terms were acceptable.

In fact on May 6, the plaintiffs in effect put forward a new offer because they sent to the town clerk a copy of the plaintiffs' valuers' J letter of April 29. The letter of May 6 concluded "We propose to dispose of contracts and enclose therefore full abstract of title to enable you to proceed". "Dispose of" I take to mean "dispense with".

The reply to that letter is the town clerk's letter of May 29. It is upon the construction of that letter that the present issue depends.

On June 7 the town clerk wrote to say that the property was unlikely to be required before 1971. Soon after he sent the draft conveyance for approval; it was approved on June 27.

Mr Ground's case for the council is that the parties were familiar with the procedure of compulsory purchase, under which a formal contract was not necessary for the purpose of producing a binding K agreement. All that was necessary was to agree the price. It is contended, rightly, that the plaintiffs had made an offer to sell at £15,000. That offer, it is said, was accepted by the town clerk, on behalf of the council, when by the letter of May 29 he agreed to dispense with contracts.

If one takes the letter of May 29, apart from the second paragraph (which accepts the proposal to dispense with contracts), it seems to me quite clear, on well-settled principles, that no binding contract of any kind resulted. That is the consequence of the words "subject to a Contract to be approved by me". The negotiating party was the council. The terms of the council's acceptance are, I must assume, correctly set out in the letter. They indicate that the L council is not to be bound except by a formal contract to be approved by the town clerk. The ordinary meaning and effect of the words "subject to contract" have been stated in many cases; *Munton v GLC (supra)* and *Chillingworth v Esche* [1924] 1 Ch 97 may be taken as examples. Looking at the matter thus far the words used in the letter of May 29 in my view exclude a binding agreement.

But that is not the end of the matter. There may, it is said, be cases in which the ordinary meaning of "subject to contract" is displaced. It is contended that in the present case, concerned as it was with a compulsory purchase, the words "subject to a Contract to be approved by me" gave the town clerk authority, if he thought fit, to dispense with formal contracts and to reach a binding agreement on M compensation. He did that, it is said, by agreeing in the second paragraph to dispense with contracts. The judge held that "it would be putting altogether too strained an interpretation upon it to say that the words 'subject to a Contract to be approved by me' mean that the council's approval of the purchase price of £15,000 had only been on terms that there would actually be a contract in a form to be agreed by the town clerk".

In considering the matter I put aside any question of what the town clerk thought the words "subject to a Contract to be approved by me" meant. All that we can do is to construe the letter.

Looking at the letter, I am unable to read it as meaning anything

A other than that the council approves the purchase of the property at £15,000 subject to contract; the form of the contract was to be approved by the town clerk but the agreement was, in essence, to be a "subject to contract" transaction in the ordinary sense. That in my view is the ordinary meaning of the language. It was a letter written, on a conveyancing matter, by the officer of the council responsible for conveyancing matters. It used familiar conveyancing language, which in my view must be given its ordinary meaning.

I come then to the second paragraph of the letter of May 29. I do not think that it advances the council's case at all. In my view the paragraph is saying no more than "There has been a notice to treat so do not let us bother with contracts". I do not think that throws
B any light upon the meaning of "subject to a Contract to be approved by me" in paragraph 1. It is in no way inconsistent with the meaning which I have attributed to those words.

Further, as to the effect of the second paragraph I make the following comments. First, the town clerk is agreeing to dispense C
with contracts relating to a disposal of land. That seems to me to be altogether remote from an intention to *create* a contract for such disposal.

Secondly, the contracting party under the agreement sought to be established is the council. It is not the town clerk and, although the town clerk may have, in practice, greater powers than a solicitor on an ordinary purchase, his authority from the council is stated in terms in the letter of May 29. In my view he has no authority to do anything except to settle the form of a contract for the purchase mentioned in the letter.

In my judgment there was never any binding agreement as to the price or anything else. Accordingly, the council has no enforceable D
agreement and the appeal fails.

The appeal was dismissed with costs. Leave to appeal to the House of Lords was refused.

ESTATE AGENTS

QUEEN'S BENCH DIVISION

April 10 1981

(Before Mr Barry CHEDLOW QC, sitting as a deputy judge of the division)

EDMONDS AND ANOTHER v ANDREW & ASHWELL (A FIRM)

Estates Gazette January 9 1982

(1981) 261 EG 53

Action against firm of estate agents by vendors of café premises for alleged breach of contract and negligence in recommending certain persons as prospective purchasers on whom complete reliance could be placed and in parting with the keys of the premises to them — Evidence that prospective purchasers had used the keys, given to them by the estate agents for the limited purpose of checking the condition of stock in the café, to enable them to take up occupation in the premises for some weeks — Held by the judge that the alleged breach of contract or negligence had not been established under either head — Even if the plaintiffs had succeeded on the issue of the keys, the judge expressed the view that there was no evidence of damage suffered — Claim against estate agents dismissed

This was an action by the plaintiffs, Mr and Mrs Edmonds, against the defendants, Andrew & Ashwell, a firm of estate agents, of Leicester and elsewhere, claiming damages for alleged breach of contract and negligence. The claim arose from the introduction by the defendants of two persons, referred to in the judgment as Jatania and Rahja, as prospective purchasers of the plaintiffs' café premises at 167 Belgrave Gate, Leicester.

K Garnett (instructed by John Parry Williams, of Saffron Walden) appeared on behalf of the plaintiffs; P H Ripman (instructed by Rutherfords, of Tamworth) represented the defendants.

Giving judgment, MR BARRY CHEDLOW QC said: This is a claim by the plaintiffs against the defendants, who are chartered surveyors, auctioneers and estate agents, arising out of the attempted sale of the plaintiffs' café premises at 167 Belgrave Gate, Leicester, in the latter part of 1973 and the early part of 1974.

Briefly, the plaintiffs' claim arises firstly because they say that the joint purchasers introduced by the defendants, namely, the persons whom I shall call for short Jatania and Rahja, were represented to them by the defendants as persons of substance and able to purchase the plaintiffs' premises, whereas in fact, say the plaintiffs, the defendants in breach of their duty to the plaintiffs, and negligently, had no grounds for so believing, and pressed the plaintiffs to accept these two persons as purchasers, even though they had no sufficient grounds for believing them able to purchase the premises.

Secondly, the plaintiffs say that, despite express instructions which they had given to the defendants by letter to retain the keys of the premises in their possession, the defendants in breach of this term, or carelessly, parted with the keys to the proposed purchasers, who then got into the premises and carried on business at them, when it was entirely improper of them so to do. The plaintiffs say that the proposed purchasers were persons of no substance, and that they subsequently, despite a notice to complete the purchase, failed to do so, and ultimately left the premises and did not go ahead with the purchase. The plaintiffs thereby say that they have been put to the loss and damage which is set out in the statement of claim.

By their defence the defendants deny that they were in breach of contract or negligent in putting Jatania and Rahja before the plaintiffs as prospective purchasers upon whom they could rely. It is admitted that they did propose these two as purchasers, but they deny that they were in any way lacking in their duty or obligation of care to the plaintiffs in so doing. The defendants admit that they handed over the keys to Jatania in order that he might check the condition of the stock and the premises which had been vacated by the plaintiffs; that it was for this purpose only and not for habitation, and they deny they were in breach of any express term in any letter, whose existence is denied. The defendants also deny in any case that the plaintiffs have suffered the loss and damage set out, and also say that in any event if there were any damage the plaintiffs have failed to mitigate their damage in the manner set out in the defence.

Thus it seems to me that there are three issues in this case. The first is: were the defendants negligent or in breach of contract in recommending or sponsoring Jatania and Rahja as potential purchasers of these premises? Secondly, were the defendants negligent or in breach of contract by parting with the keys? There is an issue here as to whether there was or was not a letter delivered to the defendants as deposed to by the plaintiffs, and furthermore, the question may arise whether, express instruction or no, the defendants were negligent or in breach of duty in parting with the keys in any event in all the circumstances. The final issue is, if either of these two issues is answered in favour of the plaintiffs, what damages flow, how the damage is to be quantified, and have the plaintiffs mitigated such damage as they have suffered?

Apart from the oral evidence which was called before me, I have been shown a large number of documents, including a bundle of about 177, to which I have been referred, and I shall in the course of my judgment refer to some of these documents. I propose to deal with the first issue, namely, whether the defendants were careless or in breach of their duty to the plaintiffs in putting forward Messrs Jatania and Rahja as potential purchasers of the plaintiffs' café premises.

The plaintiffs' case is that the defendants were urging the plaintiffs to accept them, and the defendants' support for these two was quite excessive; that they were leading the plaintiffs to believe without any proper or sufficient ground that the sale would most certainly go through and the plaintiffs contend that the defendants were making these pressing statements without any, or any sufficient or proper, grounds for making them.

Mrs Edmonds gave evidence before me, and she said, among other things, that Mr Wheelright (who is the defendants' representative, and at that time their commercial manager, now I think a partner of the defendant firm) said that Jatania was a man whom he had dealt with on previous occasions. He was a man of substance and there would be no problem in selling to him. She said he stated that there would be no trouble getting an assignment for Jatania because he was already a restaurateur. Once again he said it would be easy to sell to them and it was foolish to continue with a sale that might prove abortive. This latter statement was made by Mr Wheelright at a time when Mr and Mrs Edmonds were proceeding with the sale to another prospective purchaser called Mr Vermah. Mrs Edmonds further said that Mr Wheelright said he dealt with Jatania on many previous occasions, and he said all had gone through well and he (that is Mr Wheelright) could vouch that all was fine, "and

A he led me to believe that he (Jatania) was a man of substance". Mrs Edmonds says: "I do not say Mr Wheelright was improper in preferring Jatania or Rahja. He told me he had acted for Jatania and he was satisfactory, and Mr Wheelright was not concealing the fact that he was friendly with Jatania. He did try to persuade me to change from Vermah to Jatania and Rahja."

Well now, what is the position so far as Mr Wheelright is concerned in relation to Jatania and Rahja? Mr Wheelright in his evidence said that "George" — that is to say, Jatania — "was known to me. He was a man of diversified interests in property. I had sold him property and other properties to people he represented. He came into our offices fairly regularly." Mr Wheelright
B went on, "I have been concerned with a number of properties, including a drapery business, which I have sold to Jatania, or to those whom he represented, and I have known him for about two years. Both I (Wheelright) and Messrs Coles & Whitmore (who were the solicitors acting for Jatania and Rahja) had had no trouble with them." Mr Wheelright said that he had no misgivings about Jatania at all at that time. That is to say, towards the latter part of 1973. He said, moreover, that whereas Jatania had operated a restaurant, the other potential purchaser on the scene at the time, Mr Vermah, had not. What is more, Mr Wheelright said, whereas Mr Vermah's offer was conditional in a number of respects, the offer that Jatania and Rahja had made to purchase the property was
C the usual one of merely being subject to contract. Mr Wheelright said that he had obtained references in relation to Messrs Jatania and Rahja, which he regarded as satisfactory, and I have seen them.

Not content apparently with these references, Mr Wheelright, on behalf of his clients, the plaintiffs, had made inquiries of the British Debts Services, who apparently keep a register of debtors and judgment debtors, and neither Mr Jatania nor Mr Rahja was held on any records that British Debts Services had. Mr Wheelright, in his evidence, was quite frank about it. He said that it is quite possible he did say there would be no trouble with Jatania and Rahja, and he said frankly he felt that that was the case. He told me, "I told Mrs Edmonds that almost certainly the sale would go
D through based on the facts I then had". But he went on to say, "I did not give her a warranty, and I was not being over-enthusiastic about Jatania and Rahja".

Mr Sunderland, who was the managing clerk employed by the plaintiffs' solicitors, gave evidence before me and said as to Jatania and Rahja "We had had good references given to us. They satisfied me and, what is more, they satisfied the landlord's solicitors. I was under the impression that Jatania was a respectable man. I had no reason to think otherwise. He had a beautiful car and was well dressed, and I had no reason to think the sale would not go through." Mr Sunderland went on, "Jatania and Rahja were the only other people in the running when Mr Vermah dropped out,
E and it had been arranged that if one did not go through the other would". I ought to add here that in fact the purchaser who had originally been preferred by Mr and Mrs Edmonds, namely, Mr Vermah, had indeed dropped out because he was not prepared to sign a contract until the new year, whereas Mr and Mrs Edmonds wanted, if they could, to have a contract signed in December, and Jatania and Rahja were in fact prepared so to do.

That is not all the evidence which was given as to the first issue. But these are the considerations, or some of them, which seem to me to be substantial in relation to it. I say at once that there seems to me to be just no evidence at all that Mr Wheelright in any way behaved in breach of his contract to Mr and Mrs Edmonds, or carelessly in
F acting as he did in relation to these prospective purchasers, Messrs Jatania and Rahja. He knew them. Certainly he had done business with Jatania before, and he and solicitors found him perfectly trustworthy. The references which he had taken up were satisfactory. Neither had debts to all outward appearances. Jatania was a respectable man and a member of the Ugandan Asian community, who represented others of that community, and was a frequent attender at Mr Wheelright's offices. On the totality of all the evidence which has been presented to me, no case at all has been made out in my view, which begins to support the contention which I have called the first issue, namely, some breach of contract or negligence in Mr Wheelright in forwarding and suggesting Mr Jatania and Mr Rahja as potential purchasers of the plaintiffs' G
premises, and in my judgment this issue in the plaintiffs' case wholly fails.

I now turn my attention to the second issue, namely, the keys. It is the plaintiffs' case that on the Saturday night and Sunday morning before Christmas 1973 they were collecting their belongings from their café premises preparatory to moving to Norfolk for Christmas. What the plaintiff and her husband told me was that on the Saturday night Mrs Edmonds wrote the letter on p 66 of the bundle of correspondence, which her husband Mr Edmonds took round to the defendants' business premises and put through the letter box, with the stock list enclosed with it. Mr and Mrs Edmonds say that they then carried on packing their belongings and when they had H
finally finished and had a further load in their van, and were about to go away to Norfolk, Mrs Edmonds wrote upon one of her husband's vehicle log sheets in the cab of the lorry or van in the early hours of the Sunday morning before Christmas a further letter to Mr Wheelright, which her husband again delivered through Mr Wheelright's business door. That letter contained words to the effect, for Mrs Edmonds tells me quite frankly she cannot remember the exact words, but something like, "Herewith are the keys to The Topic Bar. Hold them until you hear from Mr Sunderland that completion has taken place." As I say, this letter together with the bunch of keys was in a separate envelope put into Mr Wheelright's letter box. A second set of keys was given to Mr Perks, J
the plaintiffs' next-door neighbour, to retain for them.

There is a stark issue about this matter. Mr Wheelright denies ever having received any such second letter at all. He says that when he came to his premises for a business appointment unrelated to this matter about, I think, December 27 1973, there was lying on his mat one letter from the plaintiffs only, and that is the letter on p 66 of the bundle, including the stock list. He says it also contained the bunch of keys, and that he is quite satisfied that there certainly were not two separate envelopes from the plaintiffs upon his mat on that particular day when he went to his business.

In the statement of claim the plaintiffs in para 4 allege that the defendants had been given express instructions to hold the keys K
until the plaintiffs' solicitors notified them that completion had taken place. A request for further and better particulars by the defendants brought forth the answer on November 11 1977 that the instructions were contained in a letter addressed to the defendants and posted through the defendants' letter box on or about December 24 together with the keys of the premises.

There was called before me Mr Matthews, of the defendants' solicitors. He told me that the defendants disclosed in their bundle of documents, among others the letter on p 66, and this was disclosed to the plaintiffs' solicitors, and indeed the latter asked for a copy of the documents, including that contained on p 66 of the bundle. At no time, Mr Matthews told me, did the plaintiffs' sol- L
icitors ever suggest that there ought to be disclosed a second or further letter of the same time as the document on p 66, and that the first day of the trial was the first time the defendants or Mr Matthews had heard it suggested that there was a second letter written on that day, delivered to the defendants in the early hours of the Sunday morning before Christmas.

The plaintiffs say that it is strange that if the letter on p 66 was the only letter no reference to keys was made in it. The defendants say that it would be surprising if two letters were delivered through the defendants' letter box within six hours or thereabouts of each other. The defendants place reliance upon the discovery point as deposed to by Mr Matthews. For my part I think it unlikely in all the M
circumstances of this case that Mrs Edmonds would have used the words which she said she used to the defendants, namely, that they were to hold them until they received instructions. Moreover, the whole tenor of the letter on p 66 seems to me to be in the nature of a document written after all the packing had been completed, and there was nothing else remaining to be done. Indeed, the letter indicates that even writing paper has been packed up. I have listened carefully to the evidence and the submissions of counsel. I think Mr and Mrs Edmonds are mistaken. I think the events of that night when they were busy and tired, and had finished packing in the early hours of Sunday morning, are confused in their minds. I do

A not think for a moment they would deliberately seek to mislead me. But I have come to the conclusion that on the totality of the evidence, and in particular the terms of the letter itself, the fact that no request was ever made for a second letter to the defendants' solicitors after the existence of the document on p 66 was disclosed, the unlikely writing of two letters within six hours of each other, and the terms that Mrs Edmonds says she used, lead me to believe that no such second letter was ever delivered to the defendants. Moreover, I prefer the evidence of Mr Wheelright that no such second letter was ever received by him.

The matter of course does not end there. It is quite clear that Mr Wheelright was instructed by Mr Sunderland, if nobody else, to B keep the keys in his possession until completion. This he did not do, but parted with them for the ostensible purpose of allowing Jatania to go into the premises to inspect them. Was this a breach of his duty to the plaintiffs? Each case must turn upon its own facts, and Mr Wheelright makes the distinction between occupied and unoccupied dwelling premises on the one hand and occupied and unoccupied commercial premises on the other. For my part, I think the distinction is a valid one. I have no doubt at all that Mr Wheelright throughout felt himself to be acting in the best interests of the plaintiffs, his clients, in doing what he did. So far as he was concerned the person who was requesting the key was a person who was going to complete the bargain within a few days, which C everyone at that time thought. He was, so far as Mr Wheelright was concerned, a reputable man. The premises were to be sold as a going concern. Yet Mr Wheelright knew that the Edmonds had left to go to Norfolk and were not coming back. He knew that there were perishables upon the premises and that the prospective purchasers might have some complaint if the premises were not run during the period certainly into the early new year. These were café premises and the chances of a prospective purchaser of such premises improperly taking possession and seeking to run them in the teeth of objections from the vendor would seem to Mr Wheelright in his experience to be so remote as to be a non-existent risk.

Finally, as Mr Ripman for the defendants points out, no expert D evidence was called before me as to the practice among estate agents in circumstances such as these. I have only got the evidence of Mr Wheelright. Can I say on the totality of the evidence that what Mr Wheelright did in parting with possession of these keys for a limited purpose, as he thought, that Jatania wanted them was a breach of any term in the contract or was negligent in all the circumstances of this case? I have given this matter anxious thought and consideration, and have come to the conclusion on all the evidence here that no such breach of contract or carelessness has been shown by the defendants in Mr Wheelright acting as he did in all the circumstances on this occasion.

I think it is very necessary to avoid the wisdom of hindsight. I take E the view that in doing what he did Mr Wheelright genuinely thought that he was acting in the best interests of his clients, and that indeed by allowing Jatania to have the keys he was avoiding rather than forwarding a breach of contract.

On this issue also in the result on the evidence before me, the plaintiffs have failed to satisfy me that in parting with the keys as Mr Wheelright undoubtedly did on this occasion, in all the circumstances of this case there was either a breach of his duty or carelessness, and on this issue also in my view the plaintiffs' claim fails.

That is enough to dispose of this action, but I ought to say that even if I were wrong upon this issue I would have the very greatest difficulty in finding that the plaintiffs have suffered any damage as a F result of it. The evidence which I have heard, and the reading of the correspondence, makes it quite clear that the prospective purchasers, certainly Jatania, went into possession as I think sometime in early January, but certainly by the end of February he had vacated them. Meanwhile, a notice to complete had been served which expired about March 4. So that the plaintiffs would have been quite unable to do anything with the premises until that time. Many of the items which are claimed by way of damages were running whether the prospective purchasers were in improper occupation or not, with the possible exception of the rates, and certainly the £46 worth of produce which disappeared. That was something which would not have been lost to them had Jatania not been in the premises. But so far as the other items are concerned, the evidence I have heard G leads me to the conclusion that the plaintiffs were not disabled from finding other purchasers in this case, by any occupation by Jatania. Indeed, Mrs Edmonds' reaction in February when she heard the sale had gone off was that she and her husband would return to the premises and run them. Then the possibility of purchase by some jeweller arose, and even during the months of April and May the possibility still existed of Jatania being able to raise some part of the purchase price and continuing with the purchase. There were, I am quite satisfied, no other purchasers in the field, and it was not until the middle of the year that Mr Chung, the ultimate purchaser, appeared on the scene. Even then it took about four months before his purchase was completed. I would have found it impossible to H have held that the extra expenses to which the plaintiffs were put between March and October 1974 could be laid at the defendants' door. It has to be remembered that the plaintiffs had forfeited the abortive purchaser's deposit of £520.

I have said enough to indicate that in my view, even had the plaintiffs succeeded on the second issue in this case, they have suffered no damage thereby. But, as I say, it is not necessary for me to come to any concluded view about this matter in the light of the view I have come to on this issue. In the circumstances this claim fails in toto, and there must be judgment for the defendants.

Judgment was given for the defendants with costs, the costs after the date of the plaintiffs' legal aid certificate not to be enforced J without leave.

OFFICIAL REFEREES' BUSINESS

November 6 1981

(Before His Honour Judge HAWSER QC)

STEPHENSON v DONALDSON & SONS

Estates Gazette April 10 1982

(1981) 262 EG 148 K

Unsuccessful libel action against firm of estate agents, surveyors and valuers — Claim by plaintiff, who had instructed the defendants in connection with the compulsory acquisition of his shop and residential premises, that a letter written by them to the local authority's estates surveyor was libellous — The letter, written in the hope of obtaining compensation for the plaintiff on the basis of total extinguishment of his business, indicated that the plaintiff's trade had suffered from increasing competition and that he was not prepared to move to premises with higher overheads or involving capital expenditure — Plaintiff sought to establish that the letter meant that he was a dishonest person, falsely pretending that his business was viable whereas it was in decline — Held that the words used L were not defamatory either in their natural and ordinary meaning or by way of innuendo — The letter was written on behalf of the plaintiff in good faith with the object of inducing the authority to make a favourable settlement — The judge discussed certain defences put forward by the defendants which it was not, however, strictly necessary to consider, as the words had been held not to be defamatory

In this action the plaintiff, Neville Stephenson, claimed damages from Donaldson & Sons, estate agents, surveyors and valuers, of 90 Dalston Lane, Hackney, London E8, for an alleged libel contained in a letter written to the estates surveyor of the London Borough of M Waltham Forest. The defendants were acting for the plaintiff in connection with the compulsory acquisition of premises occupied by him under a lease at 198 Crownfield Road, London E15.

The plaintiff appeared in person; J R K Price (instructed by Barlow Lyde & Gilbert) represented the defendants.

Giving judgment, JUDGE HAWSER said: This is an action in which the plaintiff claims damages for what he alleges to be a libel in a letter written on April 15 1980 by the defendants to Mr White, the estates surveyor and valuer of the local council.

The circumstances and background are these. There was a compulsory purchase order which included the premises, at that time

A used by the plaintiff at 198 Crownfield Road, London E15, under the terms of a lease which expired on June 24 1979. The shop and residential premises were leased to him at a rental of £920 a year (exclusive). It was evident that the new rent for the premises, had the position remained as previously, would have been in excess of £920, and I find as a fact that it probably would have been £1,500 a year (exclusive).

The plaintiff employed Donaldson & Sons, who are estate agents and valuers, to act for him in connection with the compulsory purchase order and to claim compensation for his removal from the premises. Compensation is based on two alternative situations. The first is where the council is able to find, or perhaps the client himself
B is able to find, alternative accommodation; then the compensation is limited to the costs of and occasioned by the disturbance of the business. If, however, suitable alternative accommodation is not available, then the client would be entitled to compensation on the basis of the extinguishment of the business, and usually that is a higher figure.

On May 15 1979, Donaldsons wrote to the estates surveyor and valuer of the London Borough of Waltham Forest and enclosed copies of the revenue accounts for the plaintiff's business in respect of the years ended December 31 1976, 1977 and 1978. These showed profits of £2,869 for 1976, £3,162 for 1977 and £3,498 for 1978. During this period the rent and rates had remained more or
C less static; there had been some increase in sales and (as one observes) some increase in profit, but the rate of gross profit had gone down from about 27 per cent in 1976 to about 24.7 per cent in 1977 and about 23.5 per cent in 1978. In addition there was the fact that there would have been an increase in rent as of June 1979 of the order of £580-£600 which would have diminished the profit. In real terms, having regard to inflation, the business was tending to decline during this period.

On July 16 1979, Donaldsons wrote to the estates surveyor and valuer of London Borough of Waltham Forest and stated:

We . . . would advise you that we will be prepared to recommend our client
D to accept £6,000 as compensation for this item. As you will note from the accounts, our client's profit has increased for each of the last three years.

On August 14 1979, the London Borough of Waltham Forest offered the plaintiff a lease on 76 High Road, London E15, on certain terms to the details of which I do not need to refer. Shortly afterwards this offer was turned down by the defendants on behalf of the plaintiff; details of the reasons are set out in a letter dated September 19 1979. The council indicated that they were prepared to consider negotiating on the terms. In due course the council offered other premises which were not suitable to the plaintiff: on December 12 1979 they again offered to grant the plaintiff a lease of shop premises at 76 High Road, E15, and indicated willingness
E to negotiate.

On December 17 1979, Mr Ledgett, acting on behalf of Donaldsons, wrote to the plaintiff and said:

. . . As you will see from the Council's letter, they have now offered revised terms for 76 High Road, E15 and I will be grateful if you would have another look at these premises and let me know whether you are prepared to agree. Quite frankly, I hold out little hope of you being able to find a lock-up shop only in reasonable condition and on the terms that you require and it would seem that the Council's proposal in the circumstances is reasonable.

F The council having written to say that they would negotiate on the terms, on January 7 1980 Mr Ledgett wrote to the plaintiff as follows:

. . . As I mentioned in my letter of December 17 1979, I consider that their offer of a lease on 76 High Road, together with rehousing, is reasonable, especially as Mr White mentions that he will be prepared to negotiate on the terms already quoted.

The only way that I can see to convince the Council that total extinguishment should be paid is to say that further trading in carpets on a small scale such as yours is no longer viable due to the increasing competition from multiple shops which hold a much larger stock. If this is accepted, the compensation payable will obviously be higher than their suggested method on the basis of hypothetical removals.

In reply on January 9 1980 to that letter the plaintiff stated: G

. . . Firstly I say without prejudice or malice I would prefer you take instructions rather than express opinions which would appear slanted towards bias or self-interest. No 76 High Road, E15 has been rejected by me on three separate occasions as being unacceptable and adequate [sic]. It is wrong to assume that what is reasonable is also adequate.

He then dealt with the question of relocation and went on to say:

On the question of compensation I suggest that my claim be reappraised and reiterated. I suggest a figure of £9,000. I cannot accept your suggestion on approach, it is my view too removed from the truth. I have indicated viable reasons supra.

That seems to me to be a reference to the last paragraph of the H letter of January 7 1980 in which Donaldsons had put forward their suggestion that "increasing competition from multiple shops which hold a much larger stock" was an important factor.

On January 18 1980, Mr White (who had come into the picture some months before) said, in a letter written to Donaldsons on behalf of the borough:

In my view you have not put forward any valid reason as to why my offer of 76 High Road plus rehousing is unacceptable. I take the view therefore that he is not fulfilling his requirement to mitigate his loss and accordingly I can only offer compensation for business disturbance on the basis of hypothetical removal to 76 High Road.

J

On March 24 1980, in pursuance of that, the borough offered £1,000 "In full and final settlement of your client's claim for compensation for his leasehold interest in the above property, and compensation for business disturbance," and said that if that figure was not acceptable there would be further negotiations.

On March 31 1980 Mr Ledgett, on behalf of Donaldsons, wrote back and said:

. . . Our client is not prepared to accept the compensation offered of £1,000 and will not settle for anything less than total extinguishment. Accordingly, on behalf of our client, we request any advance payment under section 52 of the Land Compensation Act 1973 pending further negotiations.

K

On April 8 1980, the London Borough of Waltham Forest wrote to Donaldsons and said:

I should be pleased to hear your client's views as to why he considers his business could not be relocated even to High Road, E15, bearing in mind that in the vast majority of cases any shop keeper taking on new premises would expect to have to carry out some alterations/adaptations/decorations etc in order to make the premises suitable for his particular use, and that compensation law requires him to make serious efforts to mitigate his loss.

On April 10 1980, Donaldsons wrote to Mr Stephenson, enclosing a copy of the council's letter of March 24. On April 14 Mr Stephenson wrote to Mr Ledgett of Donaldsons saying that he (Mr Ledgett) was free to make a claim for compensation under section L 52.

I come now to the letter which contains the paragraph in which it is alleged that the libel occurred. The letter is dated April 15 1980. Mr Ledgett says that he had a telephone conversation with Mr White the day before the letter was written. He says that Mr White told him that he had received the accounts and had formed the impression that the business was not really viable in view of the profits and the increase in rent, and "I therefore wrote to the council in the terms which I did". Mr White had no specific recollection of that telephone call. I am satisfied that it was made and did occur in the form given by Mr Ledgett. Mr White told me that before the letter of April 15 1980 was written he had decided that M he could not accept a claim for compensation based on total extinguishment: he said that had been decided in January. He had a talk with Mr Stephenson and his attitude towards relocation. His mind was already made up as a result of the accounts and the general circumstances. The letter of April 15 1980 did not affect his views on compensation.

I should read the letter fairly fully and refer in particular to the paragraph complained of. It is from Mr Ledgett to Mr White and refers to premises in the previous offer (198 Crownfield Road):

. . . Our client realises that he has a duty to mitigate his losses but, as has already been pointed out to you, he is not in the financial position which

A enables him to finance the repairs that you have required, albeit that these were negotiable.

I come now to the paragraph which Mr Stephenson complains of:

> Our client's trade is such that it has suffered over the last few years from increasing competition and he is, therefore, obviously not prepared to consider moving to premises which have higher overheads than his existing property, or where some capital expenditure is necessary to render the premises fit for trading.

That is the end of the alleged libel. The letter goes on to say:

> Finally, we must say that we have dealt with other local councils where total B extinguishment has been granted to traders of a similar age, because relocation would probably not be feasible due to increasing competition over the years. It is, of course, often the case that an established business can survive in a competitive market, but any potential is lost if the business has to relocate.

I am satisfied that Mr Stephenson was not prepared to accept the council's offers, and in particular of 76 High Road, either because of higher overheads or because of capital expenditure and that, in fact, he turned down all the offers which had been made by the council.

The response to that letter was dated April 21 1980, written by Mr White. He suggests that Mr Stephenson was unreasonable in C refusing to move to 76 High Road and also refers to the increased rent which would have applied to the existing premises had the compulsory purchase order not come into operation. He says:

> 4. You emphasise in your letter that your Client's trade had suffered in the last few years because of increasing competition. This raises the distinct possibility that the business would have been forced to close irrespective of the CPO, and one cannot help thinking that this may well have been the reason he did not wish to relocate rather than the unsuitability of other premises.
>
> 5. I am not sure what you mean by saying that an established business can survive in a competitive market. Established businesses often do not survive in the face of increasing competition, particularly where that business is a small one. In any event it is hard to regard the removal of the business to D premises close by as "disestablishing" it.
>
> 6. I find it difficult to accept that the business had any potential, particularly in the face of your information regarding the decline of the business. But if there had been any, it would in my opinion have been better served by moving to 76 High Road in view of the latter's better trading possibilities.
>
> In conclusion I regret that I am unable to change my view of the proper basis of compensation.

On May 6 1980 Mr Ledgett sent a copy of the letter of April 15 and of the reply to Mr Stephenson and said that the only alternative was to apply to the Lands Tribunal but he thought that was a waste of time and money.

E On May 12 1980, Mr Stephenson wrote to Donaldsons:

> . . . Your letter of April 15 has taken me by surprise, both as to its contents and purpose. You will recall a telephone conversation on or about January 7 in which you were orally forbidden to write in that strain, and in my letter dated January 9 you were expressly forbidden to do so.
>
> You must now realise that I have suffered damage from your actions. I look forward to receiving your account of the purpose and motive for writing this letter.

Donaldsons replied on May 15 1980:

> I am surprised that you objected to my letter of April 15 1980 as, presumably, it is the fourth paragraph thereof which gives rise to your concern. F Although I was aware of the contents of your letter of January 9, I understand that you had already said exactly the same thing to Mr White when he visited you and, indeed, it was he who brought this matter to my attention and caused me to write the letter. In the circumstances, therefore, I do not consider that you have suffered any damage from our actions, as the Council already knew this detail. . . .
>
> Despite your comments in past correspondence, I am still satisfied that you have been offered suitable alternative premises and feel that I am fighting something of a lost cause. However, I have spoken again to Mr White, who will be letting me have the Council's final offer during the next few days.

Mr Stephenson cancelled Donaldsons' instructions. A number of letters passed subsequently in which Mr Stephenson was asked whether he would negotiate through fresh agents or himself, but he G did not negotiate any further with the council and he has not applied to the Lands Tribunal.

The plaintiff says in his statement of claim:

> 5. The defendant meant and was understood to mean

by the words

> that the plaintiff was a dishonest and fraudulent person who was claiming money from the . . . Council by way of compensation under a Compulsory Purchase Order . . . by falsely pretending that the said business was viable and his [sic] rising expectations, whereas the observe was true — the business suffered from diminishing returns and was in decline and would in the foreseeable future be forced to close irrespective of the Compulsory Purchase Order. H In consequence of the alleged decline of the business the plaintiff obviously has no money to carry out even the minimum of repairs to render premises fit for trading.

A number of matters are set out as particulars of the meanings to be attached to the words. In "(vi)" of those particulars it is said:

> The Council understood the statement to mean the plaintiff was dishonest in his claim to compensation and, in view of the statement, and because of the decline, the business had no potential.

The defence as amended claims that the words were not capable of bearing any of the meanings pleaded in the statement of claim or any meaning defamatory of the plaintiff. I think that that is correct. J On the face of it, plainly the words do not, in their ordinary and natural meaning, constitute any statement defamatory of the plaintiff. They say that the plaintiff had suffered as a result of increased competition. Whether that is true of not, it is not in my view any attribution to the plaintiff of any kind of misconduct or impropriety in his business activities. It is simply suggesting that the business had tended to decline by reason of increased competition which may well be regarded as a misfortune or a difficulty with which many businesses have to contend in these days. Plainly, the plaintiff was not prepared to consider moving to premises which had higher overheads than his existing property or where capital expenditure was necessary to render the premises fit for trading. K Those two matters are alleged as being caused by the increasing competition from which the plaintiff had — allegedly, at any rate — been suffering during the last few years.

Again, in the next paragraph of the letter of April 15 1980 (which I think one should read in conjunction with the previous paragraph) the point is made that an established business can survive in a competitive market and that potential is lost if the business is to relocate. I think that in this letter Donaldsons were making an attempt (even if, perhaps, a somewhat clumsy attempt) to try to induce the council to take a view of the facts and the circumstances which would result in an offer of compensation to Mr Stephenson on the basis of the extinguishment of his business. L The council clearly had made up its mind prior to the letter of April 15 1980 that they would not grant compensation on the basis of extinguishment, and, unfortunately, the letter did not have the effect which it was perhaps hoped it might have, of persuading the council that they ought to change their mind.

Mr Stephenson has alleged an innuendo the effect of which is that the paragraph in the letter, in the circumstances, meant that he was being a dishonest and fraudulent person. I do not think that this innuendo can possibly be inferred from the words used or, indeed, from any other facts or circumstances — though it is correct to point out that no other facts or circumstances were relied upon. M

It is suggested by Mr Stephenson that the defendants, in writing this letter, intended to discredit him and intended to create in the minds of the council some sort of feeling that he was not to be relied upon: that he was in some way a dishonest and fraudulent person. Having heard Mr Ledgett, I am quite satisfied that the letter was written in good faith, and written purely and simply with the intention of trying to induce the council to adopt a more favourable attitude. In the event, unfortunately it failed to do so.

I have therefore come to the conclusion that the paragraph in question was not defamatory of the plaintiff in its natural and ordinary meaning and was not capable of bearing the meanings

A which have been alleged by the plaintiff in paragraph 5 of his statement of claim.

There are a number of other defences which have been put forward on behalf of the defendants. They allege that the words complained of were written in pursuance of the authority which was given to them and the publication was authorised by the plaintiff. I doubt, in the light of the letter of January 9 1980 from Mr Stephenson, that that allegation is justified. I think that Mr Ledgett probably went beyond any authority which he may have had, in putting forward the particular suggestion — though I have no doubt that he did so in good faith and with the best of intentions.

B Whether the defendants, in the circumstances, could rely upon the plea of qualified privilege is a matter of some doubt having regard to the authorities, because one has the rather peculiar position that if there was any defamation the defendants would have to plead that they were under a duty to publish the words and the estates surveyor and valuer had a like duty or interest to receive them. It does not seem to me that — speaking generally, at any rate — Donaldsons would have been under a duty, in pursuance of agency, contract and authority, to publish words of the plaintiff which were defamatory, even though they did act in good faith. However, this point does not really arise having regard to my finding on the original letter.

C The defendants have also pleaded that the words were true in substance and in fact. I need not say very much about this. The true picture, as is stated in paragraphs 3 and 4 of the defence, is that, over

D the years, the plaintiff's business was declining in real terms, he was not prepared to consider moving to premises which had higher overheads than his existing property or where significant capital expenditure was necessary to render the premises fit for trading. I find that if there were any libel or defamatory statement contained in the letter, it was limited to a statement that the business was declining, and I find that that was in fact true.

Finally, the defence pleads that the plaintiff has suffered no loss whatsoever as the result of the inclusion of this paragraph in the letter of April 15 1980 even assuming that it was defamatory. I am not sure whether this is a good defence, since no special damage has to be proved assuming that there was a libel of the plaintiff having

E regard to his business reputation. But I accept the proposition that the letter had no effect whatsoever upon the council — as indeed Mr White made quite plain — and that in no way, as far as the council were concerned, were Mr Stephenson's chances jeopardised by sending the letter. He is, of course, free to pursue his claim to the Lands Tribunal if he wishes. I am satisfied that this letter will have no effect whatsoever upon the Lands Tribunal or any further effect upon the plaintiff. Accordingly, I would have held, had I been in the plaintiff's favour, that the damage which he had suffered as a result of the libel was minimal damage and no more.

In the result, I have found that the statement in the letter of April 15 1980 was not defamatory in either its natural or ordinary meaning and was not capable of bearing any of the other meanings put

F upon it by the plaintiff. Therefore this action must fail.

Judgment was given for the defendants, with costs.

FEES AND COMMISSIONS

A

QUEEN'S BENCH DIVISION

February 17 1982

(Before Mr Justice KILNER BROWN)

PICKERING v SOGEX SERVICES (UK) LTD

Estates Gazette May 22 1982

(1982) 262 EG 770

Contingency fee claimed by surveyors for successful negotiations resulting in a saving of rates by securing a reduction in rateable value — Arrangement held not to offend against the ancient common law of champerty—Fee to be equal to amount of rates saved B in one year—Agreement contemplated that agreement would be reached with valuation officer before date for hearing by local valuation court and that there would be a formal decision recording the agreed valuation—The parties did not contemplate matters proceeding to a contested hearing—Negotiations resulted in a substantial reduction of rateable value and saving of rates and a formal decision of the local valuation court was recorded—Defendants refused to pay the agreed fee and tendered a much smaller sum—Among several defences the defendants submitted that the agreement was champertous and unenforceable because it involved the payment of part of the proceeds of successful pursuit in a legal process, "trafficking in litigation"—Defences rejected—As C regards the allegation of champerty, it failed on two grounds—First, negotiation, agreement and formal recording in a local valuation court did not represent litigation—Secondly, a local valuation court was not a court of law—*Attorney-General v BBC*—Even if the agreement had contemplated a contest before the local valuation court, it would not have been champertous—Obiter dictum as to proceedings before a "rent tribunal"—Case of importance to chartered surveyors' and certain other professions—Plaintiffs' claim upheld.

In this action Jeremy Pickering, trading as City Agents (a firm), sued the defendants, Sogex Services (UK) Ltd, for a contingency D fee based on an agreement to endeavour to reduce the rateable value of the defendants' premises. The fee was to be equal to the amount of the reduction in rates for the period April 1977 to March 1978. If no reduction was secured no fee would be payable. In the event the plaintiff firm claimed a sum of £22,935.41, including VAT.

Patrick A Twigg (instructed by Grangewoods) appeared on behalf of the plaintiff firm; A C L Thornton (instructed by Allen & Overy) represented the defendants.

Giving judgment KILNER BROWN J said: In this action the plaintiff firm claims a contingency fee, that is to say the amount of one year's savings on rates payable if as a result of the firm's E endeavours there was a reduction in the rateable value of the defendants' premises. The case is of importance to the chartered surveyors' profession and it may be to other professions as well, apart from solicitors and barristers, who operate under their own rules. The agreement on which the claim is based was contained in a letter dated February 17 1978. At that time the rateable value was £142,472. In November 1979 the rateable value was reduced to £116,638 and the extra rates which had been paid for the year ended March 31 1978 and which became recoverable and thereby saved amounted to £19,943.84. Value added tax on that sum came to £2,991.57. Thus the plaintiff firm claim the total of £22,935.41.

The letter on which the agreement was based was sent by Mr David Rogers of the plaintiff firm to Mr Rudge, the general services manager of the defendant company, and the relevant parts read as follows: F

> Further to our recent meeting I confirm that we would be happy to investigate the level of the rating assessment on your offices.
>
> In order to discuss the matter with the valuation officer, it is necessary to submit a formal proposal to him. As a result of these discussions, we would advise you whether we consider the assessment to be fair in relation to other property in the area. If we consider the figure to be incorrect, we are usually able to agree the matter with the valuation officer, but failing such agreement, the matter could be heard by the local valuation court and subsequently by the Lands Tribunal. In our experience, however, it is most unusual for the matter to go this far. G
>
> As I explained, if we submit a proposal before March 31 1978, the effect of any reduction in the assessment subsequently obtained, is retrospective to April 1 1977.
>
> Concerning the professional fees for these services, we would be willing to accept as a fee, the amount of rates actually saved in the year April 1977 to March 1978. As I explained, the alternative is for us to charge a fee in accordance with the recommended scales, which in round terms is a little over 2% of the rateable value and this is payable whatever the outcome of our investigation. I would add, that the above suggestion as to fees, would only apply if the matter does not have to go to court."

The acceptance dated March 10 1978 was couched in the following terms:

> As confirmed by our telephone conversation of yesterday we are pleased to request that your company file an application for reassessment of the rates paid by this company to the City of Westminster. H
>
> As previously discussed the rating assessment will be conditional as follows:
>
> 1. That City Agents agree in writing to take only the amount of rates reduction as their fee, for the period 1/4/1977 to 1/3

I think that should be 1978. It is only an interpolation, but I am quoting accurately.

> 2. That if no reduction is obtained, Sogex will be under no obligation to pay any fee, charge or service costs whatever.
> 3. That the application not be pursued further than the representation to the valuation officer. J

Pausing there, both parties recognised that if a reduction in rateable value could be achieved it was a continuing benefit and would stand for at least five years and a proportion of one year's savings out of at least five was by no means an excessive proportion of the total amount of the benefit. Furthermore the costs of a full hearing before a valuation court and on appeal to the Lands Tribunal might be considerable and therefore there would be a further financial advantage to the defendants. The arrangement was not only mutually satisfactory but would seem to be perfectly fair and reasonable. Indeed Mr Rudge, who is no longer in the employ of the defendants and who gave evidence in support of the plaintiff's claim, considered that there was no objection to the claim and K that the defendants were morally, if not legally, bound to accede to it. The plaintiff was additionally supported by Mr J W Osbourn, a chartered surveyor and partner in the firm of Garrett, White & Poland, who have not only been in business since 1898 but bear a high reputation in the City of London. He was very experienced and by a private arrangement between him and the plaintiff firm he in fact did the work on the reduction proposals.

Now the common and accepted practice where an existing rateable value is being challenged is for the objector or his profes-

A sional representative to give notice of a proposal for alteration. This is provided for by the General Rate Act 1967. Usually, as in the instant case, the notice of proposal is in general terms merely proposing a reduction in the assessment on the grounds that the present assessment is excessive and incorrect. The district valuation officer then, as in this case, serves a notice of objection. This notice also informs the objector that after the expiration of 14 days the objector is to be treated as intending to appeal against the valuation. Thereafter, as the notice sets out, a hearing will be arranged before a local valuation court unless the matter is previously settled. The procedure for compromise is set out in section 72 of the General Rate Act 1967. That is a different method of procedure. B However, in the majority of cases the service of notice is treated by the objector and the district valuation officer as the formal steps required by the 1967 Act and negotiations continue. If there is no agreement, the hearing before the local valuation court is an effective dispute and is contested. But if before the fixed date for hearing the rateable value is agreed, then the appeal hearing is a mere formality and the agreed valuation is recorded. According to Mr Osbourn, in his experience about 95 per cent of objections to proposals are dealt with in this fashion and the so-called appeal hearing is no more than a formal process of recording an agreement. This difference between negotiation and agreement to which the rating officer of the relevant local authority has to give his approval C and a fully contested hearing is well recognised by the chartered surveyors' profession. The letter of February 17 1978 was in fact suggested by and drafted by Mr Osbourn himself and is clearly understandable in the light of general knowledge and practice in this field.

As soon as the letter of March 10 1978 was received Mr Osbourn began the process of endeavouring to obtain a reduction in the rateable value and on March 21 sent the appropriate notice of proposal for alteration of the valuation list. The district valuation officer sent his notice of objection to the proposal on May 10 1978 and on May 25 a notification of appeal with the information that the clerk to the local valuation panel had been served with the D notice of proposal and notice of objection. Mr Osbourn was engaged upon the exercise of measuring up the various offices and rooms and noting unsatisfactory features, such as cramped conditions and poor lighting. As might be expected, the matter dragged on, but came actively to life again in October 1979 when the clerk to the panel gave notice to the various interested parties that a meeting of the local valuation court had been fixed for November 20 1979. On November 6 Mr Osbourn and the district valuation officer's representative made a joint inspection of the property. The clerk to the valuation panel was informed that the matter was under discussion and if not resolved would necessitate an adjournment of the fixed hearing. Over the next 10 days there was discussion E and negotiation culminating with an agreement to reduce the rateable value from £140,000 to £116,638. The rating officer of the local authority was informed and there appeared to be no other interested parties other than the defendants themselves, who willingly and no doubt happily gave their approval. All that remained therefore was for the local valuation court to be informed, and the formal decision of the court was recorded and promulgated on November 20 1979 to the effect that the valuation was £116,638. The treasurer to the Westminster City Council was informed and a repayment of the appropriate extra rates paid for the previous year was requested. This was either paid or credited to the defendants. Consequently the plaintiff firm requested payment of this sum plus F the appropriate amount of value added tax as fees due pursuant to the agreement between Mr Rogers and Mr Rudge. A Mr Ramez Irani, who was the director of administration on behalf of the defendants, declined to pay the sum requested. By letter dated January 11 1980 he tendered the sum of £5,365.34 without indicating the basis on which he calculated this to be the appropriate fee. So now there is a disputed claim before me for the full amount.

The only other relevant document is a letter dated November 5 1979 from the plaintiff firm to the defendants, which reads as follows:

> Further to your company's instructions of March 10 1978 and our telephone conversation today, I confirm that our rating consultants have G carried out preliminary rating investigations and will proceed with the appeal to reduce your rating assessment which is listed for valuation court on November 20 1979.

The first point taken by the defendants is that as the matter in fact was listed as an appeal and was adjudicated upon, albeit as a formality, it went to court. It certainly went further than "representation to the valuation officer". Therefore it is submitted that in accordance with the terms of the agreement the contingency fee did not arise and the matter fell to be dealt with upon the alternative basis of a scale fee. If the contract has to be considered with reference only to the last sentence of Mr Rogers' letter, the point 3 of Mr Rudge's letter, this would be an arguable proposition. But H the first paragraph of Mr Rogers' letter makes it quite plain that "going to court" refers to the situation where there is no agreement and the matter becomes fully contested requiring the hearing of evidence and adjudication upon a dispute. From the correspondence alone it seems to me that it was in the contemplation of both parties that "going to court" did not apply to the formality of recording an agreement. The question is put beyond doubt by the evidence of Mr Rudge, who understood perfectly well that this was the case. I reject this submission on behalf of the defendants.

The next point taken referred to the letter of November 5 1979. It was argued that this represented a variation of the original agreement as to remuneration or was a waiver by the plaintiff firm of an J entitlement to a contingency fee. That variation is said to be a widening of the original instructions as it was now contemplated that there would be an appeal after the negotiating process. I cannot accept this submission. In my opinion this letter is no more than reporting progress and the words "proceed with the appeal" have to be read in context with the previous correspondence and agreement that an appeal would only follow if there was no agreement before the day the case was listed. Nor can it possibly amount to a waiver of the agreed entitlement to a contingency fee.

Much the most interesting and important part of this case is the submission that the agreement was champertous and therefore unenforceable. I find it necessary to refer to only two authorities K out of the welter of reported case law on this subject. In *Trendtex Trading Corporation v Crédit Suisse* [1981] 3 WLR 766 and [1981] 3 All ER 520 Lord Wilberforce indicated quite clearly that whereas criminal and tortious liability for champerty was abolished by the Criminal Law Act 1967, it was expressly preserved in the context of the law of contract. Lord Wilberforce adopted the definition of champerty expressed in *Halsbury's Laws of England*, 4th ed, vol 9, para 400, in the following terms:

> Champerty is a particular kind of maintenance, namely maintenance of an action in consideration of a promise to give the maintainer a share in the proceeds or subject matter of the action.

L Such an agreement is contrary to public policy and is illegal and therefore unenforceable, since, in Lord Wilberforce's words, "it involves trafficking in litigation".

It is argued on behalf of the defendants that the arrangement in the instant case was unenforceable because it involved the payment of part of the proceeds of successful pursuit in a legal process. It is said that the doctrine extends to negotiations which contemplate an order of court and which require an order of court before an agreement which may have been reached becomes effective. Counsel for the defendants elaborated the argument by reference not only to the *Trendtex* case but also to several other authorities and also contended that if such an arrangement was by statute prohibited in the case of solicitors, then by analogy it should extend to other professions, M such as chartered surveyors or chartered accountants. It was further said that the letter of November 5, if it was not a variation or waiver, nevertheless converted an original agreement which might be said to stop short of process in court into one which extended to actual process in court and thereby became champertous and unenforceable.

As against these contentions counsel for the plaintiff took two points, either of which, it is said, would destroy the argument alleging that the agreement was champertous and therefore illegal. The first is that the actions of the plaintiff firm could not be described as trafficking in litigation. Negotiation, agreement and formal recording in a district valuation court is not litigation. There is no

A legal dispute between parties or, to use the Latin terminology, no *lis inter partes*. The fact that a local authority is an interested party or other neighbouring ratepayers may be objectors is an irrelevant consideration. It is not litigation; it is a process of administration. I agree, and on this ground alone the defence argument fails.

The other point follows naturally from the contention that there was no process of litigation. It is said that litigation in the sense in which it is used in the context of champertous arrangements requires that the process involves possible recourse to a court of law. A district valuation court is not a court of law. Support for this proposition can be found in several authorities and more particularly in the case of *Attorney-General v BBC* [1981] AC 303.
B Although that case concerned the jurisdiction to entertain proceedings for contempt of court, the status of a valuation court was fully and exhaustively examined by their Lordships in the House of Lords. In that case, although Lord Salmon was content to treat the issue in the narrow context of the law relating to contempt of court, the other four Lords of Appeal all agreed that a local valuation court is, to use Viscount Dilhorne's words, "constituted to resolve problems which arise in the course of the administration of the government of this country" and further "It is just part of the process of rating. It has to act judicially but that does not make it a court of law".

Lord Edmund-Davies and Lord Scarman in their speeches
C adverted to the case of *Society of Medical Officers of Health v Hope* [1960] AC 551, which was also cited to me. In that case Lord Radcliffe described a local valuation court as a court of limited jurisdiction, and Lord Keith of Avonholm pointed out, as also did D Viscount Dilhorne, that the creation of local valuation courts was in substitution for what used to be called assessment committees.

I have no hesitation in saying that these authoritative analyses of what is the nature and scope of a valuation court apply with equal force to the facts of the instant case. It means therefore that, even if the agreement contemplated recourse to the local valuation court, it falls outside the area where champerty might be said to arise. Furthermore it would seem to follow logically that arrangements for payment by a party of contingent financial advantage where cases are actually conducted and argued could not be said to be champertous. By the same token there could be no objection in law to a percentage arrangement where chartered surveyors E deal with a landlord's case in proceedings before a rent tribunal, because that is a tribunal which would seem to come within the same general description as to status and purpose as is the case with a local valuation court. In the instant case the plaintiff firm were content to move over to a scale fee if the matter required active representation and argument. It may be, for all I know, that the chartered surveyors' governing body have their own rules which may be merely advisory and not professionally binding, but this is a matter for the profession and not for me. All that I need say is that I cannot see any legal objection.

On the facts of this case and in the light of the law as I understand it to be, I have no hesitation in concluding that the plaintiff F firm have established their claim to the sum of £22,935.41 and are entitled to judgment therefor.

LANDLORD AND TENANT
GENERAL

A

QUEEN'S BENCH DIVISION

June 16 1981

(Before Mr Justice STUART-SMITH)

WOODTREK LTD v JEZEK

Estates Gazette February 13 1982

(1981) 261 EG 571

Action for forfeiture of lease of flat — Relief against forfeiture — Terms on which relief against forfeiture should be granted — Costs — Claim by landlords for forfeiture on grounds of non-payment of rent and service charge — Submission by tenant B **that claim in respect of service charge was misconceived — Sections 90 to 91A of the Housing Finance Act 1972 — Landlords' claim unfounded in respect of payment for estimated cost of works — Failure of landlords to supply correct details of claim, including invoices and receipts — Failure to supply satisfactory evidence continued up to date of trial — Claim so far as based on non-payment of interim service charge not substantiated — Forfeiture proceedings technically justified as tenant was in arrear in respect of rent payment — Tenant, however, entitled to relief against forfeiture and, in the circumstances, to the costs of the action subject to a deduction from taxed costs because of the arrears of rent**

C This was an action by the plaintiff company, Woodtrek Ltd, claiming forfeiture of the lease of a flat, no 5 at 12 Lindfield Gardens, London NW3, let to the defendant, Pavel Jezek, on a 99-year lease from June 24 1975.

R Powell-Jones (instructed by D B Morris & Co) appeared on behalf of the plaintiffs; R W Moxon-Browne (instructed by Vizards) represented the defendant.

Giving judgment, STUART-SMITH J said: In this action the plaintiff company claims forfeiture of flat no 5, 12 Lindfield Gardens, let to the defendant on a 99-year lease from June 24 1975 on the grounds of non-payment of rent and non-payment of a service charge. There is a claim for £150 allegedly due in respect of an interim service D charge.

The defendant concedes that at the date of the issue of the writ, the rent of £50 was in arrears and that the right to forfeiture arose, but he claims relief from forfeiture and it is accepted by the plaintiffs that that right to relief exists. It is on the terms upon which that relief should be granted that the case in the first place depends.

Secondly, the defendant says that the claim in respect of the service charge is misconceived and that the defendant should not be required to pay anything or, if anything, very little.

The lease is before me. It provides in clause 2 that it is a 99-year lease and the annual rent was payable by equal payments in advance on June 24 and December 25 in each year free of all E deductions. By clause 3 the tenant covenanted to pay the rent in the manner provided without deduction, and in clause 4(4) he covenanted to "Pay the Interim Charge and Service Charge at the times and in the manner provided in the Fifth Schedule hereto both such Charges to be recoverable in default as rent in arrear."

The Fifth Schedule to the lease lays down the contractual position with regard to the service charge and it is necessary to refer to some of those provisions. Clause 1:

> In this Schedule the following expressions have the following meanings respectively: (1) "Total Expenditure" means the total expenditure incurred by the Lessors in any Accounting Period in carrying out their obligations F under Clause 5(4) of this Lease and any other costs and expenses reasonably and properly incurred in connection with the Building including without prejudice . . .

and then various matters are set out.

Subclause (3):

> "the Interim Charge" means such sum to be paid on account of the Service Charge in respect of each Accounting Period as the Lessors or their Managing Agents shall specify at their discretion to be a fair and reasonable interim payment.

The accounting period is defined in clause 1(6) of the lease: ". . . shall mean a period commencing on the First day of January and ending on the Thirty-first day of December in any year" and in G this case the relevant accounting period is that ending on December 31 1979.

It is accepted by Mr Powell-Jones, on behalf of the plaintiffs, that the contractual rights in relation to service charges are subject to the provisions of the Housing Finance Act of 1972 as amended, in particular sections 90 and 91A, and it is material to look, I think, in the first instance, at section 91A. "A service charge shall only be recoverable from the tenant of a flat — (a) in respect of the provision of chargeable items to a reasonable standard; and (b) to the extent that the liability incurred or amount defrayed by the landlord in respect of the provision of such items is reasonable . . ." I think that I need not read any more than that. H

It is clear and is accepted that in order to recover in respect of the interim service charge — and it is common ground that the interim service charge is in fact a service charge within section 91A — four matters have to be satisfied. First, they have to be incurred in the accounting year; secondly, they have to be reasonable; thirdly, they have to have been actually incurred, and, fourthly, they have to be chargeable items within the definition in the Fifth Schedule.

The plaintiffs purchased the property and the reversion in August 1978. It was in a poor state of repair and they shortly after purchase instructed a firm called Phillips & Co to carry out a survey in respect of the work which was required. Phillips & Co reported to the plaintiffs in December 1978 that a substantial amount of work J was required to be done. As I understand it, that work was in fact never costed by builders, but Phillips & Co gave the plantiff company an estimate that it was going to cost something like £8,000 per year for 10 years.

The next thing that appears to have happened is that the landlords, the plaintiffs, sent letters to their tenants, of which examples are in the bundle before me, demanding the payment of a service charge which took account of that payment of £8,000 per year. I look, for example, at that which is addressed to the tenant of flat no 4 and it sets out:

> . . . a schedule of the works to be carried out to the property, and our client's K surveyors estimate the total expenditure for the whole building to be £8,000.00 per annum, over the next ten years. Under the terms of your lease, you are obliged to pay eighteen per cent of the landlord's total expenditure (as defined in the fifth schedule of the lease) which we calculate to be £1,440.00 per annum, and accordingly, unless we receive such sum within the next seven days, we are instructed to issue proceedings, for forfeiture of your lease.

As I understand it, similar letters were sent to all the tenants of this building, of whom there were some eight, although the one to the defendant in this case is not available.

A

That was a totally unfounded claim. The plaintiffs had no justification whatever for making a claim for that sum, or anything like it, and, not surprisingly, the tenants were somewhat apprehensive and alarmed when they received it. They formed a tenants' association in due course, although in the initial stages again the defendant does not appear to have been a member of that association.

The solicitors acting for the defendant, who were acting for the other tenants, replied to that letter on January 8 pointing out that they did not consider the charge in any way justified and on January 23 1979 they said this to the plaintiffs' solicitor:

> We are not disputing that the Lease does refer to the payment on account of service charge of an estimated amount on account of the prospective annual
> B
> expenditure, simply pointing out that under the terms of section 124 of the Housing Act 1974

that is section 91A of the 1972 Act as amended

> a service charge can only be recovered by the freeholder if it is of a reasonable amount and further only if the items upon which moneys have been expended were reasonable items, and that the moneys so expended on them were of a reasonable nature.

The next relevant thing that appears to have happened was that on December 31 1979 the plaintiffs wrote to the defendant saying this:

C

> We look forward to receiving your ground rent of £50 and interim service charges of £150 for 1979 and £1,000 for 1980, a total of £1,200, which became due on quarter day under clause 3(1) and 4(4) of your lease. We call upon you to pay the sum of £1,000 as an interim charge under clause 4(4) based on the report of Phillips & Co annexed hereto for works to be carried out in 1980. We require a further interim charge of £150 in respect of service charges for 1979 estimated to total not less than £1,200 covering inter alia, surveyors fees, legal costs, insurance (time at risk), boiler repairs and management costs, full particulars of which will be given in due course.

In fact, no particulars were ever given until after these proceedings had started.

On January 21 1980 in answer to that letter, the defendant's solicitors wrote as follows:

D

> As regards the service charge demand for 1979 in the sum of £150 per flat our clients are not able to make payments of these amounts without your first producing to them the appropriate accounts and receipts for the amount which you claim to have expended and on production of such receipts our clients will then be able to determine the extent to which the service charge is reasonable.

That was a demand which the tenants were entitled to make under the provisions of section 90 of the Housing Finance Act 1972.

It is material to notice that in the last paragraph of that letter the defendant's solicitors refer to a prospective or abortive meeting

E

which might have taken place on January 10 1980.

On February 17 1980 there is a "without prejudice" letter. The plaintiffs wrote to the defendant's solicitors saying that they had "been advised by leading counsel . . . to commence proceedings for possession for non-payment of rent and interim service charges which under the leases are to be treated as arrears of rent."

On February 29 the defendant's solicitors wrote to this effect:

> As regards the matter of service charges our clients repeat that they will only be responsible for service charges that are reasonable both as to the amount and the nature of the work incurred and to the extent that your clients are claiming service charge for work carried out or expenses incurred with regard to the building it is incumbent upon your clients to produce accounts,
> F
> invoices and other information to substantiate these amounts.

The reply to that letter was on March 6 1980 and the plaintiffs' solicitors said, among other things:

> The interim charge which is now due from the tenants is £150 per flat in respect of the accounting period ended December 31 1979" and that their clients "consider this to be a reasonable interim service charge . . .

Then, in the next paragraph, they say this:

> Furthermore a number of the tenants owe the rent due December 25 1979 of £25.00 and some owe the rent of £25.00 due June 24 1979. One or two tenants have tendered a cheque for the arrears of rent which cannot be accepted without payment in full of the outstanding interim service charge
> G
> too of £150.00. We are instructed to commence proceedings against at least one tenant in this connection which we are doing.

I should mention that the defendant had tendered a cheque for £50 for the rent outstanding on February 12 1980. That cheque was refused by the letter which I have just read, the plaintiffs making it perfectly clear that they were not going to accept it unless the full interim service charge of £150 was also paid.

The writ was issued on March 11 1980 and the cheque for payment of the rent was in fact accepted on March 27 1980 and nothing is now due in respect of rent.

The first details of the service charge that were given by the plaintiffs were sent on April 13 1981. That purported to show a

H

total, in respect of all the flats, of £1,200, but there was an error in respect of that, as the plaintiffs admit, and on June 12 they sent a fresh certificate from their accountants in which the total sum came to £795.

Mr Moschi gave evidence and sought to prove and justify the claim for £795. I regret to say that I found Mr Moschi a most unsatisfactory witness. He was glib and evasive; he was prone to enter into a diatribe against the iniquities of his tenants, real or imaginary, instead of answering the questions which were put to him, and, although I am quite satisfied that he was well aware of the material points of the questions, namely, whether or not the charge had been incurred in the accounting period of 1979 and whether or

J

not the sums expended related to such matters, he, in my view, deliberately attempted to evade answering those questions and was not prepared to answer them frankly. I am bound to say that he made a very unfavourable impression upon me as a witness and I did not find his evidence reliable.

Further than that, it is quite clear that under the Act the tenants were entitled to have production of invoices and receipts and various documents and that they had asked for such documents, yet the astonishing thing is that not one satisfactory piece of documentary evidence was produced by Mr Moschi in an attempt to support the claim set out in the second of these certificates. I find it really quite incredible that, a year or so having elapsed since the issue of the writ

K

and this matter being prepared for trial, no satisfactory documentary evidence and no other evidence was sought to be produced to reinforce Mr Moschi's evidence.

Moreover, in my judgment, Mr Moschi was clearly wrong about one matter. He sought to say that expenses which he had incurred in relation to the management of the property were incurred at an abortive visit which the surveyors made on January 10 1979. He was cross-examined on the basis of a letter, D3, which he or his office had written on January 10 1980, making it, as it seems to me, perfectly clear that the meeting was in January 1980 and not January 1979, and, as I have indicated, I think that is further made perfectly clear in the letter to which I have already referred from the

L

tenant's solicitors dated January 21 1980, yet Mr Moschi was not prepared to admit that he might have made a mistake about that. He insisted that the letter of January 10 1980 in fact referred to a meeting the year before.

For those reasons, among others, I am bound to say that I find it difficult to accept Mr Moschi's evidence unsupported by any other evidence or any documentary evidence.

Moreover, at the end of the day, Mr Powell-Jones, who has conducted this case with realism and discretion on behalf of the plaintiffs, was forced to concede that four of the items in the certificate of June 12 1981 were irrecoverable and he abandoned them, namely, no 1, no 2, no 4 and no 5. He also accepted that in

M

relation to no 6, the management charge, the most he could seek to recover in relation to that was about a half of it in respect principally of Mr Moschi attending to collect the rent and to deal with the breakdown of the boiler. The other item, which he contended was recoverable, was item no 3, which was alleged to be in respect of a boiler report. Mr Moschi said that he paid £20 in cash to a Mr Thompson, who had gone to the premises and had reported orally to him that he was unable to carry out the repairs to the boiler. In fact what happened was that the tenants themselves subsequently effected or had carried out the repairs to the boiler. I am not satisfied that Mr Moschi did in fact pay that £20. It would have been the easiest thing in the world to have produced an adequate receipt in relation to that matter and, having regard to the tenants' rights to which I have referred and the letters asking for such documents, I find it incredible that no such receipt was produced in this court.

So far as concerns the claim in respect of part of the management charges, item no 6, I do not think that the mere collection of rent comes within the chargeable items as defined in paragraph 1(1) of the Fifth Schedule to this lease. Moreover, for the reasons which I have already indicated, I am not entirely satisfied that that amount of time was expended on that aspect of the matter and, in those circumstances, I do not think that any of the items set out on the certificate of June 12 1981 fall to be recovered. I have not dealt in detail with the items which Mr Powell-Jones abandoned. They were abandoned for a variety of reasons, it becoming clear in the course of cross-examination of Mr Moschi that they were irrecoverable either as falling outside the accounting period or as not being capable of being substantiated or for a variety of different reasons.

In those circumstances, I have come to the conclusion that the claim in respect of £150, which in any event had by June 12 1981 been reduced to £99.38, is not made out. That was part of the foundation of the claim for forfeiture, because the plaintiffs, in their statement of claim, relied on non-payment of the interim service charge as founding partly the right to forfeiture. It is clear in the lease, which contains the usual forfeiture clause, that had that service charge been properly substantiated, that would have been a proper claim.

The question, therefore, is upon what terms the defendant should be granted relief from forfeiture. Mr Powell-Jones submitted to me that, as a matter of law, by virtue of section 46 of the Supreme Court of Judicature (Consolidation) Act 1925, under which relief from forfeiture is sought for non-payment of rent, and the decision of the Court of Appeal in *Belgravia Insurance Co Ltd v Meah* [1964] 1 QB 436, the plaintiffs were entitled to the costs of the action. It is clear from the judgment of the Master of the Rolls in that case that the courts, in applying the provisions of section 46 of the Supreme Court of Judicature (Consolidation) Act granting relief, follow the practice of the old Courts of Chancery subject only to the provisions of the Common Law Procedure Act of 1852, sections 210 and 212.

At p 443 the Master of the Rolls says this:

> The first thing to notice is that the ground of forfeiture was for non-payment of rent. Relief from such a forfeiture is treated differently from relief for other breaches, such as disrepair and so forth. In case of non-payment of rent, relief is still based fundamentally on the jurisdiction of Courts of Equity to grant relief, subject to the limitations imposed by sections 210 and 212 of the Common Law Procedure Act 1852. In case of other breaches, relief is based on section 146 of the Law of Property Act 1925.

Mr Powell-Jones, referring to those sections of the Common Law Procedure Act, says that the tenant here could have protected himself entirely by the payment into court or tender of the costs on the procedure for forfeiture and obtaining relief. It appears to me that the court's discretion as to the question of costs is not fettered in the way that Mr Powell-Jones submitted. I have been referred to the case of *Croft v London & County Banking Co* (1885) 14 QB 347 and, in my judgment, it is clear from the brief judgments of the Court of Appeal in that case that the court does have a discretion, notwithstanding the terms of the Common Law Procedure Act, in relation to the granting of costs. In the ordinary way, where there has been forfeiture, the right to forfeit is on the grounds of non-payment of rent and the tenant has to apply for relief. The landlord might well expect to get his costs, but there is, in my view, a discretion about the matter which can be exercised having regard to all the circumstances of the case and the way in which those costs have been incurred and the litigation has been brought about.

The fact of the matter here is that there was no letter before action in any proper sense at all. The rent was in fact tendered before the plaintiffs wrote saying that they had been advised by leading counsel to bring proceedings. It was accepted shortly after the issue of the writ, and the plantiffs, in their solicitors' letter of March 6 1980, made it perfectly clear, in my judgment, that they were not going to be content with anything less than establishing as a matter of principle and as a matter of pounds, shillings and pence the interim service charge which they alleged was due to them and it was to establish that right that in substance these proceedings have been brought.

For the reasons which I have given, the plaintiffs have failed to establish entitlement to £150 or to any sum under that head and, in my judgment, therefore, the defendant is entitled to relief from forfeiture and has succeeded on the principal issue which has been litigated before me. In those circumstances, it seems to me that the proper exercise of the discretion here is that the defendant should be granted relief from forfeiture; that he should have the costs of the action, but some deduction should be made from those costs because the defendant was in arrears and did not pay his rent and technically the proceedings for forfeiture were justified. I find it difficult to believe that they were justified in fact. This was a 99-year lease and the behaviour of the landlords, for reasons which I have attempted to outline, left a very great deal to be desired.

In those circumstances, I propose to grant relief. I propose to grant the defendant the costs of the action subject to a deduction of £150 on the taxed costs.

I should perhaps add that Mr Moxon Browne submitted to me that in fact the service charge here was irrecoverable by non-compliance with the provisions of section 90 of the Housing Finance Act 1972 and he relied upon *Anderson v Daniel Ltd* [1924] 1 KB 147. I do not find it necessary to deal with that submission in the light of the fact that I have already decided in his favour on the question of the service charge. I should only say, in deference to the argument which he addressed to me, that there appears to me to be an important distinction between the case of *Anderson v Daniel* and this case in that in *Anderson v Daniel* it was mandatory for the appropriate certificate or the appropriate invoice to be given whereas in this case it is only if the tenant exercises his rights under section 90 that the landlord is required to provide the details, and also, having regard to the provisions of section 91A which, by amendment, made the charge irrecoverable in certain circumstances, I find it a little surprising that the legislature did not say, if it intended so to do, that the service charge should be irrecoverable if the landlord failed to comply with the provisions of section 90 when required to do so, but I do not decide that point against Mr Moxon Browne. I prefer to leave the matter open.

The plaintiffs' claim was dismissed and the defendant was granted relief from forfeiture, the plaintiffs to pay the defendant's costs subject to a deduction of £150.

COURT OF APPEAL

December 2 1981

(Before Lord Justice CUMMING-BRUCE, Lord Justice TEMPLEMAN and Dame Elizabeth LANE)

SMEDLEY v CHUMLEY & HAWKE LTD (WARRELL, THIRD PARTY)

Estates Gazette February 27 1982

(1981) 261 EG 775

Landlord and tenant—Action by tenant of restaurant forming part of motel complex against landlords in respect of damage caused by alleged breach of landlords' covenant in lease — Tenant covenanted to repair interior and exterior of the demised premises, but landlords covenanted to keep the main walls and roof in good structural repair and condition throughout the tenancy — A few years after the commencement of the tenant's lease it became evident that the foundations of the restaurant were defective — There were no piles under part of the concrete raft on which the building was constructed, with the result that the raft tilted, causing damage to walls and roof — The action by the tenant concerned the liability of the landlords under their covenant — Landlords submitted that the judge was wrong in holding that, as the walls of the restaurant had become unstable owing to subsidence, the landlords were in breach of their covenant to keep the walls and roof in good structural repair and condition — The landlords argued that the imposition of such a liability amounted to requiring the premises to be improved, thus rendering them different from the premises demised — Authorities reviewed—Held that the landlords were liable—The only way to put

A **the walls and roof into a safe structural condition was to carry out such major works to the foundations as were necessary to give the walls a stable base — Appeal dismissed**

This was an appeal from a decision of Judge Mervyn Davies QC, sitting as a judge of the Chancery Division. The judge decided in favour of the tenant, Barry Thomas Smedley, in an action against the landlords, Chumley & Hawke Ltd, for breach of the landlords' covenant to keep the main walls and roof of the demised premises in good structural repair and condition. The premises demised consisted of land and buildings used as a restaurant, known as the Vandyke Restaurant, which formed part of a motel complex at
B Horning, Norfolk. The landlords appealed. The third party was Stanley George Warrell, the architect who designed the structure.

Geoffrey Jaques (instructed by Hill & Perks, of Norwich) appeared on behalf of the appellants; Michael Baker (instructed by Mills & Reeve, of Norwich) represented the respondent; John Ross Martyn (instructed by Allan Rutherford, of Norwich) represented the third party.

Giving judgment, Cumming-Bruce LJ said: By a lease dated December 31 1971 the appellant (defendant in the action), herein called the landlords, let to the respondent (plaintiff below), herein called the tenant, land and buildings for use as a restaurant (the Vandyke Restaurant) for 21 years from January 1 1972 at a rent of
C £1,300 a year, subject to upwards-only rent review. The property demised forms part of a motel complex at Horning in Norfolk. The landlords covenanted:

> 3(b) To keep the main walls and roof in good structural repair and condition throughout the term and to promptly make good all defects due to faulty materials or workmanship in the construction of the premises.

This covenant has to be read and contrasted with the tenant's covenant 2(e), whereby the tenant covenanted:

> From time to time and at all times during the term hereby granted well and substantially to repair cleanse maintain and keep the interior and exterior of the premises and the Lessors fixtures therein (if any) (unless the same shall be destroyed or damaged by fire not arising from the actual act or default of
> D the Lessee or any servant agent invitee or licensee of the Lessee) and in particular as occasion requires thoroughly to sweep all chimneys and flues in the premises and clean all windows and all cisterns boilers and water pipes and keep all ball cocks and interior pipes gas pipes and electric wires in the premises in good order and condition.

The facts found by the learned deputy judge are as follows. The Vandyke Restaurant was built together with the other buildings comprising the motel in about 1971. The landlords were responsible for its construction. It is built very near to the banks of the River Bure. In April 1976 the tenant's surveyor drew attention to the condition of the premises when he was inspecting in connection with the rent review clause. He observed that some of the floors
E sloped, and in places one had the impression of walking uphill. Secondly, he found some of the doors and windows out of true. Thirdly, he saw ruffling and crinkling of paper on some of the walls caused by the differential movement of two walls at the place of join. Later, he saw cracks in the floor. He advised that immediate action was necessary to arrest the development of the defects as subsidence was taking place and superficial repairs had been inadequate. The tenant's insurers withdrew their public liability policy, so that the tenant had to discontinue using the premises as a restaurant.

The landlords proceeded to carry out the necessary structural work necessary to put the building into a good structural condition.
F The trouble was that when the premises were built the foundations were defective. The building itself is of a timber frame. It was constructed standing on a concrete raft. The raft was supported at the river end upon piles sunk into the river bank. But there were no other piles beneath the raft. By 1976 the end of the raft furthest from the river was sinking into the ground so that the raft was tilting with consequential damage to the structure of the walls and roof. The remedial work involved jacking up the building and the concrete raft on which it stood, driving in additional piles to support the raft, placing joists upon the piles and then lowering the raft and the building down on to the joists. A new wooden floor was put on top of the original concrete raft. When the work was done, the restaurant was the same in the sense that the superstructure was the same G as it had been when it was first built, though the foundations were different.

In September 1976 the tenant brought an action against the landlords claiming breach by the landlords of covenant 3(b), and damage caused thereby. He sought an order that the landlords should carry out the works necessary to repair the defects in the structure of walls and roof, and damages for consequential loss. By their defence the landlords denied breach of covenant, and pleaded that any defects in the premises were in direct consequence of the design of the restaurant and so outside the scope of covenant 3(b). The landlords counterclaimed for breach of an oral agreement that the tenant would insure against risk of damage for subsidence. The H tenant by reply joined issue on the allegations in the defence and counterclaim. The learned deputy judge dismissed the counterclaim and no issue thereon arises in this appeal.

By the date of the hearing in the Chancery Division the work described earlier in this judgment had been completed by the landlords on the terms that the landlords' liability under covenant 3(b) should be left for the decision of the court. Thus, the issues for decision were (a) the landlords' liability under the covenant and (b) the tenant's claims for damages for breach. The landlords did not seek to recover the cost (some £6,000) of the structural work that they had carried out over a period of some seven months. The tenant claimed the cost of works that he had initially carried out, J and damages measured by the loss sustained during the period that he was unable to run his business in the restaurant, and expenditure that he had incurred on reconnecting pipes after the work had been done. The court was only asked to decide the question of liability under covenant 3(b). Determination of questions of quantum of loss was postponed.

The lease was entered into immediately after the landlord had constructed the premises. By Clause 1 of the lease:

> In consideration of the expense incurred by the Lessor in erecting the building now demised and of the rent and covenants on the part of the Lessee and the conditions hereinafter reserved and contained the Lessor demises unto the Lessee ALL THAT piece of land . . . together with the K buildings now erected thereon . . . and intended to be used by the Lessee as a restaurant . . . for the term of Twenty-One Years.

The lessee covenanted:

> 2(e) From time to time and at all times during the term . . . well and substantially to repair cleanse maintain and keep the interior and exterior of the premises . . .
>
> (f) At the determination of the said term to surrender and yield up unto the Lessor the premises with appurtenances in such state of repair as shall be consistent with the observance and performance by the Lessee of the covenants aforesaid . . .
>
> (k) To use and occupy the demised premises solely and exclusively as and for a restaurant only . . .
>
> L (l) To open and make available the demised premises for use by the public as a restaurant by a date not later than March 31 1972 and in the event of the demised premises not being open and made available as aforesaid then the Lessee shall pay to the Lessor the sum of Ten Pounds per day by way of compensation for any consequential loss that may be suffered by the Lessor . . . the sum contemplated and fairly estimated by the Lessor to be the loss he might sustain in connection with his business of letting adjacent chalets if the restaurant were not available for his tenants . . .
>
> (p) No alteration shall be made in the construction . . . of the premises or any part thereof . . . nor shall any of the principal walls or timbers be altered or cut nor any . . . substituted building whatsoever be erected or built on the premises . . .

In contrast to the lessee's covenant to repair, 2(e), is the lessor's M covenant 3(b) already set out. This covenant imposed on the lessor two obligations. One was to keep the walls and roof in the state described as good structural repair and condition. The other to remedy all defects due to faulty materials and workmanship in the construction of the premises.

This covenant imposed on the lessor two obligations. One was to keep the walls and roof in the state described as good structural repair and condition. The other to remedy all defects due to faulty materials and workmanship in the construction of the premises.

In approaching the construction of the landlord's covenant 3(b), I adopt the general approach stated by Sachs LJ in *Brew Brothers*

A *Ltd v Snax (Ross) Ltd* [1970] 1 QB 612 at 640 where in relation to a covenant "to keep in repair the demised premises", he said:

> It seems to me that the correct approach is to look at the particular building, to look at the state which it is in at the date of the lease, to look at the precise terms of the lease, and then come to a conclusion as to whether, on a fair interpretation of those terms in relation to that state, the work can fairly be termed repair. However large the covenant, it must not be looked at in vacuo.

On the meaning of the words "To keep the main walls and roof in good structural repair and condition throughout the term", I, like the learned deputy judge, derive assistance from the judgment of Fletcher-Moulton LJ in *Lurcott v Wakely* [1911] 1 KB 905 at 915.
B There this court was deciding the extent of a tenant's covenant "to well and substantially repair . . . and keep in thorough repair and good condition . . . the premises demised" in a lease for a term of 28 years of a house at least 200 years old, in which a wall had become unsafe through old age and lapse of time so that it had to be rebuilt. Fletcher-Moulton LJ first sought to give full meaning to each word of the covenant. Then he examined the effect of the covenants to keep the premises in good condition and to keep them in thorough repair as compared with the covenant to repair. The first two covenants imposed an obligation to keep the house in a certain state and he drew no wide distinction between keeping in thorough repair and keeping in good condition as both describe the condition
C of the house. If the house was no longer in that condition, the covenantor was obliged to put it into that condition, as this court had decided in *Proudfoot v Hart* (1890) 25 QBD 42. The covenant to repair by contrast imposed a duty to perform the operation of repair.

What then were the obligations accepted by the landlord and tenant respectively in relation to the Vandyke Restaurant? The landlords built it and let it to the tenant for a term of 21 years for use as a restaurant. The tenant covenanted to make the premises available to the public as a restaurant not later than March 31 1972 and in the event of their not being so available to pay the landlords a daily sum of ten pounds as liquidated damages for the loss sustained
D in their business of letting the surrounding chalets. The tenant covenanted well and substantially to repair and maintain and keep the interior and exterior of the premises and to keep all interior pipes, gas pipes and electric wires in good order and condition.

But these obligations were modified and supplemented by the landlords' obligations in covenant 3(b) in two different respects. The landlords undertook responsibility for the good structural condition of the walls and roof. They further accepted responsibility for any defects due to faulty materials or workmanship in the construction of the premises. This latter covenant likewise had the effect of cutting down the extent of the tenant's obligations to repair the interior and exterior of the restaurant. It was suggested on behalf of
E the landlords that this second covenant in clause 3(b) reduced the extent of the landlords' obligation to keep the structure of the walls and roof in good condition; it was submitted that if the walls and roof became unsafe by reason of defects in design of their footings or foundation the landlords were relieved of their obligation, as it extended only to defects due to faulty materials or workmanship. I do not agree. The two covenants expressed in clause 3(b) are independent of each other. The responsibility for keeping the walls and roof in good structural condition is unqualified, and placed firmly upon the landlords.

On the facts, the walls and roof appeared to be in good structural repair and condition when the landlords built the premises and
F when immediately after erection the tenant entered into possession. But these landlords had built the walls upon a foundation which was liable to tilt. After four years the foundation had so tilted that the walls and roof were unsafe. The only way to put the walls and roof into a safe structural condition was to carry out such major works to the foundations as were necessary to give the walls a stable base.

The learned deputy judge applied the test propounded in this court in *Lurcott v Wakely* [1911] 1 KB 905, cited the judgment of Vaughan Williams LJ at p 919, and concluded:

> The evidence shows that on September 14 1976 the walls of the Vandyke Restaurant were unstable due to subsidence. That means that the walls were not "in good substantial repair and condition" as I understand these words. G
> Accordingly, the defendant company were in breach of covenant 3(b).

In my judgment, there was ample evidence to support that finding and conclusion.

The answer of the landlords is that the learned deputy judge was wrong because the effect of the conclusion was to impose on the landlords an obligation to carry out an improvement and give to the tenant different and better premises than the premises constituting the parcels of the lease. The argument is put in two ways. There is a long line of authority to the effect that where a tenant is liable to keep in repair, he is not liable to do work that has the effect of giving the landlord a different and better house than the house that was
let. Counsel for the landlords and the third party relied upon *Lister v Lane* [1893] 2 QB 212, *Pembery v Lamdin* [1940] 2 All ER 434, *Sotheby v Grundy* [1947] 2 All ER 761 and *Brew Bros Ltd v Snax (Ross) Ltd* [1970] 1 QB 612. H

I make two observations upon those cases. It is important to distinguish the extent of the obligations where the lessor has let to the lessee an old house which has gradually deteriorated, through the inevitable effect of the passage of time, from the extent of the obligations imposed in connection with the lease of premises recently constructed. Many of the old cases are concerned with the former situation and do not assist in the analysis of the latter situation. Secondly, in order to discover whether there is an obli-
gation to do work made necessary in order to correct the effect of J
defects in design, it is necessary to examine carefully the whole lease and to decide the intention to be collected therefrom, and in this lease the intention was to place upon the landlords an unqualified obligation to keep the walls and roof in good structural condition.

Then it is contended that the work required to make the walls and roof safe was such that it involved an improvement to the premises and rendered the premises different premises from the parcels demised. Those parcels were, it is said, premises with defective foundations, so designed that the base on which the walls were built was bound to tilt over in four or five years. The work necessarily
done to make the walls and roof stable and safe involved giving the K
premises a new and different character, namely, a foundation which enabled the house to stand up instead of tumbling down. The lessor relies upon *Brew Bros Ltd v Snax (Ross) Ltd (supra)* and in particular the passage in the judgment of Phillimore LJ at p 646 where he accepted that the vital question in each case is whether the total work to be done can properly be regarded as repair, since it involves no more than renewal or replacement of defective parts, or whether it is in effect renewal or replacement of substantially the whole, a question of degree in each case.

It was contended on behalf of the landlords that the test whether the effect of the works was to render the premises something
different from the premises conveyed is to be applied by comparing L
the physical state of the premises as they were at the date of the lease with their physical state after the work had been done.

I prefer to compare the premises contemplated by the parties at the date of the lease with the premises as changed by the works actually done. Here the landlords built a complex of chalets with a restaurant for the use of the public including the persons occupying the chalets. Having built the restaurant they let it to the tenant on terms that the tenant would run it as a restaurant for the use of their licensees. The landlords accepted the obligation to keep the structure of the walls of the restaurant in good structural condition. When the work was done, they returned the restaurant to the tenant
in the state it was in at the date of the lease. The only difference was M
that the structure of walls and roof were stable and safe upon foundations made structurally stable.

I would distinguish the facts in the instant case from the facts in *Brew Bros Ltd v Snax (Ross) Ltd (supra)*. I compare and contrast the extent of the obligations imposed by the covenant to keep in repair in the context of the lease and of the circumstances in that case with the extent of the obligation, with its emphasis upon structure, in covenant 3(b). That emphasis upon structure is significant, because the structural condition of walls and roof is likely to depend on their foundations. So here I would hold that after the works were done the difference to the premises was that the walls and roof were in the condition that both parties contemplated as their necessary condition at the date of the lease.

I would dismiss this appeal.

TEMPLEMAN LJ said: I agree with the reasoning and the result. I also dismiss the appeal.

CUMMING-BRUCE LJ said that Dame Elizabeth Lane (who was not present for judgment) had asked him to state that she agreed with the judgment he had delivered.

The appeal was dismissed with costs. The costs of the third party and liability for costs as between the appellant and the third party were reserved.

COURT OF APPEAL

July 31 1981

(Before Lord Justice EVELEIGH, Lord Justice BRANDON and Lord Justice O'CONNOR)

SAMPSON v HODSON-PRESSINGER AND ANOTHER

Estates Gazette March 6 1982

(1981) 261 EG 891

Landlord and tenant — Problems created by nuisance to lower flat from excessive noise emanating from use of upper flat — Respective responsibilities of tenant of upper flat, previous landlord and present landlord — Both flats occupied by long leaseholders — Noise resulted from construction of a terrace on a flat roof which had been tiled and on which a decorative wall had been built, with access to the roof from french windows in the upper flat — Nuisance resulted from persons, such as visitors to the upper flat, treading on the tiled roof and causing an aggravating noise — Construction had taken place in the time of the previous reversioner and subsequently the reversion was purchased by the present landlord knowing of the complaints made by the tenant of the lower flat — This tenant, plaintiff in the county court action, sued the tenant of the upper flat and the present landlord — County court judge awarded plaintiff damages against the present landlord, who appealed — Previous landlord not a party to the proceedings — Held that the present landlord was in law in no better position than the previous landlord would have been and was liable both in nuisance and for breach of covenant of quiet enjoyment — Question of remedies discussed — Grant of injunctions unsatisfactory in the circumstances — Plaintiff had withdrawn claim for damages against upper tenant and preferred to uphold judgment against present landlord — Held that in the circumstances the right course was to uphold the award of damages against the present landlord and dismiss his appeal — As between the upper tenant and the present landlord the former was an innocent party despite her liability to the plaintiff

This was an appeal from a decision of Deputy Judge Raw at West London County Court. The plaintiff in the action was Robert Hamilton Sampson, tenant of flat 6 at 4 Lyall Street, London SW1. The first defendant was Miss A Hodson-Pressinger, the tenant of flat 7, the flat above the plaintiff's. The second defendant was D M Betts, the present owner of 4 Lyall Street and landlord in relation to the two flats. The deputy county court judge held that the first defendant was not liable in nuisance proceedings, but that the second defendant was liable. The judge refused an injunction but ordered the second defendant to pay the plaintiff £2,000 damages. The second defendant appealed.

R G Lamb (instructed by Franks, Charlesly & Co) appeared on behalf of the appellant: Roderick Cordara (instructed by Stoneham, Langton & Passmore) represented the respondent plaintiff; James Meston (instructed by Joynson-Hicks & Co) represented the respondent (first defendant).

Giving judgment, EVELEIGH LJ said: The plaintiff is the tenant of flat no 6 situated at 4 Lyall Street, London SW1, which he holds for a term of 99 years from June 24 1977 granted by a lease dated March 31 1978. He had previously been a statutory tenant of the flat. At one time immediately above the plaintiff's sitting-room, inner hall, kitchen and bathroom there was a flat roof which gave access to a fire escape and two water tanks.

The flat above the plaintiff's, no 7, was empty from 1974 until August 11 1978, when the first defendant moved in under a 99-year lease in similar terms to that of the plaintiff. It was granted by the same landlord. Between 1974 and 1978 conversion work was done to flat no 7. French windows were put in the wall of the sitting-room so that it opened on to the flat roof to which I have referred. That roof was tiled and a decorative wall was built using Californian bricks, so that flat no 7 had an attractive tiled terrace of some 30 ft × 19 ft.

Unhappily for the plaintiff, whenever anyone walked upon the terrace the noise of treading feet penetrated the plaintiff's flat and in particular his sitting-room. The first defendant regarded the terrace as a most desirable feature of her flat. As one would expect, she invited friends on to the terrace. She occasionally gave a party there. She was a thoroughly desirable neighbour on good terms with the plaintiff, but she insisted on the right to use the terrace in the way that any reasonable person would who was ignorant of any particular detrimental effect upon the occupier of the flat below. When she learned that the plaintiff was disturbed by the noise from the terrace she still took the view none the less that she should not be obliged to modify her use of the terrace or to put any insulating covering on it. The plaintiff complained to the first defendant and to the landlords.

In about March 1979 the second defendant showed an interest in acquiring the building no 4 Lyall Street. The plaintiff had told him about the noise from the terrace and had shown him an architect's report in which the noise was described as excessive and in which remedies were advocated. The second defendant, however, bought the property in April or May 1979 from the landlords. Thereafter he denied any responsibility for the noise resulting from the use of the terrace.

The noise complained of is that of people using the terrace, particularly when treading upon the roof. The county court judge visited the premises and heard the noise for himself. He said that, however "pleasantly" the terrace was used by the first defendant, the plaintiff could hear what was going on. He could hear conversations as a result of the tiles being improperly laid. He said that the noise was a nagging noise and interfered with the plaintiff's use and enjoyment of his premises, particularly the sitting-room. However, he held that the first defendant was not liable because she had nothing to do with the construction of the terrace. He regarded her as an innocent party in nuisance proceedings. He found the second defendant liable for the nuisance. He refused an injunction but ordered the second defendant to pay to the plaintiff £2,000 damages. These he assessed on the basis of the diminution in value of the plaintiff's flat as a result of being exposed to noise from above.

The appellant contended that the noise from flat no 7 was not sufficient to amount to a nuisance. I find this an impossible contention. The learned judge visited the premises and concluded that the noise did in fact interfere with the reasonable enjoyment of the plaintiff's flat. The appellant's contention was largely based upon the number of decibels recorded when readings were taken of the noise by an expert witness; but volume is not the sole test. The judge found the noise to be aggravating and irritating and was, in my opinion, entitled, on the evidence he heard, to reach the opinion he did.

It was also contended that the appellant could not be guilty of a nuisance, because the first defendant was using residential property in a normal way. The flaw in this argument, in my opinion, is that the property itself was not fit to be used in a normal way in so far as the terrace was concerned. It could not so be used without interfering with the reasonable enjoyment by the plaintiff of his flat. The use of the terrace put a strain upon the plaintiff that normal use in a normal building would not have done.

The first landlord is not a party to these proceedings. However, it is helpful to consider what his position would have been had he continued to be the landlord. He clearly would have been liable in nuisance to the plaintiff. He let the premises to be used for residential purposes and those purposes, in my opinion, included the use of the terrace in the way in which the first defendant used it. It must have been contemplated by the landlord that the tenant of flat no 7 would walk upon the terrace wearing ordinary footwear and that she would invite others to do so and to converse while there. He had himself been responsible for the condition of the premises which were not in fact suitable for use in the manner contemplated without causing a nuisance. It seems to me that the position of the first landlord falls

A within the principle in *Harris v James* (1876) 45 LJ QB 545. He had let the premises to a tenant for the purpose of doing an act likely to cause a nuisance. In that case the premises were let for burning lime. Any hesitation that might be felt in applying the principle to the present case lies in the fact that burning lime very plainly carries with it the risk of causing a nuisance and using a flat for residential purposes does not, at least not to the same extent. Nevertheless, the possibility of nuisance caused by treading on the floor is readily appreciated and indeed the lease in the present case contained a provision requiring the tenant to carpet the floorboards. By way of contrast, however, it was clearly envisaged that the tiles would not be covered.

B The appellant relied upon the case of *Rich v Basterfield* (1847) 4 CB 783. I do not think that that case assists the appellant for the reasons stated by Goddard J (as he then was) in *Wilchick v Marks and Silverstone* [1934] 2 KB 56 at p 65: "*Rich v Basterfield* was a case of a tenant himself creating a nuisance (in that case by smoke) by the manner in which he used the premises, which could have been used without creating a nuisance. The landlord therefore had done nothing wrong". The tenant in that case could have burned coke. He had not been authorised specifically to use the fuel he in fact used. In the present case it is clear that the use made of the terrace was the very use contemplated by the landlord and the tenant when the lease was granted and not a use for which the tenant was solely respon- C sible.

Apart from the question of common law nuisance, the plaintiff's lease contains the usual covenant for quiet enjoyment, that is, that the tenant may use the premises without interference by the landlord or those claiming under him. The contemplated use for which the original landlord let no 7 to the first defendant was one which interfered with the reasonable enjoyment of the plaintiff's flat. Consequently that landlord was, in my opinion, in breach of the covenant for quiet enjoyment. The plaintiff's enjoyment of the demised premises was unlawfully interrupted by the first defendant, a person lawfully claiming under the lessor. The plaintiff has not pleaded the case on this basis but it is a relevant consideration when I later come D to consider contribution.

It has been argued on behalf of the second defendant that even though the original landlord might be liable the second defendant is not because, it is said, he was not responsible for the construction of the roof terrace and he is unable to control the use of the first defendant's flat. I cannot accept this contention. The second defendant knew that the terrace was so constructed that its use would cause a disturbance to the plaintiff. It is a disturbance which the judge has held to amount to a nuisance. With that knowledge the second defendant has taken an assignment of the reversion and therefore taken over the role of the original landlord. He took over that landlord's responsibility for the condition of the premises and their E use. I have expressed my view as to the liability of the original landlord. I cannot see that the second defendant is in any different position. He is in receipt of rent which is payable to him on the basis that the first defendant, his tenant, is allowed by him to occupy and use the premises in the same way as she was permitted so to do by the original landlord. The second defendant is himself authorising the nuisance. He, too, is in breach of the covenant of quiet enjoyment. He is liable in nuisance and for breach of covenant.

Having reached this conclusion, I find it unnecessary to consider the effect of other terms in the two leases which have been referred to in argument. They were by no means in the forefront of the plaintiff's case.

F In these circumstances, what remedies are available? There could be an injunction against the first and second defendants. Alternatively, it is possible to award damages for prospective nuisance in exercise of the court's power first introduced by the Chancery Amendment Act 1858 (Lord Cairns' Act). In addition, damages could be granted to cover the period up to the date of trial. It is not clear upon what precise legal basis the learned judge awarded the sum of £2,000. I do not think it matters for the purpose of this appeal, provided this court is satisfied that the award can be justified in law.

To grant an injunction against both defendants is likely to produce a situation which will be very unsatisfactory from the first defendant's point of view and probably unfair to her. If the court were to grant an injunction against her an injunction against the landlord would be G unnecessary. However, the first defendant would then no doubt consider whether or not she had any remedy against the second defendant or against the original landlord. She did serve a notice in the present case claiming contribution against the second defendant in respect of any damages and costs which she might be ordered to pay to the plaintiff and in respect of her own costs of defending the action. In the event the plaintiff withdrew his claim for damages against the first defendant, hoping to remain on good terms with her, but the judge did grant her an indemnity in respect of her costs of defending the action. The contribution notice, however, is not in such a form as would enable the court to resolve such problems as might arise under the terms of the first defendant's lease if the court were to H grant an injunction against the first defendant. For example, there might be a contention that, as the injunction resulted from the act of a person claiming under the landlord (ie the plaintiff himself), the landlord was in breach of his covenant with the first defendant for quiet enjoyment. Moreover, the court has been told that the plaintiff would prefer to uphold the judgment which he has obtained rather than to seek injunctions. The figure of £2,000 may not be enough to carry out the necessary remedial works, but, while the court has not been given any details, it does appear that, at least as between the plaintiff and the first defendant, something can be done to remedy the situation. As we are told that, it seems to me that this really is a case where an award of damages would produce a satisfactory result J and achieve finality. However, on behalf of the landlord, it was contended that the court should not make such an award. It was said that the court should only grant damages for a prospective nuisance in exceptional circumstances and in very simple cases, which it was said this case was not.

A common argument against the award of damages as opposed to an injunction is that to do so would legalise or license the commission of future torts. In my opinion, while this argument may have merit when submitted by the plaintiff, it has much less strength when invoked by a defendant. In *Leeds Industrial Co-operative Society Ltd v Slack* [1924] AC 851 at p 861 Viscount Finaly quoted from the judgment of Buckley J in *Cowper v Laidler* [1903] 2 Ch 337 at p 341: K "The court has affirmed over and over again that the jurisdiction to give damages where it exists is not so to be used as in fact to enable the defendant to purchase from the plaintiff against his will his legal right to the easement". In the present case we would not be granting damages against the will of the plaintiff. That is an important consideration. In the above case the House of Lords held that it was right to grant damages as opposed to an injunction where the plaintiff's right to light was threatened by the construction of a building which had not yet reached such a height as to deprive the plaintiff's premises of light. That case also established that the power to award damages was available although Lord Cairns' Act had been repealed. I for my part then would regard the present case as an appropriate one in L which to award damages in lieu of an injunction. It is unnecessary to state a specific figure in respect of the nuisance for the period preceding the trial and to give detailed reasons to justify any particular figure in respect of prospective nuisance. This is because the evidence establishes, as the learned judge himself thought, that an appropriate figure would exceed that which could be awarded in the county court.

Strictly speaking, there could be an award in damages against both defendants. Had the plaintiff not withdrawn his claim against the first defendant I would have made an award against her. However, I regard this as a case in which she would have been entitled to a complete indemnity from the original landlord. M The original landlord had let to her premises which could not be used in the manner contemplated by the lease. These are matters which are relevant, in my opinion, when considering a claim for contribution between tortfeasors when the court in its discretion is deciding what is just as between the parties. As between these defendants I regard the first defendant as an innocent party in spite of her liability to the plaintiff.

As I would uphold an award of damages against the second defendant it follows that it would be inappropriate to grant an injunction against the first defendant. The plaintiff is accepting damages in lieu of his right to complain of nuisance in the future. There is, therefore, no basis upon which an injunction can be granted against the first defendant. I would dismiss this appeal.

A Brandon LJ: I agree.

O'Connor LJ: I agree.

The appeal was dismissed with costs against the appellant in favour of the two respondents. Leave to appeal to the House of Lords was refused.

COURT OF APPEAL

December 17 1981

(Before Lord DENNING MR, Lord Justice DUNN and Lord Justice FOX)

B
NEW ZEALAND GOVERNMENT PROPERTY CORPORATION v H M & S LTD

Estates Gazette May 22 1982
(1981) 262 EG 765

Landlord and tenant — Tenants' fixtures — Tenants' right to remove fixtures — Effect on right of renewal of surrender of old tenancy and grant of new tenancy — "All the statements in the text-books are erroneous" (Lord Denning) — Doctrine rejected that a tenant loses his right of removal if after expiry or surrender of tenancy he remains in possession by virtue of new tenancy — Point
C arose in context of a rent review clause in a lease of Her Majesty's Theatre in the Haymarket — Fixtures in this case included the seats in the auditorium of the theatre — Old lease continued under Part II of the Landlord and Tenant Act 1954 until it was surrendered by operation of law when a new lease was granted — New lease provided for the rent for the second seven years of a 21-year term to be at the "open market rental of the demised premises" — Were the fixtures to be regarded as part of the "demised premises" on which the tenants were to pay rent on the basis that the right of renewal had been lost and the fixtures belonged to the landlords? — Law as to removal of tenants' fixtures analysed — New rent should be assessed on the footing that the tenants had not lost the right to remove the
D fixtures and that the latter had not become part of the demised premises — Caution that in cases of express surrender the terms of the surrender must be considered — Judgments form an interesting excursus into the old common law — Appeal from decision of Woolf J dismissed

This was an appeal from a decision of Woolf J, on a special case stated by Judge Hawser sitting as an arbitrator appointed by consent under section 11 of the Arbitration Act 1950 and RSC, Ord 36. Judge Hawser had held that the open market rental of a new lease entered into by the tenants, H M & S Ltd, and granted by New Zealand Government Property Corporation, the landlords, should be
E determined on the basis that the tenants could only remove those fixtures annexed after the date when the material terms of the new lease had been agreed. The subject property was Her Majesty's Theatre in the Haymarket. The old lease expired by effluxion of time on October 1 1970. The new lease, executed on February 8 1973, was for a term of 21 years commencing from October 1 1970. Woolf J had held that the rent should be determined on the basis that the tenants had removed all tenants' fixtures, whether annexed during the term of the old or the new lease: (1980) 257 EG 606.

Raymond Sears QC and Anthony Porten (instructed by Allen & Overy) appeared on behalf of the appellant landlords; Ronald Bernstein QC and Kirk Reynolds (instructed by Nicholson, Graham
F & Jones) represented the respondent tenants.

Giving judgment, Lord Denning MR said: What rent is payable for Her Majesty's Theatre in the Haymarket? It gives rise to an interesting question of law about tenants' fixtures. But first for the story.

Her Majesty's Theatre is one of the most famous in London. It was built in 1898 by that great actor Sir Herbert Beerbohm Tree. The landlords were the Carlton Hotel Ltd. They leased it to him and his company, Playhouse Ltd, for 72 years from October 10 1898 to September 30 1970, at a rent of £3,700 a year. The lessees covenanted to use it as and for a first-class theatre only, to keep the demised premises (including landlord's fixtures) in good repair and so to deliver it up at the end of the term.

Many years ago the lease became vested in a company called H M G
& S Ltd., of which the leading figure is Mr Prince Littler. The ownership has become vested in the New Zealand Government.

The theatre was occupied "for the purpose of business". So the tenancy came within the Landlord and Tenant Act 1954. It meant that before the old lease came to an end the lessees were entitled to, and did, apply for a new lease under that Act. The landlords conceded that the tenants were entitled to a new lease. The only question was as to the terms of it. There were negotiations between the parties about the terms. They started early in 1970 (before the old lease expired) and continued through the next two or three years. Meanwhile, pending all the negotiations under the statute, the old lease continued in existence automatically. It did not determine on H
September 30 1970. It continued until February 8 1973 when the parties executed a new lease.

It is under that new lease of February 8 1973 that the question arises. It was for 21 years from October 1 1970 at a rent for the first seven years of £25,000 a year: and for the next seven years at the "open market rental of the demised premises". That is the point at issue. What is the "open market rental" for the second seven years? The first seven years were to end on September 30 1977. In anticipation of this, the parties negotiated for the new rental for the second seven years of the term. The landlords' surveyor put it at £55,000 a year. The tenants' surveyor did not agree. The matter went to arbitration before an official referee (Judge Lewis Hawser QC), J
sitting as an arbitrator. In the course of the arbitration a point of law arose about tenants' fixtures. There were many of them in the theatre. The tenants had put them in over many years. Typical were the seats in the stalls and auditorium. Were they to be included as part of the "demised premises" on which the tenant was to pay rent? Was the tenant to pay rent for them on the basis that they belonged to the landlord? Or was he free from paying any rent on them because they belonged to him, the tenant?

Now, before I go any further, I must observe that it is clear law that a tenant has a right to remove "tenant's fixtures" before the term comes to an end. So the tenant here had a right to remove them, not only during the old lease itself, but also at any time before the old lease as extended came to an end on February 8 1973. The old lease, K
as extended, was surrendered "by operation of law" on February 8 1973 when the new lease was granted (that has been clear for centuries: see *Thompson v Trafford* (1593) 79 ER 1131).

What was the effect of the surrender?

What, then, was the effect of the surrender? The landlords say that when the old lease was surrendered in 1973 the tenant lost his right to remove the tenant's fixtures. They became a gift to the landlord. They became part of the "demised premises". So that in 1977 the new rental was to be paid on the basis that they belonged to the landlord and the tenant was to pay rent for them just as if they were part of the structure itself. L

But the tenants say that the surrender in 1973 did not have that effect. They say that so long as they continued in possession — whether under the old lease or an extension of it, or on a new lease — they retained their right to remove them.

Tenant's fixtures and landlord's fixtures

Before I go any further, I would describe the distinction between "tenant's fixtures" and "landlord's fixtures". The term "tenant's fixtures", for present purposes, means those fixtures which the tenant himself fixed into the premises for the purpose of his trade, that is, for the business of the theatre, but which do not become part of the structure itself. Instances are the seats for the stalls, or auditorium, which are fixed by screws or bolts to the floor, wall- M
brackets for lights which are screwed on to the wall, electric transformers fixed on to the floor, and so forth. All these the tenant is entitled to remove when his term comes to an end. Whereas "landlord's fixtures" for present purposes means those fixtures which the tenant himself fixes into the premises so that they become part of the structure itself: see *Boswell v Crucible Steel Co of America* [1925] 1 KB 119. Instances are improvements made by the tenant by putting in new doors or windows in place of those that were there before, or a new frontage or a new safety curtain. These improvements become part of the structure itself. The tenant cannot remove them when his term comes to an end. All this goes back to the

A time of Sir John Holt CJ. He had before him *Poole's Case* (1703) 1 Salk 368. A soap-boiler was the tenant for years of a house in Holborn. For the convenience of his trade he put up vats and coppers and paved the back-yard. Sir John Holt CJ said:

Firstly. That during the term the soap-boiler might well remove the vats he set up in relation to trade, and he might do it by the common law (and not by virtue of any special custom) in favour of trade and to encourage industry: but after the term they become a gift in law to him in reversion, and are not removable.

Secondly. That there was a difference between what the soap-boiler did to carry on his trade, and what he did to complete the house, as hearths and chimney-pieces, which he held not removable.

B **The correct rule**

In the present case the question arises: If the term expires by effluxion of time or by surrender — and the tenant remains in possession by virtue of a new tenancy express or implied — can he still remove the "tenant's fixtures" during his extended time of possession? Or did he lose them irretrievably when his original term came to an end? There are many statements in the text-books to the effect that the tenant loses them when his original term comes to an end. Woolf J quotes the text-books at length. I think those text-books are wrong. In my opinion the tenant remains entitled to remove the "tenant's fixtures" so long as he remains in possession. That was decided in *Penton v Robart* (1801) 2 East 88. Robart was C undertenant of a yard and buildings at Battlebridge. During his subtenancy he erected a wooden shed for the purpose of making varnish. It had a brick foundation. The original term expired at Michaelmas 1800. He remained in possession *for some time afterwards,* and during that time he pulled down the wooden superstructure of the shed and carried away the utensils. The head landlord claimed that they belonged to him. The Court of King's Bench held that the tenant was entitled to remove them. Lord Kenyon CJ said (at p 90):

The old cases upon this subject leant to consider as realty whatever was annexed to the freehold by the occupier: but in modern times the leaning has always been the other way in favour of the tenant, in support of the interests D of trade which is become the pillar of the State. What tenant will lay out his money in costly improvements of the land, if he must leave everything behind him which can be said to be annexed to it . . . Here the defendant did no more than he had a right to do; *he was in fact still in possession of the premises at the time the things were taken away, and therefore there is no pretence to say that he had abandoned his right to them.*

That case was distinguished in *Weeton v Woodcock* (1840) 7 M & W 14, but not doubted. The landlords let a cotton factory to Taylor for seven years. Taylor installed a steam engine boiler, firmly fixed to the floor and walls of the engine-house. Taylor became bankrupt. His property vested in his assignees Woodcock and others. They entered into possession of the factory. The landlords then forfeited E the lease for breach of covenant by Taylor. So it came to an end. The assignees then sold the boiler and removed it. The landlords claimed that the assignees had no right to sell the boiler: because the term had already come to an end and the tenant no longer remained in possession. The Court of Exchequer held that the assignees were not entitled to remove the boiler. After the lease was forfeited, they held on — not as tenants nor by agreement with the landlords — but as trespassers. So they had no right to remove the boiler. Alderson B in giving the reserved judgment of a strong court said (at p 19):

The rule to be collected from the several cases decided on this subject seems to be this, that the tenant's right to remove fixtures continues during his original term, *and during such further period of possession by him, as he holds the premises under a right still to consider himself as tenant.*

F

The contrary view

I think that is the correct rule. But I have to recognise that there are later observations which throw doubt on it. It is said that when the tenant surrenders his existing term and takes a new tenancy, without reserving his right to remove tenant's fixtures, he loses that right altogether, even though he still remained in possession as a tenant. In the case of an express surrender, there is this dictum by Parker J in *Leschallas v Woolf* [1908] 1 Ch 641 at p 652:

In my opinion, however, if the tenant upon the surrender of his lease in order that a new lease may be granted makes no stipulation to the contrary, he does lose his right to remove tenant's fixtures.

G And in the case of a surrender by operation of law, there seems to be a decision by Scrutton J in *Slough Picture Hall Co Ltd v Wade* (1916) 32 TLR 542. To which I would add the dictum of Warrington LJ in *Pole-Carew v Western Counties and General Manure Co Ltd* [1920] 2 Ch 97 at p 122:

I think it is clear that after a surrender of the term in the land to which tenant's fixtures are attached and a subsequent lease to the same tenant, the latter can no longer remove the tenant's fixtures unless his existing right to remove them is reserved expressly or by necessary implication.

Opinion reserved

But as against all these statements, it is to be observed that on two separate occasions the Court of Appeal have expressly left it open. In H 1881 in *Re Thomas, ex parte Willoughby d'Eresby (Baroness)* (1881) 44 LT 781 at p 784 the Court of Appeal, consisting of James, Cotton and Lush LJJ, said:

If, and when the simple case shall arise of a tenant, having removable fixtures, continuing his possession under a new or extended term, we desire to hold ourselves perfectly free as to the question whether he retains his right of removal during such continuous possession, unfettered by anything said in this case.

Likewise in *Pole-Carew v Western Counties and General Manure Co Ltd* [1920] 2 Ch 97 at p 119 by Lord Sterndale MR; and at p 123 by Younger LJ.

J

Consider on principle

This means that in this court we can consider the point upon principle. I would test it by taking these illustrations:

First, a tenant takes a lease of business premises for five years, and puts in a valuable machine fixed to the floor for use in his trade. After the five years are at an end, he holds over with the consent of his landlord as tenant from year to year. Nothing is said about the machine. Then after three more years he moves out and goes elsewhere. Surely he can remove his machine and take it with him. It would be most unjust if he could not do so.

Second, the tenant does not hold over with nothing said. At the end of the five years, he takes a new lease from his landlord for K another three years to run from the expiry of the first term. But nothing is said about the machine. Then, at the end of the next three years, he moves out and goes elsewhere. Surely he can take the machine with him. There is no difference between a new lease and holding over.

Third, the tenant takes a new lease for three years from his landlord, but it is back-dated so that it starts after the first four years, and not after five. Nothing is said about the machine. The first lease is then surrendered by operation of law at the end of the fourth year. But surely at the end of the new lease, he can take the machine with him.

Fourth, the same as the last, but the tenant at the end of the fourth L year expressly surrenders the first lease and takes a grant of a new lease. Nothing is said about the machine. Surely at the end of the new lease he can take away the machine. Is it really necessary to say to him: "Did you tell the landlord that you reserved your right to take away the machine?" He would answer: "No, I did not do so, because it did not occur to me. It was my machine and I assumed I could take it away when I left".

These illustrations convince me that all the statements in the text-books are erroneous. I hold that when an existing lease expires or is surrendered and is followed immediately by another, to the same tenant remaining in possession, the tenant does not lose his right to remove tenant's fixtures. He is entitled to remove them at the end of M his new tenancy.

Improvements

But then it is said that section 34 of the Landlord and Tenant Act 1954, as amended by the Law of Property Act 1969, shows that Parliament proceeded on a different course. That section says that in fixing the rent payable under a new tenancy, there is to be taken into account any "improvements" carried out within the last 21 years, but not those carried out more than 21 years past. But the answer is that that applies to improvements made by the tenant which are "landlord's fixtures" — which the tenant is never able to remove. It does not apply to "tenant's fixtures".

A **Time of fixing rent**

Finally, coming back to our present case, the rent to be assessed for the second seven years was, and would have to be, fixed by agreement or by the judge during the period when the original lease was automatically extended by the Landlord and Tenant Act 1954. During that automatic extension the tenant would certainly have a right to remove the tenant's fixtures. So the new rent should be assessed on that basis.

"Demised premises"

The meaning of "demised premises" is also elucidated by reference to the repairing clause which is to keep " the demised
B premises and the appendages thereof", including the "landlord's fixtures" in good repair. That indicated that the "tenant's fixtures" are not part of the "demised premises".

Conclusion

After this long discussion, I think that in this court we should free ourselves from the shackles of the past. It is time that the law about removal of tenant's fixtures was brought up to date. This case gives this court the opportunity to do so. I think the rent of the Haymarket theatre should be assessed on the basis that the tenant's fixtures could have been removed by the tenant if he wished — and that they should not be regarded as part of the "demised premises" for the
C purpose of fixing the rent.

I would therefore dismiss the appeal.

Agreeing, DUNN LJ said: At common law the ancient rule was that once a chattel was fixed to the soil or the structure of the building it became part of the freehold. It was a logical step from the rule to say that if a tenant surrendered his lease, then he also surrendered the "fixture", unless in the deed of surrender he expressly reserved his right, which was also recognised at common law, to remove the tenant's fixtures. If he did not, then, since they were already part of the freehold, they were surrendered together with the rest of the demised premises.

D It was this step which Parker J took in *Leschallas v Woolf* [1908] 1 Ch 641 at p 152. But although logical the step does not seem to me to be the necessary consequence of the earlier decided cases, the *fons et origo* of which seems to be *Fitzherbert v Shaw* (1789) 1 Hy Bl 258. In that case the landlord had brought proceedings for eviction against the tenant, which were settled on terms that the tenant should surrender his lease at some date in the near future, remaining in possession meanwhile. It was held that on the true construction of the deed of surrender the tenant had lost his right to remove the tenant's fixtures. The case was expressly distinguished in *Penton v Robart* (1801) 2 East 88 on the ground that it turned on the construction of the deed of surrender. It is consistent with the view that the tenant did not lose his right to remove the fixtures until he
E gave up possession of the premises, and indeed would not have lost his right had it not been for the terms of the deed.

It seems to me that Parker J, for all his knowledge of this branch of the law, fell into error in stating the rule in the terms in which he did, which are not supported by a proper understanding of the earlier cases. This error was perpetuated in *Slough Picture Hall Co Ltd v Wade* (1916) 32 TLR 542, although the Court of Appeal expressly refused to be led into it in *Re Thomas, ex parte Willoughby d'Eresby (Baroness)* (1881) 44 LT 781 and *Pole-Carew v Western Counties and General Manure Co Ltd* [1920] 2 Ch 97.

I believe the true rule at common law to be that a tenant has the
F right to remove tenant's fixtures so long as he is in possession as a tenant, whether by holding over, or as a statutory tenant under the Rent Acts, or upon an extension of a lease of business premises under Part II of the Landlord and Tenant Act 1954. There is one exception to the rule; agricultural tenants never had the right at common law to remove tenant's fixtures (see *Elwes v Maw* (1802) 3 East 38), and this exception has by necessary implication been recognised in the provisions of the Agricultural Holdings Act 1948, which contains a compendious code for the ascertainment of "tenant's rights".

The rule is not inconsistent with the provisions of Part II of the Landlord and Tenant Act 1954 if, as I agree with the Master of the Rolls, the word "improvements" in section 34 is defined as meaning improvements to the structure and the landlord's fixtures.

If the tenant surrenders his lease and vacates the premises without G
removing the tenant's fixtures, then he is held to have abandoned them. But if he surrenders his lease, either expressly or by operation of law, and remains in possession under a new lease, it is a question of construction of the instrument of surrender whether or not he has also given up his right to remove his fixtures. If nothing is said, then the common law rule applies, and he retains his right to remove the fixtures so long as he is in possession as a tenant.

I would accordingly dismiss the appeal.

Also agreeing, FOX LJ said: There is no doubt that the common law confers upon a tenant the right to remove tenant's fixtures. The question is the extent of that right. Does it endure only so long as the tenancy exists, or will it continue so long as the tenant remains in H
possession of the premises as tenant though under a new tenancy?

The specialist text-books, on the whole, support the narrower view. *Foa,* 8th ed, para 1075, states:

> The mere fact that the tenant retains possession of the demised premises after the expiration of his interest does not extend the period during which he may sever fixtures even though the reason for his holding over is that his successor in the tenancy has failed to pay him an agreed price for them. And where the continuance in possession is under a new lease or agreement his right to carry away the fixtures is determined . . .

Foa admits the possibility of inferring where a lease is surrendered and the premises redemised to the tenant that the new demise is of the premises minus the tenant's fixtures but states that if nothing is said J
about the fixtures there would be no ground for such an inference.

Woodfall, 28th ed, at para 1—1573 warns tenants, on negotiating a renewal, to be careful to reserve the right to tenant's fixtures "as he may lose his right of removal".

Hill & Redman, 16th ed, para 425, states that a tenant must remove his fixtures during the term and that if he omits to do so they become the absolute property of the landlord "save that if the tenant remains in possession after the term in such circumstances that he is entitled still to consider himself as tenant, his right to remove fixtures continues as long as this state of things lasts . . ."

Thus while *Foa* and *Woodfall* evidently favour the narrower view, K
Foa contemplates the possible application of some less rigid rule; and *Woodfall* uses the word "may". *Hill & Redman* goes a good deal further still and specifically asserts the tenant's right to tenant's fixtures so long as he is entitled to consider himself as remaining in possession as tenant.

In my opinion the authorities binding on this court do not establish the narrower rule.

As long ago as 1852 in *Heap v Barton* (1852) 12 CB 274 we find Jervis CJ saying at p 280:

> The Courts seem to have taken three separate views of the rule. First, that the fixtures go at the expiration of the term to the landlord unless the tenant has during the term exercised his right to remove them. Secondly, as in *Penton v Robart* (1801) 2 East 88, that the tenant may remove the fixtures L
> notwithstanding the term has expired if he remains in possession of the premises after his term has expired — thirdly, that his right to remove fixtures after his term has expired is subject to this further qualification viz that the tenant continues to hold the premises under a right still to consider himself as tenant.

It was not, in the event, necessary for the court to reach any conclusion on the matter.

In *Penton v Robart (supra)* the tenancy had come to an end but the tenant remained in possession. He was permitted to remove his fixtures. Lord Kenyon at p 91 said:

> Here the defendant did no more than he had a right to do; he was, in fact, still in possession of the premises at the time the things were taken away and, M
> therefore, there is no pretence to say that he had abandoned his right to them.

In *Weeton v Woodcock* (1840) 7 M & W 14 the lease was forfeited and the assignees in bankruptcy, after the forfeiture, sought to remove tenant's fixtures. They failed. Alderson B at p 19 said:

> The rule to be collected from the several cases on this subject seems to be this, that the tenant's right to remove fixtures continues during his original term and during such further period of possession by him, as he holds the premises under a right still to consider himself as tenant.

The assignees having removed the boiler after the forfeiture, they did so at a time when they could no longer regard themselves as tenants.

A The Court of Appeal in *Re Roberts, ex parte Brook* (1878) 10 Ch D 100 seem to have felt difficulty in determining the exact limits of the principle stated in *Weeton v Woodcock* and the examples which they give (at p 109) of its operation are of limited extent but they do not disapprove it. The present is a case where, although the tenants continued at all times to occupy the premises as tenants, there was, immediately prior to the grant of the present lease in 1973, a surrender by operation of law of the 1898 lease (or rather of the statutory extension of that lease created by the Landlord and Tenant Act 1954).

There are two cases in the present century in relation to the effect of surrender upon a tenant's right to remove fixtures. The first is
B *Leschallas v Woolf* [1908] 1 Ch 641, a decision of Parker J. There was an express agreement to surrender. The judge stated, in the passage at p 652 to which the Master of the Rolls has referred, that if the tenant surrenders his lease, albeit with the view to the grant of a new lease, and does not reserve the right to remove tenant's fixtures, he loses that right altogether, even though he remains in possession as a tenant. He does, however, state at p 654:

> It may under certain circumstances be inferred that what is surrendered and redemised is not the premises in their actual condition, but the premises minus the tenant's fixtures. Possibly also, parcel or no parcel being a question of fact, evidence might be admitted under this head.

In *Slough Picture Hall Co Ltd v Wade* (1916) 32 TLR 542 Scrutton
C J followed *Leschallas v Woolf (supra)*. As regards the passage from the judgment of Parker J at p 654, which I have set out, Scrutton J said:

> But in this case nothing was said about the fixtures at all and I can find no ground for implying an exception out of the ordinary circumstances of a surrender . . .

Despite these last two authorities, it seems to me that, leaving aside the question of express surrenders, there is no authority binding upon this court which decides that, where a lease comes to an end but the tenant remains in possession under a new lease, the tenant loses his right to remove the tenant's fixtures.

In *Re Thomas, ex parte Willoughby d'Eresby (Baroness)* (1881) 44
D LT 781, the Court of Appeal on March 3 1881 gave judgment in terms which might be regarded as favouring the narrower view of the tenant's rights. Four days later, however, on March 7, the court restored the case to the list on its own motion and stated, *inter alia:*

> If and when the simple case shall arise of a tenant having removable fixtures, continuing his possession under a new or extended term, we desire to hold ourselves perfectly free as to the question whether he retains his right of removal during such continued possession . . .

In *Pole-Carew v Western Counties and General Manure Company Ltd* [1920] 2 Ch 97 Warrington J at p 122 said:

> I think it is clear that after a surrender of the term in the land to which tenant's fixtures are attached and subsequent lease to the same tenant the latter can no
> E longer remove the tenant's fixtures unless his existing right to remove them is reserved either expressly or by necessary implication — *Leschallas v Woolf (supra)*.

But Lord Sterndale MR and Younger LJ reserved their opinions upon the surrender point, which Younger LJ at p 123 described as one of great difficulty.

In these circumstances, it seems to me that this court is free to consider the matter on principle. In approaching that consideration there are two preliminary matters which I should mention.

First, I can quite see that, if a tenant upon the determination of his lease quits the holding without removing his fixtures, he may have to be taken to have abandoned them.

F Secondly, if a tenant executes an express surrender of his lease or enters into an express agreement to surrender his lease, the surrender or agreement must be construed according to its terms. It may be that the document can only be construed as a disposition of the entirety of the tenant's rights in respect of the land and will, therefore, extinguish his right thereafter to remove tenant's fixtures. Parker J in *Leschallas v Woolf (supra)* so interpreted the agreement in that case.

But when one comes to cases where the tenant, after the determination of his lease remains in possession of the property under a new lease and the determination of the old lease was either by effluxion of time or a surrender by operation of law, it seems to me that quite different considerations apply. I can see no sensible reason G
why, in these circumstances, the tenant should lose his right to remove his fixtures. The fixtures were brought in by him for the purposes of his occupation as tenant, and one would suppose that (subject to any express disposition by him) his rights in respect of them would not alter so long as he continued to occupy as tenant, whether under his original lease or a new lease taking effect upon the determination of the original lease. I do not believe that a person holding over as tenant from year to year or taking a renewal of his lease under a provision in the original agreement would imagine that any rights that he had to remove fixtures would be affected by the determination of the original lease. The Landlord and Tenant Act 1954 is designed to ensure security of tenure for business tenants. H
The Act in effect enables such tenants to obtain extensions of their leases from time to time to enable them to carry on their businesses. One would be reluctant to reach the conclusion that while tenants are secure in their tenancies they may lose their rights in respect of valuable business fixtures.

The fact that the previous lease has determined by surrender by operation of law (which is normally the case where application is made under the Landlord and Tenant Act 1954 in consequence of the statutory extension of the existing lease until the proceedings are determined) seems to me to be a mere technicality which should not affect the court's approach to the question of principle.

It is, of course, true that a tenant can, prior to the grant of a new J
lease, remove any tenant's fixtures which he attached to the property during the existing term and bring them back after the new term has been granted. But that is expensive and wasteful formalism. It should be possible for the law to preserve the tenant's right to remove fixtures in some more sensible way. In my view it would be a sensible and workable rule that a tenant should have the right to remove his fixtures so long as he remains in possession of the premises as tenant. I think it is open to the court to adopt that rule and I would do so.

It is right to consider whether such a rule would disturb property transactions entered into upon the basis reasonably supposed to exist of some other state of the law indicated in the authorities. I do not think it would. The construction of express surrenders or agreements K
for surrender would not be affected. So far as other transactions are concerned, the text-books have not been unanimous as to the limits of the law and it is clear, I think, that the two decisions of the Court of Appeal to which I have referred (ie *Willoughby d'Eresby* and *Pole-Carew*) have left the matter quite open over a long period. The law has not been settled. And, of course, if it can be shown that the parties have both acted upon the basis that particular fixtures are part of the land and not removable, effect could be given to that.

I should add that I agree with the observations of the Master of the Rolls regarding section 34 of the Landlord and Tenant Act 1954 as amended by the Law of Property Act 1969.

The result, in my view, is that the fixtures in issue in the present L
case should be regarded as fixtures which the tenant is entitled to remove and the rent should be fixed accordingly.

I would dismiss the appeal.

The appeal was dismissed with costs. Leave to appeal to the House of Lords was refused.

COURT OF APPEAL

January 28 1982

(Before Lord Justice STEPHENSON, Lord Justice ACKNER and Lord Justice GRIFFITHS) M

MORRIS GORE LTD v WESTBOURNE HILL PROPERTIES LTD

Estates Gazette May 22 1982

(1982) 262 EG 768

Application by tenants to set aside as irregular a judgment for possession, arrears of rent and mesne profits on the ground that it was entered for a larger sum than the landlords were entitled to claim — Appeal from refusal of judge (upholding refusal of master) to set judgment aside — The facts were that the landlords in their claim had

A **failed to give credit for a sum of £550 paid by the tenants, a sum for which the landlords had previously failed to give credit in an earlier action which had been compromised — Tenants failed to give notice to defend present action and judgment had been given for possession and sum claimed — Held that the judgment was not an irregular judgment which should be set aside as such — Tenants could have applied under Order 13, rule 9, to set the judgment aside in so far as it included the sum of £550, thus reducing the sum entered by that amount — However, the procedural problem was solved by the landlords agreeing to an amendment of the present judgment by a reduction of £550 — Appeal allowed to that extent**

B This was an appeal by Westbourne Hill Properties Ltd, tenants of commercial premises at 16-18 Greyhound Road, London W6, and defendants in an action for possession brought by their landlords, Morris Gore Ltd, against a refusal by Patrick Garland QC, sitting as a deputy judge of the Queen's Bench Division, to set aside as irregular a judgment entered in favour of the landlords, the present respondents.

Stephen Martin (instructed by Liebermann, Leigh & Co) appeared on behalf of the appellants; Dennis Matthews (instructed by S Farren & Co) represented the respondents.

Giving the first judgment at the invitation of Stephenson LJ, GRIFFITHS LJ said: By a lease dated March 25 1968 the plaintiffs let C commercial premises at 16-18 Greyhound Road, London W6, to the defendants for a term of 21 years. At all times material to this appeal the rent of those premises was £2,200 a year, payable by equal quarterly instalments of £550 on the usual quarter days. On September 28 1979 the plaintiffs commenced an action against the defendants in which they claimed possession of the premises on the ground of forfeiture for non-payment of rent and rent due for the whole of the years 1977 and 1978 and the first two quarters of 1979, a total of £5,500, against which they gave credit for sums of £1,200, received on November 2 1977, and £1,000, received on January 11 1978, paid in respect of arrears of rent, reducing the sum claimed to £3,300. They also claimed mesne profits.

D The plaintiffs obtained a judgment under Order 14, but on appeal to the judge that judgment was, by agreement, set aside upon condition that the defendants paid the sum claimed, namely £3,300, into court. Eventually, on December 18 1980 that action came to an end by the plaintiffs' discontinuing the action upon terms that the sum of £3,300 should be paid out to the plaintiffs but that the plaintiffs should pay the defendants their taxed costs. The defendants have now discovered, and the plaintiffs do not dispute it, that a further payment of £550 on account of arrears of rent had been paid to the plaintiffs on April 24 1978. If the defendants had been aware of this payment, and if the plaintiffs had not overlooked it, no doubt credit would have been given for it in that action.

E On February 6 1981 the plaintiffs again sued the defendants for possession of these premises, arrears of rent and mesne profits. In this action they claimed £3,300 in respect of rent, which was for the last two quarters of 1979 and the whole of 1980, and a further sum of £183.94 in respect of insurance premiums. The plaintiffs gave credit for a sum of £1,024.84 in respect of the defendants' taxed costs which they had been ordered to pay in the previous action. The plaintiffs did not, however, give credit for the £550 that they had received in respect of arrears of rent on April 24 1978. The defendants failed to give any notice of intention to defend and so, on March 10 1981, the plaintiffs entered judgment for possession of the premises and the sum claimed, namely, £2,459.08. The defendants applied to have the F judgment set aside on the ground that it was an irregular judgment because it was entered for a larger sum than the plaintiffs are entitled to claim in this action.

Both the master and the judge refused to set aside the judgment and the defendants now appeal to this court, asserting that the judgment should be set aside as irregular on the ground that it was entered for a larger sum than was actually due to the plaintiffs at the time judgment was entered.

If I may say so, Mr Martin has presented this appeal on behalf of the defendants in a most attractive and skilful manner. He submits that the plaintiffs were compelled, as a matter of law, to give credit in this action for the sum of £550 which had inadvertently been left out of account in the compromise of the earlier action. He has cited three authorities of respectable antiquity, starting in the 17th century with G *William Clun's Case* (1613) 10 Co Rep 127a; and two 19th-century cases, *Nash v Gray* (1861) 2 F&F 391, and *De Nicholls v Saunders* (1870) LR 5 CP 589. These authorities establish that where a tenant has paid a sum of money in respect of rent to his landlord before the rent is due, the money is paid subject to the implied condition that the landlord will apply it to the discharge of the rent when it falls due.

I need read only one very short passage from the judgment of Wills J in *De Nicholls v Saunders*. He says, at p 594,

> . . . payment of rent before it is due is not a fulfilment of the obligation imposed by the covenant to pay rent, but is, in fact, an advance to the landlord, with an agreement that on the day when the rent becomes due, such advance shall be treated as a fulfilment of the obligation to pay rent. H

That principle cannot avail the defendant in this case, because the £550 which was paid in April 1978 was not paid as an advance payment in respect of any rent claimed in this action; it was paid in respect of rent arrears that had accrued before April 1978. Mr Martin has therefore argued that the payment of the £550 was subject to a further implied condition that if the landlord did not apply the payment to arrears of rent, he would apply it to any future rent that might fall due. No authority is cited to support this submission; it is argued that it would be inequitable not so to hold.

I cannot accept that the payment of money due in respect of arrears of rent is subject to any such implied condition. Of course, J equity will protect a defendant in such a situation by permitting him to set up in his defence an equitable set-off in respect of the £550 if he is minded so to do. But in my view the plaintiffs were entitled to claim the arrears of rent which they seek in this action, and they were under no obligation in law to give credit in that claim for the sum in respect of arrears of rent which they received and which was paid in respect of an earlier period.

Accordingly, I have come to the clear view that this was not an irregular judgment and it is not a ground for setting the judgment aside.

However, in the course of argument Mr Matthews has very sensibly conceded that at some stage the plaintiffs would clearly have K to give credit for this sum of £550 which they had received and which they had overlooked when bringing their earlier action; so, with the consent of Mr Matthews on behalf of the plaintiffs, it appears that the sensible course to take is to amend the judgment that has been entered, and to reduce it by the sum of £550. If this is done it appears to me that there are no grounds upon which the defendants can complain that they have been inequitably treated.

There were certain other matters advanced in an affidavit sworn by Mr Randall on behalf of the defendants, but they were not urged in argument before this court and I deal with them by saying that they show no ground upon which it would be possible to set aside this regular judgment, obtained because the defendants did not give L notice of intention to defend.

To this limited extent, and with the consent of the plaintiffs, I would allow this appeal.

ACKNER LJ said: I agree, for the reasons given by my Lord, that this was a regular judgment. The defendants' proper course would have been to have applied under Order 13, r9 to set aside that judgment in so far as it included the claim for £550, thus reducing the judgment to some £1,900. In regard to that sum there was no defence either on the merits, nor was there any defence to the claim for forfeiture.

Having regard to the course that Mr Matthews has very sensibly agreed to adopt, I agree with the ultimate course that has been M decided to be adopted.

Also agreeing, STEPHENSON LJ said: If this judgment were irregular, Mr Martin's attractive argument would derive some support from some of the authorities cited in the notes in the Supreme Court Practice to Order 2, r2 and Order 13, r9. But, for the reasons given by my Lords, I cannot regard this judgment as being irregular, and I agree that the appeal should be allowed to the extent indicated.

The appeal was allowed to the extent indicated in the judgments, the respondents to have the costs of the appeal except costs of, and occasioned by, the respondents' notice.

A **QUEEN'S BENCH DIVISION**

November 9 1981

(Before Mr Justice CANTLEY)

EUSTON CENTRE PROPERTIES LTD v H & J WILSON LTD

Estates Gazette June 12 1982

(1981) 262 EG 1079

Lease or licence—Whether certain improvements should be regarded or disregarded in applying a rent revision clause in a
B **lease—Question depended on whether works had been carried out while occupation was under a licence or under an agreement for a lease of which specific performance would be granted—Works had in fact been carried out during the subsistence of an agreement, described as a "licence", which provided for a lease (in the form of a draft annexed to the agreement) to be granted—The grant was, however, subject to the fulfilment of certain conditions relating to works—Held on the evidence that occupation before the grant of the lease was under a licence only—Agreement was not such as to create an equitable lease on the principle of *Walsh v Lonsdale*—Hence improvements carried out during the occupation before the lease was granted were not to be disregarded as provided for in the lease but**
C **were to be taken into account in the evaluation of rent—Result was that the applicant property company "will reap where it has not sown", but such was apparently the intention of the parties to the agreement**

This was an application by Euston Centre Properties Ltd, the landlords, for determination of a preliminary question of law arising in an arbitration under a rent review clause in a lease of premises at 173-175 Drummond Street and 31 Hampstead Road, London NW1. The tenants, respondents to the application, were H & J Wilson Ltd.

E A Davidson (instructed by Titmuss, Sainer & Webb) appeared on behalf of the applicants; C J Lockhart-Mummery (instructed by
D Thornton Lynne & Lawson) represented the respondents.

Giving judgment, CANTLEY J said: Under the terms of a lease made on January 20 1975 the applicant demised to Tootal Ltd the business premises described in the lease for a term of 21 years from December 25 1973 at a yearly rent of £7,500. Tootal Ltd subsequently assigned the residue of this term to the respondent.

Clause 1 of the lease provides as follows for periodic revision of the rent:

> if the Lessor shall not more than twelve months nor less than six months before the expiration of the 6th 11th and 16th year of the said term give notice to the Lessee requiring the rent to be revised then and as from the expiration of the 6th 11th and 16th year the said yearly rent shall be either the rent hereinbefore reserved or such amount as shall be the market rent (but
> E disregarding any of the matters referred to in section 34(a) (b) and (c) of the Landlord and Tenant Act 1954) having regard to the rental values then current for similar property let with vacant possession without a premium and subject to the provisions of this Lease whichever shall be the greater rent.

The amount of revised rent has been referred for decision by arbitration as provided for in the lease. This motion is for determination of a preliminary question of law arising in the arbitration as to the application of the provision in clause 1 of the lease which I have just quoted.

Section 34 of the Landlord and Tenant Act 1954, as amended in 1969, deals with the determination by the court of the rent payable under a tenancy granted by order of the court under Part II of the
F Act. This section reads as follows, omitting words which are not relevant for the present purposes. Subsection (1):

> The rent payable . . . shall be such as . . . may be determined . . . to be that at which . . . the holding might reasonably be expected to be let in the open market by a willing lessor, there being disregarded . . . (c) any effect on rent of an improvement to which this paragraph applies . . . (2) Paragraph (c) of the foregoing subsection applies to any improvement carried out by a person who at the time it was carried out was the tenant, but only if it was carried out otherwise than in pursuance of an obligation to his immediate landlord . . .

Before the grant of the lease of January 20 1975 Tootal Ltd carried out improvements at their own expense and otherwise than in pursuance of an obligation to the landlord. It did so under the terms of an agreement in writing made between Tootal Ltd and the G
applicant on April 16 1974. The question I have to decide is whether at the time these improvements were carried out Tootal Ltd was the tenant of the applicant or only a licensee. If Tootal Ltd was a tenant, it is accepted that the improvements must be disregarded in ascertaining the market rent. If Tootal Ltd was only a licensee during that period, the improvements must be taken into account. The question is one of interpretation of the substance and effect of the agreement of April 16 1974, and it is necessary for me to refer in full to some of its provisions.

Throughout the agreement the applicant is referred to as the landlord and Tootal Ltd is referred to as the tenant. The agreement begins with this recital: H

> (2) It has been agreed that the Landlord will grant and the Tenant will take a Lease (hereinafter called "the Lease") of the portion comprising Shop Units Nos 64 and 65 on the Ground Floor and Units Nos 64, 65 and 66 on the First Floor of the Building which portion is shown coloured pink on the plan annexed hereto (hereinafter called 'the Demised Premises').

Clause 1 provides that it is hereby agreed that

> FORTHWITH after the date hereof the Landlord will in a proper and workmanlike manner erect or cause to be erected a party wall at First Floor level dividing the Demised Premises from the adjoining Unit No 67 and to carry out installing an anti-syphon vent pipe and upon completion thereof will cause its Architects Messrs Sidney Kaye Firmin and Partners . . . ("the Architects") to issue a Certificate ("the Architects' Certificate") to that effect. J
>
> 2. If the Tenant desires to carry out works to the Demised Premises then the Tenant shall at its own cost prepare and submit to the Landlord and its Architects for their approval four copies of a plan or plans and a specification of work which approval shall not be unreasonably withheld to be carried out by the Tenant in connection with any installation of a suitable shop front fascia and with fixtures and fittings in keeping with adjoining or neighbouring premises and the locality and for decorating and generally for carrying out such other works as are requisite to complete the Demised Premises so as to render them fit for immediate occupation and use by the Tenant . . . and forthwith upon such approval being obtained the Tenant shall apply for and use its best endeavours without delay to obtain all other consents which may be necessary for the carrying out of the Tenant's Works. K
>
> 3.(a) The Landlord will notify the Tenant in writing of the issue of the Architects' Certificate and thereupon the Tenant shall be given access to the Demised Premises and subject to the Tenant having procured all the aforesaid approvals and consents for the Tenant's Works the Tenant shall be at liberty to and shall proceed to carry out and within three calendar months thereafter complete the Tenant's Works in a good and workmanlike manner to the reasonable satisfaction of the Architects

I omit certain words

> Provided that if the Tenant shall be unable to complete the Tenant's Works within the aforesaid period of three months for any reason beyond the control of the Tenant then such period shall be extended for such period as shall be determined by the Lessors Architects . . .
>
> 4. WHEN the Landlord's Works and any works the Tenant desires to carry out and which have been approved shall have been completed in the manner L
> aforesaid the Landlord will procure the Architects to issue a Certificate to that effect and thereupon the Landlord shall serve or cause to be served on the Tenant notice in writing to that effect and within fourteen days of the service of such notice the Landlord will grant and the Tenant shall take the Lease of the Demised Premises for a term of TWENTY-ONE YEARS computed from the 25th day of December 1973 at the exclusive yearly rent of SEVEN THOUSAND FIVE HUNDRED POUNDS (£7,500.00) (such rent being subject to review at the expiration of the 6th, 11th and 16th years in accordance with the provisions on that behalf contained in the Lease) and the Lease shall be in the form and contain and reserve the several rents reservations exceptions covenants conditions and provisions as are contained in the form of Lease annexed hereto.
>
> 5.(i) The Tenant shall pay:— (a) on and from the 31st day of May 1974 M
> until the grant of the Lease a Licence rent of such daily amount and payable daily as shall be equivalent to the yearly rent to be reserved by the Lease but if the Lease shall have been granted prior to such date the Lease shall provide for the rent thereunder to commence on the before mentioned date (b) on and from the date upon which the Tenant shall be given access pursuant to Clause 3(a) hereof the insurance premiums and service charge which would be payable by the Tenant under the provisions of the Lease as if the Lease had been granted.
>
> (ii) The Tenant shall be deemed to occupy the Demised Premises subject to the same exceptions reservations covenants and conditions and to the other provisions contained in the form of Lease hereinbefore referred to so far as they are applicable and are not inconsistent herewith and so that the Landlord

A shall have and be entitled to all remedies by distress action or otherwise for recovering rent in arrear and for any breach of any of the covenants or agreements on the part of the Tenant as if the Lease had been actually granted but nothing in this clause contained shall vary or affect the application of the next succeeding clause hereof.

6. UNTIL the grant of the Lease this Agreement shall not operate or be deemed to operate as a demise of the Demised Premises nor shall the Tenant have or be entitled to any estate right title or interest in the Demised Premises or any part thereof whether erected or in course of erection or in any materials (other than materials deposited thereon by the Tenant) on the land or any part thereof . . .

9. The Tenant shall not assign mortgage charge underlet or part with the benefit of this Agreement and the Tenant (meaning the party hereto) will B accept take up and execute the Counterpart of the Lease and the Demised Premises will initially be occupied by the Tenant (meaning the party hereto) or Van Allan Ltd, and The Stretton Manufacturing Company Ltd (subsidiary Companies) and used only for the purposes authorised by the relevant Clause in the Lease . . .

13.(b) The Landlord shall have full right and liberty to stop up or alter in any way whatsoever any roads paths passages staircases drains conduits gutters pipes wires or cables whether upon the Demised Premises or upon any neighbouring property where such stopping up or alterations would not (in the opinion of the Architects) affect the Demised Premises or the Tenant's full and free use and occupation thereof.

This agreement was, among other things, an agreement for the C grant of a lease in the form of a draft lease annexed to the agreement, but it contemplated the happening of two events before the grant of the lease was to be made: (1) the execution of the works which the landlord was obliged to carry out and the issue of the certificate he was obliged to procure under clause 1; (2) the due and proper completion of works which the tenant was at liberty to carry out, subject to reasonable approval of plans and specifications for the works by the landlord and its architects and the obtaining by the tenant of any necessary consents. The tenant was not under contractual duty to carry out any works to the premises if for some eccentric reason it did not wish to do any, but once it had been notified by the landlord of the issue of the certificate under clause 1 D and it had procured the necessary approvals and consents for any works it chose to do, it was obliged to carry out those works and complete them as provided by clause 3(a) before it could insist on the grant of the lease under clause 4.

Mr Lockhart-Mummery's primary contention is that this agreement of April 16 1974 was a specifically enforceable agreement for a lease from the outset or at least became such on May 20 1974 when the landlord's certificate was issued under clause 1. I doubt very much whether it was specifically enforceable at the outset. I think there would be too many uncertainties at that stage. However, in the highly improbably event of the tenant electing to do no works and thereby choosing to do nothing to make the premises useful to the tenant for the purposes contemplated and permitted by the lease, E it may be that the agreement would be specifically enforceable once the landlord's certificate under clause 1 was issued.

Mr Lockhart-Mummery says that is enough for his purposes. He says the respondent from then on had an equitable tenancy which could only be ended by forfeiture, surrender or merger in the legal estate. I think that is a theoretical equitable tenancy which ignores the facts of this case. This case is about the improvements which the tenant in fact carried out under clause 2 and clause 3(a) before there was a grant of the lease. If at any time after it had obtained the necessary consents and approvals and before it had completed the works in accordance with the agreement Tootal Ltd had sought a decree of specific performance by the grant of a lease, its claim would F have been defeated.

The principle of *Walsh v Lonsdale* (1882) 21 ChD 9 does not create an equitable tenancy unless the agreement for the lease is specifically enforceable. The principle cannot be invoked in cases where an agreement for a lease is subject to a condition precedent, which remains unperformed by the proposed tenant and has not been waived by the landlord. (See *Cornish v Brook Green Laundry Ltd* [1959] 1 QB 394.)

Even if there had been a time when theoretically Tootal Ltd could have obtained specific performance on the basis of electing to do no works, that time had passed when the time I am concerned with began. The agreement for a lease was then under its own terms subject to a condition precedent to be performed by the proposed G tenant. In my view, Mr Lockhart-Mummery's contention that while doing the works of improvement Tootal Ltd was a tenant in equity under a specifically enforceable agreement for the grant of the lease in the form annexed to the agreement is not well-founded.

The alternative contention of the respondent is that on the true construction of the agreement of April 16 Tootal Ltd were tenants of the applicant from the time when they occupied the premises and began to pay the rent which is described as a licence rent in clause 5(i)(a) and became liable to pay insurance premiums and service charge under clause 5(i)(b).

Mr Lockhart-Mummery has pointed out various features in this agreement which he forcefully contends are more consistent with a H lease than a licence.

(1) The parties are throughout called "the landlord" and "the tenant" and the premises are called "the demised premises".

(2) What is called a licence rent is to be paid at a rate which is equivalent to the yearly rent to be reserved by the lease. Moreover, by clause 5(i)(b) the tenant is obliged also to make the payments for insurance premiums and the service charge which would be payable unde the provisions of the lease as if the lease had been granted. By clause 1 of the draft lease referred to the sums in respect of insurance premiums are to be paid quarterly by way of further rent, and by clause 2(iii) of the draft lease the sums in respect of service charge are to be paid quarterly and are recoverable as rent in arrears. By clause 3 J of the draft lease the landlord has a right of forfeiture.

(3) By clause 5 of the agreement the landlord is given the remedies of a landlord, including the remedy of distress, as if the lease had been actually granted, although it is at the same time declared that nothing in this clause is to vary or affect the application of clause 6. A right of distress can be created by agreement, although such a right created by agreement will not include a right to take the goods of third parties as a landlord's ordinary right of distress would. (See *Ex p Kennedy, re Willis* (1888) 21 QBD 384 in the judgment of Lindley LJ at p 395.)

(4) By clause 5(ii) of the agreement the tenant is deemed to occupy the premises subject to the same covenants and conditions and the K other provisions contained in the lease as are not inconsistent with the agreement. There is a right of forfeiture in clause 3(i) of the lease.

(5) By clause 4(i) of the lease there is a covenant for quiet enjoyment, although, as I read that clause, that covenant is expressly related to the lessee paying the rent and performing the covenants contained in the actual lease.

(6) By clause 13(b) of the agreement the tenant is to have "full and free use and occupation" of these premises, but the landlord reserves the right to do some stopping up and alteration on the premises if such stopping up or alteration would not affect the tenant's full and free use and occupation. This clause is consistent with the tenant having a right to exclusive possession, because if the tenant was not L given exclusive possession it would be unnecessary for the landlord to stipulate for a right to enter and do work and alteration upon the premises, and particularly work and alteration which would not affect the use and occupation to which the tenant was entitled. Work which would actually interfere with the use and occupation to which the tenant was entitled under the agreement or with the works he was entitled to do under the agreement would of course be a breach of the contract, even if the so-called tenant was a licensee. However, although the fact that an agreement gives exclusive possession is recognised as a consideration of great importance in deciding the nature of the occupation, it is not of itself decisive against a transaction being construed as a licence. The transaction must be M looked at as a whole.

It seems to me that in this agreement the landlord is at least striving very hard to get as many of the rights of a landlord as he can have without, if he can help it, overstepping the line between tenancy and licence. For this purpose clause 3(a) simply says that the tenant shall be given access to the premises and liberty to carry out and complete the approved works on the premises, and the rent which the tenant has to pay is called "a licence rent" in clause 5(ii). Clause 5(ii) is subject to clause 6, and clause 6 says firmly and clearly that the agreement shall not operate or be deemed to operate as a demise and the tenant shall not be entitled to any estate, title or interest in the

A premises. Such provisions as this are not conclusive. As Buckley LJ said in *Shell-Mex & BP Ltd v Manchester Garages Ltd* [1971] 1 WLR 612 at p 618, in determining whether a transaction is a tenancy or a licence "the court is not to have regard to the label which the parties give to the document or to the formal language of the document, but to the substance of the transaction". An express provision that nothing in the agreement shall be construed to create a tenancy will not avail if the true effect of the agreement as a whole is to create a tenancy. (See, for example, *Facchini v Bryson* [1952] 1 TLR 1386.)

I have to determine the substance of this transaction and consider the effect of the agreement as a whole. It has not been made easy by the profusion of words which the draftsman has employed. The B underlying pattern of the agreement appears to me to be as follows: (1) The only purpose of allowing Tootal Ltd to be on the premises before the contemplated lease is granted is to enable that company to execute on the premises the works referred to in clause 2, which are needed in order to make the premises fit for immediate use by Tootal Ltd or their named subsidiary companies when the lease is granted. (2) Despite the unhappy introduction of the word "underlet" in clause 9, the rights under the agreement given to the so-called tenant are personal to Tootal Ltd and cannot be assigned. (3) If Tootal Ltd within three months of notification of the issue of the architects' certificate referred to in clause 3 or within any extended period determined by the architects carry out and complete the approved C works to the reasonable satisfaction of the architects, Tootal Ltd becomes entitled to the grant of the lease. Conversely, if Tootal Ltd does not comply with these obligations in relation to the approved works, it does not become entitled to the lease. (4) Until the lease is granted Tootal Ltd is at least intended to acquire no estate in the premises. It is intended merely to have contractual rights which cannot be revoked without breach of contract so long as Tootal Ltd performs its part.

In the end it seems to me that the above is the substance of the transaction and the effect of this complicated agreement read as a whole. Accordingly, I hold that the revised rent payable under clause 1 of the lease is to be ascertained with regard to the improvements to D the property, other than tenant's and trade fixtures and fittings, carried out by Tootal Ltd before January 20 1975 under the agreement of April 16 1974. The result is that the applicant will reap where it has not sown, but that seems to me to have been the intention of this agreement when it was made by the parties.

COURT OF APPEAL

February 22 1982

E (Before Lord Justice ORMROD, Lord Justice ACKNER, and Mr Justice WOOD)

WIRRAL BOROUGH COUNCIL v SMITH AND ANOTHER

Estates Gazette June 26 1982

(1982) 262 EG 1298

Appeal by trespassers against order for possession of house—Property owned by local authority and let to a person who became ill, was admitted to hospital and later died intestate — Appellants, defendants in the action for possession, were a couple who had moved into the house after the tenant was admitted to F hospital, but had no claim of right—Local authority landlords did not serve a notice to quit on the President of the Family Division under Practice Direction (1965) 1 WLR 1237—Appellants claimed that, as the respondent council had not served a notice on the President to terminate the tenancy, the council did not have the right to immediate possession necessary to support an action for trespass—Held that council should have served a notice to quit on the President, in whom the tenancy vested on the death intestate of the former tenant until a grant of administration was made—Despite the fact that the appellants were admittedly trespassers, their appeal against the county court judge's possession order had in the circumstances to be allowed—Only a plaintiff with an immediate right to possession can sue in trespass

G This was an appeal by Barry James Smith and Laura Ann Cooper against a possession order made by Judge Collinson at Birkenhead County Court in favour of the respondents, Wirral Borough Council. The order was in respect of premises at 7 Brenig Street, Birkenhead, which were occupied by the appellants.

N A Wright (instructed by A F Russell & Co, of Birkenhead) appeared on behalf of the appellants; M Hedley (instructed by the borough solicitor) represented the respondents.

Giving judgment, ORMROD LJ said: This is an appeal by two defendants, Barry James Smith and Laura Ann Cooper, against an order for possession which was made against them by His Honour Judge Collinson at Birkenhead County Court on September 10 1981 in respect of property known as 7 Brenig Street. H

The short facts, which are agreed, are as follows. The respondents to this appeal, Wirral Borough Council, are the freeholders of this property. In July 1973 it was let on an ordinary tenancy to a Mrs Horne. She was admitted to hospital in September 1980 and, in fact, never returned to live at the premises. The first defendant, Mr Smith, moved into the property soon after Mrs Horne was admitted to hospital, ostensibly to look after it, and on December 26 1980 Mrs Horne died intestate. Some time later Miss Cooper moved in to live with Mr Smith.

On April 13 1981 the Wirral Borough Council started proceedings for possession under the summary procedure in the Birkenhead County Court. The application is: J

> Peter James Mills, Director of Administration and Legal Services, Town Hall, Wallasey for and on behalf of the applicant apply to the court for an order for recovery of possession of the premises, 7 Brenig Street, Birkenhead, Merseyside, on the grounds that the applicant is entitled to possession and that the persons in occupation of the premises are in occupation without licence or consent.

At the hearing before the learned judge an attempt was made by Mr Smith and Miss Cooper to make out that they had a right to be in the premises under the provisions of the Housing Act 1980, but that attempt entirely failed. K

The judge rejected their evidence and came to the conclusion that they were both trespassers on the property. That is now accepted by the defendants and their case in this court is that, on the facts of the case, Wirral Borough Council did not have what they must have to maintain an action for trespass, namely a right to immediate possession. The reason for this is highly technical and antiquated.

According to the law, on the death of a tenant intestate during the tenancy, the tenancy vests in the President of the Family Division and does not come to an end in law on the death of the tenant. The difficulty that Wirral Borough Council find themselves in is that they have never terminated the tenancy by notice to quit, L which of course they could do very easily by serving notice on the President, or on the appropriate person as set out in the practice direction, in which case they would automatically bring the tenancy to an end. Of course it would cease to be protected. The moment the notice to quit took effect the borough council would be entitled to immediate possession and in a position to sue the trespassers for possession and eject them. As I say, unfortunately that step was not taken.

The learned judge in the court below tried to get round this pure technicality by holding that, in spite of the fact that there was a subsisting tenancy, the owner had a better title to the premises than the trespassers. One can understand the common sense of that, but regrettably there is no authority in law to support the proposition. M

Mr Hedley, who has done his best in an impossible situation, has referred us to two cases of what he calls "competitive titles". With respect, I do not think that they have any application to the present case. We are dealing here with a fundamental piece of common law in an action for trespass which can only be maintained by someone who has a right to immediate possession. Technically the council have not got that right. It is very tiresome that they have to go through the pure formality of serving a notice to quit, but I regret that that is the state of the law. I would, therefore, allow the appeal.

Agreeing, ACKNER LJ said: It was of the very essence of the

A plaintiffs' claim that they were entitled to possession. That means, as I understand it, immediate possession. They had, in fact, disposed of their right to immediate possession when, in 1973, they granted a weekly tenancy to Mrs Horne. That tenancy did not determine automatically on her death. It would normally vest in her personal representatives, but as she died intestate it vested in the President of the Family Division. There is a simple procedure which could have been resorted to, to determine that interest, in B which case there then would have been an immediate right to possession and no possible answer to this claim. Unfortunately that course was not taken. The point taken by the appellants is a valid one and I, therefore, agree that this appeal must be allowed.

WOOD J agreed.

The appeal was allowed with costs.

LANDLORD AND TENANT
BUSINESS TENANCIES

A **CHANCERY DIVISION**

April 6 1981

(Before Mr Vivian PRICE QC, sitting as a deputy judge of the division)

U D S TAILORING LTD v B L HOLDINGS LTD

Estates Gazette January 9 1982
(1981) 261 EG 49

Landlord and Tenant Act 1954, Part II — Grant of new tenancy not opposed, but questions as to rent to be paid for agreed term of 15
B **years and as to interim rent — Shop premises occupied for business of bespoke and ready-to-wear tailors and outfitters — Zoning method of valuation adopted — Comparables considered — Correct figure to be arrived at not as a matter of arithmetical calculation but of judgment — Effect of restricted user provisions in lease reflected in a 10 per cent reduction in rental figure — Interim rent assessed by a deduction of 10 per cent to reflect yearly tenancy basis under section 24A(3) of the Act of 1954 — Interesting review of expert evidence by judge**

This was an originating summons taken out by the plaintiff tenants, U D S Tailoring Ltd. The defendants were the plaintiffs' present landlords, B L Holdings Ltd. The proceedings, under the
C Landlord and Tenant Act 1954, Part II, concerned shop premises at 204 and 206 Edgware Road, St Marylebone, London W2. The matters to be determined by the court were only the amount of the rent to be paid for the tenancy, the grant of which was not opposed, for an agreed term of 15 years, and the amount of the interim rent in accordance with section 24A of the Act of 1954.

T H Moseley (instructed by Conway Kramer, of Leeds) appeared for the plaintiff tenants; Paul de la Piquerie (instructed by Burton & Ramsden) represented the defendant landlords.

Giving judgment, MR VIVIAN PRICE QC said: These proceedings are concerned with shop premises at 204 and 206 Edgware Road, St Marylebone, London W2. The basic facts between the parties are
D agreed as follows (and I quote from paragraphs 2(i) – (v) of the originating summons taken out on behalf of the plaintiff tenant, U D S Tailoring Ltd, now known as John Collier Menswear Ltd). In that summons are stated the required particulars:

(i) The premises are known as 204-206 Edgware Road, St Marylebone, London W2.
(ii) The said premises were demised by one Maxwell & Ponting Ltd to the plaintiff, then called Price's Tailors Ltd, by a lease dated March 24 1966 for a term of 14 years from May 25 1966, the reversion is now vested in the defendant.
(iii) The annual rent reserved by the said lease is £5,250, the rateable value
E of the said premises £6,638.
(iv) The plaintiff occupies the whole of the said premises for the purposes of its business of men's and women's bespoke and ready-to-wear tailors and outfitters.
(v) By a notice to terminate dated September 10 1979 the defendants' predecessors in title, Moorage Property & Developments Ltd and Metropolitan Properties Co Ltd, being the plaintiff's landlords at that date and who hereinafter are referred to as the then landlords, gave the plaintiff notice under the provisions of section 25 of the Landlord and Tenant Act 1954, determining its tenancy of the said premises on March 25 1980, stating in the said notice the landlords would not oppose an application to the court for the grant of a new tenancy.

On October 3 1979 the plaintiff gave notice to the then landlords F
that the plaintiff would not be willing to give up possession of the said premises on the date of termination specified in the said notice of September 10 1979. The grant of a new tenancy is not opposed, save for the rent to be paid for the period of the tenancy, which has been agreed at 15 years, and save for the interim rent. A draft agreed lease has been produced before me; I do not need to refer to its terms save in one respect, namely the user restrictions, which I shall consider later.

I have been much assisted in approaching the question of deciding the open market rental value of the premises by an expert report prepared on behalf of the plaintiff tenant by Mr J A L Buckingham
[ARICS, a partner of Jones Lang Wootton] and an expert report G
prepared on behalf of the landlord — B L Holdings Ltd — by Mr Derek James Abbott [FRICS, partner of Conrad Ritblat & Co]. Both these expert valuers also gave oral evidence before me as to the assistance to be gained in deciding the open market rental value by considering the turnover of the business and, alternatively, by considering the overall average rental per square foot rather than adopting the more conventional zoning system of valuation. In the present proceedings, I am satisfied that no real assistance, on the evidence adduced before me, is to be gained from considering either of these alternatives, but I must emphasise that I have reached that conclusion on the evidence adduced before me and I
am not approaching it in any way as a matter of principle. H

However, in the present proceedings, I think that it is correct that I should proceed on the conventional zoning basis and I shall do so accordingly.

The first matter that I have to decide is which premises are to be considered as comparable. The premises, the subject of these proceedings, are in a block (known, I think, as The Parade) and it is agreed between the parties that four comparables to be considered are 222, 208, 226 and 216 Edgware Road — all in the same block, The Parade. In addition, however, the tenant considers that three additional premises should be considered, all south of the Old Marylebone Road, 198 Edgware Road and 135 and 141 Edgware
Road, the two latter being part of what is known as "The Water J
Gardens Development". I do not think that any assistance is to be derived from 198 Edgware Road since, although it is very close to the corner premises with which I am concerned — effectively being across the Old Marylebone Road — I did not find the evidence relating to it at all helpful.

Mr David Smythe, an estate management assistant working for the Legal & General Assurance Society, who gave evidence before me, was unable to give any very useful assistance on the way in which the new rent for 198 Edgware Road was computed, other than producing a document which set out the calculations (some of which were in his writing) and other calculations including, appar-
ently, the final calculations which resulted in an agreed rent of K
£11,000. Mr Smythe, however, as he pointed out, was not a valuer in any way; he was a property management assistant and only gave evidence relating to what he was told to do. As to the reasons why he was told to do that, as to the reasons why he was given no information other than the lease renewal relating to the adjoining shop, I do not know. He was unable to assist me and there was no further evidence relating to that matter. I therefore find the evidence relating to 198 Edgware Road of no assistance and I ignore it.

In my view, however, the evidence relating to 135 and 141

Edgware Road is of different quality. It is true that these premises are not in the same block as the premises I am considering. They are also in a block, south of Sussex Gardens and of the Old Marylebone Road and there was some evidence before me that they were thereby placed in a different shopping category. They are in my view, nevertheless, sufficiently proximate to the premises I am considering that the evidence relating to them should not be left out of account. The comparable premises that I shall consider are as follows: first of all, 222 Edgware Road, with a ground-floor area of 955 sq ft and I should mention that the ground-floor area of 204-206 Edgware Road is (if my arithmetic is correct) 1,531 sq ft. I should interpolate at this point that any arithmetic that I put forward in this judgment will, I hope, be checked very carefully before I am asked to make any order whatsoever pursuant to this judgment. But, as I have said, according to my arithmetic, the ground-floor area of the subject premises is 1,531 sq ft — considerably larger, I may say, than any of the comparable premises. The ground-floor area of 222 Edgware Road (the effective date of renewal being December 12 1978) of 955 sq ft, and the zone "A" rents, were calculated by Mr Buckingham at £15 per sq ft, whereas Mr Abbott calculated the zone "A" figure at, I think, £14.68. The second comparable is 208 Edgware Road (the effective date again being December 12 1978) and the ground-floor area is 736 sq ft, the overall rent £9,869. The zone "A" figure calculated by Mr Abbott at £14.59 and that calculated by Mr Buckingham at £14 per sq ft. The third, not in point of time, but in the same block, is no 226, the effective date of which is June 1979, the ground-floor area is 943 sq ft, total rent £14,500. The rent calculated by Mr Buckingham is £16.86 and that calculated by Mr Abbott £17.65. And then, finally, in the same block, no 216 where effectively the only evidence is that of Mr Abbott, that is, relating to the calculation of zone "A", because the information that was provided to Mr Buckingham in respect of these premises was, as he agreed, inaccurate. He was taken in evidence through Mr Abbott's calculations which he accepted, the effect of which is this: that for no 216 the effective date for the renewal was September 1980, ground-floor area 519 sq ft, the rent was £8,350 and the zone "A" figure was £19.10.

Those are the four comparables in the same block and I think it is correct to say that in approaching the assessment of the zone "A" figure for the premises 204-206 Edgware Road, I have taken these four premises within the same block as being more directly comparable with the subject premises than the two others that should also, in my view, be considered — that is nos 135 and 141, also in Edgware Road but in the Water Gardens development. I should also say, in respect of those two, that evidence was given before me by Mr Malcolm Donald Dalgleish, a chartered surveyor [senior partner of Dalgleish & Co], who was concerned on behalf of the tenant in the negotiations for the renewal of the leases in respect of these properties and produced before me certain documents relating thereto which basically set out the measurements of the two properties and the rent that was finally agreed in respect of them. It is true to say, however, in respect of Mr Dalgleish's evidence that, as he very fairly said when asked questions about the way in which values were put on such matters as the basements and also in respect of the disagreements that occurred during the negotiations, he was unable to remember the details of the transactions with which he was concerned. He relied upon the documents that, indeed, had been placed before me, setting out the bare facts of the measurements and the zone "A" valuations. I find that I really can rely upon those documents only, indeed, upon his evidence for no more than that — I say that without dismissing Mr Dalgleish's evidence — it seems to me that he was being as helpful as he could within the limits of his memory, but was not prepared to go outside. So I rely upon the documents that are produced by him (or, at least, to which he refers), but I also take into account Mr Abbott's expert view in relation to both these premises — that is both 135 and 141 Edgware Road — that the rents agreed for them were, in his view, on the low side. 135 Edgware Road — the effective date of the renewal was March 1979, the ground-floor area was 745 sq ft, the rent £8,700 and the zone "A" value, calculated by Mr Buckingham, was £13.75; the zone "A" calculation by Mr Abbott was £15.43, and Mr Dalgleish's figure was £15. In respect of no 141, the effective date was June 1979, the ground-floor area was 1,241 sq ft, the effective rent was £10,700, the zone "A" figure calculated by Mr Buckingham was £14, and the zone "A" figure calculated by Mr Abbott was £15.39. Mr Dalgleish's figure again was £15 for zone "A".

Very considerable discussion took place between the two experts before me as to the way in which these zone "A" figures should be approached. In his report, Mr Buckingham, who obviously was taking into account 198 Edgware Road (which, as I have said, I rejected) which he calculated as a £15 per sq ft zone "A" figure, took into account nos 135 and 141, which were lower than £15 even though Mr Dalgleish's figures were £15; he took into account the £15 zone "A" figure for no 222 and the £14 figure for no 208 and he came to the conclusion that the fair rent was of the order of £15. I think that later, and certainly in his closing speech, Mr Moseley accepted that it should be a small amount higher than £15. Mr Abbott, for the landlord, approached it in a different way; he said "Well, now, look at these figures. The most relevant ones are the ones in the block — that is in The Parade". He points out that he was concerned with negotiations that led to all four of those figures and those agreements but, in response to cross-examination by Mr Moseley, he pointed out that it was not just he with the expert valuer that was concerned in all the negotiations; there were expert valuers of high repute acting for the other side in each case, so that each of the figures which which he was concerned were figures that were the results of negotiation between expert valuers on both sides. He therefore pointed out that the most relevant valuations were the ones that were immediately contemporary in time and he points to, as can be seen from the figures I have already referred to, an increasing rate for zone "A". I think Mr Abbott accepted — and certainly counsel appearing for the landlords accepted — that perhaps the figure for no 216 was too high to be related to the block in question, that is £19 per sq ft. I think that the way it was put by counsel on behalf of the landlords is that the figure should be somewhere within the band of the two latest figures — namely £17.65 for June 1979 and £19.10 for September 1980. I take all these matters into account and all the factors that have been urged before me by Mr Buckingham. At the same time, I find that it is impossible to arrive at a satisfactory figure for the open market rental value of 204-206 Edgware Road merely as a matter of calculation. It seems to me that the appropriate way to arrive at the correct figure is to bear in mind all the factors that are adduced before me, bear in mind the figures that have been put before me, to bear in mind the very helpful expert assistance of Mr Abbott and Mr Buckingham but to weigh all those factors together as a matter of judgment, not as a matter of calculation, and upon that basis, on taking all these matters into account, I think that the zone "A" rate that should be applied to 204-206 Edgware Road, in assessing the open market rental value, should be considerably less than the £19.10 rate agreed for no 216 but considerably more than the rates for nos 222 and 208 and I assess the appropriate zone "A" rate at £17 per sq ft.

Now, Mr Buckingham asserted that the zoning method was only appropriate to nos 204-206 if a substantial discount were to be made to reflect the fact that the width and shallow depth and the consequent zone "A" area is disproportionately high. I am not satisified that Mr Buckingham, on the evidence, has made out any such proposition. I go further and say that the evidence that has been adduced before me does not support such a view in any way and I am not, therefore, prepared to make such a reduction. Before I come to the calculations with which, as I have said, I would like to have some corroborative assistance before I come to any question of making an order, it is the case that whatever the ground-floor area figure comes to, it is agreed that 5 per cent should be added (of the ground-floor rent only) to reflect the fact that the shop premises are corner premises. It is also agreed that the value to be attached to the basement should be storage — 979 sq ft @ £2, which comes to £1,958.

There are two other matters, however, to be considered before I come to the calculation: first of all, there are the restricted user provisions in the lease and, secondly, calculations were made by Mr Buckingham of a 10 per cent reduction to reflect the rebuilding clause. It has been agreed that that is not effective here in the present circumstances, so I can ignore that second reduction, but I have to consider what amount, if any, should be reflected in the open market rental value by the user restrictions contained in the lease. Now, as I have said, the terms of the lease — apart from the question of rent — have been agreed before me and I do not need to go through the lease which is, as one might imagine, a long document. The tenants' covenants with the landlord are set out in clause 2, which begins at p 2 of the lease — I do not need to consider those — but if one goes on to p 8, subclause 13 of clause 2 is in the following terms:

> Subject to the tenant obtaining all the necessary consents from statutory and other authorities to use the demised premises as a men's and women's bespoke and ready-to-wear tailors and outfitters, such trade or business to be carried on in a high-class manner provided that if, in the sole opinion of the landlord, it shall deem any application by the tenant to use the demised premises for any other retail trade or business than that hereby authorised, not to be in conflict with any trade or business already or then about to be carried on or authorised to be carried on in or upon the property forming part of the block formed by 204-256 Edgware Road and 332 Old Marylebone Road or otherwise in conflict, except as to the specified trade or business hereby authorised with any of the provisions of this lease, the landlord will not unreasonably refuse its consent to such change of use.

Now, Mr Buckingham took the view that the restrictions thereby imposed upon the tenant and accepted by the tenant, were of such a nature that it would be fair and appropriate to make a deduction of 15 per cent in the rental figure to reflect this user provision. Mr de la Piquerie, on behalf of the landlord, effectively submitted that the user restrictions contained in the clause that I have read were really no greater than those imposed upon the tenants (and, indeed, the rights given to the tenant) under the general law. Nevertheless, Mr Abbott, the landlord's expert witness, accepted that it would be appropriate to make some deduction in respect of the restrictions, whatever they were. The exact nature of the restrictions is, of course, a matter of law. I am bound to say that I am not satisfied that these restrictions are no more than coterminous with those provisions of the general law; I think that the restrictive conditions do impose a burden upon the tenant over and above that imposed by the general law and I therefore think it appropriate to make a substantial reduction. I think, however, that 15 per cent is far too high a figure for that and, again, in my assessment (which again I must emphasise cannot be a matter of calculation) the correct figure for deduction should be 10 per cent.

The calculations that I therefore make, and I will set them out so that they may be checked, are these:

In respect of the ground floor, the zone "A", which is 912 sq ft @ £17 per sq ft I calculate at £15,504.

Zone "B" — I think this figure is agreed at 619 sq ft, but again, perhaps this could be checked — 619 sq ft @£8.50 I calculate at £5,261.50, making a total of £20,765.50.

To that must be added 5 per cent to reflect the corner site, which I calculate at £1,038.28.

Again, the agreed figure for the basement, which is storage, is 979 sq ft @£2, which is £1,958, making a total of £23,761.78.

From which I deduct 10 per cent to reflect the user restrictions contained in clause 2(13) of the draft lease, which is a deduction of £2,376.18, making a total of £21,385.60.

The final matter with which I must deal is the question of the interim rents. It is quite clear upon the authorities that have been cited to me (and I do not need to quote) that in arriving at a figure for the interim rents I should take into account the fact that the interim rent is to be calculated upon the basis of a yearly tenancy and the question is as to what, if any, deduction should be made from the rent that I have just assessed in order to arrive at the interim rent in respect of these premises. Mr Buckingham considered that the due allowance that should be made to reflect yearly tenancy should be a deduction of 20 per cent from the open market rental value. Mr Abbott considered that a reduction of 10 per cent would be appropriate. He referred to a rental that had been paid in respect of 256 Edgware Road for a period of nearly two years to a retailer selling men's clothing on a series of very short tenancies without security of tenure at a quarterly rent of the equivalent of £24,700 and he has pointed out that he did not introduce this matter in order to derive a comparable rental value but to demonstrate that on a year-to-year basis a demand does exist, more so for the lessee's user, perhaps, than any other user. I think it was, indeed, agreed between the parties that the use that is made of 204-206 Edgware Road — namely as a clothing store for men's clothing, for outfitting etc — is perhaps the most popular and most valuable use that there is in the country as a whole and that, therefore, even on a year-to-year basis, it is of value. It seems to me that the figure that is put forward by Mr Abbott is, in all the circumstances of these proceedings, the appropriate figure and I therefore think that the appropriate interim rent should be the annual open market rental value that I have assessed, with a deduction of 10 per cent — namely £19,247.04.

CHANCERY DIVISION

March 10 1981

(Before Mr Michael WHEELER QC, sitting as a deputy judge of the division)

OVE ARUP INCORPORATED v HOWLAND PROPERTY INVESTMENT CO LTD

Estates Gazette January 16 1982

(1981) 261 EG 149

Landlord and Tenant Act 1954 — Summons by tenants for leave to discontinue proceedings for a new tenancy under section 24 — Whether leave to discontinue should be granted unconditionally or only on terms that tenants should not pursue any claim for statutory compensation to which they might be entitled under section 37 — Under RSC Order 21, rule 3, in contrast with county court rules, proceedings commenced by originating summons cannot be discontinued without leave of the court — Court's discretion under rule 3 — When tenants in present case decided to discontinue they were not aware that landlords had decided not to oppose grant of a new tenancy — Landlords relied on case of *Young, Austen & Young v BMA* for imposition of "no compensation" term — Case distinguished by judge on the ground that tenants here were influenced by their belief in landlords' continued opposition to grant of new tenancy — Leave to discontinue given without condition

This was a summons taken out by the plaintiffs, Ove Arup Incorporated, consultant engineers, seeking leave to discontinue proceedings commenced by them under section 24 of the Landlord and Tenant Act 1954 for a new tenancy of business premises at 88-94 Tottenham Court Road, London W1. The defendants to the summons were the landlords, Howland Property Investment Co Ltd.

John L Powell (instructed by Berwin Leighton) appeared on behalf of the plaintiffs; G N N Huskinson (instructed by Jaques & Co) represented the defendants.

Giving judgment, MR MICHAEL WHEELER QC said: This is a summons by the plaintiffs, Ove Arup Incorporated (whom I will call "the tenants") for leave to discontinue proceedings commenced by them by originating summons dated August 29 1980 under section 24 of the Landlord and Tenant Act 1954 (which I will refer to as "the Act") for a new tenancy of the third floor of business premises at 88-94 Tottenham Court Road, London W1, of which the defendants, Howland Property Investment Co Ltd, are the landlords. But for the fact that the tenancy is one to which Part II of the Act applies it would come to an end by effluxion of time on March 20 1981.

The sole question which I have to decide is whether, in the events which have happened, I should make it a term of granting the tenants leave to discontinue their application for a new tenancy that they will not pursue any claim for statutory compensation from the landlords under section 37 of the Act.

The provisions of Part II of the Act so far as relevant for the present case can be summarised as follows:

Under section 24(1) a business tenancy to which Part II of the Act applies is not to come to an end unless terminated in accor-

A dance with Part II, and the tenant under such a tenancy may apply to the court for a new tenancy (a) if the landlord has given him a notice of termination under section 25 or (b) if the tenant has made a request for a new tenancy in accordance with section 26. In the present case I am concerned with a procedure which effectively commenced with the landlord giving a notice under section 25.

Under section 25(1) the landlord may terminate the tenancy by giving notice to the tenant in the prescribed form specifying the date at which the tenancy is to terminate. Under section 25(5) the notice is not to have effect unless it requires the tenant to notify the landlord within two months whether or not the tenant will be willing to give up possession at the date of termination specified in the B section 25 notice: and under section 25(6), the notice must also state whether the landlord would oppose an application to the court for the grant of a new tenancy and, if so, the grounds upon which he would do so.

Under section 29(1) if the tenant applies to the court for a new tenancy under section 24, the court is bound, subject to the provisions of the Act, to make an order for the grant of a new tenancy at such rent and on such other terms as are provided elsewhere in Part II. But where the section 24 application to the court is made in consequence of a landlord's section 25 notice of termination, the application is not to be entertained unless the tenant has duly notified the landlord that he (the tenant) will not be willing to give C up possession at the date of termination specified in the landlord's notice (section 29(2)): and the tenant's application has to be made not less than two nor more than four months after the giving of the landlord's section 25 notice (section 29(3)).

The statutory grounds upon which a landlord may oppose a tenant's application for a new tenancy are such of the grounds set out in section 30(1) as the landlord has specified in his section 25 notice. He has a choice of all or any of seven grounds, which are set out in paragraphs (a) to (g) of section 30(1). Grounds (a), (b) and (c) are cases where, because of failure to repair, delay in paying rent or other substantial breaches of the tenant's obligations under the existing lease, he has in effect proved to be a bad tenant. Ground (d) D is where the landlord has offered a tenant suitable alternative accommodation. Ground (e) is where the existing tenancy is of part only of the property and where the landlord wants to let the whole property and thereby expects to get a substantially higher rent than he could obtain from separate lettings. Ground (f) is where the landlord wishes to demolish or reconstruct the property, and ground (g) is where he wants to occupy the property, in whole or in part, for his own business or as his residence.

Where the landlord opposes the tenant's application on such of the above grounds as he specified in the landlord's section 25 notice and establishes any of those grounds to the court's satisfaction, the court is, by section 31(1), bound to refuse to make an order for the E grant of a new tenancy. The court is also bound, under section 31(2), to refuse to make an order where the landlord's grounds of opposition are all or any of grounds (d), (e) and (f) but where, in effect, he has specified a wrong date of termination in his section 25 notice. In that event, if the landlord would otherwise have succeeded the court can substitute a different date for the date of termination of the tenancy specified in the section 25 notice but is precluded from making an order for the grant of a new tenancy: what then happens in effect is that the tenant has 14 days in which to require the court to substitute the different date for the date originally specified in the section 25 notice, and if he does, the notice then takes effect accordingly.

F The only other provision which I need mention before turning to the compensation provisions of section 37 itself is section 64, which provides for the interim continuation of a tenancy pending the determination by the court of an application for a new tenancy.

Section 37(1) provides as follows so far as material for present purposes:

> Where on the making of an application under section 24 of this Act the court is precluded (whether by subsection (1) or subsection (2) of section 31 of this Act from making an order for the grant of a new tenancy by reason of any of the grounds specified in paragraphs (e), (f) and (g) of subsection (1) of section 30 of this Act and not of any grounds specified in any other paragraph of that subsection [or where no other ground is specified in the landlord's notice under section 25 of this Act . . . than those specified in the said paragraphs (e), (f) and (g) and either no application under the said section 24 is made or such an application is withdrawn] then, subject to the provisions of this Act, the tenant shall be entitled on quitting the holding to recover from the landlord by way of compensation an amount determined in accordance with the following provisions of this section. G

The remainder of section 37 is concerned with the quantification of the compensation, which is not an issue with which I am concerned, except to notice in passing that the legislature appears to have opted for a relatively simple if somewhat rough-and-ready formula for compensation, being either the rateable value of the property concerned or twice the rateable value, dependent upon the length of time that the tenant or his predecessor in business has H been in occupation. It rather looks, therefore, as if compensation is, at least in part, for loss of goodwill, although it will be seen that a tenant cannot get compensation under ground (d) where the landlord has offered him reasonable alternative accommodation.

The words in square brackets in the extract from section 37(1) quoted above were inserted by section 11 of the Law of Property Act 1969.

It will thus be seen that under section 37(1) as originally enacted, a tenant could get compensation only if he had made an application under section 24(1) but had been opposed by the landlord on all or any of grounds (e), (f) and (g) (and on no other ground) with the result that the court had been precluded from making an order for J the grant of a new tenancy either under subsection (1) of section 30 where the landlord had, so to speak, won on the merits, or under subsection (2) of section 30 where the tenant had in effect elected to allow the section 25 notice to take effect but with a different date of termination. The 1969 amendment of section 37(1) added two further cases in which a tenant was to be entitled to compensation, namely where the only ground of opposition specified in the landlord's section 25 notice was one or more of grounds (e), (f) and (g), and, either (1) the tenant did not make an application under section 24(1) at all, or (2) he made such an application but the application was subsequently withdrawn.

It will thus be seen that the legislature clearly contemplates the K possibility of a tenant making a section 24 application, withdrawing it and getting compensation. What the legislature appears to have overlooked or ignored is the fact that the procedure for withdrawing a section 24 application differs in the High Court and in the county court. In the High Court an application under section 24 (not being proceedings commenced by writ) cannot be discontinued without the leave of the court which, under RSC O 21, r 3 may grant leave "on such terms as to costs, the bringing of a subsequent action or otherwise as it thinks just". I pause here to note that although O 21, r 3 is headed "Discontinuance", the order as a whole is headed "Withdrawal and Discontinuance", and it is clear that although section 37(1) contemplates withdrawal rather L than discontinuance, nothing turns on this distinction of language as Bridge LJ pointed out in *Covell Matthews & Partners v French Wools Ltd* [1978] 1 WLR 1477 at 1479. In the county court, however, leave of the court to discontinue a section 24 application is not required.

I now turn to the facts of the present case as extracted from the affidavit evidence of the parties. As I have already mentioned, the tenancy in the present case was due to expire by effluxion of time on March 20 1981. In the spring of 1980 the tenants heard rumours that the landlords were likely to oppose the grant of new tenancies to tenants on all floors of the Tottenham Court Road premises. Accordingly the tenants resolved at a board meeting held on March M 31 1980 to instruct their solicitors to serve a notice on the landlords requesting a new tenancy under section 26 of the Act with a view, as the tenants put it in evidence, to "ascertaining the landlords' plans so as to make contingent plans if necessary." Both sides accept that this section 26 notice was defective and inoperative. It is unnecessary, therefore, to burden this judgment with a statement of the procedure which would have been available or which would have become available if the notice had been effective. The only relevance of this notice, which was served on or about April 10 1980, is that it was the first intimation to the landlords that the tenants wanted a new tenancy.

A About this time an inspection of the premises by the landlords revealed that the property was in need of substantial modernisation and improvement, and at a board meeting of the landlords held on April 21 1980 it was resolved that section 25 notices specifying grounds (e) and (f) should be served on all tenants. At that stage the only indication of tenants' wishes was, I think, the abortive section 26 notice which the tenants had served. According to the evidence of a Mr McGee, a director of the landlords, the board contemplated that possession of the premises would be obtained at or very shortly after the termination date specified in the section 25 notice, which was March 25 1981. Accordingly the landlords instructed architects and quantity surveyors to prepare a feasibility budget. The specification B and drawings were received from the architects during August 1980 and the budget in September 1980. The budget was based on an estimated tender date of April 1981 with completion of the proposed works in July 1982.

During the period of three months between July and September 1980, five more section 24 applications were made by various tenants of parts of the Tottenham Court Road premises. I know nothing of the five applications except that two of them were apparently started in the county court and were subsequently transferred to the High Court.

The landlords were clearly alarmed by the number of these applications and sought advice from their solicitors as to the likely C date that possession of the premises as a whole could be obtained; and in view of the time which was likely to elapse before the applications were finally disposed of (including the possibility of appeals) the solicitors not surprisingly advised that the earliest possible date for commencement of the reconstruction works was October 1981 but that no guarantee could be given that the work of reconstruction could be started even then. The landlords' quantity surveyors also indicated that in view of this the original feasibility budget would need to be revised and that such a deferred and uncertain commencement date for the work of reconstruction would have a significant effect on costs.

Meanwhile, in early June 1980 the tenants had set up a working D party to look into the position generally and specifically at plans for leaving the Tottenham Court Road premises if they should have to do so. The recommendation of this working party was considered at a board meeting of the tenants held on July 21 1980. At that meeting it was decided, in the light of the working party's recommendations *and having regard to the landlords' opposition to any new tenancy* (my emphasis), to carry out alterations to other premises leased by the tenants in Soho Square whereby the Soho Square premises could be adapted so as to house the staff of the Tottenham Court Road premises, which then numbered some 50 people. The move to Soho Square, assuming a start in September 1980, was estimated to last approximately 17 weeks and to cost some £52,000. Thus, as E long ago as July 1980 the tenants had decided upon a programme which, if implemented, would culminate in the occupants of their Tottenham Court Road premises being relocated by the end of March 1981 when the existing tenancy was due to expire. At the same time, the tenants wished to protect their position with regard to the Tottenham Court Road premises and to investigate, so far as possible, the genuineness of the landlords' stated intentions. But so far as the tenants were aware, the landlords were maintaining their objections to the grant of a new tenancy.

I should here mention that the tenants are a well-known firm of consultant engineers who carry on business from a number of different premises with a staff of some 1,250 in London and 1,700 F in the UK as a whole: and it is fair to say, as the tenants quite frankly admit, that the Tottenham Court Road premises represent a comparatively small proportion of the total office space occupied by the tenants.

The tenants assert — and their assertion is not challenged — that a move of the type involved in the present case if the Tottenham Court Road lease is terminated involves a great deal of time and preparation and that it is not realistic to make and alter policy decisions of this nature on little or no notice. It is imperative therefore that contingency plans are made in good time: and the tenants stress that throughout the time that these plans were being made they believed that the landlords were genuinely opposing the grant of a new tenancy on the grounds specified in their section 25 G notice, in which event it was unlikely that the tenants would be able to remain in possession of the premises after the expiration of the current tenancy. There is no doubt that the landlords' opposition was perfectly genuine, and it was against this background that on August 29 1980 the tenants issued their originating summons under section 24 for the grant of a new lease.

By the autumn of 1980 matters were coming to a head in both camps. The tenants, in the face of continued opposition from the landlords, had incurred (and were likely to have to incur still further) considerable expense in preparing to move and in moving to Soho Square: and the compensation which they would be likely to get under section 37 of the Act was unlikely to cover the cost of H the refurbishments at Soho Square and of the move. On the other hand, the landlords, who had also incurred substantial expenditure in furtherance of their plans for reconstruction, were faced with the possibility of protracted legal opposition to their attempts to regain possession of the premises: and it so happened that each side took critical decisions at almost the same time.

At a board meeting of the landlords held on December 8 1980, after reviewing the legal position, it was agreed that it would be unprofitable for the landlords to carry out the reconstruction work originally envisaged and that it would be more expedient for them to relet the premises in their existing state. Accordingly, it was resolved that the landlords should no longer seek to oppose the J various applications for new tenancies on ground (f) and that this ground of opposition should be withdrawn. It was however agreed to continue to oppose applications for new tenancies on ground (e) where applicable. (It was common ground before me that in spite of the landlords' original section 25 notice ground (f) did not apply in the case of the tenants.) In short, as the landlords' evidence put it, they were "not prepared to take a decision to spend an unascertainable amount of money with a view to obtaining a future unascertainable rental increase".

In the tenants' camp, too, the position was being reviewed and at a board meeting also held on December 8 1980 a decision was taken to discontinue the section 24 application, in which case the K existing tenancy would terminate on March 25 1981. I am satisfied that this decision was taken as a result of the landlords' continued opposition and in the belief that the opposition was genuine and was likely to succeed.

What happened next is of particular importance. On December 16 1980 the tenants' solicitors telephoned the landlords' solicitors and informed them that the tenants intended to discontinue the application for a new tenancy and would vacate the premises on the termination of the existing tenancy. In the course of this telephone conversation the landlords' solicitors indicated they anticipated receiving instructions from the landlords to withdraw their previous grounds of opposition to the grant of a new tenancy. This was the L first occasion on which the tenants knew of this possibility. On December 17 1980 the tenants' solicitors issued the summons for leave to discontinue which is now before me. This summons was served on the landlords' solicitors under cover of a letter dated December 18 1980 and was due to come before the master on December 23 1980. But it was not until the morning of the hearing that the tenants' solicitors received a letter from the landlords' solicitors formally advising them that the landlords were withdrawing their opposition on ground (f): and even then it was only in the course of the hearing before the master that the landlords' solicitors formally withdrew the opposition on ground (e).

It will be apparent from the facts summarised above that if the M tenants had not issued their section 24 application in due time their existing lease would have come to an end on March 20 1981 and they would have been entitled to statutory compensation: but that because of the section 24 application the date of termination became wholly uncertain, although the landlords were to some extent protected by the interim tenancy provisions of section 64. Equally it is clear that the decisions taken on December 8 1980 by the tenants to withdraw their section 24 application and by the landlords to withdraw their opposition to that application were taken independently and in ignorance of the other party's change of intention.

A Such, then, are the basic facts. Do they require me (assuming that I have power to do so — a point which I shall have to touch on briefly later in this judgment) only to grant the tenants leave to withdraw their section 24 application on terms that they agree not to pursue any claim for section 37 compensation? Before attempting to answer this question it will be convenient to summarise the principal rival arguments which were presented to me: and it is right that I should emphasise at the outset that before me neither side challenged the genuineness of the motives of the other in seeking to maintain (or to resist) the claim for compensation.

In considering the relative arguments put forward on behalf of the rival parties it will be convenient to start with the arguments put B forward by Mr Huskinson on behalf of the landlords. The arguments were, of course, based solely on a case where the landlord is opposing the tenants' section 24 application on one or more of grounds (e), (f) and (g) and not on any other ground. Mr Huskinson's arguments can, I think, be summarised as follows:

(a) Under section 24 a tenant's existing tenancy is not to terminate except in accordance with the Act. The basic right which the Act is intended to assert and protect, therefore, is a right to continued occupation of the premises.

(b) Once the landlord has served his section 25 notice, the tenant has four months in which to decide whether or not to apply to the court for the grant of a new tenancy. This, says Mr Huskinson, is C the legislature's assessment of what is a reasonable time within which the tenant has to take such a decision, including time to look for suitable alternative premises: and if the tenant decides within this period not to make an application, justice is done to both sides: the landlord knows when he will get possession (as specified in his section 25 notice), and the tenant knows that he will get the statutory compensation under section 37.

(c) The tenants' decision whether or not to make a section 24 application is all-important because this can introduce uncertainty into a situation which, if no application is made, is certain (see (b) above); because the making of the section 24 application of necessity leaves the landlord in doubt as to whether he will regain D possession of the premises at all, and, if so, when. Therefore, says Mr Huskinson, if the tenant makes a section 24 application (and thus introduces this element of uncertainty) and later changes his mind, he can hardly be surprised to find that he may have jeopardised his right to statutory compensation.

(d) In considering whether or not to give leave to a tenant to withdraw his section 24 application the High Court clearly has a complete discretion under O 21, r 3 whether or not to impose terms; and leave is not given lightly or as a matter of course (see per Bridge LJ in the *Covell Matthews* decision, *supra*, at p 1485). Mr Huskinson argues that a "no compensation" term should be imposed in the present case, not merely because the tenants are E seeking an indulgence from the court, but also because the imposition of the term is necessary in order to protect the landlords' position. Their position, on the facts, is that unless the tenants are given unconditional leave to withdraw their application, they (the landlords) cannot be required to pay compensation, because the tenants will only have two courses open to them, namely (i) to accept a new lease (which they no longer want) or (ii) to allow their section 24 application to be dismissed: and neither course would qualify them for section 37 compensation. It follows, therefore, on this argument that if I were to grant the tenants unconditional leave to withdraw, the landlords would (to borrow a phrase used by Whitford J in a case to which I shall have to refer in some detail later F in this judgment) "be disadvantaged". Therefore, a "no compensation" term should be imposed in order to maintain the balance between the parties as it at present exists as a result of the tenants' deciding that they no longer wish to obtain a new tenancy.

(e) Finally Mr Huskinson relies upon the decision of Whitford J in *Young, Austen & Young Ltd v British Medical Association* [1977] 1 WLR 881: [1977] 2 All ER 884, the *ratio decidendi* of which, he says, if properly understood, effectively governs the present case.

For the tenants, Mr Powell starts by emphasising the fact that section 37(1) in terms contemplates the possibility of a tenant withdrawing his section 24 application and yet getting statutory compensation. This shows, says Mr Powell, that the mere fact of withdrawal is not in itself intended to deprive the tenant of compensation: and this in turn throws doubt on Mr Huskinson's argument G that a "no compensation" term should be imposed in order to preserve to the landlords the advantage which they have obtained as a result of the tenants no longer wanting a new tenancy; because this argument, if of general application to withdrawal cases, conflicts with — and virtually overrides — what on the face of it is a statutory right to compensation. Mr Powell stressed that the object of section 37 was to do justice between the parties and that therefore in deciding whether or not to grant unconditional leave to withdraw the court must have prime regard to the justice of the situation. The court undoubtedly has a complete discretion under O 21, r 3 to do whatever justice requires and in exercising that H discretion it should have regard to all the facts of the particular case before it: the landlords' argument based on the assumed need to preserve their accrued advantage is, says Mr Powell, an attempt to fetter the court's discretion and to substitute for that discretion something akin to a rule of law. He says that on the facts of the present case the tenants were forced to make a difficult decision (ie not to proceed with their section 24 application) at a time when, so far as they were aware, the landlords' opposition was not only genuine but might well be effective: and it was a fortnight later that the landlords' own decision to withdraw their application was formally notified to the tenants. Tenants in a case such as this cannot reasonably be expected to reverse a bona fide decision taken by J them as a result of the landlords' continued opposition merely because at a comparatively late stage the landlords themselves change their minds, even though the landlords' decision was partly due to the delay which the hearing of the six section 24 applications might involve.

I asked Mr Huskinson in what circumstances he considered that a section 24 application could be withdrawn and the tenants still get compensation, as section 37(1) clearly contemplates. He suggested that this might be a case where the landlord did not change his mind about opposing the tenant's application but the tenant thought that the landlord's opposition was likely to succeed and "wanted to get out". But if this suggestion is correct, the case postulated is not far K removed from the facts of the present case where, as Mr Rigby, a director of the tenants, deposed: "Throughout the time that these plans and decisions" (that is to say plans to move to Soho Square and to discontinue the section 24 application) "were made, the tenants believed that the landlords were genuinely opposing any application for a new tenancy on the grounds stated in its section 25 Landlord and Tenant Act notice and that accordingly it would not be possible for the tenants to remain in the premises after the expiry of the current tenancy."

I also asked Mr Huskinson how long, on his argument, the landlords could have delayed their decision to withdraw their opposition and still avoid having to pay compensation: and he suggested L that they could have withdrawn right up to, and during, the actual hearing of the present summons. But this could work manifest injustice for the tenants as the following example given by Mr Powell shows. Assume a tenant's section 24 application which the landlord opposes on ground (e), (f) or (g). During month one the tenant decides in the face of the landlord's continued opposition not to pursue his section 24 application and to apply to withdraw. During month two the tenant is offered (not by the landlord) suitable alternative accommodation but only on the terms that he enters into a binding contract forthwith, which he does. The tenant's summons for leave to withdraw comes on for hearing during month three and the landlord then for the first time announces that M he no longer opposes the grant of a new tenancy, leaving the tenant with no alternative but to allow his section 24 application to be dismissed. In circumstances such as these it would be entirely unfair, as it seems to me, to deprive the tenant of compensation.

Mr Powell points out that the uncertainty which arises in a case such as the present is the product of decisions by both sides, by the tenants to apply for a new tenancy and by the landlords to oppose the section 24 application.

If the question which I have to decide were free from authority the conclusion which I would have reached is that looking at the totality of the evidence, I could, and should, give unconditional leave to the tenants to withdraw. But if, as Mr Huskinson contends, the *Young, Austen* decision virtually answers the question in the landlords' favour, I ought loyally to follow it. I must therefore consider in some detail the facts of that case and the reasoning of Whitford J which led him to grant the tenants leave to withdraw but only on "no compensation" terms.

The procedure adopted in the *Young, Austen* case differed from that adopted in the present case, but the differences are not, I think, material for present purposes. There were two existing tenancies in *Young, Austen* in respect of each of which the landlord served a section 25 notice of termination. Then the tenants apparently served requests for new tenancies under section 26 of the Act (although I am not entirely clear how this squared with section 26(4), which provides that a tenant may not request a new tenancy if the landlord has already given a section 25 notice, but I do not think that matters), and the landlord served counter-notices under section 26(6) opposing the grant of new tenancies on ground (g) in section 30(1). The tenants applied to the High Court for new tenancies. Subsequently the landlords indicated that they were prepared to grant a new tenancy of one of the premises but continued to oppose a grant in the case of the other premises. The tenants having decided, however, that they no longer wanted a renewal of either tenancy applied to the court under O 21, r 3 for leave to discontinue the proceedings on payment of the landlords' costs. Whitford J, in an extempore judgment, held: (1) that a tenant's right to compensation under section 37(1), as amended, did not preclude the court, on the tenants' application to discontinue, from considering whether or not some term should be imposed on any claim for compensation, but (2) — and I quote from the headnote at p 881:

> that in the present circumstances, where the tenants had started proceedings in the High Court for new tenancies and now were seeking the court's indulgence because they no longer wanted those tenancies, it would not be right to give them leave to discontinue under O 21, r 3 without imposing some term; accordingly the tenants would be granted leave to discontinue on giving an undertaking not to pursue a claim for compensation under section 37 of the Act of 1954.

I observe at the outset that Whitford J imposed a "no compensation" term in respect not only of the premises which the landlords were prepared to relet but also of the other premises which they required for their own purposes and for which they were maintaining their opposition under ground (g). There is nothing in the judgment to indicate why the learned judge treated both applications in the same way except that the tenants no longer wanted new tenancies of either.

In one important respect the facts in *Young, Austen* are not clear, namely whether the tenants' decision that they no longer wanted a renewal of the tenancies was taken *before* the landlords intimated that they would relet one of the premises: or whether (as in the instant case) the tenants' decision to withdraw was taken at a time when they believed the landlords were persisting in genuine opposition to the section 24 applications. I have been referred to the reports of *Young, Austen* both in the *Weekly Law Reports* and in the *All England Reports*. There is nothing in the judgment to assist me on this point, but the language of the headnotes does lend some support for the view that in *Young, Austen* the landlords' intimation that they were prepared to relet one of the premises *preceded* the tenants' decision to apply to withdraw their applications. If this is correct, then in my judgment there is an important factual difference between *Young, Austen* and the present case. Another factual difference appears to be that in *Young, Austen* both sides seem to have been suspicious that the other party was trying to secure an unfair advantage. "In the result", said Whitford J at p 885, "a stage was reached where the tenants concluded that they would very much prefer to seek accommodation elsewhere and they have in fact . . . made preparations to move to other premises altogether. So the position now is that they no longer want a renewal of these tenancies in any event, and it is on this basis that they seek to withdraw their application."

In order to understand the *ratio decidendi* of *Young, Austen* it is necessary to quote extensively from the judgment. I start at p 885:

> I think the point made by the landlords can be put very briefly in this way: they say that if leave were not granted, and the case came on to be heard, the tenants would find themselves in the position that so far at least as the first premises are concerned, there would be no resistance to the grant of the new lease by the landlords. The only question would be the terms on which it is to be granted. So far as the second premises are concerned, there would be resistance on ground (g) of section 30(1) of the Act. In any event, however, the tenants, if the matter were to proceed, would not wish to secure new tenancies either of the first or the second premises. They would not seek an order from the court for the grant of new tenancies in respect of either of these premises.
>
> It is of course to be remembered that, even where a matter of this kind proceeds to a hearing and a new tenancy is in fact granted, the tenant has 14 days after the making of an order within which an application can be made to the court for revocation of the order, and if such an application is made and the order is revoked, then apart from specific provisions as to the date on which the existing tenancy shall be determined, the position must be that the tenant, having pursued the matter thus far and then decided not to accept the tenancy order, is not going to be entitled to compensation, for no provision in section 37 is made for compensation in circumstances such as these.
>
> So the point made on the landlords' side is that it would be quite wrong in such circumstances for leave to be given to discontinue, as if such leave were given without any undertaking with regard to the question of an application for compensation, they would be disadvantaged. As the matter at present stands, their point is that if the tenants have to proceed, they are not going to seek an order for the grant of new tenancies and in those circumstances their applications will be dismissed and no compensation will be payable to them; and it is quite wrong that they should be allowed to withdraw without some term being imposed which will preserve to the landlords the position which would undoubtedly result so far as compensation is concerned if the action were to proceed.
>
> There can be said to be some difference so far as these two premises are concerned because, of course, in relation to the first premises the landlords do not oppose the application. It was suggested to me that certain questions have to be considered, and the first was — and I shall consider the first premises initially — whether, on a withdrawal of the application, the tenants are entitled to statutory compensation. It seems to me that the answer is perfectly simple and straightforward. On a withdrawal of the application, they are entitled to compensation: section 37(1) in terms so provides. As I have said, no question arises as to the amount of the compensation. Of course, as was quite rightly pointed out on behalf of the landlords, as the matter at present stands they are not entitled to compensation because their application has not been withdrawn and it cannot be withdrawn without the leave of the court.
>
> What was said first on the tenants' side was that if the statute has provided — as I hold it does — that on a withdrawal of an application they are entitled to compensation, it would be wholly wrong that the court should purport to impose a term depriving them of compensation. This, it was said, would on the face of it be going against the provisions which Parliament has specifically enacted. It was indeed suggested to me that it was not within the power of the court to impose a term of this particular kind.
>
> Assuming however that they did not succeed on this point, it was then suggested that if the matter came to be considered whether or not a term of this kind should be imposed, then as a matter of discretion it would be quite wrong that the court should impose a term of this kind. First, it was said that it would be wrong because it would be against the apparent policy of the Act of 1954, which indicates a specific intention that there shall be a compensation upon withdrawal of an application. Secondly it was said that an absurd position would result in this way: here the sum at stake is such that the tenants have necessarily had to come to the High Court. It might well have been that the sum at issue would have been such that proceedings could have been brought in the county court. If proceedings had been brought in the county court, the tenants could have withdrawn without leave, and indeed the possibility of the imposition of any term relating to compensation simply would not have existed.
>
> I am bound to say that it does not appear to me that the court is in any way precluded from considering the question as to whether a term should or should not be imposed. The tenants come to this court, having started this proceeding, seeking an indulgence, and RSC O 21, r 3 specifically provides that in considering whether such an indulgence should be granted, the duty is thrown on the court of considering all the circumstances of the case and of considering whether, in justice, it would be right that leave to discontinue should be granted and, if so, the terms upon which such leave should be granted.

Whitford J then referred to the *Covell Matthews* decision at first instance (reported in [1977] 1 WLR 876) where, after referring to a number of cases, Graham J said this at p 879:

> The principles to be culled from these cases are, in my judgment, that the court will, normally, at any rate, allow a plaintiff to discontinue if he wants to, provided no injustice will be caused to the defendant. It is not desirable

> A that a plaintiff should be compelled to litigate against his will. The court should therefore grant leave, if it can without injustice to the defendant, but in doing so should be careful to see that the defendant is not deprived of some advantage which he has already gained in the litigation, and should be ready to grant him adequate protection to ensure that any advantage he has gained is preserved.

Then, after referring to *Robertson v Purdey* [1906] 2 Ch 615 as an example of the need to impose a term when granting leave to discontinue under O 21, r 3 in a case where if unconditional leave had been given the defendants would have been "at a grievous disadvantage", Whitford J continued:

> B The circumstances are very different, but what was submitted to me by counsel for the landlords was that the principle to be applied is clear and its application to the circumstances of this case must lead to this, that, in fairness, leave to discontinue ought not to be granted without the imposition of a term relating to compensation, as unless such a term is imposed injustice will be done to the landlords. That as I have said — and I repeat it once more — arises in this way: that, in this case, if the matter were to proceed in the circumstances which have arisen, the tenants are now no longer going to seek the fresh tenancies which they originally sought, and their action would be dismissed and their right to compensation would go.
>
> I think the submission on behalf of the landlords is a good one. It is perhaps a little troubling that, as counsel for the tenants quite rightly pointed out, if this had been a proceeding brought in the county court, the imposition of a term of the sort which I think ought to be imposed would not C have been possible. It may be that this is a matter which was never in the contemplation of those responsible for the amendment to section 37 of the Act of 1954. Prior to the amendment of that section, it will of course be recalled that the position was that there was no specific provision relating to compensation upon mere withdrawal and the right to compensation arose only after there had been a decision adverse to the tenant. It may possibly have been thought that to force the parties to a hearing in circumstances where it would be likely to be apparent to the tenants that they would not succeed in opposing a section 30(1)(g) objection would be unnecessarily onerous, and that therefore they should be granted a right to withdraw. What I think was never in contemplation was a circumstance such as this, where basically the reason why the tenants want to withdraw is not that they fear they will be unable to resist the objection of the landlords; indeed, in D one case, there is no question of objection by the landlords. Basically, the reason why they want to withdraw is that they no longer want the tenancies in question. I do not think that in those circumstances it would be right that I should grant them leave to discontinue without imposing some term. It may be that the injustice, if there be any, lies in this, that it is impossible, if proceedings be brought in the county court, for a term of this character, which I think is in fact appropriate, to be made.
>
> I am prepared to grant leave to discontinue, but only on an undertaking by the tenants that they will not pursue any claim to compensation pursuant to section 37 of the Act as amended.

Mr Huskinson claims that the *ratio decidendi* of *Young, Austen* is entirely clear and was to this effect: that by reason of the proceedings which the tenants had commenced for the grant of new tenan- E cies, the tenants had got themselves into a position where, unless they were given leave to discontinue, they could not get compensation. Therefore, the landlords — who, unlike the tenants, were not seeking the court's indulgence — were in the position where they would not have to pay compensation unless the court granted unconditional leave to withdraw: that was the landlords' existing position, and accordingly the court should only grant leave on terms which preserved that position.

Not surprisingly, Mr Powell does not accept Mr Huskinson's interpretation of *Young, Austen*. Mr Powell lays great emphasis on the factual differences between the two cases, although (as I have already pointed out) this is largely a matter of inference from F somewhat insubstantial material. He therefore urges caution against expressing the *ratio* of *Young, Austen* at a more general level than the facts of that case — as he invites me to infer them to have been — warrant. He submits that the true *ratio* of *Young, Austen* is to be found in the one sentence in the judgment at p 887: "Basically the reason why they [the tenants] want to withdraw is that they no longer want the tenancies in question." Mr Powell suggested, as a possible reconciliation between the decision in *Young, Austen* on the one hand and the approach which he invites me to adopt in the instant case, that if the tenant's decision to withdraw is their independent decision uninfluenced, or only minimally affected, by the landlords' opposition, the tenant could fairly be put on terms not to seek compensation: but that where the G tenant takes a reasonable business decision in the circumstances then prevailing and is primarily influenced in that decision to withdraw by the landlord's continued opposition the court should recognise the practical reality of the decision and should not deprive the tenant of his right to compensation.

I confess to having been puzzled as to the true *ratio* of *Young, Austen*, and I bear in mind in particular that Master Dyson clearly considered that the present case was indistinguishable from *Young, Austen*. But at the end of the day I have come to the conclusion that the basic reason for Whitford J's decision was as stated in the sentence from his judgment at p 887 which I have just quoted: because otherwise I cannot see any justification for his having H treated the two tenants' applications, one of which was no longer opposed but the other of which was still opposed, as meriting the same "no compensation" treatment.

If my assessment of *Young, Austen* is correct then, in my judgment, the decision does not preclude me from doing what, apart from authority, would have been my inclination, on the basis that what I am required to do is to hold the scales evenly between the parties in the light of the facts as a whole of the case before me (and of no other). The truth of the matter, in my judgment, is that Part II of the Act is at best a somewhat rough-and-ready method of doing justice as between landlord and tenant against the background of the general policy of Part II of the Act, which is to provide an J element of security of tenure for business tenants. I wholly endorse Whitford J's approach to the Act when he said, earlier in his judgment at p 883:

> Of course, the apparent purpose of the Act, or so it seems to me, is to endeavour to ensure, so far as possible, that justice can be done to the landlords and tenants in relation to what may be rapidly changing circumstances.

For my part I would emphasise the importance of the concluding words of the above extract, which in my judgment provide the key to the way in which the court should approach an application such as that which I have to consider. In terms, section 37(1) gives the K tenant an unqualified right to compensation if his application is withdrawn but says nothing about the circumstances in which a tenant may (in the High Court at all events) withdraw his application. This, I suggest, is because the legislature was concerned not with the procedure for withdrawal but simply with the fact of withdrawal.

Accordingly, for the reasons which I have given, I do not consider that there is anything in the *Young, Austen* decision which requires me, on the facts of the present case, to impose a "no compensation" term upon the tenants as a condition of granting the leave to withdraw their section 24 application.

There are two final points which I should mention. First, it will be L apparent from the extracts which I have cited from the *Young, Austen* decision that Whitford J considered, and expressly decided, the question whether on the true construction of section 37(1) the court had power to impose a "no compensation" term upon the tenant at all. Accordingly, before me, Mr Powell did not seek to argue that the court had no such power but he reserved the point for argument in a higher court if this case should go to appeal. Secondly, and as a safeguard against the possibility that an appellate court might hold that the court had no power to impose a "no compensation" term (in which case the only choice open to the court on an application such as the one before me would be simply to grant or refuse the application) Mr Huskinson invited me to say M which course I would take in that event. The answer is of course that on the facts of this case I would grant the application.

CHANCERY DIVISION

October 28 1981

(Before Mr Justice WHITFORD)

NEVILL LONG & CO (BOARDS) LTD AND ANOTHER v FIRMENICH & CO (UNLIMITED COMPANY)

Estates Gazette February 6 1982

(1981) 261 EG 461

Landlord and tenant — Part II of Landlord and Tenant Act 1954 — Effect of severance of reversion on rights of way granted to tenants — Before severance freeholders granted leases for 21 years (which by the date of the action had expired by effluxion of time) to the plaintiffs with rights of way over land which gave access to the rear of the properties leased to the plaintiffs — The plaintiffs' tenancies were protected by Part II of the 1954 Act — On severance of the reversion the land over which the plaintiffs' right of way subsisted became vested in the defendants — The question raised in the present proceedings was whether on the expiry of effluxion of time of the plaintiffs' terms the right of way over the part of the land running alongside the property vested in the defendants came to an end — Held that the plaintiffs continued to be protected under Part II of the 1954 Act not only in respect of the premises demised to them but also in respect of rights of way including the right of way running alongside the defendants' property — Defendants had no power to determine the plaintiffs' interest in this right of way (being only part of the interest granted to the plaintiffs) and in any case the defendants were not "the landlord" within the meaning of Part II of the 1954 Act in relation to this right of way — Plaintiffs entitled to declarations sought

In this summons the plaintiff companies, Nevill Long & Co (Boards) Ltd and S A Wise (Timber) Ltd, sought declarations against the defendant company, Firmenich & Co, an unlimited company. The declarations sought were (1) that Part II of the Act of 1954 applied to the plaintiffs' tenancies and that certain rights of way continued after the expiry of the tenancies by effluxion of time, and (2) that the defendant company was not "the landlord" for any of the purposes of Part II of the Act in relation to the plaintiffs' tenancies. The leases were of adjoining pieces of land at North Hyde Walk, Southall, Middlesex.

D K Rattee QC and R W Ham (instructed by Stilgoes) appeared on behalf of the plaintiffs; S Rich QC and J C Harper (instructed by Tarlo, Lyons & Aukin) represented the defendants.

Giving judgment, WHITFORD J said: In this matter there are two plaintiffs, Nevill Long & Co (Boards) Ltd and S A Wise (Timber) Ltd. The defendants are Firmenich & Co. The questions which I have to decide arise out of two leases in respect of adjoining pieces of land at North Hyde Walk, Southall, Middlesex. The whole of the land relevant to the matters that I have to consider at this date belonged to a company Winckley Investments Ltd and Winckley granted two leases, each of them, I think, for a term of 21 years from December 1959. I have said that Winckley were the owners of the whole of the land relevant to this dispute and it is convenient at this stage to make a brief reference to a plan attached to one of the leases, for I think for all material purposes they are in the same terms. I will take exhibit A to the affidavit of Mr Scott-Tucker sworn on the plaintiffs' side, on the third page of which we see the area the subject of this dispute. Omitting altogether a small corner of land at the bottom of the page as it is bound up in the affidavit there is next an area of land which comes out in the photocopy as being surrounded in black but should in fact be surrounded in red, which was referred to throughout the proceedings as plot 2. Above that there is another area of land, the exact limitations of which can also be seen by similar outlining by reference to exhibit B, but it comprises virtually that part of the land which is cross-hatched and that was referred to as plot 1. Finally, at the top we have an area which was referred to as plot 3. There is in the form of an inverted "L", with rather a long leg in one direction, a portion which again appears in black in the photocopy but was marked brown which is in fact a road giving access to the rear of each of these plots 1, 2 and 3; it also runs along the edge of the Grand Union Canal. At the relevant time Winckley also owned what has been referred to as plot 3.

These leases were granted, and in all material respects they are the same, to each of the plaintiffs for the term of years which I have indicated together with, so far as each of the plaintiffs is concerned, a right of way over what I will refer to as the brown land, the plaintiffs also undertaking certain obligations with regard to repairs which it might be necessary to do to this roadway. The plaintiffs are, I understand, both companies engaged in the timber trade and the terms of each of the leases appear to indicate that it may at least at one time have been contemplated that the canal would be used for the purposes of loading or unloading timber, but I was told by counsel that in actual practice the brown land constituting the roadway at the rear of the premises has in substance been used, so that access along this roadway by lorries bringing goods to or taking goods from the site in question is of importance.

So far as plot 3 is concerned, in 1967 that was sold by Winckley and eventually it came into the hands of and is now in the hands of the present defendants. So far as the other two plots are concerned, the freehold reversion in both those plots has been acquired by a company called Meek Investments Ltd, who have in fact also acquired a controlling interest in each of the plaintiff companies. But be that as it may, although Meek have secured the reversionary interest in the freehold so far as plots 1 and 2 are concerned nothing has changed so far as the plaintiffs are concerned and they still continue to occupy the land and buildings which they obtained tenancies of under their respective leases for the purposes of their respective businesses in the timber trade. The leases have now expired by effluxion of time and so far as the two portions of land outlined in red in the relevant leases are concerned no question arises which is of any concern in this present summons. No question arises likewise in respect of that portion of the brown roadway which runs alongside plots 1 and 2. The dispute which has arisen is as to whether or not in all the circumstances — and this is the first contention of the defendants — on the expiry of the term of the lease the right of way over the brown land running alongside plot 3 came to an end. There can I think be no doubt about this, that the plaintiffs must be right when they say, as they do, that the tenancies which they acquired were tenancies protected by Part II of the Landlord and Tenant Act 1954 and by virtue of the provisions of section 23 of this Act they have tenancies to which subsection (1) of that section apply, the subsection being in these terms:

> Subject to the provisions of this Act, this Part of this Act applies to any tenancy where the property comprised in the tenancy is or includes premises which are occupied by the tenant and are so occupied for the purposes of a business carried on by him or for those and other purposes.

So far as the land outlined in red is concerned — and from now on I can treat it as if I were dealing with only one of the tenancies because the same arguments apply to both of them — it might seem to anybody just reading the language of the Act, to which I have just referred, that some question might arise in relation to the right of way which was in fact granted over the whole of the brown land, but I am satisfied, having regard to the decision of the House of Lords in *Whitley* v *Stumbles* [1930] AC 544 that this tenancy is properly to be regarded as a tenancy embracing not only the land and the buildings upon it outlined in red but also the right of way, and it follows from this that so far as the interest under the right of way is concerned that would on the face of it appear to be protected just as the interest in the occupation of the land and the buildings upon the land is protected by the Act.

The difficulty that has arisen in this case was I think really put, quite forcibly so far as the defendants are concerned, in this way. It was conceded by Mr Rich on behalf of the defendants that if a situation had arisen under which, let me suppose for the moment, Winckley had never parted with their freehold interest in any of the properties in question and the tenancies had come to an end, it must undoubtedly be the case that the plaintiffs could claim to be protected tenants in respect of their interest not only of the properties but of the rights of way which had been granted; and subject to the provisions of the 1954 Act, if steps had been initiated by Winckley to bring the plaintiffs' interest to an end, they could have applied

A and could have expected to get, if their application were successful, a continuation of their existing tenancy with a continuation of the existing rights of way. As matters now stand, Meek, the company who are in fact in control of the two plaintiff companies, have secured the reversionary interest in plots 1 and 2 with the brown land adjacent, but it was submitted by Mr Rich, and I think accepted by Mr Rattee, that under no circumstances could Meek be in a position to grant to these plaintiffs upon any question of consideration of the grant of a new tenancy rights which would include rights of way over the brown land appurtenant to plot 3, which is now owned by the defendants.

The whole of the difficulty in this case has arisen out of this simple B fact, that the reversionary interest has fallen into two separate hands. I think Mr Rattee is right when he says that it seems probable that those who were responsible for the enactment of the 1954 Act never really had such an eventuality in contemplation. The authorities to which I have referred to my mind establish this beyond any possible doubt, namely, that the severance of a kind such as we have here of the freehold interest cannot possibly bring two separate tenancies into being, with the result that during the period in which the tenancies were still subsisting the plaintiffs ought properly to be regarded as being in the position of tenants of their respective portions of land edged red and the rights of way over the brown land adjacent thereto of Meek, and so far as the C brown land adjacent to plot 3 is concerned tenants only of the right of way over that particular portion of the brown land. Had that in fact been the case their position would have been one which was considered in *Land Reclamation Co Ltd v Basildon District Council* [1979] 1 WLR 767, when the Court of Appeal, considering the question of the possibility of the application of the Act to a mere tenancy of right of way came to the conclusion that it was not protected.

The two cases relied upon by Mr Rattee for the proposition that the severance of a reversion does not bring into existence two separate tenancies were *Jelley v Buckman* [1974] 1 QB 488 and *Dodson Bull Carpet Co Ltd v City of London Corporation* [1975] 1 D WLR 781. The first of these cases was a Rent Act case. It was a case which came before the Court of Appeal and in his judgment in this case Stamp LJ observed at p 498 in relation to submissions which had been put touching the possible effect of section 140 of the Law of Property Act 1925 on the question of splitting a reversionary interest in the context of the Rent Act:

> We can find nothing in the section to suggest for a moment that the legislature intended that following a severance to which the lessee was not a party he should find himself holding part of his land under one tenancy and part under another. In relation to a lease for years as opposed to a weekly tenancy the change in the law would be dramatic and had the legislature intended to create that result one would expect to find some clear ex- E pression of that intention. Not only is there an absence of such an expression of intention but we find in this section positive indications that what was assumed or contemplated was the continued existence of the pre-existing lease.

Then he goes on to deal with matters which are perhaps not quite so strictly germane.

I was referred to section 140 of the Law of Property Act by Mr Rich, who suggested, as I understood it, that it was of assistance to the claims of the defendants. He read subsection (1) of section 140, omitting certain words which he said for present purposes could be considered as immaterial, and summed it up by saying that the effect of section 140(1) was that notwithstanding severance every F condition could be enforced after severance as if that right alone had been comprised in the original lease. I do not myself see that section 140 has got anything to do with the problems with which I am concerned. There is a marginal note as to the section which reads "Apportionment of conditions on severance", and the section runs:

> (1) Notwithstanding the severance by conveyance, surrender, or otherwise of the reversionary estate in any land comprised in a lease, and notwithstanding the avoidance or cesser in any other manner of the term granted by a lease as to part only of the land comprised therein, every condition or right of re-entry, and every other condition contained in the lease, shall be apportioned . . .

G I do not need to go on. We are concerned here not of course with any question of right of re-entry nor indeed of any other condition of the lease, we are merely concerned with the effect of severance upon the undoubted right which was granted and it was a part of the right granted in a single tenancy to each of the plaintiffs in connection with their occupation and enjoyment of the use of the land which is edged in red in each of the respective leases.

If one considers the observations as to the effect of severance, it appears to me that the conclusion which necessarily follows from the judgments in the two cases to which I have referred is going to be this. What in fact the defendants are left with is a position in which there is a continuing protected interest and the only other question relevant to be determined is whether in those circum- H stances the defendants might be entitled to serve a notice determining the interest of the plaintiffs in the right of way over the brown roadway coming along next to plot 3. The whole of this question can, I think, be conveniently considered in the context of the judgment of Goff J in *Dodson Bull Carpet Co Ltd v City of London Corporation (supra)*. The facts in that case were by no means on all fours with the facts in this case. It was a case where a tenancy had been granted of two properties referred to as nos 5 and 6 and a question arose as to whether these were separate tenancies, but it was found that they were not separate tenancies and being one tenancy the question which next arose was whether a notice of determination in respect of each of the tenancies individually was J valid, and it was held by Goff J that it was not. He so held and his final determination on it I think arises from a passage which starts above the letter F on p 785. Having held that there was only one tenancy Goff J went on in these terms:

> The plaintiffs submit that there is nothing in the Landlord and Tenant Act 1954 to authorise a landlord to serve a notice of determination under section 25 as to part only of the premises comprised in the relevant tenancy, and indeed, quite the reverse, to allow him to do so would cut across and jeopardise the protection afforded by the Act. Defendants' counsel states that he cannot dispute that proposition and in my judgment he is plainly right in so doing.

Section 25 of the Landlord and Tenant Act 1954 is in these K terms:

> (1) The landlord may determine a tenancy to which this Part of this Act applies by a notice given to the tenant in the prescribed form specifying the date at which the tenancy is to come to an end (hereinafter referred to as "the date of termination").

Then there are certain provisos which I do not think for present purposes I need refer to.

I am bound to say that it appears to me that the conclusion reached by Goff J is the only conclusion which could have been come to. Mr Rich said that it was in fact reached only because upon the face of it counsel on the other side accepted the proposition; and L that one should put some discount upon the observation of Goff J that he thought counsel was plainly right in so doing. I agree that he was plainly right in so doing.

If the situation be, as in my judgment it plainly is, that at the moment the plaintiffs respectively stand in the position of being protected tenants in respect of the occupation and use of the land and premises respectively granted to them under the terms of their leases and have a continuing protected interest in the right of way, the first difficulty which is faced by the defendants is that they cannot, I think, on the plain meaning of the Act, and indeed in the light of the judgment of Goff J, determine what is in fact only a part M of the interest granted; and that, on the face of it, must be sufficient to determine this summons in favour of the plaintiffs. There is, however, another difficulty which to my mind is just as acute so far as the defendants are concerned. Section 25 in the terms which I have read out contemplates a determination of the tenancy by the landlord. There is a definition of "the landlord" in section 44 of the Act, subsection (1) being in these terms:

> Subject to the next following subsection, in this Part of this Act the expression "the landlord", in relation to a tenancy (in this section referred to as "the relevant tenancy"), means the person (whether or not he is the immediate landlord) who is the owner of that interest in the property

A comprised in the relevant tenancy which for the time being fulfils the following conditions, that is to say —

(a) that it is an interest in reversion expectant (whether immediately or not) on the termination of the relevant tenancy, and

(b) that it is either the fee simple or a tenancy which will not come to an end within 14 months by effluxion of time and, if it is such a tenancy, then no notice has been given by virtue of which it will come to an end within 14 months or any further time by which it merely continued under section 36(2) of section 64 of this Act,

and is not in reversion expectant (whether immediately or not) on an interest which fulfils those conditions.

Mr Rich submitted that the relevant tenancy in this particular
B case is the right of way. The defendants are not in my judgment the persons who are the owners of that interest and they can fall neither within (a) nor (b). They are the owners in fee simple of the land which is encumbered by the existing right of way and I can well understand that either they might think that this was a valuable property or they might desire that they should be in a position to stop passage across it. But none of this is really of any relevance if they cannot, as I would hold, properly claim to be a landlord within the meaning of the Act.

I should refer to one other case which was referred to by both counsel. It was a case which was heard in the same year as that in which Goff J gave his decision in the *Dodson Bull* case. It was a case
C which came before Judge Edgar Fay sitting as a deputy judge of the High Court and it is reported in [1975] 1 QB 361, being the case of *Skelton (William) & Son Ltd v Harrison & Pinder Ltd*. I do not propose to go into this case in detail. It is a case in which the judge was referred to *Jelley v Buckman* among a number of other authorities. It is a case in which he gave careful consideration to the effect of the provisions of section 140 of the Act of 1925. It was a case in which in the whole of this context he considered, in the light of the Act of 1954, three situations affecting the method of terminating the teanancy: first, a tenancy never protected; secondly, a tenancy protected which none the less had lost protection; thirdly, a tenancy continuing to be protected. He was specifically concerned
D with the rights of free entry. Mr Rattee submitted, and I think there was substance in the submission, that in some parts at least of this judgment it may be found that there is a conflict with the findings of Goff J in the case to which I have referred, given a few months later in the same year. I follow the judgment of Goff J.

In my judgment the plaintiffs are entitled to the declarations which they seek in the summons, namely a declaration that Part II of the 1954 Act applies to the tenancies created by the leases and that the rights of way included in such tenancies continued after December 1980 in accordance with the provisions of section 24 of the 1954 Act. They are also entitled to a declaration that the defendant company is not the landlord in relation to such tenancies
E for the purposes of Part II of the Act and is not entitled to serve notice terminating such tenancies under section 25 of the 1954 Act.

Judgment was given for the plaintiffs, with costs.

CHANCERY DIVISION

December 10 1981

(Before Mr Justice SLADE)

INTERNATIONAL MILITARY SERVICES LTD v CAPITAL & COUNTIES PLC

F
Estates Gazette February 27 1982

(1981) 261 EG 778

**Landlord and Tenant Act 1954 — Tenants' entitlement to compensation under section 37 when court is precluded from granting a new tenancy — Original basis of compensation the rateable value or twice the rateable value as the case might be — Amendment to basis of compensation made by Local Government, Planning and Land Act 1980, section 193 and Schedule 33, para 4, implemented by Landlord and Tenant Act 1954 (Appropriate Multiplier) Regulations 1981 — New basis the rateable value multiplied by 2¼ or, as the case may be, twice the rateable value multiplied by 2¼ — Ques-
G tion as to whether in the circumstances of the present case the old or the new basis applied — The landlords had served a section 25 notice on December 17 1980, the regulations came into operation on March 25 1981 and the tenants quitted on June 24 1981 — Landlords contended that the new basis did not apply in cases where the landlords' section 25 notice had been served before March 25 1981, the date when the regulations came into force — Presumption against retrospective legislation invoked — Held, rejecting various arguments put forward on behalf of landlords, that the new enlarged basis of compensation applied as from March 25 1981 to every tenant who quits thereafter, provided that the qualifying conditions in section 37 of the 1954 Act are satisfied — Compensa-
H tion in present case was equal to the amount of the rateable value multiplied by 2¼**

This was an originating summons in which the plaintiffs, International Military Services Ltd, tenants of office accommodation at St Andrew's House, 40 Broadway, London SW1, sought, in addition to other relief, a determination of their entitlement to compensation under section 37 of the Landlord and Tenant Act 1954, as amended by the Local Government, Planning and Land Act 1980, on the basis of the rateable value of the premises, £45,180, multiplied by 2¼, ie in the sum of £101,655. The multiplier of 2¼ was prescribed by the Landlord and Tenant Act 1954 (Appropriate Multiplier) Regulations 1981 (SI 1981 No 69). The landlords,
J defendants to the summons, were Capital & Counties PLC.

J Cherryman (instructed by Clifford-Turner) appeared on behalf of the plaintiffs; J S Colyer QC and M R King (instructed by Debenham & Co) represented the defendants.

Giving judgment, SLADE J said: In this originating summons the plaintiffs are International Military Services Ltd. The defendants are Capital & Counties PLC, formerly called Capital & Counties Property Co Ltd. They were also formerly the plaintiffs' landlords in respect of certain premises in London.

The Landlord and Tenant Act 1954 ("the Act of 1954") in certain circumstances gives a tenant the right to receive compensation on leaving the premises. By a notice of December 17 1980
K served pursuant to section 25 of the Act of 1954 the defendants gave notice terminating the plaintiffs' tenancy of the premises on June 24 1981. It is common ground that the plaintiffs quit the premises on June 24 1981 and then became entitled to compensation. Effectively, the principal issue in the case now is whether such compensation falls to be assessed in accordance with a statutory scale of compensation which was in force on June 24 1981, as the plaintiffs assert, or in accordance with a scale which was in force on December 17 1980, as the defendants assert.

By a lease of May 13 1976 ("the 1976 lease"), made between the defendants of the one part and Four Millbank Nominees Ltd of the other part, certain office premises comprising the fourth floor at St
L Andrew's House, 40 Broadway, London SW1, were demised by the defendants to Four Millbank Nominees Ltd for a term from February 23 1976 to March 25 1978 at a yearly rent of £32,000. By a deed of variation and supplemental lease ("the 1977 deed") dated April 18 1977, and made between the same parties, the 1976 lease was varied so that it should thenceforth be construed as granting a term from February 26 1976 to June 24 1981, instead of a term expiring on March 25 1978. The defendants thereby demised office premises comprising the second-floor annexe and the third floor at St Andrew's House to Four Millbank Nominees Ltd from February 7 1977 for the residue of the term of the 1976 lease as varied by the 1977 deed, at the additional yearly rent of £34,000. By an assign-
M ment dated July 19 1978, Four Millbank Nominees Ltd assigned the second, third and fourth floor premises to the plaintiffs for all the residue of the term created by the 1976 lease and 1977 deed. I will call these premises collectively "the premises".

At all material times the rateable value of the premises has been £45,180. It is common ground that the tenancy is one to which Part II of the Act of 1954 applies.

At this point it will be convenient to summarise some of the history of the legislation relevant to the present case. Section 24(1) of the Act of 1954, as amended, provides:

A tenancy to which this Part of this Act applies shall not come to an end unless terminated in accordance with the provisions of this Part of this Act;

A It further confers on the tenant under such tenancy the right, subject to certain conditions, to apply to the court for a new tenancy (a) if the landlord has given notice under section 25 to terminate the tenancy, or (b) if the tenant has made a request for a new tenancy in accordance with section 26. Section 25(1) confers on the landlord the right to terminate a tenancy to which Part II of the Act applies "by a notice given to the tenant in the prescribed form specifying the date at which the tenancy is to come to an end", though this subsection has effect subject to later provisions in the Act. Section 25(2) provides:

> Subject to the provisions of the next following subsection, a notice under this section shall not have effect unless it is given not more than 12 nor less
> B than six months before the date of termination specified therein.

Section 25(5) provides:

> A notice under this section shall not have effect unless it requires the tenant, within two months after the giving of the notice, to notify the landlord in writing whether or not, at the date of termination, the tenant will be willing to give up possession of the property comprised in the tenancy.

Section 25(6) requires a landlord's notice under the section to state whether he would oppose an application to the court under Part II of the Act for the grant of a new tenancy and, if so, to state on which grounds mentioned in section 30 he would do so. Section 29(1)
C provides:

> Subject to the provisions of this Act, on an application under subsection (1) of section 24 of this Act for a new tenancy the court shall make an order for the grant of a tenancy comprising such property, at such rent and on such other terms, as are hereinafter provided.

Section 29(2) provides:

> Where such an application is made in consequence of a notice given by the landlord under section 25 of this Act, it shall not be entertained unless the tenant has duly notified the landlord that he will not be willing at the date of termination to give up possession of the property comprised in the tenancy.

D Subsection (3) provides:

> No application under subsection (1) of section 24 of this Act shall be entertained unless it is made not less than two nor more than four months after the giving of the landlord's notice under section 25 of this Act or, as the case may be, after the making of the tenant's request for a new tenancy.

Section 30(1) begins with the following words:

> The grounds on which a landlord may oppose an application under subsection (1) of section 24 of this Act are such of the following grounds as may be stated in the landlord's notice under section 25 of this Act or, as the case may be, under subsection (6) of section 26 thereof, that is to say:

E There then follow seven subparagraphs, of which I need read only subparagraph (g):

> subject as hereinafter provided, that on the termination of the current tenancy the landlord intends to occupy the holding for the purposes, or partly for the purposes, of a business to be carried on by him therein, or as his residence.

Section 31(1) provides:

> If the landlord opposes an application under subsection (1) of section 24 of this Act on grounds on which he is entitled to oppose it in accordance with the last foregoing section and establishes any of those grounds to the satisfaction of the court, the court shall not make an order for the grant of a new tenancy.

F

Section 37(1), in its amended form, provides:

> Where on the making of an application under section 24 of this Act the court is precluded (whether by subsection (1) or subsection (2) of section 31 of this Act) from making an order for the grant of a new tenancy by reason of any of the grounds specified in paragraphs (e), (f) and (g) of subsection (1) of section 30 of this Act and not of any grounds specified in any other paragraph of that subsection, or where no other ground is specified in the landlord's notice under section 25 of this Act or, as the case may be, under section 26(6) thereof, than those specified in the said paragraphs (e), (f) and (g), and either no application under the said section 24 is made or such an application is withdrawn, then, subject to the provisions of this Act, the tenant shall be entitled on quitting the holding to recover from the landlord by way of compensation an amount determined in accordance with the following provisions of this section. G

The words in this subsection, reading from "or where . . ." to ". . . is withdrawn" were inserted in it by section 11 of the Law of Property Act 1969 as from January 1 1970. I pause to comment that no reliance has been placed by either side in the present case on the phrase "subject to the provisions of this Act". These words appear to refer to section 38 of and paragraph 5 of the Ninth Schedule to the Act of 1954, and possibly to other provisions. Section 37(2), up to March 25 1981, when it was amended in circumstances which I will describe later, read:

> The said amount shall be as follows, that is to say, — H
> (a) where the conditions specified in the next following subsection are satisfied it shall be twice the rateable value of the holding.
> (b) in any other case it shall be the rateable value of the holding.

It is common ground that this is a paragraph (b) case, because the conditions of section 37(3) are not satisfied.

Section 37(5)(a), so far as material, provides:

> For the purposes of subsection (2) of this section the rateable value of the holding shall be determined as follows:
> (a) where in the valuation list in force at the date on which the landlord's notice under section 25 . . . is given a value is then shown as the annual value (as hereinafter defined) of the holding, the rateable value of the holding shall be taken to be that value; J

On November 13 1980 the Local Government, Planning and Land Act 1980 (which I will call "the Act of 1980") received the Royal Assent. Part V amended the General Rate Act 1967 so as to abolish the statutory requirement for quinquennial revisions of rateable values. The legislature considered that this would or might necessitate certain changes to the provisions governing the appropriate multiplier for the purpose of section 37(2)(a) and (b) of the Act of 1954. Section 193 provided that the enactments specified in Schedule 33 to the Act should have effect subject to the amendments specified in that Schedule. Paragraph 4 of Schedule 33, which was headed "Landlord and Tenant Act 1954", read as follows: K

> (1) In subsection (2) of section 37 of the Landlord and Tenant Act 1954 (compensation where order for new tenancy precluded on certain grounds) the words "the product of the appropriate multiplier and" shall be inserted after the word "be" in paragraphs (a) and (b).
> (2) The following subsections shall be added after subsection (7) of that section:
> "(8) In subsection (2) of this section "the appropriate multiplier" means such multiplier as the Secretary of State may by order made by statutory instrument prescribe.
> (9) A statutory instrument containing an order under subsection (8) of this section shall be subject to annulment in pursuance of a resolution of either House of Parliament." L

In the absence of any further provision, the provisions of paragraph 4 of Schedule 33 would presumably have come into operation as soon as the Act of 1980 received the Royal Assent. However, subsection (5) of section 47, which was the last of the sections in Part V of the Act of 1980 dealing with rates, made it clear that such was not the intention. Section 47(5) provided:

> The provisions of Schedule 33 to this Act which give the Secretary of State power by order to prescribe multipliers and which are specified in subsection (6)(a), (b) and (c) below shall not have effect until he exercises the power conferred by them.

M

Section 47(6), so far as it is material, provided:

> The provisions of Schedule 33 mentioned in subsection (5) above are —
> (a) paragraph 4; . . .

The wording of section 47(5) is not, perhaps, entirely happy. First, the use of the words "which give the Secretary of State power by order to prescribe multipliers" are a little puzzling, since the provisions of Schedule 33 specified in subsection (6) do much more than give the Secretary of State "power by order to prescribe multipliers". Secondly, it is not entirely clear what meaning is to be attached to the phrase "until he exercises the power conferred by

A them" if, as in the event which happened in the present case, the Secretary of State makes an order prescribing multipliers but declares by his order that it is not to come into operation until a later date. In my judgment, however — and I do not think there is really any substantial dispute about this — the only proper sense that can be given to section 47(5) is to construe it as meaning that the amendments to the various specified statutory provisions relating to multipliers, including section 37(2) of the Act of 1954, which are to be effected by Schedule 33, are not to have effect until the date on which an exercise of the Secretary of State's power to prescribe multipliers is expressed by his order to have operative effect. Any other construction of section 47(5) could result in a twilight period
B between the making of a statutory instrument and the date on which the relevant regulations came into operation, during which there was no appropriate multiplier at all for the purpose of section 37(2) of the Act of 1954.

I now revert to the history of the present case. On December 17 1980, that is more than a month after the Act of 1980 received Royal Assent, the defendants served a written notice of that date on the plaintiffs terminating their tenancy of the premises on June 24 1981. The notice, as usual, required the tenants within two months after the giving of the notice to notify the landlords in writing whether or not they would be willing to give up possession of the premises on the specified date. It further stated that the defendants
C would oppose an application to the court under Part II of the Act for the grant of a new tenancy, on the ground that, on the termination of the current tenancy, the defendants intended to occupy the premises for the purposes of a business to be carried on by them in the premises, that is to say, on the ground specified in section 30(1)(g) of the Act of 1954.

On January 21 1981 the Secretary of State for the Environment, in exercise of the new powers conferred on him, made the Landlord and Tenant Act 1954 (Appropriate Multiplier) Regulations 1981 (which I will call "the regulations of 1981"). These provided:

1. These regulations may be cited as the Landlord and Tenant Act 1954
D (Appropriate Multiplier) Regulations 1981 and shall come into operation on March 25 1981.
2. The appropriate multiplier for the purposes of paragraphs (a) and (b) of section 37(2) of the Landlord and Tenant Act 1954 shall be 2¼.

These regulations were subject to annulment in pursuance of a resolution of either House of Parliament within 40 days of February 2 1981, which was the day upon which the regulations were laid before Parliament: (see paragraph 4 of Schedule 33 to the Act of 1980, and section 5 of the Statutory Instruments Act 1946). The regulations never were annulled and so came into operation on March 25 1981.

E In the meantime, by a letter of February 6 1981, the plaintiffs gave the defendants notice that they would not be willing to give up possession of the premises on the date specified in the defendants' section 25 notice. The plaintiffs, however, made no application to the court for a new tenancy. The last date upon which it would have been open to them to do so would have been April 18 1981. They let that date pass by, so that by April 19 the defendant landlords had acquired an indefeasible right to possession of the premises on the following June 24 1981. By a letter of April 16 1981 the defendants had claimed, first, that the plaintiffs were not entitled to any compensation at all under section 37 of the Act of 1954 and secondly that, even if some compensation were payable, the scale of compen-
F sation should be ascertained by reference to the date on which the section 25 notice had been given.

The plaintiffs duly quit the premises on June 24 1981. In the light of the information as to the defendants' attitude which they had received, they issued an originating summons on June 30 1981, in which they sought the following relief:

(1) That it may be determined whether, under and by virtue of the provisions of section 37 of the Landlord and Tenant Act 1954 (as amended) and in the events which have happened, the Plaintiff became entitled on quitting the premises on June 24 1981 to recover from the Defendant by way of compensation an amount determined in accordance with the provision of the said Act.

(2) If the answer to question (1) is in the affirmative, that it may be G
determined whether the said amount which the Plaintiff became entitled on quitting the premises on June 24 1981 to recover from the Defendant was:
(a) £45,180 (being the rateable value of the premises) or
(b) £101,655 (being two and one quarter times such rateable value).
(3) Payment of the amount determined in answer to question (2) above.
(4) Interest pursuant to the Law Reform (Miscellaneous Provisions) Act 1934 on the said amount from June 24 to the date of Judgment.
(5) Further or other relief.
(6) Costs.

To complete the history, by a letter of October 2 1981 the defendants accepted that compensation is payable to the plaintiffs under the Act of 1954, and said that in the circumstances the only H
issue was one as to quantum.

I now turn to consider the nature and extent of the plaintiffs' right to compensation. Very briefly, they contend that question 2 of the originating summons should be answered in the sense of alternative (b). The defendants claim that it should be answered in the sense of alternative (a).

The effect of the regulations of 1981, when read in conjunction with sections 47(5) and 47(6)(a) of the Act of 1980, was in my judgment that as from March 25 1981 both the regulations themselves and the amending provisions of paragraph 4 of Schedule 33 to the Act of 1980, relating to multipliers, came into full force and effect. The consequences were as follows. First, section 37(2) of the J
Act of 1954 thenceforth took effect in its amended form so as to read:

The said amount shall be as follows, that is to say, (a) where the conditions specified in the next following subsection are satisfied, it shall be the product of the appropriate multiplier and twice the rateable value of the holding; (b) in any other case it shall be the product of the appropriate multiplier and the rateable value of the holding.

Secondly, as from March 25 1981, the "appropriate multiplier" for the purpose of these two paragraphs became 2¼.

According to the terms of section 37(1) of the Act of 1954, as amended, there are a number of conditions precedent (some of K
them alternative) which have to be fulfilled if a tenant is to be entitled to compensation on quitting the holding. The relevant conditions which have been satisfied in the present case are the following:

(1) The landlord has served on the tenant a notice under section 25;
(2) the notice has specified no other ground other than one of the grounds specified in paragraphs (e), (f) and (g) of section 30 of the Act of 1954; and
(3) the tenant has made no application for a new tenancy.

Nevertheless, though conditions of the nature specified in section 37(1) have to be satisfied before a tenant's right to compensation can arise, the only event which entitles the tenant to compensation L
is the actual quitting of the holding. If for any reason whatever he does not quit, he does not become entitled to compensation. The relevant wording of section 37(1) of the Act of 1954 is:

. . . then, subject to the provisions of this Act, the tenant shall be entitled on quitting the holding to recover from the landlord by way of compensation an amount determined in accordance with the following provisions of this section.

The word "then", in my judgment, merely means "in that event", that is to say, if the earlier conditions specified in the subsection are fulfilled. The entitlement arises on the quitting of the holding (*and* M
not before). In my judgment, therefore, it is quite plain that the amount of the entitlement must be assessed in accordance with the law as it stands at the date of the quitting. I am not sure that this point is really in dispute.

The argument of Mr Colyer on behalf of the defendants, if I have understood it correctly, in essence involves the submission that the amendments to section 37 of the Act of 1954, which were introduced by the combined effect of section 193 of the Act of 1980 and paragraph 4 of Schedule 33 to that Act, and of the regulations of 1981, which altered the scale of compensation to which tenants quitting the premises were to be entitled, should be read with some implied limitation to the effect that they were not to apply in respect

A of tenants who quit their premises following the service of a landlord's section 25 notice which had been served before March 25 1981. The latter is, of course, the day when the regulations of 1981 came into operation. Any other reading of the relevant statutory provisions would in his submission involve giving them retrospective effect. He submitted that it is to be presumed that a statute, or statutory instrument, of a substantive rather than a procedural nature is not to have retrospective effect; and that the Act of 1980 and the regulations of 1981 read together can only have retrospective effect if a clear legislative intention of this nature has been demonstrated. Such intention, in his submission, has not been shown.

B In support of his submissions, Mr Colyer relied heavily on a decision of Brightman J in *Re 14 Grafton Street, London, W1* [1971] Ch 935. The headnote to that case, omitting immaterial references to authorities at the end, reads as follows:

> On September 27 1969, the landlords gave notice under section 25 of the Landlord and Tenant Act 1954 terminating a business tenancy of certain premises on April 1 1970. The notice required the tenants to notify the landlords within two months whether they would give up possession or not; it also stated that any application for a new tenancy would be opposed by the landlords on the basis that the premises were to be reconstructed and that the tenants could not claim compensation unless an application was made to the court for a new tenancy. The tenants' intention to vacate the premises
> C was communicated to the landlords by a letter dated October 13 1969. On October 22 1969, the Royal Assent was given to the Law of Property Act 1969 which came into force on January 1 1970. Section 11 of that Act amended section 37(1) of the Act of 1954 with the effect that a tenant was entitled to compensation without having first applied to the court for a new tenancy. On February 11 1970, the tenants inquired about compensation. The landlords denied liability. On the tenants' summons seeking compensation to be paid:
>
> *Held*, (1) that in order to be entitled to compensation for disturbance a tenant must, first, have served a counternotice on the landlord under section 29(2) of the Act of 1954, stating that he was unwilling to vacate the premises and secondly, under section 37(1) of the Act (before amendment), he must have applied to the court for a new tenancy; that a counternotice which expressed willingness to quit was irrevocable; and that, accordingly,
> D after October 13, the tenants could not give a notice of unwillingness to quit and had lost their right to apply to the court and the landlords, in turn, had acquired an indefeasible right to obtain possession on April 1 1970, without paying compensation.
>
> (2) That on its true construction the amendment to section 37(1) of the Act of 1954, not being procedural, could not have retrospective effect so as to revive the tenants' right to compensation; and that, accordingly, the tenants were not entitled to compensation.

I accept Mr Colyer's submission that this case is authority for the proposition that the amendment to the law which was effected by the relevant statutory provisions in the present case was substantive rather than procedural. This point has not been in dispute. Like-
E wise, the case is authority for the proposition that enactments are not to be construed so as to have a retrospective operation unless their language is such as to require such a construction. Brightman J quoted with approval a number of passages from judgments in earlier cases which established this principle. I need read only one.

In *Lauri v Renad* [1892] 3 Ch 402, Lindley LJ said at p 420:

> It certainly requires very clear and unmistakeable language in a subsequent Act of Parliament to revive or re-create an expired right. It is a fundamental rule of English law that no statute shall be construed so as to have a retrospective operation unless its language is such as plainly to require such a construction; and the same rule involves another and subordinate rule to
> F the effect that a statute is not to be construed so as to have a greater retrospective operation than its language renders necessary.

Brightman J (at p 944F) pointed out that, in the events which had happened, the landlords on any footing "had an indefeasible right to recover possession on April 1 1970, without the payment of compensation . . ." His conclusion, on the retrospectivity point, is to be found summarised in the following words (at p 947):

> Under the Act of 1954, by virtue of the unopposed notice served by the landlords, they had acquired the right before the commencement of the Act of 1969 to future possession of business premises without compensation. If the Act of 1969 is construed in such a way as to deprive them of that right, that Act would in my view be retrospective. Under the Act of 1954, by virtue of the notice served by the landlords, the tenants at one time had a G
> contingent right to compensation, ie a right to compensation contingent upon the landlords successfully opposing an application made by the tenants for an order for the grant of a new tenancy. The tenants lost that right before the Act of 1969 came into operation, by reason of their failure to take certain steps open to them. If the Act of 1969 is construed in such a way as to revive that right, that Act would in my view be retrospective.

Having then decided that the Act of 1969 could not properly be described as procedural, Brightman J concluded (at p 949):

> In the result I dismiss the application, on the ground that the Act of 1969 did not deprive the landlords of their right to recover possession without payment of compensation and did not revive the tenants' lost right to compensation. H

I pause to comment that there is one obvious point of distinction between the facts of the *14 Grafton Street* case and those of the present case. In the former case the effect of the construction contended for by the tenants was to revive a right to compensation which, before the relevant legislation, had been wholly lost. In the present case, there is no question of any right of the tenants to compensation ever having been wholly lost. At very most, what has happened is that the introduction of the new legislation increased the quantum of a prospective right which the tenants had at all relevant times. On the facts of the present case, as Mr Cherryman pointed out, the landlords did not, on any footing, obtain an indefeasible right to possession of the premises until April 19 1981, that J
is to say, several weeks after the relevant statutory provisions came into effect. On the present facts, therefore, the relevant statutory provisions, even if applied according to their letter, will not have any retrospective effect at all in the strict sense. When they came into operation, even if they are construed in the manner for which the plaintiffs contend, they operated as between the plaintiffs and the defendants merely for the future and not for the past.

Mr Colyer suggested that they did at least deprive the defendants of an inchoate right, in the sense that, as from the service of their section 25 notice, the defendants could reasonably have expected that, if the plaintiff tenants made no application under section 24, the defendant landlords would both be able to get possession on K
June 24 1981 and would only have to pay compensation according to the tariff in force on December 17 1980. I do not find this argument compelling. First, the section 25 notice was served after the Act of 1980 had received the Royal Assent, so that anyone interested had been warned that a new "appropriate multiplier" might be introduced in the near future. Secondly, Mr Colyer referred me to no authority in support of his proposition that the presumption against retrospectivity applies in respect of such vague inchoate rights as those of the defendants as at December 17 1980 on which he relies for this purpose.

I attach rather more weight to another of his points. He observed that, though on the timetable of the present case the landlords had L
not yet acquired an indefeasible right to possession by March 25 1981, they could have done so, if their section 25 notice had been served rather earlier; in this contingency, the landlords before March 25 1981 might have acquired an indefeasible right to recover possession at a date later than March 25 1981. On this hypothetical timetable, the coming into operation of the regulations of 1981 would still not have deprived the landlords of an indefeasible right to possession, so that the case would not even then have been on all fours with the *14 Grafton Street* case. Nevertheless, the regulations would have altered the scale of compensation which the landlords would have had to pay to the tenants, as the price of this indefeasible right, if and when the tenants left the holding. As Mr Colyer M
pointed out, the service of the landlord's section 25 notice is the first of the steps which begins the process of determining the tenant's rights to compensation; when he serves it, he may be influenced by (*inter alia*) consideration of the amount of the compensation which he may in due course have to pay to the tenant on quitting. I therefore accept that, by analogy with the *14 Grafton Street* case, the plaintiffs' construction of the relevant statutory provisions, which makes no distinctions between cases where landlords' section 25 notices have been served before March 25 1981 and those where they have been served after that date, could be said to involve a degree of retrospective effect in the hypothetical circumstances

A postulated by Mr Colyer. For this reason I also accept that they should not be construed so as to have greater retrospective effect than the language renders necessary.

Nevertheless, subject to the points arising on section 47(5) of the Act of 1980, with which I have already dealt, the language of the relevant statutory provisions seems to me plain and unambiguous. When they are read together, sections 47 and 193 of the Act of 1980, paragraph 4 of Schedule 33 to that Act, and the regulations of 1981, in my judgment make it quite clear that as from March 25 1981 the new amendments to paragraphs (a) and (b) of section 37(2) of the Act of 1954 are to apply in the case of every tenant who thereafter quits a holding, provided only that the other conditions B of section 37(1) have been satisfied. It would have been easy for the legislature in the Act of 1980, or for the Secretary of State in the regulations of 1981, expressly to provide that the amending provisions should not apply in cases where the landlord's section 25 notice had been served before March 25 1981. No such exception was made, and I can see no sufficient grounds for implying one.

On the facts of the present case, the plaintiffs on quitting the premises on June 24 1981, for the first time acquired the benefit of what Brightman J in the *14 Grafton Street* case described as "a debt created by statute, upon which the tenant may sue in other proceedings if necessary": (see [1971] Ch at p 942 C). For the purposes of ascertaining the amount of that debt one must, in my judgment, C look at the provisions of section 37 of the Act of 1954 as they stood in their reamended form at June 24 1981 when the right arose. In this reamended form they were, in my judgment, unequivocal. When they are so looked at, it is plain that the tenants' compensation falls to be calculated with the application of an appropriate multiplier of 2¼ to the rateable value of the holding.

This, I think, is really the end of the defendants' case, but in deference to Mr Colyer's argument I should refer to one or two other points. He drew my attention to a number of suggested capricious results or anomalies which might follow if the plaintiffs' construction of the relevant statutory provisions were correct. In particular, he pointed out that the plaintiffs' construction involves taking the D rateable value of the premises as at December 17 1980, but the "appropriate multiplier" as at June 24 1981, when the landlords' notice expired and the plaintiffs quit the premises. This, he submitted, means that like would not be related to like. He contended that one would expect the provisions for an "appropriate multiplier" to be prescribed so as to ensure that in any given case the relevant multiplier would be ascertained as at the same day as that which falls to be multiplied is ascertained.

I think that, so far as it goes, this comment is a fair one. On the other hand, in my judgment it has to be accepted that the statutory purpose of the introduction of the new system of "appropriate multipliers" was, as Mr Cherryman submitted, to effect from time E to time a rough and ready uplift in the amount of the statutory compensation to be given to tenants under section 37 of the Act of 1954, in place of the automatic uplift that they would otherwise have enjoyed by virtue of the system of automatic quinquennial rating valuations which was abolished by the Act of 1980. The introduction of a new "appropriate multiplier" was, I think, bound to produce some anomalies, particularly in the early stages. Nor do the anomalies appear to me all to lie in one direction, when the competing constructions are considered. If Mr Colyer were right in submitting that the relevant statutory provisions have no application in any case where the landlord's section 25 notice has been served before March 25 1981, this would mean that landlords who F chose to serve a section 25 notice during the period between January 21 1981 (when the regulations of 1981 were made) and March 25 1981 (when they came into effect) would escape the liability to pay compensation on the higher scale, even if the tenant quit the premises long after March 25 1981. It is hard to see any statutory purpose reflected in this result.

Mr Colyer did raise some other points, but I hope he will forgive me if I do not refer to all of them. For, once it is accepted that the relevant amendment of section 37(2) of the Act of 1954 and the introduction of the new "appropriate multiplier" came into full force and effect on March 25 1981, as I think it must be accepted, it seems to me inevitably to follow, on the clear wording of the section as amended, that the plaintiffs' rights fall to be determined in G accordance with the tariff which represented the law when they quit the premises. The words of the subsection as amended are entirely apt to apply to the plaintiffs' situation on June 24 1981, and the defendants, in my judgment, can point to no statutory provision which enables them to escape from their full effect.

In answer to the questions raised by paragraphs 1 and 2 of the originating summons, I therefore propose to answer question 1 in the affirmative sense and question 2 in the sense of alternative (b).

After counsel had made submissions as to payment of interest, SLADE J said: It is common ground that, on the basis of my judgment, the plaintiffs are entitled to interest on the amount of H £101,655 from June 30 1981 to the date of judgment. The plaintiffs have not asked for interest in respect of the period from June 24 to 30, I understand, because, although they had formally quit the premises on June 24, they did continue to have certain use of the premises for certain purposes for the next six days.

The question arises as to the rate of interest which should be awarded to the plaintiffs in the discretion of the court. In this context I have been assisted by a reference to the recent decision of Forbes J in *Tate & Lyle Food & Distribution Ltd v Greater London Council* [1981] 3 All ER 716. The facts of that case were very different, but he made some pertinent observations as to matters of principle. In particular, he pointed out (for example at p 722) that J there are considerable differences between personal injury cases and cases involving injury to business concerns. He said (at p 722 d):

> ... I do not think the modern law is that interest is awarded against the defendant as a punitive measure for having kept the plaintiff out of his money. I think the principle now recognised is that it is all part of the attempt to achieve *restitutio in integrum*. One looks, therefore, not at the profit which the defendant wrongfully made out of the money he withheld (this would indeed involve a scrutiny of the defendant's financial position) but at the cost to the plaintiff of being deprived of the money which he should have had. I feel satisfied that in commercial cases the interest is intended to reflect the rate at which the plaintiff would have had to borrow K money to supply the place of that which was withheld.

Forbes J went on to say:

> I am also satisifed that one should not look at any special position in which the plaintiff may have been; one should disregard, for instance, the fact that a particular plaintiff, because of his personal situation, could only borrow money at a very high rate or, on the other hand, was able to borrow at specially favourable rates. The correct thing to do is to take the rate at which plaintiffs in general could borrow money. This does not, however, to my mind, mean that you exclude entirely all attributes of the plaintiff other than that he is a plaintiff. There is evidence here that large public companies of the size and prestige of these plaintiffs could expect to borrow at 1 per cent over MLR, while for smaller and less prestigious concerns the rate might be L as high as 3 per cent over MLR. I think it would always be right to look at the rate at which plaintiffs with the general attributes of the actual plaintiff in the case (though not, of course, with any special or peculiar attribute) could borrow money as a guide to the appropriate interest rate. If commercial rates are appropriate I would take 1 per cent over MLR as the proper figure for interest in this case.

This was the rate which, subject to certain qualifications, Forbes J awarded.

Mr Cherryman has told me on instructions, and this has not been disputed, that the plaintiffs are a wholly-owned subsidiary of the Ministry of Defence but, nevertheless, run their business as a commercial business. There is, I think, no doubt at all that the defendant M company similarly runs its business. This can, I think, fairly be described as a commercial dispute in which, on the basis of my judgment, the plaintiffs have been kept out of the greater part of a debt which has been due to them since June 24 1981. I say "the greater part", because in October a payment was made by the defendants on account. In all the circumstances of the case, I think that commercial rates are appropriate to apply for the purpose of assessing the amount that should be paid to the plaintiffs. I think that, on the facts which I know, it is right to treat these particular plaintiffs as being plaintiffs of standing and prestige, who, with the benefit of the attributes, would be able to borrow at a more favourable rate than smaller and less prestigious concerns. I am told that

A Minimum Lending Rate ceased to have effect on August 20 1981 and that from then on the equivalent rate has been the Clearing Bank Base Rate.

In all the circumstances, following the guidance of the *Tate & Lyle* case, I propose to award the plaintiffs interest on the sum of £101,655 from June 30 1981 to judgment at the rate of 1 per cent over Minimum Lending Rate from June 30 to August 20, and at 1 per cent above the Clearing Bank Base Rate from August 21 onwards until judgment. Credit, however, is to be given to the defendants for the interest which has already been paid by them on the sum of £45,180 to the plaintiffs.

I think that it may be desirable that there should be some further
B definition of the Clearing Bank Base Rate, for the purpose of the order which I am making. Subject to further submissions, I think a convenient way of dealing with this would be for the plaintiffs' counsel and the defendants' junior counsel to agree and sign a minute embodying these provisions in clear workable form.

Liberty was given on the defendants' application to apply for a certificate under section 12 of the Administration of Justice Act 1969 to enable a "leapfrog" appeal to be brought to the House of Lords.

C

HOUSE OF LORDS

February 18 1982

(Before Lord HAILSHAM OF MARYLEBONE, Lord Chancellor, Lord WILBERFORCE, Lord KEITH OF KINKEL, Lord SCARMAN and Lord BRANDON OF OAKBROOK)

O'MAY AND OTHERS (PRACTISING AS INCE & CO) v CITY OF LONDON REAL PROPERTY CO LTD

Estates Gazette March 27 1982
(1982) 261 EG 1185

D **Landlord and Tenant Act 1954, Part II — Appeal by landlords of office accommodation in the City of London from a decision of the Court of Appeal allowing an appeal from a decision of Goulding J — Dispute as to terms of new tenancy to be granted to tenants under Part II of the Act of 1954 — Whether, as landlords contended, the terms of the tenancy should be converted on the grant of the new tenancy into a "clear lease" — The object of the "clear lease" arrangement was to shift to the tenants, in return for an adjustment in the fixed part of the rent the burden of fluctuations in the costs of supplying services and providing for the repair and decoration of the exterior and common parts of the building — Tenants would bear, by means of a service charge, appropriate proportions according to**
E **their interests of the cost of these items, so that the landlords would receive an income less subject to fluctuations due to outgoings than before — Held that the onus was on the landlords to justify the alteration in the terms of the tenancy required by the "clear lease" proposal — Although the proposed adjustment of the fixed element in the rent might appear to be an adequate compensation for the transfer of the burdens to the tenants, the tenants were being asked to accept risks which were unpredictable and inconsistent with the nature of their interest in the land — Tenants should not be required to accept the grant of a new tenancy on such terms — Appeal from decision of Court of Appeal dismissed**

F The appellants, City of London Real Property Co Ltd, were the owners of Knollys House, an office building situated at 9-12 Byward Street and 14-15 Seething Lane, EC3, in the City of London. The respondents, who practised as solicitors under the style of Ince & Co, were the tenants of accommodation forming part of the building. The proceedings which gave rise to the present appeal were an application by the respondents as plaintiffs for the grant of a new tenancy under Part II of the Landlord and Tenant Act 1954. The appellants did not oppose the grant but proposed that the new tenancy should contain terms of the "cleared lease" type. Goulding J granted a new tenancy on such terms ((1979) 249 EG 1065), but the Court of Appeal reversed his decision on this point ((1980) 255 EG 151). In the present appeal to the House of Lords the appellants sought to have the "cleared lease" principle G restored.

Ronald Bernstein QC and Benjamin Levy (instructed by Nabarro Nathanson) appeared on behalf of the appellants; Derek Wood QC and Jonathan Gaunt (instructed by Ince & Co) represented the respondents.

In his speech LORD HAILSHAM LC said: This appeal raises questions as to the proper approach to be taken by the courts on applications by business tenants for a new lease under Part II of the Landlord and Tenant Act 1954 as amended in 1969.

The appellants are the owners of an office building known as Knollys House situated at 9-12 Byward Street and 14-15 Seething Lane, EC3, in the City of London. The respondents are a well- H known firm of City solicitors practising under the style of Ince & Co. They are currently the tenants of the fifth floor of Knollys House together with some spaces for the parking of cars in the basement. The demised premises are occupied for the purposes of the professional practice of Ince & Co, and for the purposes of Part II of the Landlord and Tenant Act 1954 the respondents are business tenants.

The proceedings which are the subject of this appeal were commenced on January 20 1977 by the respondents as plaintiffs against the appellants as defendants by way of originating summons for the grant of a new tenancy under Part II of the Landlord and Tenant Act 1954 as amended together with ancillary relief. The appellants J do not oppose the grant of a new tenancy in principle, and as will be seen comparatively few questions are in issue between the parties. In the events which happened, by a judgment dated December 18 1978 Goulding J granted a new tenancy on the terms put forward by the appellants.

The respondents appealed against this judgment and on June 19 1980 the Court of Appeal (Civil Division) consisting of Shaw, Brightman and Buckley LJJ unanimously allowed the appeal and granted the new lease on the modified terms demanded by the respondents. In either case the new lease will in fact terminate on March 24 1982, but there are other proceedings pending which may result in the respondents' tenancy being prolonged on terms which K will no doubt be influenced by the judgment of your Lordships in the instant appeal.

The tenancy of the respondents derives from two separate leases dated April 25 1972 dealing respectively with the office accommodation and the parking spaces. In 1976 the area occupied by the respondents was increased by two deeds of variation dated May 27 1976 and the rent correspondingly increased. The term of the new tenancy is agreed and the dispute revolves around the rent to be paid and the terms to be embodied in its operative covenants. The real dispute relates to the covenants.

The origin of the dispute derives from the desire of the appellants to convert the terms of the tenancy, which did not previously L possess this characteristic, into what is known as a "clear lease". The effect of giving effect to this proposal would, according to the evidence, be to enhance the value of the appellants' reversion by a sum somewhere between one and two million pounds, and at the same time to render it more readily marketable partly owing to the increasing part played (amongst others) by pension funds and life insurance companies in purchasing office property as an investment. The purpose of a "clear lease" is to render the income derived from the rent payable by the tenants as little subject to fluctuation in respect of outgoings as may be possible. The method proposed by the appellants in the present case is to transfer in effect the risk of fluctuation of the items in the covenants which in the M nature of things will be executed by the landlord to the respondents by providing that the appellants should be fully reimbursed in respect of the fluctuating elements by provision for fluctuations in what has been referred to, perhaps inaccurately, as the service charge, in return for a flat diminution in the fixed element in the rent. In return for the transfer of risk, the appellants are prepared to accept a fixed reduction in the amount of the fixed rent calculated as a matter of figures at a sum of 50p a square foot, in actual fact reducing the fixed rent component of the total rent which would otherwise be £10.50 per square foot to £10.00 if a clear lease were granted. The main bone of contention between the parties is that

A the respondents are unwilling to be insurers of the risk of fluctuation, and would prefer instead to pay the full fixed rent of £10.50 a square foot in place of the reduced rent of £10.00.

It goes without saying that with an office block in multiple occupation by a number of different business tenants the management of the common parts and the exterior, and the various services from lifts to lighting, must normally be part of the responsibility of the landlord. This was so under the terms of the old tenancy in the instant case, and remains so under the appellants' proposed terms for the new. The mechanism for the conversion of the new lease to a clear lease under the appellants' proposed terms for the new lease was contained in a series of provisions forming part of the first and B second Schedules to the draft, the effect of which would be to transfer the risk of fluctuation which had been a burden borne by the landlord to the respondents as tenants. It is necessary at this stage to consider in a little more detail how, under the appellants' proposed terms, this was to be done. For convenience, I extract the details and the effect from the judgment of Brightman LJ in the Court of Appeal:

> The term of the lease is three years from March 1979. There are expressed to be three rents: first, a fixed rent of £100,700 a year. Secondly, by way of further or additional rent, a proportionate part of certain insurance premiums paid from time to time by the lessors in respect of the building of which the demised premises form part. This is called an insurance rent. C Thirdly, by way of further or additional rent, a service charge calculated and payable in the manner set out in the First Schedule. Clause 2 contains a covenant by the lessees to keep the interior of the demised premises in repair and decorated. Clause 3 contains a number of covenants on the part of the lessors which can be summarised as follows: (1) to provide heating; (2) to clean the common parts; (3) to keep the exterior and interior of the building in repair, other than the demised premises; and (4) to provide the services mentioned in paragraphs 1 and 11 of Part I of the Second Schedule; this is sub-clause (1)(iv), and I shall have to refer to it later.
>
> The function of the First Schedule is exclusively to calculate the service charge, as indeed the opening words of the First Schedule proclaim. Put shortly, the schedule provides that the service charge is to consist of the fair and reasonable proportion, to be certified by the lessor's surveyor, of the D total of five items: item (a), the actual or estimated cost in any year of the term of providing the services mentioned in paragraphs 1-11 of Part I of the Second Schedule; these are the services which the lessors covenanted to provide in clause 3(1)(iv) of the body of the lease; item (b), a management fee, upon which nothing turns; item (d) (i), an amount, to be revised annually by the lessors at their discretion, towards the estimated cost of replacing lifts, boilers and other equipment on the basis of certain life expectancies, and item (d) (ii), a similar annual amount towards the estimated cost of the decoration of the exterior and common parts of the interior of the building calculated on the footing that such work will be carried out at the end of specified periods. Paragraphs 1 to 11 of the Second Schedule, Part I, which represent item (a) above, comprise various items of routine expenditure but most importantly include also "*repairing, maintain-* E *ing*, cleaning, painting and decorating *the structure exterior* and common parts of the building . . . plant and equipment . . . and all drains, pipes, wires etc."
>
> The scheme of the draft, therefore, is to impose on the landlord the obligation to provide all the services to the building and to repair, maintain and decorate the building, both externally and internally (except the demised premises); but to entitle the lessors to recover, as additional rent, a proportion of the money which is spent on services and on external and internal repairs, plus a yearly sum which can best be described as estimated provision for depreciation of lifts, boilers and other equipment and for external and internal decorations. The service rent so collected from the tenant, though gauged by reference to actual or estimated expenditure, becomes in the hands of the lessors their absolute property to deal with as F they please. If all leases are on the same basis, which is not yet the case, the effect will be to extract from the tenants in each year sums which are calculated to recover or recoup the whole financial burden of the lessors' compliance with the covenants of the lease, in order to leave the fixed rent in the hands of the lessors as clear profit exonerated from the burden of all outgoings.

I turn now to the statutory provisions of Part II of the Landlord and Tenant Act 1954. These, as amended by subsequent enactments are contained in sections 23-46 and are conveniently set out in the volume of "Statutes in Force" relating to the Law of Landlord and Tenant at pp 19-41, of which I shall quote the most relevant sections only.

G By way of preface, I would point out that these provisions are examples of that class of statute referred to by Lord Simon of Glaisdale in *Johnson v Moreton* [1980] AC 37 at pp 65-68 in the context of the Agricultural Holdings Acts, in which he reflected how in recent years Parliament has thought it right to reverse the tendency remarked in the famous generalisation in Maine's Ancient Law that progressive societies had regularly moved from status to contract by limiting the freedom of contract in certain areas either for general political and social reasons of policy or in order to protect a class of persons who were in a less strong negotiating position from those with whom they had to negotiate. It is profitless to specify examples, because, as I pointed out at p 60 of H the same case, there are few of them *in pari materia*, but, out of the large number of such enactments particularly notable are those relating to employer and employee and to landlord and tenant, and in this field notably in tenancies of residential accommodation, agricultural tenancies, and business tenancies. A beginning was made with business tenancies in the Landlord and Tenant Act 1927, now largely repealed. The Act of 1954 followed chronologically on the report of the Jenkins Committee (Cmd 7982), but did not exactly embody its recommendations. The effect was to limit the grounds on which the landlord of business tenancies might recover possession at the end of a contractual tenancy and to substitute for the landlord's freedom of action a right on the part of J the business tenant to secure a new lease, or in certain other cases to claim compensation for the loss of his security of tenure. The relevant provisions affecting the terms of the new tenancy are sections 32 which deals with the property to be comprised, 33 which deals with the length of the tenancy, 34 which prescribes the rent, and 35 which governs the terms of the new lease. Of these I need only quote part of section 34 and section 35 as relevant to the present appeal. Section 34 reads as follows:

> 34(1) The rent payable under a tenancy granted by order of the court under this Part of this Act shall be such as may be agreed between the landlord and the tenant or as, in default of such agreement, may be determined by the court to be that at which, having regard to the terms of the tenancy (other K than those relating to rent), the holding might reasonably be expected to be let in the open market by a willing lessor there being disregarded . . .

There follow four disregards irrelevant for my present purpose, and two further subsections, also irrelevant, respectively dealing with the definition of improvements, and conferring on the court, in appropriate cases, power to provide for variations in the rent. Section 35 is as follows:

> 35. The terms of a tenancy granted by order of the court under this Part of this Act (other than terms as to the duration thereof and as to the rent payable thereunder) shall be such as may be agreed between the landlord and the tenant or as, in default of such agreement, may be determined by the L court; and in determining those terms the court shall have regard to the terms of the current tenancy and to all relevant circumstances.

From these sections I deduce three general propositions.

(1) It is clear from section 34 that, in contrast to the enactments relating to residential property, Parliament did not intend, apart from certain limitations to protect the tenant from the operation of market forces in the determination of rent:

(2) In contrast to the determination of rent, it is the court and not the market forces which, with one vital qualification, has an almost complete discretion as to the other terms of the tenancy (which, of course, in turn must exercise a decisive influence on the market rent to be ascertained under section 34); and

M (3) In deciding the terms of the new tenancy, as to which its discretion is otherwise not expressly fettered, the court must start by "having regard to" the terms of the current tenancy, which *ex hypothesi* must have been either originally the subject of agreement between the parties or themselves the result of a previous determination by the court in earlier proceedings for renewal.

A certain amount of discussion took place in argument as to the meaning of "having regard to" in section 35. Despite the fact the phrase has only just been used by the draftsman of section 34 in an almost mandatory sense, I do not in any way suggest that the court is intended, or should in any way attempt, to bind the parties to the terms of the current tenancy in any permanent form. But I do

A believe that the court must begin by considering the terms of the current tenancy, that the burden of persuading the court to impose a change in those terms against the will of either party must rest on the party proposing the change, and that the change proposed must, in the circumstances of the case, be fair and reasonable and should take into account, amongst other things, the comparatively weak negotiating position of a sitting tenant requiring renewal, particularly in conditions of scarcity, and the general purpose of the Act, which is to protect the business interests of the tenant so far as they are affected by the approaching termination of the current lease, in particular as regards his security of tenure. I derive this view from the structure, purpose, and words of the Act itself. But, if I required B confirmation of it, I would find it in the passages cited to us in argument from the judgment of Denning LJ in *Gold v Brighton Corporation* [1956] 1 WLR 1291 at p 1295 and of Widgery LJ in *Cardshops Ltd v Davies & Another* [1971] 1 WLR 591 at p 596 (also cited with approval by Shaw LJ in the instant case). The point is also emphasised by the decision in *Charles Clements (London) Ltd v Rank City Wall Ltd* (1978) 246 ESTATES GAZETTE 739, where the court rejected an attempt by the landlord as a means of raising the rent to force on a tenant a relaxation of a covenant limiting user which would have been of no value to the particular tenant, and *Aldwych Club Ltd v Copthall Property Co Ltd* (1963) 185 ESTATES GAZETTE 219 where the court rejected an attempt by the tenant to C narrow the permitted user with a view to reducing the rent.

A further point which was canvassed in argument, and with which I agree, is that the discretion of the court to accept or reject terms not in the current lease is not limited to the security of tenure of the tenant even in the extended sense referred to by Denning LJ in *Gold v Brighton Corporation (supra)*. There must, in my view, be a good reason based in the absence of agreement on essential fairness for the court to impose a new term not in the current lease by either party on the other against his will. Any other conclusion would in my view be inconsistent with the terms of the section. But, subject to this, the discretion of the court is of the widest possible kind, having regard to the almost infinitely varying circumstances of individual D leases, properties, businesses and parties involved in business tenancies all over the country.

This brings me directly to the judgments of the courts below, and I begin with that of Goulding J. I was at first of the opinion that, in view of the width of the discretion conferred by section 35, it would be extremely difficult to justify interference by an appellate court with the exercise of discretion by a trial judge of wide experience who had taken infinite pains with the detail of the case and expressed his views in a judgment rightly characterised by the Court of Appeal as "pellucid". I must now describe how I have come to change my mind on this point and now consider the decision of the Court of Appeal as embodying the correct conclusion.

E The learned judge admittedly arrived at his decision by the application to the facts of the instant case of four tests or the answers to four questions, after cautiously and, in my opinion correctly, making it plain that "I do not regard them as a correct scheme of analysis for all similar cases".

The four tests, or questions, were as follows:

(1) Has the party demanding a variation of the terms of the current tenancy shown a reason for doing so?

(2) If the party demanding a change is successful, will the party resisting it in principle be adequately compensated by the consequential adjustment of open market rent under section 34?

F (3) Will the proposed change materially impair the tenant's security in carrying on his business or profession?

(4) Taking all relevant matters into account, is the proposal, in the court's opinion, fair and reasonable as between the parties?

These tests were accepted as valid by the Court of Appeal, and although, as will shortly be seen, I have my own reservations as to the appropriateness of this method of approach, for the moment I am content to say that it is only at the fourth and last that the learned judge enters into the realm of what he wittily and aptly describes as:

> the realm of pure discretion, where it is hardest to condense in words the dialogue of the intra-cranial jury room.

G It follows, I believe, inexorably from this mode of approach that if, as the Court of Appeal ultimately concluded, the answers proposed by the judge to any of the first three questions are erroneous, either because the questions themselves are in some degree defective or because the answers are in some degree mistaken, the matter of discretion becomes clearly at large for an appellate court, since the learned judge had entered the "intra-cranial jury room" only on the basis of the answers he had given to the first three questions he had formulated. In the event the Court of Appeal concluded, and, as will be seen, I agree, that the affirmative answer proposed by the learned judge to the second question was one which they could not accept. In the words of Shaw LJ who, at the request of Buckley LJ, delivered the leading judgment: H

> It is here that, regretfully, I find myself parting company with the learned judge. The advantage conferred on the landlord by a clear lease is the freedom from factors which are of uncertain outcome and which elude absolute quantification and lend themselves to what at best is a conjectural assessment, however informed the source. That freedom removes an element of risk from the landlord's financial situation; but the element of risk is not obliterated. It is transferred to the tenant. In the outcome, the diminution in the rent, though "fair compensation" when in prospect, may prove in actuality to be wholly inadequate as the years go by. If the risk is disadvantageous and therefore unacceptable to the landlord at a commensurable rent why should it be more acceptable and less disadvantageous to the tenants albeit at a reduced rent? This is the more so since they are not in the business of property management and property dealing. J

Brightman LJ put the same point as follows:

> I agree entirely with my Lord, Shaw LJ, that this case has to be considered under section 35 and not under section 34, because the substantial issue is not the amount or calculation of the rent but the incidence of certain unusual financial burdens which will have the effect of controlling the rent. Approaching the case on this basis, and applying the three tests enunciated by Goulding J, I entirely agree with Shaw LJ that a short-term tenant such as the appellants, is not adequately compensated by a small reduction in rent for the assumption of the financial risks implicit in the maintenance of the structure of an office block. Those risks are proper to be borne by the owner of the inheritance or of a long term of years but are not appropriate to be borne by one who is in possession for three years and has no further interest K save a limited statutory security of tenure. Such risks are indeterminate in amount and could prove to be wholly out of proportion to the very limited interest held by a short-term tenant. No doubt they can be quantified annually when averaged over a number of years, and better still when averaged over a number of properties. But they are not capable of being fairly allowed for, by way of a reduction of rent, when the tenancy has only a few years to run. The tenant in such a case would be made the insurer of the landlord against unknown contingencies of undefined cost. I wholly agree with the approach of Shaw LJ, and I respectfully adopt the entirety of his reasoning. I would accordingly allow the appeal.

Buckley LJ in a succinct, but none the less powerful, judgment said:

> The effect of the proposed modification of the terms of the existing tenancy L in the present case is to make the lessee an insurer of the lessor against risks connected with the structure of the building which under the terms of the existing lease have to be borne by the lessor. They are risks which are inherent in the ownership of the building. The reduction in the rental which is said to compensate the lessee for assuming these risks is far from being an indemnity against them. The lessee might find himself at any time unexpectedly and unforeseeably saddled with very heavy capital expenditure for which the reduction in rent would by no means compensate him, except possibly in the very long run; but, as has already been pointed out, a particular tenant may fill that role only for a relatively short period and compensation which is only compensation on a very long-term basis would be no comfort to him.
>
> The proposed alterations in the terms of the lease, in my judgment, alter M the mutual relations of the lessor and the lessee so drastically, and so adversely to the lessee, that the court could not in the exercise of its discretion properly impose them on the lessee under the Act in the absence of some more cogent reason than can be discovered in this case.
>
> For these reasons, and for those already given by my Lords, I agree that this appeal should be allowed.

For reasons which I will shortly express in my own words I myself agree with this criticism of the learned judge's proposed answer to the second question, and it follows from this, for the reason I have already given, both that the point was open to the Court of Appeal and that the appeal to your Lordships' House must in consequence be dismissed.

A I wish, however, slightly to widen my reservations to the approach of the learned judge. By formulating the questions and answering them in the way he did, I believe that he had logically, so to speak, painted himself into a corner before he came to the exercise of any discretion at all. Take for instance, since I will be coming in due course to the crucial question, the way in which the first and third questions are formulated and answered. Admittedly the landlord's desire to vary the terms of the current lease in such a way as to maximise the value and marketability of his reversion is a perfectly valid and respectable reason from the landlord's point of view for seeking to secure the tenants' agreement to vary the terms of the tenancy. So, in my view, in *Gold*'s case (*supra*) the desire of
B the corporation to improve the tone of the street of which it was an important landlord was a perfectly valid reason in itself for seeking to secure the tenant's agreement to vary the terms of the *Gold* tenancy. So also in the *Clements* case (*supra*) was the desire of the landlord to increase the rent, and so in the *Aldwych Club* case (*supra*) was the tenants' desire to diminish the rent. But in my view it was precisely at that stage and not only at the fourth and last that the "intra-cranial dialogue" should have begun, and the interests of the other, and unwilling, party to the proposal should have been taken into consideration, and were ultimately so taken into consideration by the court in the cases cited. The improvement in the tone of the street was a perfectly valid aim of the landlords. But if
C done at the expense of the tenant's current business it was not a reasonable term to introduce into a lease against the tenant's wishes under an Act of Parliament the purpose of which was to protect the tenant's interest in carrying on his existing business. In the other two cases the desire to increase or diminish the rent was a perfectly legitimate negotiating objective for the landlord or the tenant respectively, but not, as the court held, by forcing on the opposing party an unwanted advantage which, in the circumstances, would have conferred no real benefit on him and to which he did not agree.

When I come to the third test, the same kind of consideration applies. The security of tenure which is protected by the Act does
D not simply mean protection from the danger of eviction by the landlord. The judgment of Denning LJ in *Gold*'s case is enough to establish that. The intra-cranial dialogue which the judge conducts must in my view enter in at every stage of the analysis, and the question should be asked: granted that the negotiating objective of the party proposing a variation is perfectly legitimate, is it reasonable in the circumstances to impose it willy nilly on a resisting landlord or tenant?

This brings me to the crucial second question. Here, too, the intra-cranial dialogue is vital. To some extent the learned judge actually conducted it, since the dialogue is in a sense implicitly involved in the qualifying adverb "adequately" before the word
E "compensated" in the question. But in my view the word "adequately" itself conceals an ambiguity, and it seems to me that, at the critical point in his judgment, the ambiguity misled the judge into a question-begging assumption. Of course, in a sense, the testimony of the experts was agreed. But the transcript of the evidence is in my view conclusive to the effect that all that was agreed was *if* (emphasis mine) a fluctuating obligation was to be replaced by a fixed figure or insurance premium calculated either at market value (or an estimate of the chances) made at the commencement of the new term, the sum of 50p deducted from the rent was adequate compensation. But the real question is contained in the protasis and not the apodosis of the above conditional clause. The point at issue
F was whether it was reasonable under section 35 to impose the fluctuating burden on the tenant against his will in return for a fixed figure valid throughout the new tenancy and calculated as at the beginning of the tenancy. I am not satisfied that in his answer to the second question the learned judge really considered this point adequately, and perhaps not at all, and I believe his failure to do so was due at least in part to a misunderstanding at the critical point of the exact nature of the concession made in cross-examination of the tenants' own expert witness. The learned judge reported this in the following terms:

> Nevertheless he (ie the witness) was satisfied that the agreed rent reduction . . . was a fair compensation in all the circumstances.

In two questions and answers which were properly brought to our attention the witness had said in the first: G

> Q. Does it not follow that if my Lord decides to approach this matter in this way: what would the rent be if the terms were those of form 16; answer £105,000 odd. If I decide after hearing all the evidence and argument the new tenancy ought to be on form 20, and if I accept as I presumably must the figure that the two surveyors have agreed of £100,000 odd, will the tenant be fairly compensated for the transfer of burdens resulting from the change from form 16 to form 20? The answer to that question ought to be in the affirmative.
>
> A. It ought to be in the affirmative, I accept that. *But the risk involved in the transfer of burden over the period of the lease I still maintain cannot be fully anticipated now. All the market can do is take a view, and it takes a view in the* H
> *way it has in terms of the two rents we were talking about.*

The emphasis in the above answer is mine.

The second question and answer was:

> Q. Is £4,950 per annum a figure which taken over the three years this form will operate, is it likely to produce a net benefit to the tenant or a net loss to the tenant. If £4,950 is not the right figure, what is the right figure?
>
> A. I think in terms of the discussions that Mr Perkins and I had, my view was that the difference of £4,950 was the right difference bearing in mind that we had previously fixed the rental value in the market. *Whether that adequately compensates the tenant over the term of the lease bearing in mind the transfer of the burden is something I do not think anybody can necessarily say* J
> *today. We would have to wait and see until the end of the lease.*

Again the emphasis added to the answer is mine.

That the learned judge did so misunderstand the evidence appears not only from the judgment, but from an interposition by the learned judge recorded on the following page of the transcript, which, for the reason that it involved quite a considerable discussion with counsel, I forbear to quote at length.

The matter thus being clear for an appellate court, I now give my own reasons for agreeing with the Court of Appeal. It is obvious that in the case of an office block in multiple occupation by different tenants the actual management of the exterior parts, common parts, K
lifts, boilers, and ancillary services will ordinarily rest in the hands of the landlords who will ordinarily covenant to provide them. Some of these items are very readily calculable or may readily be made the subject of insurance. Some may fluctuate enormously and the extent of fluctuation will, as the tenants' witness said, only ultimately be ascertainable at the end of a lease. Obviously it is to the advantage of the landlord to transfer the financial risk of fluctuation to the tenant, and there can be no possible reason why, if the tenant agrees (and the evidence was that many do), he should not do so. But the crucial question is, if the current lease does not so provide and the tenant does not agree, by what possible reasoning should the court impose the burden on the tenant against his will as
a condition of his receiving a new tenancy under Part II of the L
Landlord and Tenant Act 1954? It may be granted that the transfer of the risk from the landlord to the tenant is a perfectly legitimate negotiating aim for the landlord to entertain. But the argument is two edged. It is equally a legitimate negotiating aim of the tenant to resist the change. Granted that a reduction in the rent of 50p from £10.50 per foot to £10.00 per foot is, in the limited sense described, an adequate estimate of the compensation which a landlord will offer if the risk is to be transferred. But the argument is again two edged. It may equally be argued that an additional 50p is the adequate estimate of the rent rightly payable to the landlord if the risk is to be kept where it is under the current lease. But neither of the two statements assists to answer the question where, in the new M
lease, is the risk of fluctuation to lie. If I am correct that the inference from the authorities is that the language of section 35 requires that the party (whether landlord or tenant) requiring a change must justify as reasonable a departure from the current lease in case of dispute about its terms, the answer must be that *prima facie* it must lie where the current lease provides, and that a mere agreement about figures based on either or both of two rival hypotheses does not shift the burden in any way.

For completeness I must deal with one remaining point. Canvassed before Goulding J was a dispute between the parties which I have not hitherto mentioned including one about the so-called

A "challenge provisions" and one about a term referred to by Goulding J as the "funding provisions". This dispute was not dealt with in the judgments of the Court of Appeal, though arguments relating to it are contained in the respondents' printed case, and at the hearing before the Appellate Committee the appellants sought, and obtained, leave to file a supplemental case canvassing the arguments in detail. If effect were given to them separately it would have been open to the appellants, driven back from their main case, to adopt an intermediate position, reverting to the order of Goulding J for certain minor purposes, but accepting defeat at the hands of the respondents on the main bone of contention. It became apparent, however, that such an intermediate position is not really tenable.
B The case has been conducted throughout, and, in particular, was conducted in the Court of Appeal on the footing that what the appellants wanted was a "clear lease", and it was clearly accepted that, if they failed on that, in effect they failed altogether, since there was neither material on which the intermediate position could be formulated with adequate precision nor room at all for effect to be given to it as a separate argument. As part of the total package designed to impose a clear lease the individual details formed a legitimate part of the case. As a separate entity they had no real independent existence, and to give effect separately now would be to give effect to a new case, not argued in the Court of Appeal, and, so far as one could judge from junior counsel's contemporary note,
C effectively disclaimed as a separate and alternative case in the exchanges before the Court of Appeal.

In the result my opinion is that the appeal fails and should be dismissed with costs.

Agreeing that the appeal should be dismissed, LORD WILBERFORCE said: The facts of this case, and the relevant statutory provisions, have been fully stated by the Lord Chancellor and I shall not repeat them. The general purpose and policy of the Act of 1954 is clear. It was to provide security of tenure for those tenants who had established themselves in business in leasehold premises so that they could continue to carry on their business there. This objective was identified in the Final Report of the Leasehold Committee of
D 1950 (Cmd 7982), as the principal, indeed the only, objective then recommended to be achieved by legislation. There was no suggestion, nor did the Act, when enacted, contain any provisions to the effect that business tenants required any greater protection than that necessary to enable them to continue their business: no protection, nor did the Act, when enacted, contain any provisions, to the contractual terms. I accept therefore the landlord's contention that, in principle, tenants are not to be protected from market forces; as regards rent, indeed, this is expressly laid down in section 34, which requires that the rent is to be the open market rent, subject to certain specified "disregards". The same underlying principle ought to be applied to the determination of other terms in the new
E lease, subject, however, to the guidelines laid down in section 35 — as to which see below.

The Act, in the portion (Part II) of it which deals with business tenancies, is in the main a discretionary Act, giving wide powers to the judge to grant and settle the terms on which the business tenant is to have a new lease. This applies particularly to sections 33 and 34, which relate to the duration of the tenancy and the rent. The crucial section, for present purposes, is section 35, which relates to the terms of the tenancy, other than terms as to duration and rent. This section contains a mandatory guideline or direction to "have regard to" the terms of the current tenancy and to all relevant circumstances. The words "have regard to" are elastic: they compel
F something between an obligation to reproduce existing terms and an unfettered right to substitute others. They impose an onus upon a party seeking to introduce new, or substituted, or modified terms, to justify the change, with reasons appearing sufficient to the court (see *Gold v Brighton Corporation* [1956] 1 WLR 1291 — on "strong and cogent evidence" per Denning LJ; *Cardshops Ltd v Davies & Another* [1971] 1 WLR 591, 596 per Widgery LJ.)

If such reasons are shown, then the court, applying the words "all relevant circumstances", may consider giving effect to them: there is certainly no intention shown to freeze or, in the metaphor used by learned counsel, to "petrify" the terms of the lease. In some cases, especially where the lease is an old one, many of its terms may be out of date, or unsuitable in relation to the new term to be granted. G
If so or for other good reasons shown, the court has power to order a modification by changing an existing term or introducing a new one (eg a break clause, cf *Adams v Green* (unrep) March 20 1978 CA*). Before doing so it will consider any objections by the tenant, and where there is an insoluble conflict, will decide according to fairness and justice.

The question at issue in this case does not relate to duration or rent. The duration of the new lease is agreed at five years from March 25 1977: the appropriate rent follows from determination of the terms. According to whether the terms are fixed according to the proposals of the landlord, or to those of the tenant, the amount of the rent follows, the experts being agreed as to the amount. What H
the experts were not agreed upon is the prior question whether the tenants can, in principle, be compensated by any reduction in rent for the greater burden sought to be imposed on them by the landlord.

The nature of that burden can be briefly summarised as follows. Instead of the landlord being responsible (as under the old lease) for repairs, maintenance, and decoration of the exterior and common parts of the building, and for providing and maintaining lifts and other plant (including boilers), the tenants are to bear, by way of service charges, a proportion, attributable to their holding, of the cost of these items. These costs are to be ascertained by certificate of the landlord's surveyor which the tenants have only a limited J
right to challenge. There is further proposed a funding provision under which the tenants are to pay annually an amount based on the assumption that work is done at intervals or on assumed life expectancies. I do not detail these provisions, because at this stage the landlord's proposals must be regarded as one packaged whole and, if the main provision for shifting the burden is unacceptable, must be rejected however fair other provisions taken by themselves might be. Mr Bernstein QC did endeavour to invite us to consider separately one portion of them — namely that relating to lifts and equipment — but that we cannot do. We can only deal with the case as presented to and considered by the judge; we cannot exercise an original discretion upon part of it. Correspondingly, no decision by K
the House upon the package as a whole can prevent, or prejudice, a later or separate application as regards any severable part of the package.

The landlord's case for a shift of the burdens, as I have indicated, rests upon two broad foundations. The first is that such a shift is, for quite genuine and respectable reasons, in the interest of the landlord. I accept this; I think that it is shown by evidence that in the present and foreseeable state of the property market, a freehold interest in commercial property, at least in London, commands a higher price if let on "clear leases" — ie leases in which the tenants bear all the costs and risks of repairing, maintaining and running the building of which their demised premises form part, so that the rent L
payable reaches the landlord clear of all expenses and overheads. Thus the landlord demonstrates a genuine interest in departing from the existing terms of the lease. Secondly, the landlord asserts that new leases of property, such as that with which we are concerned, are now granted and accepted by tenants as "clear leases". This, too, is supported by the evidence. But, in my opinion, though a relevant circumstance it is not decisive. There is no obligation, under section 35 of the Act, to make the new terms conform with market practice, if to do so would be unfair to the tenant. And there is no inherent necessity why the terms on which existing leases are to be renewed should be dictated by those of fresh bargains which tenants may feel themselves obliged to accept. The court has to M
compare the advantage desired by the landlord with the detriment to be suffered by the tenants and to consider whether any monetary compensation offered against that detriment ought fairly to be imposed upon the tenants in exchange for the acceptance of that detriment. That money is not necessarily fair compensation for a change in existing rights is obvious in itself and is well recognised by the law in relation, for example, to compulsory acquisition and to the granting of damages instead of an injunction.

There can be no doubt in my opinion that this detriment is real

*(1978) 247 EG 49.

A and serious. Considering only the obligation to bear a proportion of the cost of maintaining and repairing the exterior and common parts of the building, to impose this upon the tenants is something which they may most reasonably resist. They risk incurring a liability which is unpredictable and which may, in the event of a structural defect, be very great. They have no power of precautionary inspection or survey, since they only have access to part of the building. They have no means of verifying that work for which they are charged was necessary at the time, or was truly repair and not improvement, the cost of which ought not to be put on the tenants, nor of controlling what work has been done or how it has been done. As tenants, carrying on a solicitor's business, they have no B staff capable of performing these tasks, whereas the landlord, as the large property company with an interest in over 200 buildings in the City of London, has. If work can be ordered and effected by persons, other than those who have to bear the cost, risks of extravagance and misdirection of effort may be created. The same separation of responsibility necessitates a system of charging by certificate and preventing challenges, except in rare cases, which is of its nature onerous and prejudicial.

The character of the two parties' interests in the land — the landlord's an indefinite one by freehold, the tenants' a limited one over a comparatively short period, even though capable of renewal if the tenant so wishes, is such as to call for the assumption of long C term risks by the former: his benefit too is long term and will not, according to the evidence, emerge till the 1990s. Transference of these rights to the leaseholders, accompanied, as is inevitable, by separation of control, creates a risk disproportionate to their interest. If it is reasonable for the landlord to wish to get rid of these risks in exchange for a fixed payment (reduction in rent) it must be equally, or indeed more, reasonable for the tenants to wish, against receipt of that reduction, to avoid assuming it. The tenants are being asked to bear all the risks of property management, a business which they have not chosen, being management by others in the interest of those others. The present distribution of burdens is that freely and contractually agreed upon so recently as in 1972. To recast it involves a D serious departure from the terms of the current lease. In my opinion, a court which has to have regard to the terms of the current lease ought not to sanction such a departure, and such other circumstances as should fairly be taken into account — the landlord's wishes and the increasing acceptance by others, in different situations, of clear leases — are insufficient to give grounds for so doing.

The balance to be struck, and the result of striking it, are to my mind clear, and my only hesitation as to the result of this appeal comes from the fact that the learned trial judge, whose discretion it was, in a very well reasoned judgment, came to the conclusion that the landlord's proposals should be accepted. But I have been per- E suaded that he did so partly at least upon a misconception of the effect of some of the evidence and partly through losing sight at the critical point of the vital question whether, even for monetary compensation which arithmetically might be fair, the case for compelling the tenants to accept such compensation in exchange for the transference of burden was made out. The noble and learned Lord on the Woolsack has developed the argument on this point in a manner with which I wholeheartedly agree, and I need not contribute any additional reasons. I have no doubt that the Court of Appeal, which was, of course, well aware of the respect to be given to the exercise of discretion by the trial judge, was entitled and right to take a different view. I respectfully agree with all three judg- F ments.

I would dismiss the appeal.

Lord Keith of Kinkel, Lord Scarman and Lord Brandon of Oakbrook agreed that the appeal should be dismissed for the reasons given in the speeches of the Lord Chancellor and Lord Wilberforce.

G

COURT OF APPEAL

November 30 1981

(Before Lord Justice STEPHENSON, Lord Justice DONALDSON and Sir David CAIRNS)

WATKINS v EMSLIE

Estates Gazette March 27 1982
(1981) 261 EG 1192

Landlord and tenant — Action for possession of a field — Estoppel H — Tenancy to which Part II of Landlord and Tenant Act 1954 applied — Field used for purposes of running livery stables and a riding school — Tenant's notice under section 26 of 1954 Act requesting a new tenancy invalid because her tenancy was not for a term of years certain exceeding one year but was a periodic tenancy — Landlord submitted that, because of tenant's service of a request for a new lease and her solicitors' insistence that she was entitled to apply for a new lease, he must be taken to have refrained from serving a notice to determine under section 25 and thereby suffered a detriment — Held that there was no evidence that landlord refrained from serving such a notice because he believed that tenant's request was valid — *Greasley v Cooke* distinguished — In any case it was J doubtful whether an estoppel could be relied on to enable the court to make an order for possession when the tenancy had not in fact been determined in any of the ways laid down by Part II of the 1954 Act — Landlord's appeal against county court judge's order in favour of tenant dismissed

This was an appeal from a decision of Judge Tibber at Edmonton County Court dismissing an action for possession by Ernest John Watkins, the plaintiff and present appellant, against Mrs Christine Emslie, the defendant and present respondent. The proceedings related to the possession of a field at Enfield in Middlesex of which the respondent was the tenant and the appellant the landlord.

K J Lindsay QC and R C I'Anson Banks (instructed by Kenneth Shaw & Co) appeared on behalf of the appellant; K H Zucker QC and M Warwick (instructed by Tyrer Roxburgh & Co) represented the respondent.

Giving the first judgment at the invitation of Stephenson LJ, Sir David Cairns said: This is an appeal from two judgments of His Honour Judge Tibber, sitting in the Edmonton County Court. Both judgments were given in an action whereby the plaintiff, Mr Watkins, sought an order for possession of a field at Enfield in Middlesex against the defendant, Mrs Emslie. The first judgment was given on February 16 1981 on a preliminary point, which the learned judge decided in favour of the defendant; the second judgment was given L two days later, when he dismissed the action, and the plaintiff appeals against both those judgments.

There was a previous action between the same parties, the defendant being at that time unmarried and sued in her maiden name of Healy. In that action the plaintiff contended that the defendant was a trespasser on the field, but the same judge on March 3 1980 held that she was a tenant and dismissed the action for possession.

In the second action the plaintiff, by his particulars of claim, alleged that the defendant was a tenant up to May 31 1980 but that the tenancy was determined on that day and she had continued using the field as a trespasser. The reason why he said that the tenancy had determined on that date was this: on September 29 M 1979 her solicitors on her behalf had served on the plaintiff a notice under section 26 of the Landlord and Tenant Act 1954 requesting him to grant her a new lease as from June 1 1980. Section 26 is in Part II of the 1954 Act, which of course applies to business tenancies, and the relevant provisions of the section are subsection (1) and subsection (5).

Subsection (1) reads:

> A tenant's request for a new tenancy may be made where the tenancy under which he holds for the time being (hereinafter referred to as "the current tenancy") is a tenancy granted for a term of years certain exceeding one year, whether or not continued under section 24 of this Act, or granted for a term of years certain and thereafter from year to year.

A Subsection (5) provides as follows:

> Where a tenant makes a request for a new tenancy in accordance with the foregoing provisions of this section, the current tenancy shall, subject to the provisions of subsection (2) of section 36 of this Act and the provisions of Part IV of this Act as to the interim continuation of tenancies, terminate immediately before the date specified in the request for the beginning of the new tenancy.

Neither section 36 nor anything in Part IV was applicable. So if the defendant was a tenant, and if it was a business tenancy and if she held for a term of years certain exceeding one year, the effect of her notice was to determine the tenancy on May 31 1980.

B By her defence the defendant denied that the tenancy had been within the Landlord and Tenant Act, and averred that it was a tenancy of an agricultural holding within the Agricultural Holdings Act 1948 as amended. As an alternative, she averred that if it was not within the 1948 Act it was within the 1954 Act.

The plaintiff, by his reply, contended that the defendant was estopped from denying that the tenancy was protected by the 1954 Act by reason of the notice given in September 1979 requesting a new tenancy, and by repeated assertions by her solicitors on her behalf that the tenancy was so protected, and by her pleading in the first action that it was protected by the 1954 Act, and by the judge having found that it was. On similar grounds he contended that she
C was estopped from saying that the September 1979 notice was not a valid notice.

Now, it was indeed correct that the learned judge had said, at the end of his judgment in the first action: "My findings are therefore that Mr Healy and now Miss Healy are tenants. They were tenants for the running of livery stables and a riding school for the purpose of the Landlord and Tenant Act 1954".

The preliminary issue that the judge heard on February 16 1981 was as to whether that finding created an estoppel. He held that it did not, because the question of what type of tenancy it was was not an issue in the first action.

Then, in his second judgment, on February 18, dealing with the
D other estoppels alleged, he held that none of the representations created an estoppel because none of them was relied upon by the plaintiff.

The plaintiff appeals to this court solely on the basis that the learned judge was wrong in his conclusions as to estoppel. The defendant, by a respondent's notice, seeks to support the judgment on the additional ground that the tenancy was under the Agricultural Holdings Act, and that therefore the learned judge had no jurisdiction to order possession; alternatively that, even if the tenancy was within the Landlord and Tenant Act, it was not one which entitled her to obtain a new lease, because the notice under section 26 was a nullity.

E That was how the matter stood until the early days of last week, but the defendant's solicitors on her behalf told the plaintiff's solicitors about a week ago that the defendant would no longer contend that her tenancy was not within Part II of the 1954 Act. Mr Zucker, who appeared in this court on behalf of the defendant, confirmed to us that that was so, saying that in his submission the evidence which she gave in February of this year, and which the judge accepted, established that the tenancy was indeed a business tenancy within that Act. It is therefore now common ground that this was the character of the tenancy. In case anybody should hereafter wish to resile from that common ground, I may say that I have considered the evidence, and am quite satisfied that it was indeed a tenancy to which Part II of the 1954 Act applies.

F It is accepted by Mr Lindsay, on behalf of the plaintiff, that the defendant's tenancy was not in fact for a term of years certain exceeding one year; it was probably a weekly or a monthly tenancy. Accordingly, he concedes that the only basis on which he can succeed in this appeal is on the ground of estoppel. The alleged estoppel, which was the subject of the preliminary issue, is now unnecessary and unhelpful to the plaintiff – unnecessary, because it is now agreed that there was a tenancy to which Part II of the 1954 Act applies; unhelpful, because it does not relate to the term or period of the tenancy.

What is still relied upon is the service of the notice requesting the new tenancy, the putting of that notice in evidence at the hearing of the action and the insistence by the defendant's solicitors on her G
behalf of her right to claim a new tenancy in pursuance of that notice. Mr Banks, junior counsel for the plaintiff, drew our attention in particular to the fact that after the judgment had been given on March 3 1980, by which the plaintiff's contention that there was no tenancy at all had been rejected, the defendant's solicitors continued to rely on the request made in September 1979 for a new lease.

The argument on behalf of the plaintiff is that it was because of the service of the request for a new lease and of the insistence of the defendant's solicitors that she was entitled to apply for a new lease that the plaintiff must be taken to have refrained from serving a notice to quit under section 25, and thereby suffered a detriment in H
delaying the date at which he might be able to obtain possession.

There was in fact no evidence by the plaintiff that he ever placed any reliance on the request or on the defendant's solicitors' letters, and indeed it is clear from the way in which his case was presented in the first action and from the evidence that he gave at the hearing of that action, that the reason why he did not serve a notice to quit under section 25 was that it was his case that the defendant was never a tenant at all.

It is true that after March 3 1980 he could no longer rely on that view; but on March 17 1980 the plaintiff's solicitors wrote to the defendant's solicitors giving reasons why a new tenancy should not be granted and saying "We will undoubtedly be instructed to serve J
notice to quit on your client". But they did not do so, and on April 25 1980 the defendant's solicitors pointed out that there had never been such a tenancy as entitled the defendant to apply for a new lease. Still no notice to quit was served, and the plaintiff started the new action in July 1980. In the circumstances I find it impossible to say that there was any evidence that the plaintiff refrained from serving a notice to quit because he believed, or his legal advisers believed, that the request for a new tenancy was a valid request – and indeed, Mr Lindsay realistically did not contend that there was any such evidence.

However, Mr Lindsay relied on the decision of this court in *Greasley v Cooke* [1980] 1 WLR 1306. In that case a woman had K
lived in a house, first as a living-in maid to the owner and his sons, and later, after the death of the father, as an unpaid housekeeper to the family, and later still as the sole occupant of the house. An action for possession was brought against her, which she defended, and she counterclaimed for a declaration that she was entitled to occupy the house for the rest of her life if she wanted to. Her case was that two of the sons had given her assurances which led her to believe that she could regard the house as her home for the rest of her life, and her counsel's contention was that she acted to her detriment by continuing to render services to the family without payment as a result of her belief that she would be able to remain there. The contention on her behalf was that she was entitled to L
succeed in her defence and counterclaim on the ground of a proprietary estoppel.

The county court judge who heard the case, while accepting her evidence, held that there was no estoppel, because she had failed to prove that it was as a result of her belief that she continued to render the services. The Court of Appeal allowed her appeal, holding that the court should infer that it was because of the assurances given to her that she had so continued.

The Master of the Rolls, at p 1311, in the course of his judgment said:

> These statements to Miss Cooke – that is, the defendant, and the statements were the assurances that I have referred to – were calculated to influence her M
> – so as to put her mind at rest – so that she should not worry about being turned out. No one can say what she would have done if Kenneth and Hedley had not made those statements. It is quite possible that she would have said to herself: "I am not married to Kenneth. I am on my own. What will happen to me if anything happens to him? I had better look out for another job now; rather than stay here where I have no security".
>
> So, instead of looking for another job, she stayed on in the house looking after Kenneth and Clarice. There is a presumption that she did so, relying on the assurances given to her by Kenneth and Hedley. The burden is not on her, but on them, to prove that she did not rely on their assurances. They did not prove it, nor did their representatives. So she is presumed to have relied on them. So on the burden of proof it seems to me that the judge was in error.

A Waller LJ, at p 1312, from G to H, said:

It was a difficult question – not a simple question – which the judge had to consider; and one cannot help being sympathetic in that he had to consider it in the absence of argument from the plaintiffs, and, as an explanation for his getting the onus of proof wrong in this particular case, that is probably the main reason. However, I am satisfied that he did, unfortunately, and that he was in error in the way in which he put the onus of proof.

Dunn LJ agreed; at the foot of p 1313 he said:

The judge considered his judgment and found for the defendant on every point except one, and that is the only point that we can consider on this appeal. It is a narrow point. There is no doubt that for proprietary estoppel
B to arise the person claiming must have incurred expenditure or otherwise have prejudiced himself or acted to his detriment. The only question before us is as to the burden of proof of the detriment. The judge thought that the onus lay on the claimant to prove it. I agree that in that he fell into error for the reasons given by Lord Denning MR and I also would allow this appeal.

The present case is not one of proprietary estoppel. It is not one of any assurances being given. It is not one in which the plaintiff is alleged to have acted in a particular way in reliance on what was said or done by or on behalf of the defendant, but only that he refrained from acting; that is to say, from serving a notice to quit under section 25. However, the plaintiff here knew perfectly well that there had been no letting for a term of years of more than a year; he
C was advised by solicitors, and I very much doubt if the service of the notice, or the letters written by the defendant's solicitors, could be said to be calculated to make him believe, despite the clear words of section 26(1), that the notice was a valid one.

Further, since sections 24 to 27 lay down strict rules as to how a tenancy within Part II of the Act can be terminated, I am by no means satisfied that an estoppel can be relied on to enable the court to have jurisdiction to make an order for possession when the tenancy has not in fact been terminated in any of those ways.

If the plaintiff can clear all those fences – and if the onus was on the defendant to establish that it was not by reason of any representation on her behalf that the plaintiff did not serve a notice to quit – I
D am of the opinion that the onus was discharged. The judge found as a fact that the plaintiff did not refrain from granting a new lease because of the defendant's notice, but because he was alleging that there was no tenancy. In my judgment, the judge rightly so found. He made no reference to what occurred after March 3 1980, but it is abundantly clear that the plaintiff's solicitors were persuaded that the tenancy was a business tenancy, not because of the defendant's notice, or her solicitors' letters, but because of the judge's finding. On March 17 they were taking the view that the plaintiff was entitled to serve a notice to quit under section 25. This was inconsistent with the defendant's notice being a valid notice, because section 26(4) expressly provides that no notice to quit shall be given
E after the making by the tenant of a request for a new tenancy.

For those reasons I would dismiss this appeal.

DONALDSON LJ: I agree.

STEPHENSON LJ: I also agree.

CHANCERY DIVISION

May 22 1981

(Before Mr Justice SLADE)

F LLOYDS BANK LTD v NATIONAL WESTMINSTER BANK LTD

LLOYDS BANK LTD v CITY OF LONDON CORPORATION

Estates Gazette April 3 1982

(1981) 262 EG 61

Landlord and Tenant Act 1954, Part II — Motions by tenants seeking leave to discontinue proceedings for grants of new tenancies — Leave of court required under RSC Order 21, rule 3 — Court may impose such terms as it thinks just — Landlords, relying on decision in *Young, Austen & Young Ltd v BMA*, submitted that tenants should be given leave to discontinue, if at all, only on giving undertakings not to pursue any claims for compensation under section 37 of the 1954 Act as amended by the Law of Property Act 1969 — Substantial amounts of compensation involved in the present two cases — Judge's detailed analysis of arguments for and against imposition of term requiring the forgoing of compensation — *Young, Austen* decision not accepted by judge as authority for a general principle that where a tenant has made an application to the High Court for a new tenancy, and then wishes to withdraw the application because he no longer wants the tenancy in question, he should be allowed to discontinue only on giving an undertaking not to pursue a claim to compensation under section 37 — In special cases justice might require such a term, but not in either of the present cases — Leave to discontinue given in both cases subject only to appropriate conditions as to costs — Landlords' summonses under section 24A (interim rent) directed to stand (to preserve them from falling on discontinuance of actions) G H

These were two motions by the plaintiffs, Lloyds Bank Ltd in both cases, in proceedings against National Westminster Bank Ltd and the Corporation of the City of London respectively, under Part II of the Landlord and Tenant Act 1954. By the motions the plaintiffs sought the leave of the court under Order 21, rule 3 to discontinue the proceedings. The proceedings had been brought by originating summonses. The only question was as to what terms, if any, should be imposed on the plaintiffs if leave to discontinue were given. J

Jonathan Gaunt (instructed by Cameron Markby) appeared on behalf of the plaintiffs; M Rich QC and J Male (instructed by Herbert Smith & Co) represented National Westminster Bank Ltd; Michael Essayan QC (instructed by the Comptroller and City Solicitor) represented the City Corporation.

Giving judgment, SLADE J said: This is a judgment in two motions. In one of them, Lloyds Bank Ltd seeks an order that it may be at liberty to discontinue certain proceedings which it has instituted against the Corporation of London ("the corporation"), asking for the grant of a new tenancy, pursuant to Part II of the Landlord and Tenant Act 1954 ("the Act"), of part of certain K
premises known as 3 Broad Street Place, London EC2. I will call this motion, these proceedings and these premises, respectively, "the Broad Street motion", "the Broad Street action" and "the Broad Street premises".

In the other motion, Lloyds Bank Ltd seeks an order that it may be at liberty to discontinue certain proceedings which it has instituted against National Westminster Bank Ltd ("National Westminster"), asking for the grant of a new tenancy, pursuant to Part II of the Act, of certain premises at first floor, Berkeley Square House, Berkeley Square, London W1. I will call this motion, these proceedings and these premises, respectively, "the Berkeley Square motion", "the Berkeley Square action" and "the Berkeley L
Square premises".

It is common ground that the hitherto subsisting tenancies in respect of the Broad Street premises and the Berkeley Square premises are tenancies to which Part II of the Act applies.

The court's leave to discontinue is required because RSC Order 21, rule 3 provides that, except as provided by rule 2, a party may not discontinue an action (whether begun by writ or otherwise) without the leave of the court and the right to discontinue conferred by rule 2 does not apply to proceedings begun by originating summons, such as these.

Order 21, rule 3 further provides that the court hearing an application for the grant of such leave may order the action to be M
discontinued "on such terms as to costs, the bringing of a subsequent action or otherwise as it thinks just".

The plaintiffs, Lloyds Bank Ltd, accept that, if they are to be given the leave sought in respect of either action, this should be on the terms, at least, of paying any costs incurred by the defendants up to the date of the plaintiffs' application to withdraw. If they are given leave, no question arises of their bringing of any subsequent action because, having regard to the provisions of the Act, any such action would be out of time. The question therefore is what terms, if any, "otherwise" should be imposed on the plaintiffs, as the price of giving the leave sought, if leave is given at all?

The defendants in each action, in reliance on the decision of Whitford J in *Young, Austen & Young Ltd v BMA* [1977] 1 WLR 881, claim that the plaintiffs should be given leave, if at all, only on giving an undertaking not to pursue any claim for compensation under section 37 of the Act. The plaintiffs, while not disputing that the court would have jurisdiction to impose such a term, contend, in effect, that on the facts of these two cases the imposition of such a term would be neither appropriate nor just.

The discretion conferred by RSC Order 21, rule 3 is a wide one. Apart from the *Young, Austen* decision, two cases have been cited to me relating to its exercise. The earlier of them was *Stahlschmidt v Walford* (1879) 4 QBD 217. In that case, an action had been referred to an arbitrator to state a special case. He had then made findings of fact with regard to all but a very small part of the claim. These findings had been in the defendant's favour. The plaintiff had then applied for leave to discontinue the action. The Court of Appeal, reversing the decision of Field J, who had granted the plaintiff leave, refused him such leave. Cockburn CJ said (at p 219):

> The defendant, as it seems to me, is in justice entitled to the fruits of these proceedings and we ought not to interfere to deprive him of them.

Mellor J, after referring to the relevant rule, which was in substantially the same form as the present Order 21, rule 3 said (at p 219):

> But the discretion thus given must be exercised within certain limitations, and so as not to take away from the defendant any advantage to which he is fairly and reasonably entitled.

In *Covell Matthews & Partners v French Wools Ltd* [1977] 1 WLR 876, Graham J, after referring to certain authorities, including the *Stahlschmidt* decision, echoed the language of Mellor J in saying this (at p 879):

> The principles to be culled from these cases are, in my judgment, that the court will normally, at any rate, allow a plaintiff to discontinue if he wants to, provided no injustice will be caused to the defendant. It is not desirable that a plaintiff should be compelled to litigate against his will. The court should therefore grant leave, if it can, without injustice to the defendant, but in doing so should be careful to see that the defendant is not deprived of some advantage which he has already gained in the litigation and should be ready to grant him adequate protection to ensure that any advantage he has gained is preserved.

On the particular facts of that case, which did not raise any dispute concerning compensation under section 37 of the Act, Graham J gave leave to a tenant to withdraw an application for a new tenancy under the Act, on terms designed to ensure that the landlords were not prejudiced. The Court of Appeal in due course ([1978] 1 WLR 1477) upheld his decision, expressing no criticism of the principles as formulated by him. Bridge LJ, however, added (at p 1485D):

> It is perfectly clear from authorities to which we have been referred that the granting of leave to discontinue is no mere formality, that an applicant for leave to discontinue may be put in a position, by the terms which the court is minded to impose upon him as a condition of ordering discontinuance of his proceedings, which will be so onerous that he would not wish to accept them.

Buckley LJ, after referring to Order 21, rule 3, said (at p 1487):

> That rule does empower a court to whom application is made to discontinue an action upon terms, but those terms must, I agree, be terms which are imposed upon the applicant.

In the circumstances, my approach to the present cases is intended to follow the principles stated by Graham J and the Court of Appeal in the *Covell Matthews* case, subject to two points. First, while Graham J referred specifically merely to the need for the court, in considering whether to grant a plaintiff leave to discontinue, to preserve for the defendant any advantage that he had gained in the litigation, I am sure that he would have regarded it as no less important to consider the need to compensate the defendant for any proven loss that he might have suffered through the institution of the proceedings. This, I would think, is the reason why plaintiffs are frequently given leave to discontinue only on the terms of paying defendants' costs. Secondly, the advantage gained in the litigation to which Graham J was referring, in my judgment, means an advantage which a defendant has *already* gained by resisting the proceedings. In other words, it covers what Cockburn CJ called "the fruits of these proceedings".

The relevant legislation

Before examining the facts of each of the two present cases, I think it will be convenient to refer to the relevant legislation. Section 24(1) of the Act (as amended) provides that a tenancy to which Part II of the Act applies shall not come to an end unless terminated in accordance with the provisions of Part II. It further confers on the tenant under such a tenancy the right, subject to certain conditions, to apply to the court for a new tenancy if the landlord has given notice under section 25 to terminate the tenancy or the tenant has made a request for a new tenancy in accordance with section 26.

Section 24A enables a landlord who has given a section 25 notice or has received from the tenant a section 26 request to apply to the court to determine a rent, which it would be reasonable for the tenant to pay while the tenancy continues by virtue of section 24, and gives the court the power to determine such a rent.

Section 25(1) confers on the landlord the right to terminate a tenancy to which Part II of the Act applies, by a notice given to the tenant in the prescribed form, specifying the date at which the tenancy is to come to an end, though this subsection has effect subject to the provisions of section 64.

Section 25(5) provides that a notice under that section shall not have effect unless it requires the tenant, within two months after the giving of the notice, to notify the landlord in writing whether or not, at the date of termination, the tenant will be willing to give up possession of the property comprised in the tenancy.

Section 25(6) requires a landlord's notice under the section to state whether he would oppose an application to the court under Part II of the Act for the grant of a new tenancy and, if so, to state on which of the grounds mentioned in section 30 he would do so.

Section 26(1) confers on the tenant the right to make a request for a new tenancy, if his current tenancy is a tenancy granted for a term of years certain exceeding one year. The tenant, however, does not have this right if the landlord has already given notice under section 25 to terminate the tenancy: see section 26(4).

Section 26(6) provides:

> Within two months of the tenant's request for a new tenancy the landlord may give notice to the tenant that he will oppose an application to the court for the grant of a new tenancy, and any such notice shall state on which of the grounds mentioned in section 30 of this Act the landlord will oppose the application.

Section 29(1) provides:

> Subject to the provisions of this Act, on an application under subsection (1) of section 24 of this Act for a new tenancy, the court shall make an order for the grant of a tenancy comprising such property at such rent and on such other terms as are hereinafter provided.

Section 29(3) provides:

> No application under subsection (1) of section 24 of this Act shall be entertained unless it is made not less than two nor more than four months after the giving of the landlord's notice under section 25 of this Act or, as the case may be, after the making of the tenant's request for a new tenancy.

Section 30(1) of the Act begins with the following words:

> The grounds on which a landlord may oppose an application under subsection (1) of section 24 of this Act are such of the following grounds as may be stated in the landlord's notice under section 25 of this Act or, as the case may be, under subsection (6) of section 26 thereof, that is to say:

There then follow seven subparagraphs, (a) to (g). I need read only subparagraph (f):

> That on the termination of the current tenancy the landlord intends to demolish or reconstruct the premises comprised in the holding or a substantial part of those premises or to carry out substantial work of construction on the holding or part thereof and that he could not reasonably do so without obtaining possession of the holding.

A Section 31 of the Act, subject to certain conditions, precludes the grant of a new tenancy by the court where the landlord opposes an application for a new tenancy on grounds on which he is entitled to oppose it, in accordance with section 30 and which he establishes to the satisfaction of the court.

Section 37(1) (in its amended form) provides:

> Where on the making of an application under section 24 of this Act the court is precluded (whether by subsection (1) or subsection (2) of section 31 of this Act) from making an order for the grant of a new tenancy by reason of any of the grounds specified in paragraphs (e), (f) and (g) of subsection (1) of section 30 of this Act and not of any grounds specified in any other paragraph of that subsection, or where no other ground is specified in the landlord's notice under section 25 of this Act or, as the case may be, under B section 26(6) thereof, than those specified in the said paragraphs (e), (f) and (g) and either no application under the said section 24 is made or such an application is withdrawn, then, subject to the provisions of this Act, the tenant shall be entitled on quitting the holding to recover from the landlord by way of compensation an amount determined in accordance with the following provisions of this section.

The words in this subsection, reading from "or where" to "withdrawn" were inserted in it by section 11 of the Law of Property Act 1969, as from January 1 1970. I pause to make two comments on this subsection. First, the reason why the legislature, from the very beginning, included a reference only to paragraphs (e), (f) and (g) of section 30(1) is, I think, fairly clear. Paragraphs (a), (b) and (c) C all deal with cases where the tenant is, in one way or another, in breach of his obligations. Paragraph (d) deals with the case where the landlord has offered and is willing to provide or secure suitable alternative accommodation for the tenant on reasonable terms. In a case where the landlord can establish one or more of the grounds referred to in paragraphs (a), (b), (c) or (d), statutory compensation may therefore not appear very appropriate. However, it was the legislature's clear intention that, in a case where the landlord successfully resisted an application to the court solely by reference to one or other of the grounds referred to in paragraphs (e), (f) or (g), the tenant should be entitled to compensation.

Secondly, however, as section 37(1) was originally enacted, the D tenant, in order to secure his compensation in such circumstances, not only had to serve a counternotice on the landlord stating that he was not willing to give up possession, but also had to make and pursue an unsuccessful application for a new tenancy: (see *Re 14 Grafton Street, London W1* [1971], Ch 935).

Now, however, in its amended form, section 37(1) gives the tenant who has a tenancy protected by Part II of the Act the statutory right to compensation in a case (among other cases) where all the following conditions are fulfilled:

(1) The landlord has served on the tenant a notice under section 25 or section 26(6); and

(2) the notice thus served has specified no ground other than one E or more of the grounds specified in paragraphs (e), (f) and (g) of section 30 of the Act; and

(3) the tenant either has made no application at all under section 24 or has withdrawn any such application after it has been made; and

(4) the tenant has quitted the holding.

In a case where these four conditions are fulfilled, section 37(1) (as now amended) confers a clear and unqualified statutory right on the tenant to compensation. The amount of compensation to be awarded under the section is specified in section 37(2), as amended by paragraph 4 of Schedule 33 to the Local Government, Planning and Land Act 1980 and in the Landlord and Tenant Act 1954 F (Appropriate Multiplier) Regulations 1981. Nothing much turns on these provisions for present purposes, save to say that if compensation were to be due to the plaintiffs in either case at all, it would be substantial, having regard to the rateable value of each set of premises. I think it is common ground that in the case of the Broad Street premises the compensation (if any) would be £65,776. In the case of the Berkeley Square premises, there is some dispute whether it would be £21,734 or £48,900.

The Broad Street Motion

I now propose to turn to the facts of the Broad Street motion, which I regard as the more difficult of the two motions before me. G These facts appear from an affidavit sworn by Mr W F Abbott, a chartered surveyor who is employed by the plaintiffs in the estates department of their premises division, an affidavit sworn in answer by Mr Anthony Wheaton, a principal in the office of the City Solicitor, and an affidavit sworn in reply by Mr Abbott. From these affidavits the following facts appear.

The plaintiffs hold the Broad Street premises under three leases dated June 1 1905, August 31 1910 and March 31 1917, the terms granted by which were all limited to expire on December 18 1980. The reversioner under these leases is the Corporation of London ("the corporation"). The rateable value of the Broad Street premises is £32,888.

H Until March 31 1981 the plaintiffs occupied the Broad Street premises as their Finsbury Circus branch. By January 1980 they had received information that the building of which the premises formed part was to be the subject of major works of refurbishment and renovation and that, if these works were to be carried out, they would necessitate the vacation by the plaintiffs of the Broad Street premises. The plaintiffs, however, did not at that time know whether new leases would be granted to them when their current leases expired in December 1980.

In these circumstances, the plaintiffs' solicitors, on February 8 1980, pursuant to section 26 of the Act, served on the defendants' solicitors requests for the grant of three new tenancies beginning on December 25 1980. On April 1 1980 the plaintiffs' solicitors J received from the City Solicitor three counternotices, dated March 31 1980, saying, in effect, that the corporation was not willing to grant new tenancies and would oppose any application to the court for such tenancies on the grounds set out in section 30(1)(f) of the Act. These counternotices were served pursuant to section 26(6) of the Act.

On June 5 1980 the plaintiffs issued three originating summonses seeking the grant of three new tenancies pursuant to Part II of the Act. These three originating summonses constitute the Broad Street action. They were consolidated by an order of Master Dyson made on June 26 1980. On that occasion he gave directions for the trial of the preliminary issue, whether new tenancies should be K granted, having regard to the landlord's grounds of opposition.

At a meeting of July 14 1980 the corporation informed the plaintiffs that it would not oppose the application for new leases. The corporation's decision not to oppose the application was confirmed by a letter of July 22 1980, written by the City Solicitor to the plaintiffs' solicitors and in other correspondence.

At the meeting of July 14 1980 and in subsequent correspondence in July, the plaintiffs had indicated to the corporation that it was their present wish and intention to obtain new leases and to continue to occupy the premises. However, by August 6 1980 the plaintiffs had changed their minds in this regard. On August 6 they L reached the decision to close their Finsbury Circus Branch in the spring of 1981 and to move it to their Stock Exchange Branch, and to vacate the Broad Street premises accordingly.

On August 7 1980 they instructed their solicitors to inform the corporation of their intention to vacate, but in the event revised these instructions, because they considered that the decision had to remain confidential until the customers and staff of the Finsbury Circus Branch had been informed of the impending move.

In due course, however, by a letter of October 20 1980 the plaintiffs' solicitors informed the City Solicitor that the plaintiffs would be vacating the Broad Street premises in the spring of 1981. After some intermediate correspondence, the plaintiffs' solicitors, M on January 23 1981, wrote to the City Solicitor, informing him that they would be ceasing business at the premises at about the end of February and would be finally vacating them about a month after that. They drew attention to the fact that their clients were entitled to statutory compensation and asked for confirmation that it would be paid to them immediately on their finally vacating the premises.

The date for vacation was subsequently extended by agreement between the plaintiffs and the corporation to March 31 1981. The plaintiffs have, I understand, now vacated the premises, though no surrender of the leases has been formally accepted by the corporation, so that the plaintiffs would be in a position to resume possession if they should desire to do so.

A On March 9 1981 the corporation applied for the determination of an interim rent under section 24A of the Act. On March 24 1981 the City Solicitor wrote to the plaintiffs' solicitors stating the terms on which the corporation was prepared to consent to an application by the plaintiffs for leave to discontinue the Broad Street action. One of these terms was that the plaintiffs should undertake not to seek compensation under section 37 of the Act.

In these circumstances, the Broad Street motion was issued on April 6 1981, by which the plaintiffs seek leave to discontinue the Broad Street action.

Mr Essayan, on behalf of the corporation, has submitted that, if the plaintiffs are to be given leave to discontinue at all, this should
B only be on the terms that they undertake not to seek statutory compensation because on no possible view, he says, could they now receive compensation if they were compelled to continue with their application to the end. He points out that ever since July 14 1980 the corporation has not relied on any ground of opposition. He invites me, therefore, to consider what would happen if I were to refuse the plaintiffs leave to discontinue. Their application for a new tenancy would then come before the court with no opposition from the corporation. One of two alternative possibilities might then occur. In the one alternative, the plaintiffs might ask for their application to be dismissed, in which event they would obtain no statutory compensation. In the other alternative, the plaintiffs
C might ask for a new tenancy. If they did so, section 29(1) of the Act would oblige the court to grant a new tenancy because no ground of opposition would have been established. It would be its duty to decide the terms of the new tenancy. Once the order had been made, the plaintiffs could either accept these terms or, if they considered the rent fixed by the court unacceptably high, they could ask the court to revoke its order under section 36(2) of the Act. Whether they accepted the new tenancy or not, however, they would not get compensation, because the conditions of section 37(1) would not have been satisfied.

Mr Essayan's analysis of the position that would arise, as a practical matter, if I were to refuse the plaintiffs leave to discon-
D tinue, appears to me correct and I proceed on this basis. From this hypothesis, he went on to submit that, since it is thus a certainty that the plaintiffs will not get compensation if they proceed with the Broad Street action, they cannot or should not now be placed in a better position by seeking the court's indulgence, so as to enable them to discontinue these proceedings. In his submission, the corporation has gained an advantage from the litigation, within the principle enunciated by Graham J, of which it would be wrong to deprive it.

In submitting that the court's discretion should be exercised in this manner, Mr Essayan naturally relied strongly on the decision of Whitford J in the *Young, Austen* case. I will return to that decision a
E little later, but think that I should first attempt to consider the submissions put forward on behalf of the corporation, without reference to authority.

First, and I think foremost, Mr Essayan submitted that for the court to refuse the desired leave to discontinue would be entirely consonant with the policy of the Act. He pointed out that, even under section 37(1) as amended, it is only in very limited circumstances that a tenant is entitled to compensation on quitting the premises. The policy of the Act, he submitted, is that the tenant shall receive compensation only in a case where he is deprived of his new tenancy through circumstances entirely beyond his control and for which he is not responsible. This, he submitted, was the reason
F why the legislature, in enacting section 37(1), singled out paragraphs (e), (f) and (g) of section 30. It did not, in his submission, contemplate that the tenant should get compensation if he merely decided that he did not want a new tenancy.

On reading section 37(1) in conjunction with section 30(1), I accept that the legislature contemplated that the tenant who received a landlord's notice under section 25 or section 26(6) should not receive compensation on quitting his holding, either if he was in such breach of his obligations under the tenancy that he would not have obtained a new tenancy if he had asked for it or if the landlord had made an offer of alternative accommodation of the nature referred to in section 30(1)(d). I cannot, however, accept the wider submission that the legislature contemplated that the tenant G
should never get compensation if his motives for failing to press his possible rights to a new tenancy were merely that he did not want a new tenancy. Section 37(1), as amended, says nothing whatever about the motives which may prompt a tenant either to omit to apply for a new tenancy or to withdraw his application after it has been made. The motives prompting a tenant to take either of these courses, after he has received a landlord's notice under section 25 or section 26(6), relying on one or more of the grounds specified in paragraphs (e), (f) or (g) of section 30, may be many and mixed. I cannot impute an intention to the legislature to withhold compensation from a tenant who complies with all the other conditions in section 37(1) merely because his motives for omitting to apply to H
the court, or for withdrawing an application when made, may be of a particular nature. The rights to compensation given by the subsection, in circumstances such as those of the plaintiffs in the present case, to a party who withdraws his application and then vacates the premises, are in terms absolute, not qualified by reference to any considerations of motive.

Perhaps it might be suggested that the legislature, in amending section 37(1), would have had in mind that the court might from time to time, in the exercise of its discretion and in appropriate circumstances, effectively deprive a tenant of his right to compensation, by requiring him to abandon this right as the price of allowing him leave to withdraw his proceedings. However, I think there is a J
conclusive answer to any such suggestion. As the legislature must have been well aware, many applications under the Act fall to be brought in the county court, where it is open to a tenant to discontinue his application merely by notice to the landlord, without obtaining the leave of the court. It cannot therefore be inferred that the legislature, in amending section 37(1), specifically contemplated that, in practice, the exercise of a tenant's right to statutory compensation might be controlled and fettered through an exercise of the court's discretion.

For these reasons, in so far as Mr Essayan submits that the court would be giving effect to the true intentions of the legislature if it were to withhold the leave to discontinue now sought by the plain- K
tiffs against the corporation, I must, without hesitation, reject this submission.

In support of the corporation's case, however, Mr Essayan put forward a number of further points which did not directly depend on any presumed intention of the legislature. He accepted that, if the plaintiffs had indicated that they were prepared to vacate the premises as soon as they received the corporation's counternotices of March 31 1980, they would have been entitled to statutory compensation, whatever their motives for such acquiescence might have been. However, he submitted that, by applying to the court, the plaintiffs radically altered the respective positions of themselves and the corporation, to the benefit of themselves and the detriment L
of the corporation. If the plaintiffs, on receipt of the corporation's counternotices, had not applied to the court, they would have been obliged to vacate the Broad Street premises when their original tenancies came to an end in December 1980. The corporation would, in these circumstances, have had the advantages not only of earlier possession but also of knowing for certain, at latest by June 1980, that it would be entitled to possession in December of that year; for a date in June was the latest date on which the plaintiffs would have been entitled to apply for new tenancies: see section 29(3) of the Act. With this certain knowledge, Mr Essayan submitted, the plaintiffs could either have taken steps during the last six months of 1980 to relet the Broad Street premises as from M
December 1980 at a full market rent or could instead, if they had preferred, have made arrangements with contractors to effect alterations to them.

In his submission, the corporation have thus suffered detriment through the plaintiffs' reliance on their statutory rights to apply for new tenancies, in the form of delay in obtaining possession of the premises, uncertainty as to their position and loss of rent. Though the corporation will no doubt receive the award of an interim rent on its section 24A application, this will not, he suggested, be the same as a full market rent. On the other hand, he suggested, the corporation has gained an advantage in the litigation by the very

A institution of the Broad Street action, simply because this action cannot be withdrawn without the leave of the court. In these circumstances, he submitted, even apart from the *Young, Austen* decision, it is only just that, if the court is to deprive the corporation of this advantage by giving the appropriate leave to the plaintiffs, it should do so by giving the corporation compensation in the form of the undertaking which it seeks.

With all respect to Mr Essayan's very persuasive argument, it seems to me that it involves some begging of the crucial question. The so-called advantage to the corporation gained in the litigation is of a quite different nature from the "fruits of the litigation" referred to by Cockburn CJ in the *Stahlschmidt* case. It can only be a
B real advantage if there is a good reason why the court should deprive the tenant of the compensation to which he is otherwise entitled by statute; but this is the very question which the court now has to decide. I have already expressed the view that any deemed intention of the legislature constitutes no good reason for depriving the tenant of his compensation in such a case. One is therefore left with the question whether, on the particular facts, justice requires that the plaintiffs should be required to forfeit the compensation to which they are *prima facie* entitled, as the price of obtaining the desired leave.

There are, in my judgment, several good reasons why justice does not so require, on the facts of the present case. First, as Mr Gaunt
C has pointed out on behalf of the plaintiffs, it is not and could not be suggested that the amount of the statutory compensation would bear any relation in monetary terms to any financial loss which the corporation may have suffered. Secondly, in its evidence the corporation has not asserted any special damage or given any particulars of financial loss. Thirdly, if and in so far as the corporation may have suffered financial loss through delays and uncertainties caused by the plaintiffs' reliance on their statutory rights, such delays and uncertainties are likely to arise in any case where a tenant relies on the system of statutory protection afforded to him by the Act. These, however, are inherent in the system. It has not been asserted in the present case that the plaintiffs acted in any way improperly in
D relying on their statutory rights or that they abused the process of the court, in issuing their applications for new tenancies. Fourthly, in so far as the corporation alleges that it has lost rent through the exercise of the plaintiffs' statutory rights, it will have the benefit of the remedies given to it by section 24A of the Act which, it is to be inferred, were regarded by the legislature as providing adequate recompense to landlords in such circumstances.

For all these reasons, at least in the absence of authority, I would not be minded to impose any conditions on the plaintiffs, save in regard to costs, in giving them the leave to discontinue their proceedings against the corporation which they now seek. I must, however, turn to the decision in *Young, Austen & Young Ltd v*
E *BMA* [1977] 1 WLR 881, on which Mr Essayan replies. The facts of that case, as summarised in the headnote to the report, were as follows:

> Following notices to terminate two tenancies served by the landlords under section 25 of the Landlord and Tenant Act 1954, the tenant served requests for new tenancies under section 26 of the Act. Counternotices were served by the landlords under section 26(6) opposing the grant of new tenancies under section 30(1)(g) on the ground that on termination of the existing tenancies they intended to occupy the premises for the purposes of their own business. The tenants applied to the High Court for new tenancies pursuant to Part II of the Act on terms to be agreed. Subsequently, the landlords indicated that they were prepared to relet one set of premises. The tenants having decided, however, that they no longer wanted a renewal of
> F the tenancies, applied to the Court under RSC Order 21, rule 3 for leave to discontinue the proceedings on payment of the landlords' costs.

Whitford J (at p 885) summarised the submissions made on behalf of the landlords as follows:

> So the point made on the landlords' side is that it would be quite wrong in such circumstances for leave to be given to discontinue, as if such leave were given without any undertaking with regard to the question of an application for compensation, they would be disadvantaged. As the matter at present stands, their point is that if the tenants have to proceed, they are not going to seek an order for the grant of new tenancies and in those circumstances their applications will be dismissed and no compensation will be payable to them;
> G and it is quite wrong that they should be allowed to withdraw without some term being imposed which will preserve to the landlords the position which would undoubtedly result so far as compensation is concerned if the action were to proceed.

On the tenant's side (see at p 886 of the report) substantially two principal submissions were made. First it was contended that it was not within the power of the court in granting leave to discontinue to impose a condition of the kind which, on the face of it, would be going against the provisions which Parliament had specifically enacted. Secondly, it was contended that in any event, it would be wrong for the court, as a matter of discretion, to impose such a condition.

H Whitford J rejected the first of these submissions, holding (at pp 886 and 887) that the court on any such application had the duty to consider all the circumstances of the case and whether, in justice, it would be right that leave to discontinue should be granted. I respectfully agree with this part of Whitford J's decision; and, indeed, Mr Gaunt, on behalf of the plaintiffs, has not challenged its correctness. His challenge is directed to the second part of the decision, which related to the exercise of the court's discretion.

In this context, Whitford J (at p 886) referred to the passage from Graham J's judgment in the *Covell Matthews* case, which I have already read. He observed (at p 887) that, if the matter were to proceed in the subsisting circumstances, the tenants were no longer
J going to seek the fresh tenancies which they originally sought, their action would be dismissed and their right to compensation would go. The kernel of his decision as to the question of discretion is to be found in the following two paragraphs of his judgment (at pp 887 to 888):

> I think the submission on behalf of the landlords is a good one. It is perhaps a little troubling that, as counsel for the tenants quite rightly pointed out, if this had been a proceeding brought in the county court, the imposition of a term of the sort which I think ought to be imposed would not have been possible. It may be that this is a matter which was never in the contemplation of those responsible for the amendment to section 37 of the Act of 1954. Prior to the amendment of that section, it will of course be recalled that the
> K position was that there was no specific provision relating to compensation upon mere withdrawal and the right to compensation arose only after there had been a decision adverse to the tenant. It may possibly have been thought that to force the parties to a hearing in circumstances where it would be likely to be apparent to the tenants that they would not succeed in opposing a section 30(1)(g) objection would be unnecessarily onerous, and that therefore they should be granted a right to withdraw. What I think was never in contemplation was a circumstance such as this, where basically the reason why the tenants want to withdraw is not that they fear they will be unable to resist the objection of the landlords: indeed, in one case, there is no question of objection by the landlords. Basically, the reason why they want to withdraw is that they no longer want the tenancies in question. I do not think that in those circumstances it would be right that I should grant them leave to discontinue without imposing some term. It may be that the
> L injustice, if there be any, lies in this, that it is impossible, if proceedings be brought in the county court, for a term of this character, which I think is in fact appropriate, to be made.
>
> I am prepared to grant leave to discontinue, but only on an undertaking by the tenants that they will not pursue any claim to compensation pursuant to section 37 of the Act as amended.

Essentially, therefore, the *ratio* of Whitford J's decision on the question of discretion, as I understand it, was that the court should require the tenants to abandon their statutory right to compensation as the price of being allowed to withdraw their application, because otherwise they would be receiving compensation in circumstances which Parliament would never have contemplated. In this manner,
M Mr Essayan submitted, Whitford J filled a lacuna in the Act by doing what was fair and equitable.

Any case in which the exercise of the court's discretion is involved must depend on its particular facts and I express no view on the exercise of the court's discretion on the particular facts of the *Young, Austen* case. I do not, however, accept the submission of Mr Essayan that the passing reference by Bridge LJ to that case in his judgment in the *Covell Matthews* case [1978] 1 WLR 1477 at p 1485 amounted to specific approval of that decision, in so far as the exercise of the discretion was involved. The question at issue before the Court of Appeal in the *Covell Matthews* case was a very different one.

A However, in the present case it has been suggested that the *Young, Austen* decision established a general principle that, in any case where a tenant has made an application to the High Court for a new tenancy under the Act and then wishes to withdraw the application because he no longer wants the tenancy in question, he should be allowed to discontinue only on giving an undertaking not to pursue any claim for compensation under section 37 of the Act.

In so far as the decision can be read as establishing any such general principle, I would respectfully dissent from this principle for a number of reasons, which more or less appear from what has already been said in this judgment, but can be summarised as follows: (1) I am not satisifed that there is any lacuna at all in section B 37(1) of the Act as amended. The amendment gives a tenant who, after receiving a landlord's notice of the nature referred to in the section, forbears to make an application to the court or withdraws his application after it has been made, the clear and unqualified right to receive compensation on quitting the holding. I am far from convinced that, even if the point had been put to it, the legislature would have intended to make any distinction between tenants who withhold or withdraw an application to the court with one motive and those who do so with another motive, particularly since motives may be very mixed and any such distinction would, I think, have given rise to formidable difficulties both of statutory drafting and of evidence. (2) Even if there be a lacuna in section 37(1), in my C judgment it is one for the legislature to fill and not for the High Court when exercising its discretionary jurisdiction under RSC Order 21, rule 3.

I accept that special cases could arise when the court, in exercising its discretionary jurisdiction, might properly consider that, quite apart from any presumed intention of the legislature, justice to the landlord demanded that the tenant should be required to give an undertaking of the nature sought by the landlords in the present case. The Broad Street motion, however, is not in my view one of them. The court should not force the plaintiffs to continue with the Broad Street action merely for the purpose of ensuring that they do not receive the compensation which Parliament has said they D should have on withdrawing their application. I therefore propose to give the plaintiffs the leave to discontinue the Broad Street action which they seek, subject to an appropriate condition as to costs. I will hear submissions in due course as to its precise form.

So that the corporation's summons under section 24A of the Act will not fall if and when the plaintiffs' Broad Street action is discontinued, Mr Essayan has asked for a direction that this summons under section 24A should stand as a counterclaim: (see the notes in the Supreme Court Practice 1979 under RSC Order 97, rule 9A). I propose to give such a direction.

The Berkeley Square motion

E This motion concerns certain office premises on the first floor of Berkeley Square House, Berkeley Square, which have been used by the plaintiffs since 1961 as their Greater London (North) regional head office premises. The plaintiffs hold them by virtue of an underlease dated August 1 1961, made between Berkeley Square Holdings Ltd and the plaintiffs, the reversion of which is now vested in National Westminster.

Clause 1 of the underlease excepted and reserved to the landlords certain rights to enter the premises for the purposes of effecting repairs and alterations to the structure of Berkeley Square House and other similar rights. The term limited by this underlease was due to expire on March 25 1981. The plaintiffs, in fact, occupy F certain other parts of Berkeley Square House.

The evidence before me on this motion consists of three affidavits, namely (1) an affidavit of Mr N P Draper, a chartered surveyor who is employed by the plaintiffs in the estates department of their premises division; (2) an affidavit sworn by Mr J A Glass, on behalf of National Westminster; and (3) an affidavit sworn in reply by Mr Draper on behalf of the plaintiffs.

From this evidence I think that the following facts appear. As long ago as 1978 National Westminster instructed project managers in connection with certain works of improvement to Berkeley Square House which it was then contemplating. In April and May 1980 Mr Glass, who is a project manager with the firm in question, made contact with all the tenants of the building with a view to G seeking their cooperation in the proposed works of improvement and discovering whether, on the expiry of their various interests, they would be interested in continuing the occupation of their existing premises or in taking additional accommodation. In particular, as Mr Glass explains in his affidavit, it was convenient for the purposes of the works for National Westminster to have possession of the plaintiffs' Berkeley Square premises and essential for them to have at least the cooperation of the plaintiffs.

On May 21 1980 Mr Glass met Mr Draper. At this meeting, Mr Draper indicated that the plaintiffs might be willing to move out of the Berkeley Square premises. No decision, however, had yet been reached by the plaintiffs and indeed none was reached, it seems, until August 20 1980. H

Against this background, on July 31 1980, National Westminster served on the plaintiffs a notice dated July 30 1980, under section 25 of the Act, terminating the plaintiffs' tenancy on March 25 1981. The notice stated in effect that National Westminster would oppose an application to the court for the grant of a new tenancy on the grounds set out in section 30(1)(f) of the Act. A covering letter of July 29 1980 explained that the notice was served "in order to maintain maximum flexibility having regard to the landlords' proposals to undertake various works at and within the building and the extent to which access will be required to your offices to complete the project, which will commence in the near future". J

Mr Glass, in paragraph 6 of his affidavit, suggested that Mr Draper at all times appreciated, first, that the works proposed on the Berkeley Square premises were such that they did not amount to demolition or reconstruction of a substantial part of the building; secondly, that they did not amount to substantial works of construction of the holding; and, thirdly, that much of the work was work which National Westminster was undoubtedly entitled to carry out under the terms of the underlease. In other words, Mr Glass suggested that the section 25 notice, though indicating National Westminster's reliance on the section 30(1)(f) grounds of opposition, was specifying grounds which Mr Draper well knew were ill-founded in law and in fact. K

Mr Draper, in his affidavit in reply, specifically denied that he had any such knowledge as alleged by Mr Glass. For the purpose of the present application I do not propose to examine all the evidence in this context. I will merely say that, on its present state, I do not think it would be right to find or assume in favour of National Westminster that Mr Draper had any such knowledge or indeed that National Westminster itself was serving a section 25 notice which specified grounds of opposition to a new tenancy, which it knew to be wholly without substance in law or in fact.

On receipt of this notice, the plaintiffs gave further consideration to the matter and, in due course, on August 20, decided that both their Greater London (North) regional head office and Greater London (South) regional head office, which occupied other nearby premises, should be relocated in premises elsewhere. L

On September 5 1980, however, to protect their position, the plaintiffs' solicitors gave notice to National Westminster that they were not willing to give up possession of the Berkeley Square premises on the date mentioned in the notice of July 30.

On October 14 1980 Mr Draper wrote to Mr Glass telling him of the plaintiffs' decision to vacate the Berkeley Square premises, but asking him whether National Westminster would allow the plaintiffs to keep them for up to three months after March 25 1981 to facilitate their move. By November 20 Mr Draper still had not received confirmation that the plaintiffs would be permitted to hold M over after March 25 1981. On that day, the plaintiffs, again to protect their position, issued the Berkeley Square action, seeking the grant of a new tenancy pursuant to Part II of the Act. In view of the time-limit imposed by section 29(3) of the Act, these proceedings had to be issued, if at all, by the end of November 1980.

After the service of this summons, further negotiations took place between the parties which, so far as the plaintiffs were concerned, were principally designed to enable them to remain in the Berkeley Square premises for a few weeks after March 25 1981 and, so far as the National Westminster were concerned, were principally designed to enable them to obtain the necessary access

A to the premises for the purpose of carrying out the contemplated works. As a result, the plaintiffs, without objection from National Westminster, remained in the premises for a few weeks after the end of March. They have now surrendered the keys to the premises, though they have not formally surrendered possession. A summons under section 24A of the Act was issued by National Westminster on March 4 seeking interim rent.

In all the circumstances, the plaintiffs wish to withdraw their application for a new tenancy, but National Westminster have proved unwilling to consent to its being withdrawn, save on the terms that the plaintiffs abandon their claim to statutory compensation.

B I hope that Mr Rich, who appeared on behalf of National Westminster, will not regard it as discourteous if I deal quite briefly with his attractive argument. Essentially, I think the same reasons apply for rejecting it as apply in rejecting the corporation's argument, but *a fortiori*.

He submitted that when National Westminster served its notice of July 30 1980, relying on ground (f) set out in section 30 of the Act, it, in effect, made an offer to the plaintiffs that, if they accepted it and vacated the premises accordingly, National Westminster would pay them statutory compensation. Instead, however, the plaintiffs elected to take proceedings and thereby gain the advantage of a continuation of their tenancy, by virtue of section 64 of the
C Act. The landlords, on the other hand, suffered the correlative disadvantages of uncertainty as to the date of determination of the tenancy and delay. In these circumstances, Mr Rich submitted that the court could reasonably take the attitude that a tenant who has taken advantage of section 64 in such circumstances should either continue with his proceedings until the end or, if he does not wish to continue with them, should be allowed to withdraw at a date of his choice only on terms which do not leave him compensation for a removal which he has elected to make and has not been forced on him. Why, he asked rhetorically, should the plaintiffs complain in such circumstances, if the court now holds that they should not have compensation for a loss which, he submitted, they have not suffered
D and which, he submitted, they would not get if they took their proceedings to their conclusion?

This question is thus based on two hypotheses of fact, namely, that (1) the plaintiffs have suffered no loss through National Westminster's reliance on its statutory rights and (2) the plaintiffs would in the end inevitably fail to obtain compensation if they were compelled to pursue the Berkeley Square action.

I answer Mr Rich's rhetorical question in this manner. Even if the two hypotheses of fact upon which it is based are correct, the plaintiffs could, I think, justifiably complain if the court were to take this course, for much the same reasons as I have indicated in the part of this judgment relating to the Broad Street motion. The
E court could not, in my view, by reference to any supposed intention of the legislature, justifiably deprive the plaintiffs of the statutory compensation to which they are *prima facie* entitled. Nor, in my view, are there sufficient special facts to necessitate such a deprivation in the interests of justice to National Westminster. Though Mr Glass, in his affidavit, says that the delay in the vacation of the premises has imposed "an extremely tight programme" on National Westminster's contractors, it has not asserted any special damage or given any particulars of financial loss. It now has the access to the premises which it requires for the purpose of carrying out its works. It has not been asserted that the plaintiffs abused the process of the court in issuing their application for a new tenancy or
F acted in any way improperly in relying on their statutory rights.

An element of tactical manoeuvring is inevitably introduced by the system of the Act; and, as I see the position, the plaintiffs' application to the court in this case was no more and no less of a tactical manoeuvre than the landlords' section 25 notice. I do not criticise either of them. In so far as National Westminster may have lost rent through the exercise of the plaintiffs' statutory rights, it will have the benefit of interim rent assessed under section 24A of the Act in accordance with the measure which the legislature has deemed appropriate in such circumstances.

Even on the two hypotheses of fact Mr Rich invites me to make, I thus see no sufficient grounds for imposing the desired condition,
G particularly when the amount of the statutory compensation, of which he seeks to deprive the plaintiffs, bears no relation in monetary terms to any pecuniary loss which National Westminster may have suffered.

I should, however, add that on the facts of this case I am by no means convinced that either of these two hypotheses of fact is necessarily correct. As to the first, though Mr Rich invited me to infer that National Westminster's section 25 notice was not the cause of the plaintiffs' decision to vacate the premises, I am not satisfied on the evidence that the notice may not have been at least one of the contributory causes. Inquiries of this sort into motivation, particularly on the basis of affidavit evidence, are an exercise
H of very doubtful value; this is an issue of fact which I regard as being an open one on the evidence before me. I feel equal doubts as to the correctness of the second hypothesis of fact put forward by Mr Rich. It is at least theoretically possible that, if I were to refuse the leave now sought, the plaintiffs would reconsider their position and pursue their application for a new tenancy. Mr Gaunt made it plain that they were reserving their rights to take this course. It is no less theoretically possible that, in this event, National Westminster would oppose the grant of a new tenancy on the grounds set out in its section 25 notice and oppose it successfully, in which event, the plaintiffs, on vacating, would qualify for statutory compensation. Mr Rich suggested, in effect, that the section 25 notice should not
J be taken seriously, that National Westminster would not want to oppose the grant of a new tenancy because, under the existing tenancy, it would have all the rights which it needed to do the additional works to Berkeley Square House and because it could not establish the section 30(1)(f) ground anyway. These submissions may or may not be well founded. They would be relevant matters for consideration if and when an application by the plaintiffs for a new tenancy came before the court for hearing. However, they are not, in my view, suitable to be prejudged on this application by the plaintiffs for leave to discontinue and in any event, as I have already indicated, are not in my view clearly established by the evidence before me.

K In these circumstances, on the facts of this case, I am not satisfied that the inevitable outcome of refusing the plaintiffs the leave now sought would be that they would in the end be left without statutory compensation. On this ground, the Berkeley Square motion is, in my view, distinguishable from the *Young, Austen* case on its facts. For this additional reason, I can see no sufficient grounds for imposing on the plaintiffs in the Berkeley Square motion an obligation to give the suggested undertaking not to pursue their claim for statutory compensation, as the price of giving them leave to discontinue the Berkeley Square action.

On all these grounds I propose to give the plaintiffs leave to discontinue the Berkeley Square action, subject to an appropriate
L condition as to costs. I will, at the same time, direct that National Westminster's summons under section 24A shall stand as a counterclaim in the Berkeley Square action.

COURT OF APPEAL

December 21 1981

(Before Lord Justice LAWTON, Lord Justice BRIGHTMAN and Lord Justice OLIVER)

W J BARTON LTD v LONG ACRE SECURITIES LTD M

Estates Gazette May 29 1982

(1981) 262 EG 877

Landlord and Tenant Act 1954, Part II — Interlocutory appeal by tenants, bakers and confectioners, from order for discovery made by county court judge, requiring tenants to disclose "all documents relating to trading" — Order made in course of tenants' application for a new tenancy — Order involved searching for and disclosing such information as delivery notes, cash rolls, manufacturing costings, wage slips and so on — Whether information required by order was relevant — Order related both to the subject premises and to another set of retail premises occupied by the tenants rented

from the same landlords — *Harewood Hotels Ltd v Harris* and a passage in *Woodfall* discussed — Special cases where production of trading figures for the purposes of Part II of the Act of 1954 may be relevant and admissible, eg hotels, petrol-filling stations, theatres, racecourses — Such considerations do not, however, apply in the ordinary case of shop premises with no peculiar features in an area where there are plenty of comparable premises from which the open market rents can be deduced — In the present case there was nothing to suggest that the discovery sought by the landlords would be of any assistance in determining the open market rent — Indeed the information sought might be positively confusing — Appeal from judge's order allowed

This was an appeal by W J Barton Ltd, multiple bakers, tenants of business premises at 302 Green Lanes, Palmers Green, London N13, from an order for specific discovery made by Judge Tibber at Edmonton County Court in proceedings brought by the tenants under Part II of the Landlord and Tenant Act 1954. The landlords, respondents to the present appeal, were Long Acre Securities Ltd.

John Foy (instructed by Barker, Gooch & Swailes, of Enfield, Middlesex) appeared on behalf of the appellants; Robin Belben (instructed by Stafford Clark & Co) represented the respondents.

Delivering the judgment of the court at the invitation of Lawton LJ OLIVER LJ said: This is an interlocutory appeal with leave of the judge from an order for specific discovery made in the Edmonton County Court by His Honour Judge Tibber in proceedings under the Landlord and Tenant Act 1954 on October 5 1981. The point raised is, on the face of it, a fairly short and simple one, but since it has potentially far-reaching implications as regards other applications under the Act we thought it advisable to take a little time for consideration. The applicants (the appellants in this court) are the tenants of business premises at 302 Green Lanes, Palmers Green, N13, holding over under a lease for five years granted in 1976 by the Prudential Assurance Co Ltd. On May 29 1980 the landlords gave notice to terminate the tenancy on March 25 of this year and a counternotice under the Act was duly served by the appellants on June 19 1980. That was followed on September 4 1980 by an application in the Edmonton County Court by the appellants for a new lease, in response to which the landlords stated that they would not oppose the grant of a new tenancy but objected to the rent and certain of the other terms proposed by the appellants. Thereafter, no agreement having been arrived at, the matter proceeded in the usual way save that, in some way which is not entirely clear to us but which does not matter for present purposes, the present respondents, Long Acre Securities Ltd, became substituted for the original respondents. Valuers were appointed by both sides and their reports have been exchanged. To all intents and purposes the matter might have appeared to be ready for trial. The premises consist of a shop with a flat above and the lease contains a covenant restricting the user to the business of a baker and confectioner. At all material times the appellants, who are multiple bakers, have carried on there the business of retail bakers, confectioners and pastrycooks. The reports of the experts which have been exchanged indicate that there is no shortage of comparable premises in the vicinity although they differ about what premises are comparable and, no doubt, about the degree of comparability. It is not, however, suggested by the landlords that there is any lack of comparable premises and indeed this was conceded by Mr Belben, at the outset of his submissions.

There is one further circumstance which requires to be mentioned, since it has assumed an importance in the argument on behalf of the respondents. The appellants are, as has been said already, multiple bakers and among the several premises at which they carry on business they have a shop about 7 miles away from the premises in suit. The shop in question is 163 Fortess Road, Tufnell Park, London N7. Those premises are held by them under a lease from a company which is a member of the same group as that to which the respondents belong. That lease has, as we understand it, been renewed pursuant to the Act at a rent arrived at by agreement. There is not, on the face of it, any reason why those premises should be considered to have any relevance at all to the issue between the appellants and the respondents. It is true that the appellants carry on there a similar business to that carried on at the premises in suit, but they are in a totally different area both geographically and socially and Mr Belben was unable to suggest any reason why the coincidence that the landlords happen to belong to the same group as the respondents should be of any significance.

On August 27 1981 the respondents issued an application before the registrar for an order that the appellants disclose by way of further discovery all accounts for the past three years in connection with the premises in suit and the Tufnell Park premises. The registrar evidently took the view that the accounts sought were irrelevant to any issue in the proceedings, because he dismissed the application. The respondents appealed to the judge and on October 5 1981 His Honour Judge Tibber allowed the appeal and made the order now appealed against. It is in astonishingly wide terms and directs the appellants within seven days to disclose "all documents relating to trading" at the two premises without any limitation at all of the period covered by the documents. Mr Belben, however, has told us that the absence of any limitation to the past three years (which was, of course, mentioned in the application) was due to a slip in drawing up the order. The mutation from "all accounts" to "all documents relating to trading" was, we have been told, attributable to the fact that counsel then appearing for the respondents stated, on instructions, that no separate accounts of the appellants' individual businesses were kept and it seems that the learned judge's intention in making the order in the wide terms mentioned was to compel disclosure of such trading records as would enable the respondents to ascertain the amount of the turnover of the appellants at each of the two premises and the amount of the expenses and outgoings attributable to each.

Whether the terms of the order are apt for that particular purpose need not, however, be further considered at this juncture because what is in issue is not so much the ambit of the order but whether an order of this type ought to have been made at all.

From that order the appellants appeal to this court and the ground of appeal is that the order is onerous when the evidential value of the documents relating to trading is non-existent or very small having regard to the provisions of section 34 of the Landlord and Tenant Act 1954. The onerous nature of the discovery sought is, as Mr Foy very fairly conceded, difficult to sustain in the absence of any affidavit on behalf of the appellants indicating what is involved, but it is, to some extent, self-evident since, on the face of it, it involves searching for and disclosing such minutiae as delivery notes, cash rolls, manufacturing costings, wage slips and so on. But Mr Foy's principal attack is on the relevance of the discovery which the learned judge has ordered.

The inquiry upon which the court is directed to embark in applications for new tenancies under the Act is set out in section 34(1) of the Act, the relevant provisions of which are as follows:

> The rent payable under a tenancy granted by order of the court under this Part of this Act shall be such . . . as, in default of agreement, may be determined by the court to be that at which, having regard to the terms of the tenancy (other than those relating to rent), the holding might reasonably be expected to be let in the open market by a willing lessor, there being disregarded — (a) any effect on rent of the fact that the tenant has or his predecessors in title have been in occupation of the holding, (b) any goodwill attached to the holding by reason of the carrying on thereat of the business of the tenant (whether by him or by a predecessor of his in that business), . . .

There are two additional matters specified in the section which fall to be disregarded, but they are irrelevant to any issue raised by this appeal and can be ignored for present purposes.

So what the court has to look for is the open market rent of the premises simply as premises at which a business of the type carried on by the tenant can be carried on, but that rent is not to be enhanced, reduced or otherwise affected by the tenant's own actual occupation of the holding or by any goodwill created as a result of the business which the tenant has carried on. To put it broadly, the rent is to be arrived at on the hypothesis that the premises are empty and without regard to the tenant's previous trading.

This immediately raises the question in one's mind — of what relevance to such an inquiry are the tenant's trading results? The court is not concerned with the tenant's ability to pay rent but with the rent which a willing lessor could command for these premises in

A the hypothetical open market and there is a perfectly well-recognised way of arriving at that by reference to the rents payable for similar premises in the vicinity.

Indeed, if one is to take into consideration the results of the tenant's trading as a relevant factor in arriving at the open market rent, the elimination from that consideration of any effect on rent from the tenant's occupation and from the goodwill involves an extraordinarily difficult practical exercise.

In his judgment the learned judge observed: "It is highly material to know, in order to ascertain the open market rent, what the trading position is"; and a little later he said: "In a case of this sort where the open market rental value is in dispute, evidence of B trading is relevant and admissible to consider and show what the open market value is".

We confess that, for our part, we are entirely unable to follow this in the case of a property such as this where there are, as it is conceded that there are, plenty of comparable premises in the vicinity from which the open market value of premises of this type can be deduced. No doubt evidence of the tenant's trading would indicate whether his business had been successful or unsuccessful and so might be a pointer to the rent which this particular individual tenant might be prepared to pay in order to spare himself the disruption of moving to other similar premises in the area, but that has nothing to do with the open market rent which the court is C directed by the Act to ascertain. That it is for this purpose that the discovery is sought is, we think, tolerably clear from the statement of the landlord's valuer submitted by Mr Belben, which contains a frank avowal that if this tenant, in the light of its turnover in its Tufnell Park shop, has been prepared to pay a particular rent there, the disclosure of that turnover and of the turnover of the business in the premises in suit will be an indication of the sort of rent which it might be prepared to agree to for these premises.

Mr Belben sought originally to justify the learned judge's order, not on any general principle that the tenant's trading records are relevant in every application for a new tenancy under the Act, but on the special ground that they were relevant in the instant case D because of the fact that the tenant carried on a similar business in another area in premises rented from, in effect, the same landlord.

It is, however, difficult to see why the trading records of a similar business carried on in a totally different area, miles away from the premises in suit, should be thought to be any guide at all to the open market value of the latter. Why, it may be asked, stop at Tufnell Park? Why are the trading results of businesses carried on in Croydon or Piccadilly or Burnham-on-Crouch not equally relevant?

Ultimately, Mr Belben felt compelled to fall back on a general proposition that the accounts of a tenant's trading in business premises — and, we suppose, also in other premises — are always E relevant to any application under the Act as providing a guide to the open market rent. Support for such a proposition is said to be found in the 28th edition of *Woodfall, Landlord and Tenant*, vol 2 at para 2-0740, where it is said that "Evidence of the trading accounts of the existing business is admissible for the purpose of showing the earning capacity of the premises but not for any other purpose". The authority cited in support of this proposition is the decision of this court in *Harewood Hotels Ltd v Harris* [1958] 1 WLR 108. That case was concerned with some rather unusual hotel premises in Tunbridge Wells catering largely for retired people and consisting of three adjoining houses in a terrace of six. It is not entirely clear from the report, but it rather looks as if the only com- F parable premises in the area were the three adjoining houses belonging to the same landlords which were, of course, not strictly comparable, since they remained as private houses and had not been adapted for use as an hotel. It was in these circumstances that the tenant tendered and the judge received evidence of the tenant's trading for the purpose of showing what sort of profit a tenant of this hotel might expect to make and thus what level of rent a prospective tenant would be likely to pay. The landlord appealed on the ground that the evidence was irrelevant and that, in any event, the judge was precluded from considering it by the provisions of section 34(1)(a) and (b) to which reference has been made above. This court dismissed the appeal. We cannot, however, read the G decision as supporting any general proposition that evidence of this type is relevant and admissible in every application under the Act. It was relevant in that case because of the absence of any comparable premises and of the nature of the business under consideration; and the effect of the decision appears to us to be only this, that where such evidence is required in order to establish the open market rent there is nothing in section 34 which prohibits its reception for this limited purpose. But to extract from that some more general proposition that evidence of trading is always to be admitted seems to us to be a quite impermissible extension of the ambit of the decision.

Mr Belben relies particularly on the following passage from the H judgment of Lord Evershed MR (at p111):

> If the evidence was led for the purpose of showing that these tenants ought to be granted some concession because of some particular hardship that they had suffered, that might be another matter; and I agree also that the terms of the paragraph serve to exclude the consideration that a tenant might be expected to be willing to pay rather more than an outsider because he would not wish to be disturbed in his occupation. But, in my judgment, it is plainly legitimate for a judge to hear evidence which bears upon the question which he has to decide, namely: what would the particular holding reasonably be expected to be let at in the open market? Plainly, I should have thought, in arriving at a conclusion upon that question, it is legitimate to hear evidence of what similar premises which are being let for a particular purpose (as the one in suit is) can be expected to earn for a potential J lessee in the market in the place where the premises are; and, if so, then similar evidence is, in my judgment, admissible for proving the same point about the premises in suit. In other words, if the purpose of the evidence of the figures was for that limited objective, then I think for my part that they were perfectly admissible and that no objection can be made to them.

Now that, on the face of it, appears to be a statement which at least is capable of general application, but it has, in our judgment, to be read in the context of the particular facts which rendered the evidence relevant in that case, namely, that the premises were capable of being used only as an hotel and that the only physically comparable premises appear to have been private houses. This, we think, appears rather more clearly from the judgment of Romer K LJ, who said at p115:

> But I cannot find anything in the language of those provisions which renders it irrelevant to look at such material as the accounts of this company as part of the material on which the judge must make up his mind as to what rent might reasonably be expected to be obtained for the premises in the open market. I should have thought that one of the first things that anybody who was going to set up an hotel business in this type of premises in this locality would want to know would be what prospects he had of making a good thing out of it, and therefore, what rent he would be prepared to pay for the premises — if, indeed, he decided to take a tenancy of the premises at all. I cannot see any ground, either on common sense or the language of the section, upon which that consideration should be relegated to the realm of the irrelevant. I do not think that the judge would L be entitled to look at the accounts for the purpose of seeing what the company could afford to pay. Mr King-Hamilton suggested that that really was what the judge did in the present case — that he looked to see what profit they made in one year and what loss they made in another year, and so on, and said that this company, having regard to those figures, ought not to be expected to pay more than a particular rent. I should be disposed to agree that that was wrong, if the judge did it; but I do not think that he did it at all. He came to the conclusion that this hotel was being run efficiently; he came to the conclusion — and indeed it was agreed on all hands — that the premises could only be used as an hotel, and then, for the purpose which I have mentioned of seeing what a new person, an outsider, would pay as rent for these premises, he looked into the question of what kind of prospects, having regard to the law of supply and demand, this hotel would M have.

Certainly the decision shows that there may be cases where the production of trading figures may be both relevant and admissible and that where that is so section 34 does not inhibit the consideration of such evidence for the narrow limited purpose described even though the exclusion of any consideration of the effect of the tenant's occupation and of goodwill may present the judge with a very difficult task.

It is, we think, clear that there are several types of premises, of which an hotel is only one example, where the ascertainment of an open market rent may depend upon an assessment of the likely profitability of the business for which the premises are peculiarly

A adapted. Other instances might be a petrol-filling station, a theatre or a racecourse, in all of which the market rent may well depend upon the average takings. It is both unnecessary and would be unwise to seek to define or limit the categories of premises where such evidence would be relevant for the limited purpose described in the *Harewood Hotels* case. But that is a far cry from saying that such evidence is always relevant and, in our judgment, considerations of this sort do not apply in the ordinary case of shop premises with no peculiar features in a business area such as that with which the instant case is concerned and where there are plenty of comparable premises from which the open market rents can be deduced.

B There is no peculiarity in the premises themselves or the business carried on there which would lead to the conclusion that trading records would be of any assistance and it cannot, in our judgment, make any difference that the landlord, for his own purposes, seeks to restrict the user of the premises by a covenant in the lease. Such a restriction may, no doubt, decrease the obtainable rent by restricting the market to those engaged in the particular specified trade, but it cannot alter the nature of the relevant evidence from which the open market rent is to be ascertained.

In the ordinary way, the best evidence of the open market rent is what traders carrying on business in the area are prepared to pay for premises of this type, and evidence of the successful or unsuccessful
C nature of the particular tenant's business would, in our judgment, generally be not only irrelevant but also positively confusing. Nothing in the circumstances of this case suggests that the discovery sought by the respondents could be of any assistance in determining the open market rent. It would, in our judgment, be very unfortunate and would greatly increase the expense and length of proceedings under the Act if the statement which we have referred to in *Woodfall* — which is perfectly accurate so far as it goes in relation to the case cited in the footnote — were thought to justify a general proposition that the tenant's business accounts are always relevant material or the notion that orders for discovery of the type sought in this case ought as a matter of course to be made in applications under
D the Act for new tenancies. In the circumstances of this case we are in no doubt that the registrar came to the right conclusion in dismissing the respondents' application and the appeal should accordingly be allowed.

The appeal was allowed with costs in the Court of Appeal and below.

COURT OF APPEAL

January 11 1982

E (Before Lord Justice STEPHENSON, Lord Justice ACKNER and Sir John WILLIS)

TURONE AND ANOTHER v HOWARD DE WALDEN ESTATES LTD

Estates Gazette June 19 1982
(1982) 262 EG 1189

Landlord and Tenant Act 1954, Part II — Terms on which new lease of gentlemen's hairdressing premises should be granted — Landlords' appeal against county court judge's decision — Decision challenged on the ground that there was no evidence to justify it —
F Various issues had been raised as to the quantum of rent, namely, whether the appropriate approach was a "global" approach or by a zoning method, what "comparable" should be used, whether an ante-room should be in zone A or lower zone B, and what effect on rent resulted from landlords' proposal, after the case was called on, to substitute a 7 years' for a 10 years' term — Criticism of judge's decision and reasoning rejected — There was evidence on which the judge was entitled to arrive at his conclusions — Landlords' appeal dismissed — Some discussion of valuation submissions

This was an appeal by the landlords, Howard De Walden Estates Ltd, from a decision of Judge Honig at Bloomsbury and Marylebone County Court on an application for a new lease by Vincenzo Turone G
and Carlo Barbato, tenants of premises at 1a Moxon Street, London W1, used as a gentlemen's hairdressing establishment. Judge Honig fixed the rent for a new tenancy of seven years' duration at £2,550 per annum, with an interim rent of £2,400.

N Patten (instructed by Baileys, Shaw & Gillett) appeared on behalf of the appellants; D Guy (instructed by Colombotti & Partners) represented the respondents.

Giving the first judgment at the invitation of Stephenson LJ, ACKNER LJ said: Howard De Walden Estates Ltd, the appellants, are the landlords of 1a Moxon Street, London W1. They are appealing against an order made by His Honour Judge Honig on October 15 1980 at the Bloomsbury County Court on the hearing of H
an application for a new tenancy under Part II of the Landlord and Tenant Act 1954. It was an application by the respondents to this appeal, Vincenzo Turone and Carlo Barbato, who were the tenants of the premises.

The premises, 1a Moxon Street, London W1, comprise a gentlemen's hairdressers. Moxon Street is a small side street running eastwards from Marylebone High Street and the premises comprise a ground-floor shop with an area known as an ante-room off it, and a basement. The tenants were occupying the premises under a lease for 10 years which was dated May 14 1970 and which reserved initially a rent of £550, but at the end of which had risen to £1,000 per annum.

The application for a new lease was not contested; the only issue J
which the learned judge had to decide was on what terms the new lease should be granted.

Mr Patten, in a careful and well-formulated argument, was at pains at the outset frankly to accept that his appeal raises essentially one short point concerning the judge's findings as to the rent to be reserved in the new lease. Mr Patten properly accepted that it was not open to him to challenge the judge's findings of fact; he would have to establish an error of law and he sought to do so, and he so amended his notice of appeal, by alleging that there was no evidence available to the learned judge upon which to base his decision that the right rent for the new lease was £2,550.

Essentially there were four issues concerning the fixing of the K
correct rent. So far as the term is concerned, initially it had been common ground that the new lease should be for a period of 10 years, but when the case was called on, and not until the case was called on, the landlords indicated that their consent to 10 years no longer continued and that they were going to contend for a period of seven years because as a matter of good estate management that was, according to their point of view, the appropriate period, since the lease would then fall in at the same time as that of adjoining property.

The four issues concerning the quantum of rent were these: firstly, what was the appropriate approach in fixing the market rent? The rival views were these: on behalf of the tenant it was contended that L
the global approach was the right approach, and by that you increased the rent by a percentage figure appropriate to the extent to which rents had become inflated since the rent was last fixed. The landlords, however, contended for the zoning method; the zoning method involves dividing the premises into zones A, B and C, the best area being zone A, and you consider what is the proper figure for that, and then for the next portion of the property, zone B, you halve that figure and for the remaining portion, zone C, you halve that.

The next issue was: what were the comparable properties? I need spend little time on this because the learned judge took the view that the property to which he should apply his mind was 3b Marylebone Street. M

The next issue was whether the ante-room, to which I referred when describing the property, should be considered as zone A and therefore be included with the shop, or whether it should be looked upon as property of far less value and therefore be considered as zone B, attracting a rent of half that of the shop.

Finally, the fourth point at issue was whether there should be any discount for the reduction, if the judge approved it, of the term from 10 years to seven years.

I can now deal with those issues quite shortly. So far as the approach was concerned, the learned judge considered that the zoning approach was the more appropriate one and this is what he

A adopted. I have already indicated the comparable property which in his view he should pay regard to, which was 3b Marylebone Street. It was the property which had been chosen by the landlords' surveyors; they had been involved in the negotiation of a rent review in March of 1979. The property was the subject-matter of a lease for a term of some 18 years granted in 1975 and expiring in March of 1993. So on the first two issues the evidence of the landlords' expert was preferred.

I now deal with the third issue, the ante-room. The use being made of the ante-room by the respondents, the tenants, was not for the purpose of the gentlemen's hairdressers, but as a place where tea could be made, towels could be dried and occasionally the
B hairdresser himself could take a break there and have some rest. The landlords' expert contended that that was not the appropriate use that should be made of the premises and, albeit that it would involve extra expense, it could be thrown into the shop premises. The learned judge rejected that view; he was of the opinion that the use which was being made of the ante-room was a perfectly reasonable one and he therefore zoned the ante-room as zone B, which therefore attracted half of whatever was the appropriate rent for the shop.

So far as the fourth issue, the discount, was concerned, this only fell to be considered, as I have already indicated, when the case was being heard. Criticism was made by Mr Patten about the tenants' expert's evidence, who said in terms, when he gave that evidence,
C that he had not considered the matter before that day. I do not think that is a fair criticism to make. If the landlords had given advance notice that they were going to change their minds and ask for the lease to be limited to a term of seven years rather than 10, there could have been some substance in this criticism.

The tenants' surveyor said that if a seven-year term is substituted for 10 years the rent should be lower by 25% than the current valuation. This was disputed by the landlords' expert, who said that it should make no difference at all. It seems to me that as a matter of common sense one would expect some discount. There is clearly the advantage of having the certainty of a term of 10 years rather than the lesser term with the possibility, albeit the probability, of a right to
D renewal under the Act. That right, of course, is not absolute. The landlord might wish, when the time came for the grant of a new lease, to contend that he required possession for his own occupation, which is no doubt unlikely; or that he wished to gain possession in order to replan, remodel and alter the whole premises. He might have disposed of them in time for a successor to be entitled to have been the landlord for the appropriate five years and himself to ask for possession in order to occupy; so there is no certainty that, come seven years, a new lease would be granted. So that as a matter of ordinary business approach to the situation, one would expect that a lease of 10 years, specially one which can be assigned, as this lease could be with the consent of the landlord, would be more valuable
E and would attract a higher rent.

The learned judge, when he came to deal with that particular matter, said that in the interests of good estate management the landlords were contending that the letting of the subject premises should be for seven years and not for 10 years; he said that Mr Broomfield had said in evidence that a lease of seven years would not really be less advantageous because of the applicants' rights to apply for a new lease, and he went on: "This seems to me a wholly fallacious argument. To have to return to the court to obtain a renewal is not the same as getting 10 years originally. However, I see a certain force in the landlords' argument based on good estate management in this case, and I conclude that it would be reasonable
F to grant a seven-year term but I must take account of the fact that it puts the tenant in a worse position when I decide the rent payable thereunder."

I turn now to the figures which the learned judge applied when fixing the rent. So far as 3b Marylebone Street, the comparable premises, was concerned, the evidence which had been given by the landlords' surveyor was that if one applied the zoning approach, the zone A figure was £10 per sq ft; accordingly the zone B figure was £5 per sq ft and zone C £2.50 per sq ft. He went on to maintain, in his report and in his evidence, that as the rent had been fixed in March of 1979, a further increase had to be made to take into account the 18 months which had expired, and that accordingly he would increase those figures to: £12 for zone A, £6 for zone B and it would follow G
that zone C would be up to £3.

The learned judge had a contest before him as to whether in fact these premises were inferior or superior to those the subject-matter of the litigation, and he said that he did not think there was much in it on the evidence. But he went on to say: "However, I don't know enough about the various terms of the lease (of the comparable) to say how they compare with the grant of a seven-year lease with a review after four years. What I have decided to do is to divide the premises into three parts, zone A, zone B (the ante-room) and the basement" (zone C). He concluded his judgment by saying: "I conclude, therefore, that in the case of 1a Moxon Street the proper figure for the shop area should be £8 per sq ft, for the ante-room £4 H
per sq ft and for the basement £2 per sq ft", and thus the total rental produced was the figure of £2,546, which he rounded up to £2,550 per annum, and he fixed the interim rent, to which there is no criticism if the earlier figure which I have mentioned is right, at £2,400.

The learned judge did not say in terms how he arrived at the £8 per sq ft. Mr Patten accepts that if the learned judge had said: "I do not think, however helpful the comparable is, that I should follow it slavishly because there has been a contest, not only as to whether it is superior or inferior property, but also I know little about the other terms of the lease than those to which reference has been made, and accordingly I have decided to take £10 per sq ft for zone A, the figure J
analysed by the landlords' expert as being appropriate in 1979", it would have been extremely difficult to fault the learned judge. If he had then said in terms that the tenants' surveyor says that the discount should be 25% for the reduction from 10 years down to seven years, and that he proposed to take the appropriate figure of 20%, thus producing £8 per sq ft, the learned judge could not have been faulted in that regard — certainly not in this court. There would have been material upon which his decision could be justified, although there could be counter-arguments as to whether it was the right ultimate decision; but in this court it would have been unappealable. But, says Mr Patten, that has not been spelled out in terms, and it would be wrong for this court so to do. K

I do not take that view. This was an extempore judgment. The learned judge is not in those circumstances to be expected to identify each and every process in his reasoning. He has in this case made his approach clear, namely, the zoning approach, and what is the comparable which he is prepared to have regard to; a comparable does not mean a facsimile which he has to follow in each and every respect. He has indicated why he has downgraded the ante-room contrary to the landlords' surveyor's evidence, and he has said in terms that he thinks a discount should be made and that he intended to make a discount in the figure at which he has arrived.

It seems to me to be highly probable that he did take £10 as being the zone A figure, and he did discount it in the way in which I have L
suggested. Mr Patten did not suggest any other approach which seems to have been likely to have been adopted by the learned judge, except that he totally disregarded the comparable and plucked a figure, so to speak, out of the air. I do not think that that fits in with the learned judge's very careful judgment; nor is it consistent with the sort of reasoned approach that we have learned to expect from this judge, with his considerable experience of this type of litigation.

I therefore conclude that Mr Patten cannot achieve what he set out to do, namely, to establish an error of law in that there was no evidence to justify the learned judge's decision.

I should perhaps, for the sake of completeness, mention one final point which was made by Mr Patten, although realistically, on the M
basis that it was not to be put in the forefront of this appeal. The learned judge excluded from the new lease a clause which was to prohibit absolutely assigning or subletting part of the premises. He did so because both surveyors were of the same opinion, namely, that as a matter of practice there could be no question, having regard to the way in which the premises were laid out, of their being the subject-matter of any such assigning or subletting; it just was not physically on. Accordingly, the judge decided — and it seems to me, with respect, very sensibly — why bother to include a clause in the lease which could have no relevance at all to the premises? So he excluded it.

In the notice of appeal it is contended that the learned judge failed to take into account the position of any assignee from the applicants. That point, understandably, was not pursued by Mr Patten, because it has no relevance to the point that the premises were not physically likely ever to be the subject-matter of such a subletting. Mr Patten said: "Ah, but there is always the question of sharing, which is something short of actual subletting". Significantly, not only was this point not taken in the notice of appeal, but it was never raised in the court below, and for my part I do not consider that the point is open to be argued here. It appeared, at the end of the hearing of the matter before the learned judge, to be common ground that there was no need to consider further subclauses 22(b) and 22(c) and accordingly the learned judge, in my judgment rightly, concluded that there was no need to burden a new lease with it.

For these reasons I would dismiss this appeal.

SIR JOHN WILLIS: I agree; there is nothing that I can usefully add.

STEPHENSON LJ: I also agree.

The appeal was dismissed with costs.

COURT OF APPEAL

November 20 1981

(Before Lord Justice LAWTON and Lord Justice TEMPLEMAN)

MANCHESTER CITY COUNCIL v NATIONAL CAR PARKS LTD

Estates Gazette June 26 1982

(1981) 262 EG 1297

Agreement for the use of land in the centre of Manchester as a car park — Whether agreement created a tenancy to which Part II of the Landlord and Tenant Act 1954 applied or whether it granted a licence only — Agreement was framed in the terms of a licence and restricted the use of the land to car park use from one minute past midnight to 2 am and from 7 am until midnight — There was, however, evidence that tickets were issued for parking on a 24-hour basis — Held that the terms of the agreement and the surrounding circumstances indicated that the transaction was a licence only and the fact that the licensors did not enforce the restriction on the hours of use was not sufficient to justify an inference to the contrary — Appeal by licensees dismissed

This was an appeal from a decision of Vice-Chancellor Blackett-Ord, sitting as a judge of the High Court, on an application by the plaintiffs (respondents to the appeal), Manchester City Council, for summary judgment under Order 14 of the Rules of the Supreme Court. The defendants (the present appellants) were National Car Parks Ltd and they appealed against the Vice-Chancellor's order granting the respondents an injunction and ordering an inquiry into damages. The order restrained the appellants from using as a car park an area of land fronting Lower Mosley Street, Manchester, following a notice from the respondents requiring possession of the land to be relinquished. The appellants, who had refused to give up possession, claimed that they had a tenancy to which Park II of the Landlord and Tenant Act 1954 applied.

Bruce McIntyre (instructed by A E Hamlin & Co) appeared on behalf of the appellants; Nigel Howarth (instructed by R M W Taylor, Town Hall, Manchester) represented the respondents. Mr Howarth was not called on.

Giving judgment, LAWTON LJ said: This is an appeal by the defendants, National Car Parks Ltd, against an order made by His Honour Vice-Chancellor Blackett-Ord, sitting as a High Court judge, on the hearing of the plaintiffs' application for summary judgment under Order 14 of the Rules of the Supreme Court, whereby the learned Vice-Chancellor adjudged that the defendants, whether by themselves, their directors, officers, servants or agents or otherwise be restrained from using as a car park the land fronting Lower Mosley Street in Manchester edged in red on the plan annexed to the judgment and from entering thereon or from trespassing thereon in any manner whatsoever. Further it was ordered by the learned Vice-Chancellor that an inquiry be made by the district registrar as to what damages the plaintiffs had suffered by reason of the defendants' continued use of the land for the purpose of parking vehicles thereon from and after March 25 1980 to the date of possession; and that the defendants' counter-claim be dismissed.

The plaintiffs became owners of the land in question in December 1979. The previous owners had been a company called Garrard Securities Ltd. By an agreement in writing dated October 25 1976 and made between Garrard Securities Ltd of the one part and the defendants of the other part Garrard Securities Ltd granted the defendants the right to use the land in question, together with other adjoining land edged in blue on the plan annexed to the statement of claim, from July 19 1976 for a term of six months between the hours of one minute past midnight and 2 am and 7 am and midnight daily for the purposes of a public car park and for no other purposes whatsoever. The defendants, who are a very well-known organisation, went into possession of the land pursuant to that agreement and they have been in possession of it ever since. It is clear from the agreement to which I have referred that Garrard Securities Ltd at the time when they made the agreement had in mind the possibility that they might want to develop the land. As I have already said, in December 1979 Manchester Corporation bought the land from Garrard Securities Ltd and they at once gave the defendants notice to give up possession of the car park.

The defendants have not given up possession and they have contended, both before the learned Vice-Chancellor and in this court, that what they were granted by Garrard Securities Ltd was a tenancy to which the benefits of the Landlord and Tenant Act 1954 apply. It follows, as they have submitted, that the Manchester Corporation had no legal power to turn them out otherwise than under an order of the court made pursuant to the provisions of the Landlord and Tenant Act 1954.

The question in this case turns upon the agreement which was made on October 25 1976. If, on its true construction, it is clear that a licence was granted, then this appeal must fail. The agreement was, as I have already stated, in writing. It purports to be an agreement granting a licence to the defendants, but, as is well established, the label which parties put upon an agreement of this kind is not conclusive. That does not mean, however, that the label should be wholly disregarded. In this case, having regard to the situation of the land and the purpose for which the use of the land was being granted to the defendants, there was clearly a good commercial reason why Garrard Securities Ltd would not want to grant a tenancy. This land is in the centre of Manchester and is clearly, if developed, very valuable land indeed. That is a factor which has to be considered. Another factor which has to be considered as forming part of the matrix of this case is this: it is well known now that, if car parks are open to the public for 24 hours every day, during the night undesirable activities are likely to go on there. Broken-down vehicles are dumped and criminal activities of various kinds may take place. It follows, therefore, as a matter of convenience to everybody, agreements relating to car parks can properly be, and often should be, so worded as to ensure that they are not open all through the night.

The parties to this agreement seem to have had both the points which I have mentioned in mind. Clause E, paragraph 1, of the agreement reads as follows:

> The Licensee shall have the right to use the site during the licence term between the hours of 00.01 and 02.00 hours and 07.00 and 24.00 hours daily for the purposes of a public vehicle park and for no other purpose whatsoever.

It seems to me that that clause, on the face of it, did not grant, and was not intended to grant, the defendants exclusive possession of the land.

There is another clause in the agreement which entitled Garrard Securities Ltd to give a short notice terminating the licence if they wanted the land for development. That again fits in with what I would have expected in a case of this kind.

Mr McIntyre, on behalf of the defendants, has had to concede that, on the face of this document, there was a grant of a licence and not the grant of a tenancy. He has submitted, however, that subsequent events showed that the original parties intended to

A dress up a tenancy as a licence and that it was the intention of Garrard Securities Ltd when they entered into this agreement to grant exclusive possession of the land. In other words, he says that the written agreement was a sham and the real agreement was something entirely different.

He bases that submission on the facts set out in an affidavit of Mr Raymond Scott Durrant, who is a chartered surveyor employed by the defendants. In paragraph 2 of his affidavit, which was sworn on July 29 1981, he said:

> I note that clause E of the agreement dated October 25 1976 purported to restrict the times at which the defendants could use the site. In fact at all material times the defendants left a kiosk on the land together with ap-
> B propriate signs setting out the tariff. In addition we have continuously advertised and let season tickets for parking on the land on a 24-hour basis. I believe it was well known to Garrard Securities Ltd and the plaintiffs herein that that was the position. To that extent the terms of the agreement dated October 25 1976 do not reflect what was happening on the ground.

Of the two points taken in that paragraph, one is clearly bad, because the licence agreement envisaged that some kind of portable kiosk could be left on the land. As to the other point taken, all it C amounts to is that Garrard Securities Ltd, and the plaintiffs later, did not enforce their rights. That is a very long way from saying that, right from the beginning, Garrard Securities Ltd never intended to enforce their rights and that the plaintiffs, when they acquired the land from Garrard Securities Ltd, were willing to allow a similar situation to go on, namely, that that which purported to be a licence should be treated as a tenancy and that they were knowingly taking over from Garrard Securities Ltd as the defendants' landlord.

I find it impossible to spell out any collateral agreement of that kind from such facts as are before the court. It seems to me that the parties, very sensibly, in October 1976, agreed that there should be D a licence and not a tenancy. I would dismiss the appeal.

TEMPLEMAN LJ, agreeing, said: For the reasons given by the Vice-Chancellor and by my Lord, it seems to me that there is no tenable defence in this action and, accordingly, it is right that we should uphold the order below on appeal.

The appeal was dismissed with costs.

LANDLORD AND TENANT
RENT ACTS

A
COURT OF APPEAL

July 24 1981

(Lord Justice EVELEIGH, Lord Justice BRANDON and Lord Justice O'CONNOR)

GRIFFITHS v ENGLISH

Estates Gazette January 23 1982
(1981) 261 EG 257

**Rent Act 1977, section 12 — Whether resident landlord provisions excluding a tenancy from being a protected tenancy applied — Sole question in present case was whether the tenant's dwelling-house
B formed "part only of a building" within the meaning of section 12(1)(a), in which case the tenant had a restricted contract only and not a protected tenancy — Held that the whole property, in spite of having a centre portion with extensions on both sides, having no internal connection with the centre, constituted one building — *Bardrick v Haycock* considered and principles applied, although on the facts that decision went the other way — Appeal from decision of county court judge, who made an order for possession in favour of landlords, dismissed**

This was an appeal by the tenant against an order for possession of a dwelling-house which constituted the left-hand extension of a
C property known as "Wessex Manor", described by Eveleigh LJ as a "quite small" property. The structure of the property, with the main house in the centre and extensions on each side, is described in the judgment of Eveleigh LJ. The appellant (defendant in the court below) was Richard English; the respondent (plaintiff) was Thomas Percy Griffiths.

K F Wylie (instructed by Jasper & Vincent, of Eastleigh, Hants) appeared on behalf of the appellant; J A H Haig-Haddow (instructed by Bell, Pope & Bridgwater, of Southampton) represented the respondent.

Giving judgment, EVELEIGH LJ said: The property which we have to consider in this case appears clearly from the photograph before
D the court. Looking at the photograph we see the front of the property — the centre of which can be described as neo-Georgian — with a centre doorway fronted by a porch and one room on either side of that doorway. To the left, as we look at the photograph, there is a single-storey extension which goes along for approximately 20 ft and then turns at an angle of about 60° to the rear for another 10 ft or so. The exact measurements are not of importance. The front of the main house itself is some 25 ft, or so it would appear. To the right of the property there is another extension which travels for about 5 ft. That, too, is single storey and that extension then turns at a right angle to the main house and goes towards the road a distance of some 25 or 30 ft or thereabouts. That
E extension has a gabled roof. The extension to the left of the property has a flat roof. That flat roof is used as a terrace. The extension itself on the left is probably some 15 ft in depth and so the terrace is a substantial one. It is accessible from the main house and it is surrounded by an attractive modern terrace rail. The extension to the left of the property has a main doorway approximately in the centre of the wall before it turns off at 60°, as I have said.

There is no communication between that extension and the main house. There is, indeed, no communication between the extension on the right and the main house. There had been a doorway and that has been blocked up.

The extension on the left is occupied by the defendant. It is his F
dwelling-house. The main house itself is divided into flats and the terrace is available for use by the occupant of a flat. The extension on the right, as we look at the building, with the gabled roof is occupied by the landlord, the plaintiff.

The property was not always in the condition in which we now see it, for immediately to the left of the house there originally existed a garage. It was attached to the house, and away to the left of the garage there was an air-raid shelter. Reconstruction work was undertaken, which connected the air-raid shelter to the garage. Windows and doors were put in and rooms were formed within those exterior walls. The air-raid shelter was turned into a bathroom and kitchen. These reconstructed premises were now available to be let as a dwelling-house. There is no longer a garage there. G

The learned county court judge made an order for possession, although the defendant had claimed that his was a protected tenancy. The order was made on the basis that the premises were not protected by virtue of section 12 of the Rent Act 1977. That section lays down a number of conditions which have to be fulfilled in order to take a tenancy out of the Act for the purposes of security of tenure, but the relevant one with which this case is concerned is contained in section 12(1)(a). It reads as follows:

> The dwelling-house forms part only of a building and . . . the building is not a purpose-built block of flats.

No question of a purpose-built block of flats has been raised in this H
case and the sole question is whether the defendant's dwelling-house forms part only of a building.

Perhaps it is not quite accurate to put it as simply as that, because it is also necessary for it to be established that the landlord himself occupied the building; but the question was raised before the learned judge in the simple form, as indeed was appropriate. Is the whole of the property that I have described a building or does it consist of more than one building?

Counsel for the appellant has recognised that this involves a question of fact and the learned county court judge visited the premises and came to the conclusion that they were one building.
There is no appeal to this court on a question of fact in a case such as J
the present because its rateable value is such that it does not come within the provisions of section 109 of the County Courts Act, which allows an appeal on a question of fact where the rateable value of the premises is £500 or over. The rateable value in this case is less than that. But it has been submitted on behalf of the defendant that the learned judge misdirected himself in a number of respects which are set out in the grounds of appeal.

Before turning to those specific allegations I would consider the matter of fact myself — although, indeed, it is not necessary to do so unless one comes to the conclusion that the learned judge did misdirect himself. This property, looking at it, did it not have the centre
door to the defendant's premises, could easily be regarded as fol- K
lows. Here we have a house which has a single-storey extension running away to the left: an extension that might well be providing a garden room or something like that to be used in connection with a terrace and garden at the back. To the right the property gives the appearance of a substantial extension to the main house for use as domestic offices, possibly for use as storerooms, but also for use for personal occupation. But in my judgment, the appearance is that of one property, and I would go further and say one building. I would not readily recognise any part as being that of a separate

A building. One may add, in addition to the description I have already given, that the front forecourt of the property is contiguous with the whole of the premises: the defendant's dwelling-house, the centre house and the wing on the right. In a similar way the garden is contiguous with the perimeter of the property at the back. It should be emphasised, I think, that this property is really quite small. It is called "Wessex Manor" and that might lead one to think that the property is of substantial dimensions. It is not.

Now, I myself would have come to the same conclusion as that of the learned judge and, saying that, it is really unnecessary or academic to consider the particular criticisms made of the judgment in the course of the appeal. Suffice it to say that, in my opinion, they B do not demonstrate that the learned judge misdirected himself. It is no disrespect to the very careful and helpful argument of counsel for the appellant if I do not deal with each one specifically; but take as an example the ground of appeal which complained that the learned judge had regard to an expression of opinion of the plaintiff himself when he said that he would not have let the property if he thought that it was other than a single building and thus entitled to the benefit of the provisions of section 12 of the Act. It is true the learned judge did refer to that expression of opinion, but he was doing so after having formed and expressed his own opinion and, in my judgment, he merely stated it, as it were, as lending some support generally to the view that he independently had formed. It C was an unnecessary remark but this was an extempore judgment. Judges sometimes do make such remarks; and it seems to me that the other grounds of appeal, broadly speaking, are explicable upon the same basis.

These cases are very much questions of fact and for the decision of the learned judge. This is made quite clear in a case which has been quoted to this court, *Bardrick v Haycock* (1976) 31 P & CR 420. Cairns LJ said this at p 425:

> When an Act of Parliament uses the word "building" without defining it there must be some structures or pairs of structures which as a matter of law could be said to be two buildings within the meaning of the Act and some which as a matter of law could be said to be one building. If a judge held, for D example, that two quite separate houses constituted one building, he might be said to have erred in law. On the other hand, if a judge held that two floors in the same house constituted separate buildings, that might be said to be erroneous in law. There must, however, be some borderline cases as to which it is a question of fact whether there is one building or two. All the grounds of appeal here are to the effect that the judge erred in law. Some guidance as to what Parliament meant by "building" in the particular Act may be gained by considering the context in which the word is used. I do not find in the context of the Act here any such indication as would enable me to say that where a new habitable structure is built adjoining an existing house which is divided into flats, but with no internal communication, the main structure together with that house constitutes one building.
>
> I consider that this is one of those borderline cases where the question is E one of fact, as with a pair of semi-detached houses: see the decision of the Divisional Court in *Humphrey v Young* [1903] 1 KB 44 to which Scarman LJ has referred.

The facts of that case were mentioned to the court because the decision went the other way; but it is emphasised in the judgments delivered that the question is one of fact. As I have said, I myself in this case would have taken the view that this was one building and, that being so, I hope I am right in saying that that deprives counsel of the advantage of the further ground of appeal on which he relies that, upon the evidence in this case, no reasonable person could come to the conclusion that this was one building.

F I would dismiss this appeal.

Agreeing, BRANDON LJ said: I would just mention that the provisions of section 12 of the Rent Act 1977 under the wording of that section apply only to tenancies of dwelling-houses granted on or after August 14 1974; but, with the assistance of counsel, we were able to see that the provisions were made to apply to tenancies granted earlier by Schedule 24, paragraph 6, to the 1977 Act. There was, therefore, no dispute below and no dispute here that these furnished tenancies were governed by the provisions of section 12.

I would direct myself entirely by the judgment of Cairns LJ in *Bardrick v Haycock*, which my Lord, Eveleigh LJ, read. I think it clear that there may be some cases where no reasonable judge at G first instance could conclude that premises constituted a single building, and other cases where it would be quite impossible for him reasonably to conclude that they did not constitute a single building; and in these cases it could be said that, if the judge comes to a conclusion which is wholly unreasonable, then he has erred in law. But in between those cases there must be a band of cases where different views can perhaps be taken reasonably by different judges. That being so, in so far as a judge in what might be described as a borderline case comes to a conclusion one way or the other as to whether there is one building or more than one building, then the matter is one of fact and no appeal to this court is available in respect of his decision.

H I think this case comes within that concept. It is a case where there were plausible arguments capable of being put forward that the defendant's flat was a separate building, and there were plausible, and perhaps more than plausible, arguments capable of being put forward that it was all part of the whole structure known as Wessex Manor. The judge decided that the flat was part of the whole structure; that was a decision of fact and, in my view, we would have no power to interfere with it even if we disagreed with the conclusion to which the learned judge came. Like my Lord, Eveleigh LJ, however, if I were asked to re-try the case as if I were a judge at first instance, on the material before us, I think that I should come to the same conclusion as did the learned judge.

J It was contended in the notice of appeal that the learned judge had regard to a number of matters which he should not have taken into account and had failed to take into account matters which he should have taken into account. Of course, if that were established, that would show an error of law on the part of the learned judge; but I find it difficult to see how these complaints can be justified.

In paragraph 2 of the notice of appeal it is complained that:

> The learned judge was wrong in treating as irrelevant the fact that a substantial part of the defendant's dwelling was at one time part of an air raid shelter standing separate from, independent of and predating the main part of Wessex Manor.

K In my view, the previous history of the building and evidence that at one time it was in a different form from the form in which it is now were not relevant to the decision whether the present structure is a single building or not.

Then it is said that:

> The learned judge was wrong in treating as irrelevant the fact that at no time was the defendant's dwelling internally connected to the rest of Wessex Manor.

I do not know that there is any reason to suppose that the judge did treat it as irrelevant. What I think that he did say, or meant to say, was that it was not a conclusive matter. It could not by itself outweigh all the other matters which he thought pointed in a L different direction.

Then it is said that:

> The learned judge was wrong in treating as irrelevant the history of the construction of the parts of Wessex Manor.

Here again, it seems to me that what the judge has to deal with was the building as it now stands and not the history of how it came into being.

Then it is said that:

> The learned judge was wrong in taking into account facts irrelevant to the question of whether the defendant's dwelling was part only of a larger M building, namely
> (a) the fact that it had no separate garden;
> (b) the fact that the plaintiff had a greenhouse and shed at the back of the defendant's dwelling;
> (c) the fact that the whole complex known as Wessex Manor was once in single occupation; and
> (d) the fact that the plaintiff had not expected to grant a protected tenancy to the defendant.

To my mind, in deciding the question of fact which the learned judge had to decide, a very large number of circumstances can be relevant and all of them should be taken into account. I certainly think that the absence of a separate garden and the presence of a

A greenhouse and shed at the back of the defendant's dwelling and the fact that the whole complex of buildings had once been in single occupation were all material facts that the trial judge was entitled to take into account. With regard to the last matter, the fact that the plaintiff had not expected to grant a protected tenancy to the defendant, I agree with what my Lord, Eveleigh LJ, has said. What I think the learned judge was intending to say when he referred to that was that the attitude which the plaintiff had adopted was consistent with the intention of the legislature in respect of this provision as explained by Scarman LJ (as he then was) in *Bardrick v Haycock*. He said at p 424:

> B Therefore, the mischief at which the section was aimed was the mischief of that sort of social embarrassment arising out of close proximity — close proximity which the landlord had accepted in the belief that he could bring it to an end at any time allowed by the contract of tenancy.

All the judge was saying was that the plaintiff did regard the tenancy of this flat, if it was a protected tenancy, as giving rise to the sort of close proximity which, according to Scarman LJ's judgment, this provision was intended to avoid.

Then, in paragraph 6 of the notice of appeal, it was suggested that:

> The learned judge was wrong in holding that the respective dwellings of the
> C plaintiff and the defendant, parts of each of which were freestanding buildings before the central section of Wessex Manor was built, could become parts of the same building within the meaning of section 12 of the Rent Act 1977 by the gradual construction of the said central section.

I find this a ground of appeal that is totally untenable. It seems to me that, if you have two buildings which are separated and you then join them together with another building between the two in a manner which ordinary people would regard as creating a single building, then what you have got left is a single building.

I think that I have dealt with all the points in the notice of appeal except the final one, which is:

> D . . . the conclusion was one that could not reasonably be reached, and was inconsistent with the evidence heard and accepted by the learned judge.

Since both Eveleigh LJ and I consider that we would have arrived at the same conclusion as the learned judge, it follows that neither he nor I can possibly regard that ground as having any substance in it at all.

For the reasons which I have given and those already given by Eveleigh LJ, I agree that the appeal should be dismissed.

O'CONNOR LJ said: I agree with both judgments which have been delivered and have nothing to add.

The appeal was dismissed with costs up to the date of the legal aid
E certificate, the costs to be referred to the registrar of the Southampton County Court, both as against the appellant and the Law Society. Further stay of execution for 28 days was granted.

COURT OF APPEAL

July 8 1981

(Before Lord Justice ORMROD, Lord Justice O'CONNOR and Mr Justice SHELDON)

F

PATEL AND ANOTHER v PATEL

Estates Gazette January 30 1982
(1981) 261 EG 361

Rent Act 1977 — Claim for possession under Case 9 in Schedule 15 — Claim by personal representatives who had no beneficial interest in house — Claimants were administrators of estates of two persons who had been the owners of the house and who had been killed in a car accident — Claimants became guardians of the orphaned children of the deceased and later adopted the children — Object of possession proceedings was to provide a home for the claimants and the children — Was the claim barred by a rule that personal representatives cannot be landlords for the purpose of Case 9 unless they had a beneficial interest in the house? — *Sharpe v Nicholls, Parker v Rosenberg and Harrison v Hopkins* considered — No rule that personal representatives cannot be landlords for this purpose — True explanation of the cases is that a personal representative normally cannot claim relief under Case 9 because in so doing he would be acting in breach of trust — In present case claimants were certainly not acting in breach of trust in seeking to live in the house with the children as a family home — Appeal from county court judge allowed and order for possession made G

This was an appeal by the plaintiffs, M M Patel and his wife I M Patel, from a decision of Judge Tumin at Willesden County Court refusing their claim for possession of a house at 9 Garratt Road, Edgware, from the defendant, B Patel (no relation), who was in occupation as a statutory tenant. H

P H Morgan (instructed by Herbert & Gowers & Co) appeared on behalf of the appellants; the respondent, B Patel, appeared in person.

Giving the first judgment at the invitation of Ormrod LJ, O'CONNOR LJ said: This is an appeal by the plaintiffs from an order of His Honour Judge Tumin, sitting in Willesden County Court, refusing their claim for the possession of a house at 9 Garratt Road, Edgware.

J The facts giving rise to this case are tragic and simple. In December 1977 Mr and Mrs H M Patel, the brother and sister-in-law of the male plaintiff, were killed in a motor-car accident. They left two small children, then aged five and three. They were the joint owners of 9 Garratt Road. The plaintiffs became guardians to the two children in July 1978. They became administrators of the estates of both the deceased in April 1979 and in March 1980 they adopted the two children, who are now aged eight and five. By an agreement in writing, dated June 18 1979, they let 9 Garratt Road to the defendant (who is not a relative, although he has the same surname) for one year from April 1 1979 until March 30 1980. That agreement expired by effluxion of time and, subject to a matter I will deal with under the respondent's notice, the defendant held over as a statutory tenant, as he was entitled to do. Thus it was that the plaintiffs brought these proceedings for possession of the house, founding themselves on Case 9 of Schedule 15 to the Rent Act That provision reads so far as is material: K

> Where the dwelling-house is reasonably required by the landlord for occupation as a residence for:
> (a) himself, or

the court may make an order for possession. That provision appears in the 1977 Act, which is a consolidating statute, effectively re-enacting paragraph (h) in Schedule 1 under section 3 of the 1933 Rent Act. L

In an admirable and careful judgment the learned judge directed himself that the issues in the case were:

> (1) Are the plaintiffs landlords for the purposes of Case 9: if they are not, that is the end of the plaintiffs' case;
> (2) if the plaintiffs are landlords for Case 9, have the plaintiffs established that they reasonably require the property for themselves?;
> (3) if the plaintiffs reasonably require the property for themselves, has the defendant established greater hardship?;
> (4) is it reasonable to order possession?

First of all the learned judge considered the last three topics and he found in favour of the plaintiffs on all three heads. M

The defendant, who has appeared in person (and, if I might say, has conducted his case quite admirably) does not seek to challenge the findings under (2), (3) and (4) to which I have referred.

In answer to the first question the learned judge, holding himself bound by the authority of two cases in this court, found against the plaintiffs, and it is against that that the appeal is brought.

Having looked at the cases I, for my part, have concluded that the learned judge did not ask himself the right quesion. Question (1) should have stopped at "Are the plaintiffs the landlords of the house?"

A The reason why he asked the question as he did is undoubtedly due to the way in which the two cases upon which he founded himself, namely, *Sharpe v Nicholls* [1945] 1 KB 382 and *Parker v Rosenberg* [1947] 1 KB 371, have been dealt with in the textbooks. The proposition has found favour with the textbook writers from those cases that personal representatives cannot be landlords unless they have a beneficial interest in the house. As I propose to show in my judgment, that is not what the cases decide. The question which should be asked is: are the plaintiffs landlords? And the next question should be: have they established that they reasonably require the property for themselves?

Let me get two matters out of the way before I look at the
B authorities. The provision in Case 9 that the landlord reasonably requires the dwelling-house for occupation as a residence for himself, "himself" includes his spouse and children: see *Smith v Penny* [1947] KB 230. Secondly, "landlord" is to be interpreted, where there are joint owners, as "landlords": see *Baker v Lewis* [1947] 1 KB 186.

The two cases of *Sharpe v Nicholls* and *Parker v Rosenberg* are most conveniently considered in a later decision of this court, *Harrison v Hopkins* [1950] 1KB 124. That case was concerned as to whether a personal representative tenant was a tenant for the purposes of the Rent Act. Cohen LJ (as he then was) dealt in his judgment with the problems raised by *Sharpe v Nicholls* and *Parker*
C *v Rosenberg*. His judgment sufficiently sets out the facts of the case and I shall read an extract from it. At p 132 he said:

> In these two cases, *Sharpe v Nicholls* and *Parker v Rosenberg*, this court had to consider the rights of a landlord to possession. The question in each case, however, arose not under section 12(1)(f) of the Act of 1920, but under section 3 of the Rent and Mortgage Interest Restrictions (Amendment) Act 1933, and para (h) of Schedule I thereto. In the first case [that is *Sharpe v Nicholls*] proceedings were commenced by the widow and nephew of the deceased landlord as personal representatives to recover possession of premises within the Rent Restriction Acts for the occupation of the widow. There was no evidence to prove that the widow had any beneficial interest in the estate of the deceased landlord. The actual decision in the case did not turn on whether personal representatives were landlords within the mean-
> D ing of para (h), but Morton LJ expressed an opinion on the point, saying:
>
> "The plaintiffs' whole case, as I have shown from their pleadings, is based upon the allegation that they were the owners as personal representatives. I cannot find any admission at the trial in any shape or form that the widow had any beneficial interest in the house. I do not know in the least who was entitled to the house. For all I know Mr Sharpe may have made a will leaving it to a nephew or to any other person. In those circumstances one must consider whether it can be said that the 'dwelling-house is reasonably required by the landlord for occupation as a residence for himself or herself', when the plaintiffs are legal personal representatives suing in that capacity and one of them wants to live in the house. In my opinion, such a case is not within the terms of para (h) of Sched I at all. Strange results would follow if that were not so. For instance, you might
> E have four legal personal representatives, none of whom was related to the testator at all, and one of them might require the house as a residence for himself or herself, having no beneficial interest whatsoever in the property. I am clearly of opinion that such a case could not possibly be within the terms of para (h). It is also to be observed that the words 'himself or any son or daughter of his . . . or his father or mother' seem to refer to a person who is the landlord not in the sense that he or she is one of several personal representatives, but in the sense that he or she is the sole owner of the property subject to the tenancy. It seems to me, therefore, that the plaintiffs entirely failed to prove that they came within para (h); and that if they failed to prove that they came within para (h) the words of section 3 of the Act absolutely precluded the county court judge, and would preclude this court, from making any order for the recovery of possession."
>
> F These observations were obiter; but in the second case, that of *Parker v Rosenberg*, they were approved and applied to a case where legal personal representatives were seeking to recover possession of premises for the occupation of a sister of the testatrix who was the tenant for life under her will. Tucker LJ said: "It is clear that neither the definition section of the Act of 1920 nor para (h) of Sched I to the Act of 1933 confer on anyone any right to an order for possession which he does not otherwise possess. Para (h) is designed to relax in certain cases the previously imposed statutory restrictions on the right of recovery. In the present case Miss Marsh, apart from the Rent Restrictions Acts, would not have been entitled to sue in ejectment. She was not a party to the lease and was not entitled to the reversion. There is nothing in para (h) of Sched I to the Act of 1933 enabling her to sue or be added as a plaintiff. As the proper plaintiffs, the trustees, cannot bring themselves within para (h) and Miss Marsh cannot G herself sue, it follows that no order for possession could properly be made in the present case."
>
> Those decisions lend some colour to Mr Stabb's argument, but I would observe that the question in each case before the court was not as to the meaning of the expression "landlord" in the abstract, but as to the right of the landlord to possession under para (h) of Sched I to the Act of 1933, and the landlord had therefore to prove not only that he was a landlord but also that he required the premises for his own occupation, and that greater hardship would be involved in refusing him possession than in depriving the tenant of possession, a burden which plainly an executor or administrator without a beneficial interest in the property would be unable to discharge. These considerations are not expressly mentioned in the judgments in the cases cited, but they afford a means of reconciling them with the earlier H
> opinion expressed by Morton LJ in *Lawrance v Hartwell* [1946] KB 543.

It will be seen that, in considering these cases, Cohen LJ was saying, in my judgment correctly, that they are not decisions construing "landlord" in the abstract. The real reason why it would not be reasonable to make an order in favour of personal representatives who are landlords is because, in the ordinary course of events, if they claimed possession for their own occupation, they would be acting in breach of trust. This proposition will be found in the same case in the judgment of Asquith LJ (as he then was) at p 135 where he said:

> Further, it is contended, that, on the alternative view, a personal representa- J
> tive who is a trustee would necessarily, in circumstances such as those of the present case, make a profit out of his trust.

In my view, the true ratio of those cases is that a personal representative normally cannot claim relief under Case 9, as it now is, because, in so doing, he would be acting in breach of trust. This is a wholly exceptional case. In the present case the plaintiffs, under the terms of the agreement, are quite plainly the landlords of 9 Garratt Road and, in my judgment, the defendant is not entitled to challenge that status. There is nothing in the Act which says that personal representatives cannot be landlords. Indeed, section 152 leads one to the contrary conclusion as one would expect. As landlords they are personal representatives in this case and they are K
trustees for the two children. In the present case, in considering whether it is reasonable to make the order on that particular ground, they are not acting in breach of trust in seeking to live in the house with the children. The children are the beneficial owners of the house and those exceptional circumstances take this case out of the decisions in *Sharpe v Nicholls* and *Parker v Rosenberg*.

In my judgment, the learned judge understandably fell into error in considering that he was bound by those decisions to find against the plaintiffs. On the main ground of the appeal, therefore, I am satisfied that these plaintiffs were entitled to rely on the provisions of Case 9 and, as all the other requirements were found in their favour, unless there is anything in the respondent's notice I would L
set aside the judgment and order possession of the house.

The defendant has submitted in his respondent's notice that the learned judge found facts from which we should infer that he had a further contractual tenancy which had never been determined. It was his case that, in March 1980, just before the written agreement expired by effluxion of time, he agreed with the male plaintiff that he should stay on in the house indefinitely at the same rent. The learned judge dealt with that proposition in his judgment in this way:

> I find that the facts as to what happened in March and April 1980 do not require a complete finding. The first plaintiff did give an indication that the defendant would not be required to move out at once. The first plaintiff's M
> recollection is that he went to America on business and does not recall seeing the defendant. I also find that at some time about then there was some mention made of the defendant buying the premises and of the premises being sold. It took some time for the first named plaintiff to make up his mind and discuss it with his family. These matters do not affect the first plaintiff's evidence.

In the evidence it is quite plain that the defendant said he went to see the male plaintiff on Sunday March 30 and gave him a cheque for £150 for the next month of tenancy and that it was accepted by the male plaintiff. The male plaintiff, in his evidence — and the learned judge held that he was an honest witness and that he accepted his evidence — said that, on the morning of March 30 he left for the United States of America by air, that he never saw the defendant, and never received any cheque. The fact is that the plaintiffs' solicitors sent a cheque dated March 31 back to the defendant in a letter early in April. The evidence does not show as to how that cheque came into their possession; it may be that it was put through the letter box by the defendant, because he lives within a short distance of the plaintiffs' address at 1 Deansbrook.

For my part I can see no reason for inferring that the learned judge found any facts from which it is possible for this court to say that the defendant had acquired a contractual tenancy by agreement with the male plaintiff by tendering rent and its acceptance, and I would not hold that he is able to get that submission on its feet. For these reasons I would allow this appeal and order that the plaintiffs have possession of the house.

Agreeing, SHELDON J said: I agree that the appeal should be allowed and that the respondent's appeal should be dismissed. The plaintiffs let these premises to the defendant. They created the tenancy agreement, whatever their title, and even if at the time they had none, by general principles, using the words in their ordinary meaning, as between themselves and the defendants, they became his landlord. The learned judge, after careful examination of all the relevant facts, found that they reasonably required the premises for occupation as a residence for themselves and their family. He also decided that the defendant had not succeeded on the issue of greater hardship and held that it was reasonable to make an order for possession. At first glance, therefore, it is difficult to understand why such an order should have been refused.

The mistake made by the learned judge, and the error into which he fell, resulted, in my opinion, from a misunderstanding of the real basis on which the cases of *Sharpe v Nicholls* and *Parker v Rosenberg* were decided. Nor was such a misunderstanding inexcusable, having regard to some of the observations made in the course of the judgments in question. Thus, in *Parker v Rosenberg* at p 376 Tucker LJ stated in terms that the earlier case had decided that "personal representatives having no beneficial interest in the dwelling in question cannot avail themselves of paragraph (h) of Schedule I to the Act of 1933:" — in other words, that they could not claim possession of the dwelling-house on the grounds that it was "reasonably required by the landlord . . . for occupation as a residence for himself".

Factually in those cases, although *Sharpe v Nicholls* was a case which was sent back to the county court to be reheard, that may have been so — maybe in most cases the result would be the same — but, in my opinion, Tucker LJ's observation is not properly to be understood as stating a proposition of law. In my view it is not correct to say as a matter of law that personal representatives, having no beneficial interest in the property, could never be the landlord within the meaning of paragraph (h) of Schedule I to the 1933 Act — a provision which corresponds in terms to Case 9 of Schedule 15 to the Rent Act 1977.

In my opinion, subject only to one comment, the correct approach to these cases and to what they decided is set out in the judgment of Cohen LJ in *Harrison v Hopkins* to which reference has already been made but which, for completeness, I will repeat. The learned lord justice there said that in those cases of *Sharpe v Nicholls* and *Parker v Rosenberg*

> the question . . . before the court was not as to the meaning of the expression "landlord" in the abstract, but as to the right of the landlord to possession under para (h) of Schedule I to the Act of 1933, and the landlord had therefore to prove not only that he was a landlord but also that he required the premises for his own occupation, and that greater hardship would be involved in refusing him possession than in depriving the tenant of possession, a burden which plainly an executor or administrator without a beneficial interest in the property would be unable to discharge.

My only criticism of this passage is that the comment that this was "a burden which plainly an executor or administrator without a beneficial interest in the property would be unable to discharge", though doubtless in most cases is true, was an unnecessary addition to the general proposition and, as the facts of the present case have demonstrated, was not necessarily correct. In my opinion, the question of whether or not personal representatives can bring themselves within the provisions of Case 9 is one that has to be answered by applying normal principles of law, using normal canons of construction and in the light of all the facts of each particular case. In the circumstances I have no doubt whatever that, on the findings of fact made by the learned judge, the plaintiffs are entitled to the order they seek.

I would adopt, indeed, the direct approach of the court in *Stratford v Syrett* [1958] 1QB 107, reflecting the closing passage of Lord Evershed MR at p 114. I can see no answer to the simple proposition that, being the defendant's landlord, they have succeeded in establishing a ground which entitles the judge to make an order for possession, and that he has thought it reasonable to do. That order, therefore, in my opinion, should have been made.

Also agreeing, ORMROD LJ said: I entirely agree with both the judgments which have been delivered and only add one thing. This case is very exceptional in that the personal representatives, the landlords, are in a position to show that they require the premises for occupation by themselves, they are able to show that they reasonably require the premises because, in occupying them, they will not in the particular circumstances of this case be in breach of trust, and, consequently, they are able to deal with the other matters as well. They did succeed in showing that it was reasonable for the court to make an order and also they were able to rebut the suggestion that greater hardship fell on the tenant rather than on themselves. If they were in breach of trust, of course, they could not do any of those latter things. They could not show that they reasonably required it or that it was reasonable to make an order, and they would not have any standing at all on the hardship issue.

For those reasons I agree that the appeal should be allowed and an order for possession made.

The appeal was allowed with costs.

QUEEN'S BENCH DIVISION

July 16 1981

(Before Mr Justice WEBSTER)

DUNDON v KENNA AND OTHERS

Estates Gazette January 30 1982

(1981) 261 EG 367

Rent Act 1977 — Appeal against decision of a rent assessment committee — Complaint by landlord that he had not received notice of a hearing to which he was a party — Rent Assessment Committees (England and Wales) Regulations 1971 — The papers received by the panel office from the rent officer had shown that the appellant's address was either "22 Brooklands Road" or "22 Brooklands Road East" (in fact the correct spelling was "Brookland") — Certain interlocutory notifications were sent to the appellant addressed to "22 Brooklands Road East" and were acknowledged without any indication that the address should have been different — Subsequently, notifications of the date of the committee hearing were sent to "22 Brooklands Road East, Liverpool L13 3EL" (a correct postal description except that "Brookland" should be singular) — It appeared that, in connection with an earlier case, the appellant had made an oral request that the decision should be communicated to him at a different address, 4 Normanton Avenue, Liverpool 7 — The appellant said that he had not received notice of the hearing in the present case and was not aware that it had taken place until some time later — Held that the committee or the panel were entitled to infer that 22 Brookland Road East was the appellant's usual or last-known address for the purpose of regulation 11 of the 1971 regulations and there was no ground for setting the committee's decision aside — The judge also dismissed a complaint that the committee had wrongly exercised their discretion in dealing with six flats at one day's sittings — Court's view that if there was a power to rehear (as to which no opinion was expressed) it would be sensible and fair to do so in presence of appellant

This was a statutory appeal under the Tribunals and Inquiries Act 1971 by the landlord, Mathew Dundon, from a decision of a rent

A assessment committee of the Merseyside and Cheshire Rent Assessment Panel concerning premises at 115 and 121 Ullet Road, Liverpool, consisting of six flats.

The appellant, Mathew Dundon, appeared in person; Stephen Archer (instructed by Hill, Dickinson & Co, of Liverpool) represented Mrs G M Howarth, a respondent.

Giving judgment, WEBSTER J said: This is an appeal by Mathew Dundon from a decision of the rent assessment committee of the Merseyside and Cheshire Rent Assessment Panel which was made On October 29 1980 in relation to premises at 115 Ullet Road and 121 Ullet Road, Liverpool, on Merseyside, the premises containing between them six flats. The appellant before the court is appealing
B from that decision and, although a number of grounds are relied upon, this court can take cognisance of two of them, namely that the contention that the appellant was not given notice of the hearing date on which that decision was made and in consequence was not present at the hearing, and that it was not his wish that the six flats should be dealt with at one hearing. I will assume for the purpose of this appeal that he is able in principle to challenge the decision of the committee to conduct the hearing in relation to the six flats on that one day.

The regulations that govern the matter are the Rent Assessment Committees (England and Wales) Regulations 1971 (SI 1971 no 1065). Regulation 3(3) provides that notices of the date, time and
C place of any hearing shall be given by the committee not less than 10 days before the date of the hearing to, *inter alia*, the landlord. The appellant in this case, Mr Dundon, is the landlord of the premises which were the subject-matter of the decision under appeal.

Regulation 4 provides that "At the hearing — (a) the parties shall be heard in such order, and subject to the provisions of these regulations, the procedure shall be such as the committee shall determine; . . .".

The only other regulation to which I need refer is regulation 11, which provides:

> Where any notice or other written matter is required under the provisions of these regulations to be given or supplied by the committee . . . it shall be
> D sufficient compliance with the regulations if such notice or matter is sent by post in a prepaid letter and addressed to the party for whom it is intended at his usual or last known address. . . .

The facts about the giving of notice have been sworn to by Mr Myers, who is the Secretary of the Merseyside and Cheshire Rent Assessment Panel, and although there is one matter sworn to in Mr Myers' affidavit which Mr Dundon does not accept (although he is not able specifically to challenge it), there are other matters that he would add.

The facts which I am about to recite are effectively unchallenged and indeed unchallengeable. They are these. When the files were received in the panel office from the rent officer, it was noticed that
E the appellant's six applications for the registration of fair rents gave as his address 22 Brooklands Road on five of the applications and 22 Brooklands Road East, Liverpool 13, on the sixth. 22 Brooklands Road, Liverpool 13, was the address entered in the rent register. Mr Myers subsequently ascertained the correct spelling of the road referred to as "Brookland" (in the singular). Brookland Road East and Brookland Road West were a continuous thoroughfare. The even house numbers in Brookland Road East run from 2 to 24 and continue in Brookland Road West from 26 to 86, so that 22 Brookland Road East is the correct address.

Two notifications, dated August 8 and 18 respectively, that the appeals against the rent officer's determination had been referred
F to the rent assessment panel by the rent officer, were sent by the panel to 22 Brooklands Road East. Two reply forms with a slip attached, each dated September 8 1980, were received by the panel on September 11 1980, and the acknowledgements did not contain any indication of a different address for the appellant. As a result it was assumed — it was a reasonable assumption or inference to make or draw — that it would be appropriate to send any further communication to the same address, namely 22 Brooklands Road East. In consequence two notifications of the hearing, each dated October 14 1980, were sent to the appellant by ordinary first-class post to 22 Brooklands Road East, Liverpool L13 3EL, which is the postal code, on October 15 1980.

Mr Dundon, when addressing the court, made two points of G
relevance to this particular question. First of all he said that he had told the committee, or the panel, at a previous oral hearing that his address was 4 Normanton Avenue, Liverpool 17, from which he wrote subsequently the letter of November 12 1980. But in my judgment that cannot displace the inference the committee or the panel were entitled to draw as to his usual postal address. In my judgment the facts which I have recited were sufficient to establish that, for the purpose of regulation 11, 22 Brooklands Road East was Mr Dundon's last-known address. Secondly, Mr Dundon when addressing me said that although the outward post book showed that the letter had been posted, it did not show the address to which it had been posted. Although for that reason he does not accept that H
the letter was posted to that address, I have to accept the unchallenged fact deposed to by Mr Myers in his affidavit that it was sent to that address. In my judgment the words of regulation 11 were satisfied and there is no reason to set aside the decision on the ground that the appellant was not given notice of the hearing in accordance with the regulations.

The only other ground which, as I have said, raises on the face of the documents an arguable ground for setting aside the decision is that the committee wrongly decided to deal with the six flats at one sitting. In the notice of hearing sent to the appellant of October 14 1980, he was notified that the hearing with regard to flats A, B and C at 115 Ullet Road would take place respectively at 1.30, 1.50 and J
2.10 pm on October 29 and that the hearing with respect to flats 1, 3 and 5, 121 Ullet Road would take place respectively at 2.30, 2.50 and 3.10 pm. I can see no basis on which it can be contended that the committee failed to exercise the discretion which they clearly had as to the time of these hearings.

Accordingly I have to dismiss this appeal.

I only add this. When he was arguing the appeal Mr Dundon made two things clear. One was that he thought that he had applied to the committee for a rehearing and had been refused it; and the other was that he thought that that was a matter that he was entitled to canvass in this appeal. It is not, in my judgment, a matter that he was entitled to canvass on this appeal, not merely for any technical K
reason — there is a technical reason in that the decision to refuse to rehear is not referred to in the notice of appeal — but really for a reason of greater substance, that in the documents, such as they are, there is not included any document containing a request for a rehearing and a refusal of one, nor is there any specific reference to any such request or refusal. I do not know whether there is a power to rehear or not. I am not able to express a view one way or the other. If there is power to rehear, I would have thought it would be sensible and fair for the committee to rehear the matter in the presence of Mr Dundon. But that is not a matter with which I can deal on this appeal.

No order was made as to costs. L

COURT OF APPEAL

September 24 1981

(Before Lord Justice LAWTON, Lord Justice BRANDON and Lord Justice TEMPLEMAN)

METROPOLITAN PROPERTIES CO LTD v GRIFFITHS

Estates Gazette February 13 1982

(1981) 261 EG 565 M

Landlord and tenant — Rent Act 1977 — Action by landlords against statutory tenant for possession of flat on the ground of breach of covenants — The particular breach alleged was occupation of the flat by named persons, in addition to the tenant, contrary to a covenant not, *inter alia*, to assign, charge, underlet or part with possession or occupation or share the occupation of the flat or any part of it — The landlords' case depended on the inference, if any, to be drawn from a letter to their solicitors from the electoral registration officer of the City of Westminster — This letter showed that a named individual, in addition to the tenant, was on the register of electors for 1980 for the address of the flat and had therefore been

A **resident in the flat on October 10 1979 for the purposes of the Representation of the People Act 1949 — Held that the entry in the register of electors did not constitute sufficient evidence that the presence of the named individual was of a nature and quality which created a breach of the tenant's covenants — Form A, which is addressed to occupiers, might not have been completed or might have been completed under a misapprehension as to facts or requirements, the identity of the person who completed it, if it was completed, was not known, and in any case the categories of persons entitled to be registered were very wide — There were too many imponderables — Landlords' appeal from decision of county court judge refusing possession order dismissed**

B

This was an appeal by landlords, Metropolitan Properties Co Ltd, from a decision of Judge Warde at Bloomsbury and Marylebone County Court dismissing an action for possession against Miss Mary Griffiths, tenant of flat 107 St Mary's Mansions, St Mary's Terrace, Westminster.

David Neuberger (instructed by J Memery & Co) appeared on behalf of the appellants; R Spearman (instructed by Moss, Beachley) represented the respondent.

Giving the judgment of the court at the invitation of Lawton LJ, TEMPLEMAN LJ said: This is an appeal from a decision of His Honour Judge Warde given in the Bloomsbury and Marylebone
C County Court on March 17 1981, whereby he dismissed an action for possession of the flat 107 St Mary's Mansions, St Mary's Terrace, Westminster, brought by the plaintiff landlords, Metropolitan Properties Co Ltd, against the defendant tenant, Miss Mary Griffiths.

Miss Griffiths became tenant of the flat on the terms of a tenancy agreement dated November 4 1965 for a period of three years. After the expiry of that term she remained in possession of the flat as statutory tenant. By particulars of claim dated August 21 1980 the landlords sought possession on the grounds that Miss Griffiths was in breach of tenant's covenants in the agreement, namely clause 2(9) and clause 2(11). So far as material those covenants required
D Miss Griffiths to use the flat as a private dwelling-house only in the possession of one family, not to accommodate boarders, lodgers or paying guests and not to assign, charge, underlet or part with possession or occupation or share the occupation of the flat or any part of it, either furnished or unfurnished.

The particulars of claim alleged occupation of the flat by specific persons at various times from 1967 onwards, including in those persons a Miss Jean Rimmer and, in or about 1980, a Mr Murray Cooper.

By her defence Miss Griffiths did not admit any breach of covenant, pleaded waiver of any breach of covenant and in any event pleaded that it would not be reasonable in all the circumstances for
E the court to make an order for possession.

It was conceded that the landlords could not complain of the occupation of Miss Rimmer, who has been sharing the flat for a number of years to the knowledge of the landlords. The landlords called oral evidence about the other alleged breaches of covenant, but the learned judge was not impressed.

At the conclusion of the landlords' witnesses, counsel representing Miss Griffiths decided not to call any evidence on her behalf. In the circumstances, as the learned judge held, the landlords' case depended on the inference, if any, to be drawn from a letter dated September 18 1980.

The letter was addressed to the landlords' solicitors and was
F expressed to be signed by the electoral registration officer of the City of Westminster. The letter stated that "the names which appear in the register of electors in my custody for the years 1965 to 1980 inclusive in respect of the address 107 St Mary's Mansions are as follows". Miss Griffiths appeared in 1966 and every subsequent year except 1974. Miss Rimmer appeared in 1967 and every subsequent year. The entry for 1980 specified Mr Murray Cooper, Miss Griffiths and Miss Rimmer in that order.

Sections 1 and 2 of the Representation of the People Act 1949 provide that a person entitled to vote must be resident in the relevant electoral area on the qualifying date and must be registered in the register of electors for that area. Section 6 provides for the appointment of a registration officer for each parliamentary G
constituency. By section 7 a registration officer must prepare and publish combined registers of parliamentary electors and local government electors for the electoral areas included in his constituency. By section 9, with a view to the preparation of the register, the registration officer shall have a house-to-house or other sufficient inquiry made as to the persons entitled to be registered and must prepare and publish electors' lists showing the persons appearing to him to be entitled to be registered, together with their qualifying addresses. By section 39 the register of electors shall, for the purposes of Part I of the Act, be conclusive on the question whether or not a person registered therein was on the qualifying date resident at the address shown. H

It is common ground that the qualifying date for the year 1980 was October 10 1979. For the purposes of Part I of the Representation of the People Act 1949, therefore, the admitted registration of Mr Murray Cooper proves conclusively that he was resident at the flat on October 10 1979.

To determine whether the registration of Mr Murray Cooper under the Act was any evidence at all against Miss Griffiths and that Mr Cooper was resident at the flat in such manner as to constitute a breach of the relevant tenant's covenants, it is necessary to examine the machinery whereby the register was compiled.

Section 42 of the 1949 Act gave power for provision to be made by regulations with respect to the procedure to be followed in the J
preparation of the register.

The relevant regulations are now the Representation of the People Regulations 1974 (SI 1974 No 648). Some slight amendments have been made to those regulations but they are not material to this appeal. By regulation 25 the registration officer may require any householder or person owning or occupying any premises within the area for which he acts, or the agent or factor of any such person, to give information required for the purposes of his registration duties. By regulation 69(1) the Form A to the regulations shall be used for the purpose for which it is expressed to be applicable. By regulation 74(1) if any person fails to comply with or gives false information in pursuance of any requisition of the registration officer to give information required for the purposes of his K
registration duties pursuant to regulation 25, that person is liable on summary conviction to a fine not exceeding £50.

Form A is delivered or posted addressed to the occupier of each residence within the area of the registration officer. Form A is headed "Return by occupier as to Residents". It specifies the qualifying date, which is October 10 in every year. It informs the recipient that the Form A is sent by the electoral registration officer. It says *inter alia* "I have to compile and publish an up to date Register of Electors . . . To do so, I need information which you, as occupier, are obliged by law to supply . . . The notes within tell you how to fill up the form, but if you need further help, I shall be glad to L
give it . . . Please complete and sign this form, and return it to me now . . .". There is then a table where the occupier must specify the full address of the flat, room, floor or house of which he is the occupier. There is a second table to be completed of "Residents eligible to be included (see notes 1 and 2)". The table requires the surname, title and full christian names of each eligible resident to be inserted. There then follows the question "Is any part of your house/flat separately occupied by persons not entered above?" with a direction to answer Yes or No. There follows a declaration in these terms "I declare that to the best of my knowledge and belief the particulars given above are true and accurate . . .". Under the declaration the occupier who is completing the form must sign. M

There then follow the notes. Note 1 specifies, so far as material "You are required to enter all British subjects . . . who will be resident at your address on October 10" in the relevant year . . . "Including (a) Those who normally live at your address and are temporarily away . . . (b) Resident guests (but not short stay visitors). (c) Lodgers and resident domestics. (d) Anyone who is away working, unless his absence will be for more than six months". There then follow other categories and instructions not material to this case, and finally there are instructions to return the form.

The letter dated September 18 1980 is conclusive for the purposes of Part I of the Registration of the People Act 1949 that Mr

A Murray Cooper was resident at the flat on October 10 1979. For the landlords it was argued that the letter constituted *prima facie* evidence that Mr Murray Cooper was present at the flat at any rate on October 10 1979 and that his presence was of a nature and quality which constituted a breach by Miss Griffiths of the tenant's covenants, and that in the absence of any evidence to the contrary the learned judge was bound in law to find that breach of covenant had been proved by the landlords.

The landlords rely on section 1 of the Evidence Act 1845 and section 14 of the Evidence Act 1851, but even if these sections apply they only cover the same ground as the concession made by counsel for Miss Griffiths that the letter dated September 18 1980
B was evidence that a Mr Murray Cooper was registered in the register of electors for the year 1980 in respect of the flat, a concession which avoided the necessity for calling the registration officer or producing the register to prove this fact.

Section 9 of the Civil Evidence Act 1968 provides, *inter alia*, (subsection (2)c) that any rule of law "whereby in any civil proceedings public documents (for example public registers and returns made under public authority with respect to matters of public interest) are admissible as evidence of facts stated therein" shall continue to have effect.

No authority was cited to us that there is or was any rule of law that entries in the register of electors for the purposes of the
C Representation of the People Act 1949 are admissible evidence of the occupation of any premises, let alone of the quality of occupation, of any person, as against any party to litigation, without at any rate production of the relevant Form A and some evidence linking the signature on Form A with the litigant.

It is doubtful whether section 39 of the 1949 Act, which for the purposes of Part I of the Act makes the entry in the register conclusive of the fact that the person registered was on the qualifying date resident at the address shown, is any evidence at all for any other purpose.

But if the entry in the register was evidence, the learned judge was not in our judgment bound to accept that it was sufficient
D evidence of breach of covenant by Miss Griffiths. Not only was no notice to admit served on Miss Griffiths but no explanation was given of the failure to produce a relevant Form A and then to prove the contents and signature of Form A. The learned judge accepted that the letter dated September 18 1980 was written in good faith but said "It is not enough, on the balance of probabilities, to show that Mr Murray Cooper was actually resident on the premises . . .".

We agree. There are too many imponderables. Form A may not have been completed; if it was completed the identity of the person who completed it is not known, and there is no evidence that any form was completed to the knowledge of Miss Griffiths. Form A may have been completed under a misapprehension as to the facts
E or as to the requirements to be satisfied for entry on the form. Moreover, the categories of persons who are entitled to be registered are defined so widely by the Act that it is impossible, in the absence of evidence, to be certain that any inference to be drawn from the register of electors that a Mr Murray Cooper was present at the flat constituted presence of a nature and quality which created a breach of the tenant's covenants.

Accordingly, the landlords' contentions fail and this appeal must be dismissed.

The appeal was dismissed with costs.

F

QUEEN'S BENCH DIVISION

November 23 1981

(Before Mr Justice FORBES)

KOVATS v CORPORATION OF TRINITY HOUSE

Estates Gazette May 1 1982
(1981) 262 EG 445

Rent Act 1977 — Statutory appeal under Tribunals and Inquiries Act 1971 from decision of rent assessment committee on a point of law — Appeal by tenant who complained that committee went G **wrong in law because, as a fair rent had previously been registered, all that the committee should have done was to revise the existing registered rent by reference to the effect of inflation — Fallacious argument derived from section 67 (3) of the 1977 Act — In any case rents do not move precisely in step with the movement of inflation — Committee in this case acted sensibly by looking at comparative properties and rejecting the argument that the only proper factor to consider was the cumulative rate of inflation since the previous registration — No error of law on part of committee — Appeal dismissed.**

This was an appeal by Laszlo Joseph Kovats against the decision H of a rent assessment committee of the London Rent Assessment Panel. Mr Kovats was the tenant of three floors of a four-storey terrace house at 64 Trinity Church Square, London SE1, of which the landlords were the Corporation of Trinity House.

W Bowring (instructed by Kingsley, Napley & Co) appeared on behalf of the appellant; P Goldsmith (instructed by Freshfields) represented the respondent corporation.

Giving judgment, FORBES J said: In this case Mr Kovats appeals, as he is entitled to do, against a decision of the rent assessment committee sitting at Newlands House, Berners Street, London W1, and the decision, which they gave on August 19 1980, was in relation to the fair rent for a dwelling-house of which the appellant is J the tenant.

The situation is this, quite briefly, that in 1976 the rent officer registered a fair rent of, I think, £820 a year for this property. In January 1980 there was a further application to the rent officer to register a new fair rent and the rent officer did so in May 1980. The appellant appealed to the rent assessment committee as he was entitled to do. From their decision, of course, an appeal lies to this court on a point of law and one has to look to see what is the point of law on which this appeal arises. Well, broadly, as I understand it, it is this, that the rent assessment committee went wrong in law because, as there was already registered in 1976 a fair rent for this very property, all they were entitled to do in arriving at a new fair K rent was to apply to that rent the cumulative rate of inflation since the date of the application to register that rent; because that would have been the relevant date under subsection (5) of section 67 of the Rent Act 1977, which is the relevant statute for these purposes.

That is really what is being argued and Mr Bowring, who stuck to his point with great tenacity, suggests that, under section 70, the procedure there for determining a fair rent in effect only applies to the first determination, because if all the matters which are properly taken into account under section 70 have been taken into account on the first registration of a fair rent, nothing remains to be taken into account unless it can be shown that circumstances have changed, such as those set out in section 67(3), and, he says, no L such circumstances were shown to have changed in this case.

Section 67(3) is wholly irrelevant to any argument that arises in this case. That is a section which provides, as it were, a hurdle or a gate, or whatever analogy you like to draw, in the path of the landlord applicant if he wants to apply for a fresh determination of a fair rent less than three years since the last determination. If he wants to do that, he has to satisfy the rent officer that one or other of the matters set out in subsection (3) (and they end up with "any other circumstances taken into consideration when the rent was registered or confirmed") have been so changed as to make the registered rent no longer a fair rent. All those matters provide, as I say, a hurdle to be surmounted or a gate to be gone through by the M applicant landlord or the tenant before either of them can ask within the three-year period for a fresh registration of a new rent. If they do not bring themselves within those provisions then they fail. Their application will not be entertained. If they do pass through the gate it is quite clear, from the case to which Mr Bowring has drawn my attention, *London Housing & Commercial Properties* v *Cowan* [1977] 1 QB 148, that once you have gone through the gate then all the matters which are now set out in section 70 fall to be considered by the rent officer and he would be wrong, as the case showed, if he considered only the matters set out in section 67(3). In other words, what is set out in that subsection are merely qualifying matters which the applicant has to satisfy before his application can be entertained. Once he qualifies under that subsection then all matters have to be looked at again as required by section 70. But in any event these considerations apply only to "mid-term" applications, ie applications within the three-year period.

Now, as I say, I cannot see that there is anything in section 67(3) which bears the slightest relevance to the argument, particularly having regard to the case which Mr Bowring has just brought to my attention and which is binding on me. I cannot see that section 67(3) has any relevance to any matter which has to be taken into account under section 70 and, of course, if it really is suggested that a rent officer or rent assessment committee can only take into account the incidence of inflation once they have determined a fair rent, and that has been registered in the past, then, it seems to me, that would be a really very curious arrangement for Parliament to have intended.

The rate of inflation — and Mr Bowring acceded in the end to this definition — is a statement of the percentage rise in the weighted average of a number of items in a chosen price index over the period since the last time on which a statement was made. If the rate of inflation is a monthly rate, then what happens is you tot up all the items in the price index, then you apply a weighted average to the total of all the rises and falls which may occur over the whole spectrum of the items in the price index and, in the end, as I say, you get an inflation figure in percentage terms. But one has only got to define the rate of inflation in those terms for it to be seen that you simply cannot, with any form of commonsense, justice or logic, apply a weighted average of that kind to one item like rent.

It does not at all follow that rents will go up precisely in step with the rate of inflation. In fact they frequently do not. It has been apparent for a long time that, for instance, building costs do not keep in line with inflation rate, and building costs, to some extent, are one of the many matters which may together determine what is a rent chargeable for a particular property. There is no evidence at all, it seems to me, which any sensible rent assessment committee could take into account that the rate of inflation is a proper multiplier to apply when considering what should be a fair rent for property which already has a registered fair rent. The way in which rents vary, differ, rise or fall, differs all over the country and they can differ very markedly within a very short geographical distance. All those matters have to be taken into account and are properly taken into account by looking at comparable rents, which is precisely what the rent officer did in this case. I cannot see that there is the slightest ground for suggesting here that the rent assessment committee were wrong in law. They seem to me to have operated in accordance with the sensible way in which you would expect them to operate by looking at comparable cases and rejecting any argument that the only proper thing to look at was the cumulative rate of inflation since the last registration. In my view they acted perfectly correctly and it cannot be said that they went wrong in law and this appeal therefore fails.

The appeal was dismissed with costs.

COURT OF APPEAL

March 12 1982

(Before Lord Justice CUMMING-BRUCE, Lord Justice EVELEIGH and Lord Justice MAY)

METROPOLITAN PROPERTIES CO LTD v CRONAN AND ANOTHER

Estates Gazette June 12 1982

(1982) 262 EG 1077

Rent Act 1977 — Position of former wife after divorce where no application had been made by her under section 7 of the Matrimonial Homes Act 1967 — Appeal by ex-husband and ex-wife against possession order granted to landlords — Husband and wife and a son of the marriage lived in a flat, no 26, of which the husband was the tenant, until matrimonial difficulties arose and the husband moved out — He took a tenancy of another flat, no 66a, from the same landlords — When tenancy of flat 26 was about to expire, the husband took a further three-year tenancy of it, while still continuing to live at no 66a — Wife and son continued to live at flat 26 — Eventually, after the contractual tenancies of both flats had expired, the wife obtained a divorce, an order for joint custody of the son being made, with care and control to the wife — Landlords sought possession of flat 26 — It was the wish of both the husband and the wife that she should remain in flat 26 and they both appealed against the county court judge's order for possession — Court's review of case law beginning with *Skinner v Geary* — Submissions by appellants that, although the husband had been the nominal tenant, the wife was the real tenant, and that there should be an "abandoned child" exception to the *Skinner v Geary* rule, analogous to the "deserted wife" exception, both rejected — In the absence of an order under section 7 of the Matrimonial Homes Act 1967, the wife had no claim to remain in occupation after the divorce — Appeal against possession order dismissed

This was an appeal by John Barry Cronan and his ex-wife, Mrs Cronan, from a decision of Judge Bolland at Westminster County Court granting an order for possession to the respondents (plaintiffs in the action), Metropolitan Properties Co Ltd, in respect of a flat at 26 Carlisle Mansions, Carlisle Place, London SW1.

Peter Millett QC and Daniel Pearce-Higgins (instructed by Piper, Smith & Basham) appeared on behalf of the appellants; David Neuberger (instructed by Memery & Co) represented the respondents.

Giving the first judgment at the invitation of Cumming-Bruce LJ, MAY LJ said: This is an appeal from the order made by His Honour Judge Bolland in the Westminster County Court on May 1 1981 that the respondents should recover possession against each of the appellants of the flat at 26 Carlisle Mansions, Carlisle Place in London.

The brief facts of the case are as follows. The respondents were and are the owners of a block of residential flats known as Carlisle Mansions, which is situated behind Victoria Street in south-west London. On May 19 1971 the first appellant, Mr Cronan, took a lease of the flat numbered 26 Carlisle Mansions. This was for a term of three years. At some stage during this demise the second appellant, Mrs Cronan, began to live with Mr Cronan and they were ultimately married in April 1973. On some date between then and February 8 1974 a son was born to them.

Unfortunately marital difficulties developed between Mr and Mrs Cronan and on February 8 1974 the former took another three-year lease of 66a Carlisle Mansions. He moved out of no 26 and has ever since lived in no 66a as his residence. It is quite clear that thereafter he had no intention of returning to no 26.

Be that as it may, when the three-year term of no 26 granted to him in 1971 was about to expire, he was offered and accepted a further three-year lease of this flat from June 1974. Although this fresh lease contained a covenant on his part to use the flat only as a private residence occupied by him and his own immediate family and not for any other purpose, there is no doubt that ever since 1974 Mr Cronan was living in 66a Carlisle Mansions and his wife and son have been living in no 26. Further, it is also clear, and the learned county court judge so held, that by their servants or agents the respondents were at all material times well aware that this was the situation.

This continued not only during the currency of the contractual tenancies of both 26 and 66a Carlisle Mansions but for a time thereafter. Towards the end of 1977 Mrs Cronan instituted divorce proceedings against her husband and the decree nisi was made on November 18 1977. At the same time an order for joint custody of the son was made, with care and control to Mrs Cronan. On January 3 1978 the decree nisi was made absolute.

At some stage thereafter more senior members of the respondents' organisation became aware that Mr Cronan was the nominal tenant from them of both 26 and 66a Carlisle Mansions and the respondents accordingly instituted proceedings against him for possession of no 26. These were originally on the basis that as he was living in no 66a and not in no 26, there was clearly a breach of the covenant concerning occupation to which I have already referred. As I have said, however, at all material times other servants of the respondents more directly concerned were well aware that Mr Cronan was not living in no 26 and consequently, having received rent in those circumstances there was no doubt that the respondents had waived their right to rely upon any such breach and that they would not have been entitled to possession on this ground alone.

However, the true state of the relationship between Mr and Mrs Cronan and the fact that a decree absolute had been granted only became fully known to the respondent landlords when the appellants served their defence in the county court proceedings for possession. I should add that although quite clearly at the time of the decree Mrs Cronan could have made an application under section 7 of the Matrimonial Homes Act 1967 for the transfer to her of the then existing statutory tenancy of no 26, for some reason or other no such application was made.

Upon receipt of the defence in the proceedings, the landlords promptly amended their particulars of claim to allege that even though Mr Cronan, and thus Mrs Cronan, might have been entitled to remain in occupation of no 26, at first under a contractual tenancy, and then under a statutory tenancy until January 3 1978, the latter determined upon the grant of the decree absolute and that accordingly they were entitled to possession on this basis also.

Notwithstanding various arguments which were addressed to the learned county court judge on behalf of the appellants and have been repeated before us, to which I shall hereafter refer, upon these facts he came to the conclusion that neither Mr nor Mrs Cronan, nor indeed their son, had any right to remain in possession of no 26 and made the possession order accordingly. It is against that possession order that the appellants now appeal herein. I should also add that when questions of maintenance came to be decided between Mr and Mrs Cronan on their divorce, he undertook to continue to pay the rent of no 26 and to allow his wife and son to remain in occupation of that flat. In consequence a nominal order for maintenance in favour of Mrs Cronan was made, coupled with one for periodical payments of £17 per week in respect of their infant son. It is quite clear that both Mr and Mrs Cronan intended that Mrs Cronan should remain in no 26, where she had in fact been living since some date between 1971 and 1973.

In so far as the law is concerned, ever since the well-known judgment of Scrutton LJ in *Skinner v Geary* [1931] 2 KB 546, the basic principle governing the operation of the Rent Acts and the entitlement to security from dispossession which they give is that the latter can be enjoyed only by an occupying tenant and only for so long as he is in occupation of the relevant dwelling-house as his home. See also, for example, per Lord Wright in *Hiller v United Dairies (London) Ltd* [1934] 1 KB 57, and the Court of Appeal in *Cove v Flick* [1954] 2 QB 326n.

Over the last 50 years a limited number of apparent exceptions to this general rule have been developed by the courts as "empirical solutions to prevent injustice being done", per Lord Wilberforce in *National Provincial Bank Ltd v Ainsworth* [1965] AC 1175 at p 1252. One of these apparent exceptions, for example, is the case of temporary absence — the sea captain who is away on a voyage but intends to return to the relevant dwelling-house. Analogous to this is the case of the man who can properly be said to have two homes, each of which attracts the protection of the Rent Acts — the home in the country and the flat in London which he occupies during the week in order to carry on his business. Another apparent exception to the general rule developed by the courts is that which can briefly be described as that of the "deserted wife": in such cases it has been held that the husband tenant, though absent, must be deemed to be occupying the dwelling-house by his wife with whom he has a duty to cohabit and to whom he owes a duty to maintain. The security of tenure which such a wife enjoys only lasts at least so long as she remains the tenant's wife: once the marriage has been ended by a decree absolute of divorce, the protection ceases: *Robson v Headland* (1948) 64 TLR 596, approved and applied by the Court of Appeal in *Heath Estates Ltd v Burchell* (1979) 251 EG 1173.

There may also be an apparent exception to the general rule in a case where the landlord knows that the tenant is merely nominal and the real tenant is somebody else. In these circumstances the real tenant may be protected. However, having regard to the basic principle governing the Acts and in the light also of the decisions in *S L Dando Ltd v Hitchcock* [1954] 2 QB 317 and *Cove v Flick* I think that this last apparent exception, if indeed it can ever be made good in law, applies only where the landlord grants the tenancy to the nominal tenant as a sham and for the purpose of being free of the burdens of the Rent Acts to which he would otherwise be subject in respect of the "real" tenant. Where by a genuine transaction, albeit that the true facts are known to the landlord, the latter lets a dwelling-house to one person, for instance a father, for the purpose of its occupation by another, for instance that child of that father, then when the contractual tenancy expires I do not think that the occupying child is entitled to the protection of the Rent Acts. It might appear that in such circumstances the landlord was having the benefit of both worlds; during the tenancy he looks to the father for the rent and the due performance of the covenants; at the end of it he is entitled to vacant possession: but one must not forget that *prima facie* the demised premises belong to the landlord to make what profitable or other use of them he may wish; he is restrained only to the extent that the statutes may restrict his freedom of action in this way. Given that the landlord proposes to let the dwelling-house in any event, and thus in all probability be encumbered by a statutory tenant of it in due course, he can, if the opportunity offers and by due negotiation and agreement, let it to one person at the behest of another and still be able to look to that other for the rent and the due performance of the covenants, if the tenant should default, for instance by obtaining that other's guarantee of the tenant's obligations.

That this is the basic principle, but that there have grown up these apparent exceptions, is clear from the provisions of subsection (1)(a) of section 2 and subsection (3) of the same section of the consolidating Rent Act 1977, which read as follows:

> (1) Subject to this Part of this Act —
> (a) after the termination of a protected tenancy of a dwelling-house the person who, immediately before that termination, was the protected tenant of the dwelling-house shall, if and so long as he occupies the dwelling-house as his residence, be the statutory tenant of it . . .
> (3) In subsection (1)(a) above and in Part I of Schedule I, the phrase "if and so long as he occupies the dwelling-house as his residence" shall be construed as it was immediately before the commencement of this Act (that is to say, in accordance with section 3(2) of the Rent Act 1968).

In these circumstances, and subject to the two contentions of counsel for the appellants, to which I shall refer, the analysis of the relationships between the various parties in this case in law at relevant times must be this. In June of 1974 the respondents granted a contractual tenancy of flat no 26 to Mr Cronan. The mere fact that they then knew that he was separated form Mrs Cronan, that it would be she and their son who would be living in no 26 and that he would be living in flat 66a, of which they had also granted him a tenancy four months earlier, cannot alter what was the legal position between the respondents and Mr Cronan during the three-year term of that tenancy, namely that they were the landlords and he the tenant pursuant to that contractual demise. When that three-year term ended and until Mrs Cronan ceased to be Mr Cronan's wife upon the grant of the decree absolute in January 1978, by the apparent exception to the general principle governing the Acts applicable to the case of a deserted wife, Mr Cronan was deemed to remain in occupation of no 26 by his wife and consequently she could not be evicted. Once the decree absolute was pronounced, however, Mr Cronan could no longer be deemed to be in occupation of no 26 and neither he nor she could claim any security of tenure of that flat under the Rent Acts. Mrs Cronan could have preserved her position in and in relation to no 26 had she or those representing her taken the appropriate steps under section 7 of the Matrimonial Homes Act 1967, but neither of them did so.

Against this apparently inevitable result on the facts, which were hardly in dispute, counsel for the defendants advanced two arguments with which I shall deal in the reverse order to that in which he presented them.

Counsel submitted that in the very special circumstances of this case, although Mr Cronan was what he described as the nominal tenant, nevertheless the actual tenant throughout was Mrs Cronan. He supported this submission by a reference to the covenant which

A there was in the June 1974 tenancy agreement that the tenant should use the flat only as a private residence occupied by him and his own immediate family and not for any other purpose. Quite clearly both sides knew that this would immediately be breached as soon as the tenancy agreement was signed, if Mr Cronan was not in occupation of no 26, and thus the reality of the contractual tenancy between the respondents and Mr Cronan in June of 1974 must have been and, intended to be that Mrs Cronan was the "actual" or "real" tenant and that her husband was a mere nominee for her.

I confess first to an unworthy suspicion that when the tenancy was granted in June 1974 it was made on the respondents' usual printed form and that neither party applied their minds either to the content
B or effect of the covenant.

Even if they did, they may have read the covenant disjunctively — although I do not think that this is an acceptable construction of it — and thus thought that the factual situation thereafter obtaining constituted no breach. Alternatively, if the occupation of no 26 by Mrs Cronan and her son alone after June 1974 was *prima facie* a breach of the covenant, I have no doubt that, being fully aware of the then facts, the plaintiffs were prepared to accept this and did in truth waive the ex hypothesi breaches of covenant from the start.

In the end, counsel for the defendants effectively accepted that in order to make good this first basic submission, namely that from June 1974 to 1977 it was Mrs Cronan who was the "actual" or
C "real" tenant, he had to contend either that the tenancy agreement of June 1974 was a sham, in the sense that to the knowledge and with the agreement of both parties it embodied an apparent legal and contractual relationship contrary to the true facts, or that throughout in June 1974 Mr Cronan was acting as his wife's agent. In my opinion not only was there no finding by the learned county court judge that the transaction had been a sham in this sense but also there was no evidence to support any such finding. On the issue of agency, the learned judge below made a specific finding against the defendants. Not only was there material before him upon which he was entitled to make such a finding but also, I think, that any decision to the contrary would have been perverse.

D Although by virtue of subsection (3) of section 2 of the Rent Act, 1977, re-enacting in somewhat different words subsection (2) of section 3 of the 1968 Rent Act, we have to accept and apply to the basic principle which started with *Skinner v Geary* the various apparent exceptions which the courts had worked out and upheld prior to 1968, I think that the mere fact of the enactment of the Matrimonial Homes Act 1967 makes it more difficult in this or any other case to argue that the apparent and relevant contractual tenancy was in truth otherwise than between landlord and husband. A deserted wife has her rights under the Rent Acts on the decided cases so long as she remains a wife. Coincident with her ceasing to be the wife by virtue of a divorce, she can continue to protect her
E position by an appropriate application under the 1967 Act. There is no need to postulate, nor I think is there any justification for postulating, the sham or agency situation similar to that for which counsel contended in the present case.

Counsel's second submission was that if Mrs Cronan lost any security of tenure which she previously had as a result of the decree absolute, nevertheless on similar reasoning which gave rise to the apparent exception to the general principle in favour of the "deserted wife", the courts should uphold a similar exception in favour of the "abandoned child". Just as the decisions in favour of the deserted wife rested on the obligation of her husband to maintain her and to provide her with a roof over her head, so also should an
F infant son be entitled to rely upon his father's common law obligation to maintain him to support the contention that his, the son's, continuing occupation of premises within the Rent Acts should in law be considered to be that of his father, and thus attract the protection given by those Acts to a tenant (the father) in occupation of the relevant premises (flat 26) by the son (whom at common law he was bound to maintain).

I do not think that such a contention can be validly supported for a number of reasons. First, the mutual duties and rights of parent and child are different from those of husband and wife. Although there is a duty on the husband and father to support his wife and children, there is, in my opinion, absent from the father-child relationship the mutual right and duty of both husband and wife to cohabit. G
Although the right of a wife to remain in occupation of the matrimonial home even as against her deserting husband is incapable of precise definition and may change overnight by the act or behaviour of either spouse (see per Lord Upjohn in the *National Provincial Bank Ltd* case at p 1233) it certainly exists as an aspect of the status of marriage.

On the other hand, at least in the case of an adult child, he or she is correctly described as a licensee in the parents' home and such a licence can be withdrawn. Although the court is expecially slow to grant an injunction which will exclude even an adult child from that home, there is power to do so: see *Waterhouse v Waterhouse* (1905) 95 LT 133, *Stevens v Stevens* (1907) 24 TLR 20 and *Egan v Egan* H
[1975] 1 Ch 218. I cannot think, and at the least we were shown no authority to the contrary, that the legal nature of a child's position in the parents' home changes from one of status to that of licensee so soon as the child obtains his or her majority. As at present advised, I think that in law a licence to remain in the parents' home can be withdrawn even in the case of a minor child, though of course one hopes that this would occur only in very special cases. If such a course of action were adopted, then no doubt the appropriate care and other proceedings would follow under the general legislation relating to children.

For present purposes I think that there is therefore a significant difference between the position of a deserted wife on the one hand J
and an abandoned child on the other *vis-à-vis* the matrimonial or family home.

Secondly, so far as I am aware, this is the first occasion upon which an apparent exception to the basic principle in favour of an abandoned child has ever been contended for. For my part, I would be averse to upholding it unless I were satisfied that not only in this case but in others with similar, though not identical, facts such an exception was necessary to prevent a general injustice. I am not so satisfied. In most cases no injustice will be done to a child, because he or she will throughout be in the care of either the father or the mother in his or her home and if not still a contractual tenant, then a statutory tenant in their own right. Further, the succession K
provisions in the Acts, now contained in the First Schedule to the 1977 Act, amply protect a child living with its statutory tenant parent if the latter should die. In truth, Mr and Mrs Cronan's son has been disadvantaged in the present case only because of the omission to make an appropriate application under the Matrimonial Homes Act 1967, and I am sure that I must be on my guard to prevent my natural sympathy for him from leading me to be a party to making bad law.

Thirdly, although a father can no doubt in law be in beneficial occupation of premises for rating purposes by the presence therein of an abandoned child (compare *Cardiff Corporation v Robinson* [1957] 1 QB 39), ex hypothesi he is not in occupation of those premises as *his* residence, or *his* home, and thus has no protection L
under the Rent Act 1977, nor would have had under any of its predecessors following *Skinner v Geary.*

Although it was found by the learned county court judge, and has not been disputed before us, that at least after June 1974 the plaintiffs must have known or at least have the knowledge imputed to them that Mr Cronan lived in no 66a and Mrs Cronan with their son lived in no 26, this is not enough to enable the appellants to resist the claim made against them for possession of no 26. Counsel told us that he could not rely upon any form of estoppel based upon this knowledge. In the result, when the contractual term granted in June 1974 in respect of no 26 expired, Mrs Cronan was entitled to statutory protection under the Rent Act only so long as Mr Cronan M
must in law be considered to have been occupying that flat and doing so as his residence or home. So long as he and she remained married, this was so; as soon as they were divorced, his notional occupation and thus her protection ceased.

In *S L Dando Ltd v Hitchcock, supra* at p 325, Lord Goddard CJ summarised the basic principle governing the operation of the Rent Acts and the anomalous position of the deserted wife in a passage to which we were referred and which I think has particular relevance to this case. He said:

> . . . I think that the Acts are intended and designed to protect tenants and tenants only. That there has been an inroad into that principle in the cases of

A husband and wife is no doubt true. I cannot help thinking that in those cases the shadow of the old common law doctrine that husband and wife are one in law has possibly, consciously or unconsciously, affected the courts. The Acts put very considerable difficulties in the way of landlords and circumscribe their legal rights to a very great extent. I do not think that we ought by decisions to enlarge the difficulties of landlords or to go further than the declared object and policy of the Acts dictate — that is, to protect tenants — and I think that it cannot be denied that that means tenants who live in these houses. Where there is a personal tenant who does not live in the house, never intends to live in the house, and declares that his intention is never to live in it, I can see no reason why his tenancy should be protected to enable him to keep in the house a manager, or a partner, or anyone else whom it may be convenient to have there.

B For these reasons I would dismiss this appeal.

EVELEIGH LJ: I have had the advantage of reading both the judgments prepared by my Lords; I agree with them both.

Agreeing that the appeal should be dismissed, CUMMING-BRUCE LJ said: The appellants failed to avail themselves of the right conferred by section 7 of the Matrimonial Homes Act 1967 to apply for a transfer of Mr Cronan's tenancy of 26 Carlisle Mansions to Mrs Cronan. The ingenious submissions presented on behalf of the appellants emerge on analysis as no more than a desperate attempt to persuade the court to develop a new extension of the statutory protection conferred by the Rent Act 1977 in order to relieve the C appellants of the consequences of that omission.

For the reasons stated by May LJ there is no principle to be discovered in the case law upon the scope and effect of section 2(1) of the Rent Act which entitles the court to extend the statutory protection in the manner now claimed.

I would add only one consideration to those already explained. Counsel invited the court to erect a new exception upon the foundation of the former husband's right to joint custody where there was in existence what has been described as a "split order", granting care and control of the child of the family to the mother with joint custody to the mother and father. Such an order by necessary implication deprives the father of the right to claim that his D child should reside with him save during periods of staying access. The home of the child is the home of the mother, who has the right and duty to provide the child with care and control. If the parties wanted the mother to maintain her right to live with the child in the former matrimonial home after she ceased to be Mr Cronan's wife, all that they had to do was to apply before decree absolute to the court by which the decree nisi was granted for an order under section 7 of the Matrimonial Homes Act. Where the court has dissolved a marriage, the Matrimonial Causes Acts, the Matrimonial Proceedings and Property Act, and the Matrimonial Homes Act 1967 together constitute the statutory code which prescribes the rights and duties of the parties to provide for all aspects of the E maintenance, including accommodation, of a child of the family. The only provision in this code for continuing statutory protection of a tenancy of a home in which the father formerly resided with his family for the purpose of affording protection for a child in the care and protection of the mother is to be collected from section 7 of the Matrimonial Homes Act 1967.

The appeal was dismissed with costs against both appellants, the order for costs against the second appellant not to be proceeded with without an order of the court. It was ordered that possession should be given to the respondents three months from the date of judgments in the Court of Appeal. Leave to appeal to the House of Lords was refused.

F

G

QUEEN'S BENCH DIVISION

February 26 1982

(Before Mr Justice FORBES)

R v A RENT ASSESSMENT COMMITTEE FOR LONDON AND OTHERS, EX PARTE ELLIS-REES AND OTHERS

Estates Gazette June 26 1982

(1982) 262 EG 1298

Rent assessment committee's refusal to adjourn hearing challenged by tenants — Application in judicial review proceedings for an order of certiorari to quash committee's decision on fair rents and for an H order of mandamus directed to rent officer to expunge registrations made in consequence of committee's determinations — Tenants' complaint that committee wrongly refused a lengthy adjournment of a hearing to enable them to be represented by a QC of their choice who was himself a tenant of a flat in the block (although not personally involved as such in the appeals to the committee) and who had offered his services free of charge — Further complaint that committee, in refusing the adjournment requested, had allowed only a short adjournment of a few hours until the afternoon of the hearing day — An earlier request for a postponement of the hearing had been refused in an interlocutory decision by a committee of vice-presidents of the rent assessment panel — Tenants had been J informed by vice-presidents that they could apply on the day of the hearing, to the committee appointed to conduct the hearing, for an adjournment, but were warned that they should come fully prepared to present their case if the adjournment were refused — Landlords had throughout opposed an adjournment — Committee's discretion considered — Judge's review of case law relating to the exercise of the court's supervisory jurisdiction when asked to interfere with a subordinate tribunal's discretion as to adjournment — *R v S W London Supplementary Benefits Appeal Tribunal ex parte Bullen* and *R v Thames Magistrates' Court, ex parte Polemis* — Doctrine in *Bullen* and *Polemis* extended principle of *Associated Provincial Picture Houses Ltd v Wednesbury Corporation* — Question in such K cases was whether refusal of adjournment amounted to a denial of justice because applicants were not given a reasonable opportunity to present their case, the applicants not being themselves to blame — Held that in the circumstances it was impossible to say that the committee had exercised their discretion wrongly — Application dismissed

This was an application for judicial review by a number of tenants of St George's Court, a block of flats situated over shops in Gloucester Road, London SW7. The first respondents were a rent assessment committee of the London Rent Assessment Panel, the second respondent was the rent officer for the Royal Borough of Kensington and Chelsea, and the third respondents were the owners L of the flats, St George's Court Ltd.

W Henderson (instructed by Winckworth & Pemberton) appeared on behalf of the applicants; Simon Brown (instructed by the Treasury Solicitor) represented the first respondent, the rent assessment committee; Nicholas Patten (instructed by Pickering & Kenyon) represented the third respondents; the second respondent, the rent officer, was not represented and took no part in the proceedings.

Giving judgment, FORBES J said: In this matter Mr Henderson moves for an order of certiorari to bring up and quash determinations of fair rents made by a London rent assessment committee on August 7 1981 and for an order of mandamus directing M the rent officer in effect to expunge the rents registered on that day as a result of that determination.

The decision of the rent assessment committee which is being challenged, and on which the applicants here base the whole of their case, was a decision given on August 6 1981 on a preliminary point not to adjourn an appeal by the landlords against a determination by a rent officer of fair rents for certain flats in St George's Court, Kensington.

St George's Court, as I understand it, consists of a block of 52 flats over shops in Gloucester Road, midway between Gloucester Road underground station and Kensington Gardens. The size and amenities of those flats are perhaps best indicated by the rents registered in the end by the rent assessment committee as fair rents. They were all, save for one much smaller flat occupied by a Miss Stuart — who appears in the history later — around £3,000 a year with amounts, in most cases, larger than £500 as additional payments for services.

The flats are owned by St George's Court Ltd, who manage them through the well-known firm of surveyors and estate agents, Chestertons. There is a very active tenants' association, whose chairman is Lady Burford and whose vice-chairman is a Mrs Baker. The latter also appears later in the history and she was an estate agent for some 27 years before what she describes as her "semi-retirement" in 1979.

The history, I think, can briefly be taken in this way. On December 1 1980 there was an application by the landlords for registration of fair rents. That was received by the rent officer on December 17. Notification of that was given to the tenants on December 18, and the hearing of that application by the rent officer took place on February 10 1981. His decision was given on February 19, and on March 9 there was a meeting of the tenants' association. I might perhaps add that one cannot go through the papers in this case without gleaning the suspicion that all is perhaps not well between the tenants' association and the landlords in this block of flats, and perhaps has not been for some time.

There was a meeting of the tenants' association to decide whether or not to appeal the rent officer's decision. As I understand it, the tenants decided not to do so, but on March 16 the landlords lodged an objection against the rents determined by the rent officer and that, as I understand it, was received by the rent officer on March 18. The papers went to the panel in the normal way and on June 23 1981 notification of the hearing date was given to the parties by the rent assessment panel. That date was August 6 1981.

It so happened that among the tenants of St George's Court is Mr Richard Rougier QC, who was of course not only a tenant but also an eminent leading counsel. He had, before this, offered to appear on behalf of the tenants free of charge. Unfortunately, August 6 is, of course, in the long vacation and Mr Rougier had other fish to fry, including having to be in the United States of America, as I understand it. The situation was that he would not be back until the week beginning September 28 and the most convenient date for him was October 5.

Becoming aware of that, Lady Burford telephoned, I think originally on June 29 1981, but certainly on July 8 1981, to Sir Alan Dawtry. Sir Alan Dawtry was the president of the London Rent Assessment Panel, and I think Lady Burford had met him on some occasion. She rang him up in order to explain the position and ask him what should be done. As I understand it, he said that questions of adjournment were dealt with ordinarily by a meeting of two vice-presidents of the rent assessment panel, and there was going to be a meeting of that panel and he would put the matter before them. In order to make doubly sure, Lady Burford wrote on July 15 1981. I do not think I need read the letter but it was a letter asking for the adjournment and asking for the case to be postponed until October 5, which was the only date in effect that Mr Rougier could be there, because after that he was going to be heavily engaged in other matters.

The request for an adjournment was mentioned to Chestertons and on July 21 they wrote refusing to agree to any adjournment. The committee of two vice-presidents met on July 23 and decided not to grant the adjournment, and a letter communicating that decision was sent to Lady Burford on July 24. That letter is exhibited to the affidavits in this case. It starts off with a reference to Lady Burford's letter of July 15 and goes on:

> A committee have considered the matter and I regret to have to say they have decided that the application should not be granted. Their reasons for making this decision are that it would mean a further delay of two months before the hearing involving a total period of nearly 11 months between the date of the landlords' application to the rent officer and the decision by the committee; also, while the tenants are entitled to have whom they wish to represent them, the panel cannot be expected to agree to a postponement of as long as two months so that a particular counsel could appear. In refusing the application the committee also took into account the fact that the landlords' agents opposed it.

Then there follows a short paragraph which is of the greatest importance, it seems to me:

> It is possible for the tenants' association to make another application for adjournment at the hearing but if this is refused by the committee they must be prepared to present their case.

That was the formal letter from the London Rent Assessment Panel in relation to that determination by the committee of two vice-presidents.

Mr Henderson does not seek to challenge that decision. He does not say there was anything wrong with the two vice-presidents at that stage taking that decision in the way they did. I pass on then to see what went wrong thereafter. I should say, perhaps, that Lady Burford was not one of the 14 tenants involved, but she was, as chairman of the tenants' association, taking an active part in dealing with their case on their behalf at this stage. She telephoned Sir Alan Dawtry again, I think on July 30 and certainly again on August 4. At one or other, probably the earlier, of those calls there was a conversation about the adjournment again. I think it was probably the earlier, and I put it in that way because there is some difference between Lady Burford's affidavit and Sir Alan Dawtry's about the exact content of these conversations and the exact dates on which they took place. Frankly, I do not think anything turns on that at all, but it was probable that on July 30, in the conversation between Lady Burford and Sir Alan Dawtry, Sir Alan in all probability said that the committee of two vice-presidents was meeting again on August 4 and that they would perhaps consider it again.

In fact, on August 3 the committee which was going to hear this matter had already been appointed. The way in which the panel works is that applications for adjournments — interlocutory matters, as it were — are dealt with by the committee of two vice-presidents until an actual committee to hold the hearing is appointed. Once that committee is appointed all interlocutory matters are dealt with by that committee and not by the committee of two vice-presidents. So when the vice-presidents met on August 4 the matter was out of their hands.

There was another conversation, as I have indicated, on August 4 between Lady Burford and Sir Alan. I think it was the result of that conversation which has caused the sense of indignation which is apparent to the tenants in this case. It is clear that as a result of that conversation, Lady Burford understood that the application which would have to be made on August 6 to the committee for an adjournment would be a mere formality, and she in fact phoned Mr G H W Ballantine, who was the partner of Chestertons dealing with the matter, and went so far as to indicate to him that it really was not worth his attending on the 6th because the adjournment would inevitably be granted. Mr Ballantine deposes to that.

However, Lady Burford in her last affidavit generously puts the matter in this way:

> Even after the telephone conversation which I had with Sir Alan on August 4, when he told me that since the committee for August 6 had already been appointed he no longer had any jurisdiction to grant an adjournment, I still believed that an adjournment would be given on the 6th and I acted in that belief when I telephoned Mr Ballantine on August 4. With the benefit of hindsight it may be that the strength of my belief was not fully justified on the basis of what Sir Alan actually said to me on August 4. As far as I can recall the words used by Sir Alan on August 4 were to the effect that he and his vice-presidents saw no reason why in the light of the undertaking to back-date any increases that the rent assessment committee might impose an adjournment should not be granted, but that the matter was out of his and their (the vice-presidents') hands; that there was a possibility that the committee would not grant an adjournment on the 6th and therefore that in case the worst came to the worst the tenants should be prepared to argue the substance of their case on the 6th. But he also said that he would be very surprised if an adjournment was not granted though not necessarily for the six weeks that I had suggested.

In view of that paragraph I do not think it is necessary to go to Sir Alan's affidavit for his recollection of the conversation. It is not quite the same but it is not fundamentally different.

That seems to me to have been the position on August 4 and the cause of Lady Burford's expectation that the adjournment would be granted. On August 6 Lady Burford turned up on behalf of the 14 tenants — in fact, one tenant had by then withdrawn — confident that her application for an adjournment would be granted. Mr

A Ballantine was there and opposed it. After some deliberation the committee refused to adjourn to a day of the week beginning September 28, which is what was being asked for, and decided to proceed with the hearing but on condition that copies of Mr Ballantine's case should be given to the tenants and the hearing would be postponed from the morning until 2pm.

Nobody was really available for the tenants at that stage, but they managed to get Miss Stuart to come along at two o'clock in the afternoon. It will be remembered that she was the lady who occupied the smallest of the flats in St George's Court. She was wholly unprepared to deal not only with her own flat but with any of the others, but she was there and the committee heard Mr Ballantine and
B Miss Stuart. They viewed the premises, deliberated on the matter and, on August 7, they made their determination of fair rents which should be registered. It is that decision not to adjourn in those circumstances which is now challenged. But, of course, if that decision is quashed the result would be that the substantive decision would go as well and the matter would have to be submitted to a fresh committee for determination.

Mr Henderson, who presented the case for the applicants most ably, takes several points. First of all, he says that there was considerable prejudice to the tenants in not being represented, and he points to the fact that in a subsequent application in relation to six other flats in the block, when the tenants were represented by Miss
C Baker — who, it will be remembered, had some considerable expertise as an estate agent — the rents registered by that rent assessment committee were considerably below the level of those registered in relation to the 14 flats on August 7. I do not know if it is safe to assume that that result was wholly due to the fact of representation, though it may have been, but I do not think that that matters. I am prepared to assume — in fact, I am astonished if anybody came to any other conclusion — that it would considerably prejudice the tenants if they were not able to be represented by a trained advocate and *a fortiori* by one who was of the standing of Mr Richard Rougier. So I do not really need to say any more on Mr Henderson's first point, and I do not think anybody has suggested
D from first to last in this case that it was not of considerable prejudice to the tenants that they were unable to have counsel of their choice in that way — effectively not being represented at all.

The second point which Mr Henderson takes is this. He says that the extent of that prejudice to the tenants so far outweighs any prejudice to the landlords that no reasonable tribunal could have failed to grant an adjournment. On this aspect he invites me to consider a number of matters concerned with what in fact was the prejudice to the landlords. He suggested that on an analysis there was very little prejudice; it only amounted in effect, he said, to a postponement of the receipt of any increases of rent for the time of the adjournment and this would be balanced by further increases of
E rent which would have occurred by the time of the adjourned hearing due to inflation. In so far as Mr Ballantine had suggested that it would be necessary to rewrite his case, Mr Henderson said that was an exaggeration because his case was only based on the rents of other comparable flats in St George's Court and there would be no change in those over the period of the adjournment.

I should perhaps say that under the Housing Act 1980 increases of rent take effect from the date of registration — which in a case such as this would be the date on which the rent assessment committee determined the new rent — and that was the reason why the offer that Lady Burford was making that any increase of rent should be back-dated to August 6 failed to get off the ground. It was not an order
F which the rent assessment committee could make, and if it had been tendered voluntarily and accepted it was a matter which was declared to be illegal under the Rent Acts. So that suggestion could not be acted on.

There is a statutory provision for adjournment in the rules governing the procedure of rent assessment committees. It is regulation 8 of the Rent Assessment Committees (England & Wales) Regulations 1971. Of course, as there is a statutory provision governing adjournments, it is that statutory provision which obviously governs the situation in which this rent assessment committee found itself on August 6. It reads in this way:

> The committee at their discretion may of their own volition or at the request of the parties, or one of them, at any time and from time to time postpone or G adjourn a hearing but they shall not do so at the request of one party only unless, having regard to the grounds on which and the time at which such request was made against the parties, they deem it reasonable to do so. Such notice of any postponed or adjourned hearing which is reasonable in the circumstances shall be given to the parties by the committee.

I might add one additional comment about the rent assessment committee and that is this — and I hope I am right — I have been unable to discover that they have any power to award costs and, of course, very frequently the award of costs is a matter which can take care of injustices when an adjournment is asked for.

That is the provision which governs this committee, it seems to me, in determining whether or not to grant an adjournment. But if this H court is to review such a determination it would, I think, be against the background of the well-known reluctance of courts generally to interfere with matters of discretion when exercised by subordinate courts. In addition to that there is another factor in this case which seems to me to be of vital importance. This court does not exercise an appellate jurisdiction: it exercises a supervisory jurisdiction, and so rules which apply to courts considering discretions in their appellate jurisdiction do not necessarily apply when this court is considering prerogative rights, in which case we have purely a supervisory jurisdiction.

Mr Henderson referred me to two cases: *Scutt v Scutt* [1950] WN J 286 and *Walker v Walker* [1967] 1 WLR 327. Both those cases were instances of the Divisional Court of the Probate, Divorce & Admiralty Division exercising an appellate jurisdiction from decisions of justices. Neither of them were decisions of this court exercising, on the contrary, a supervisory jurisdiction. I emphasise the difference because there are circumstances in which an appellate court, though usually with extreme reluctance, will interfere with the discretion of a lower court if the appellate court thinks that the lower tribunal has in fact exercised its discretion wholly wrongly.

I have been referred also to the case of *Hanson v The Church Commissioners* [1978] 1 QB 823. I do not want to spend too much time on this case. The facts were very briefly these. This was a rent assess- K ment committee case but the tenant had asked the clerk whether he could withdraw his application or his appeal, and the clerk — erroneously as it turned out — had said he could. The tenant thereafter assumed that the thing was dead. The landlords' surveyors, however, turned up at the hearing and they argued that whereas they had, in fact, not worried so much about the rent which was registered, they wanted to take the present opportunity to ask the rent assessment committee to increase it. The rent assessment committee proceeded to do so. They then went to have a look at the premises and when they got there the tenant was not there but his wife was. They asked his wife if they could come in and look but his wife demurred a bit and said she had better ring up the tenant. When the L tenant heard about it he said: "No, you jolly well do not come into my house without me, but what is all this about anyway?" The chairman explained that they had already decided to increase the rent, or words to that effect, and the tenant said: "Hold on a minute, give me half an hour and let me come and argue the matter." The chairman in effect said: "No, we cannot wait any longer. We have got a lot of other work to do. The hearing has already taken place and the case will have to go on without you. We will just have to record that you refused the committee access for inspection", which is what was done.

On those facts, if I may say so, perhaps it is not surprising that any court would come to the conclusion that the tenant had been denied M any proper opportunity of putting his case. That in effect is what the court decided in that case. The Master of the Rolls referred to all the errors — I am not going to read it in detail — starting at the bottom of p 833, and I think the important part of that judgment appears at p 834 C, where he said:

> There was a further error, I fear. When the chairman telephoned in the afternoon, the tenant told him that he understood the appeal had been withdrawn and that was the reason he had not appeared at the hearing. On being told that, I think the chairman should have realised that there must have been a misunderstanding and that the tenant ought to be given an opportunity of being heard, at any rate before increasing the rent from £800 to £900.

A Those errors were, in my opinion, such as to vitiate the decision of the committee. If I may summarise it, the landlords agreed at the statutory consultation before the rent officer to a figure of £800. Yet the landlords before the rent assessment committee applied to increase it to £900 in the absence of the tenant and at a time when he had no warning that they were seeking an increase. He had no opportunity of dealing with it. It is one of the cardinal principles of natural justice that a matter should not be decided adversely to a man unless he has had a fair warning of the case against him and a fair opportunity of dealing with it.

It is interesting to see that the Master of the Rolls is there putting it squarely on that principle of natural justice which is often referred to as the *audi alteram partem* rule.

B Roskill LJ (as he then was) had this to say at p 838, and as Mr Henderson relied on it particularly I think I should read it:

I am very conscious of the irritation as well as the extra expense which last minute adjournments can cause, particularly with tribunals of this nature, and I recognise the desire of any tribunal to avoid such adjournment wherever possible. But sometimes, if justice is to be done, adjournments are essential. I think this was such an occasion unless either the tenant with full knowledge of the position had refused to accept an adjournment or had been in a position to come along within half an hour as he had expressed himself willing to do. In those circumstances, much as I sympathise not only with the chairman but above all with the landlords who have at no time done anything wrong, I think this is one of these unfortunate cases where a muddle has arisen, and I would
C therefore allow certiorari to issue to quash the determination of the committee

and he thought it should be remitted to another committee. That deals with the case of *Hanson*, and it seemed to me that that case was not strictly a question of revising the discretion of a lower tribunal to adjourn. The ratio of the case was, as I have indicated from the passage from Lord Denning's judgment, the fact that it is a fundamental principle of natural justice that you cannot alter a man's position without giving him a proper opportunity of meeting whatever it is he has to meet and a fair hearing of whatever it is he has to say.

D The principles to be applied, I think, are clearly set out in a case called *R v South-West London Supplementary Benefits Appeal Tribunal, ex parte Bullen*. That was a decision of this court on May 7 1976. The court consisted of Lord Widgery CJ, Donaldson J (as he then was) and Mars-Jones J. I am not sure that I need go too far into the facts of the case but the question was whether the applicant should be deprived of supplementary benefit under certain provisions of the Supplementary Benefit Act 1978. Under that Act it appears that if it appeared to the commission that a person was refusing or neglecting to maintain himself, they could make a report to the Appeal Tribunal, and if they did so the function of the Appeal Tribunal, as the court found, was to look at the report submitted by
E the commission, listen to what the individual in question had to say at the hearing, for which an opportunity must be given under the section, and then decide whether to approve the proposed restriction on the applicant's rights or not. That was the function of the Appeal Tribunal.

In *Bullen's* case, the tribunal gave the applicant only 48 hours' notice, and that was a shorter period than was customary. According to the applicant, the day chosen for them to listen to him was the very day on which, after many years or certainly a very long time, he had finally got an interview for a job. He got in touch with the tribunal and explained the position to them. He also got a social worker to ring them up and talk to them about it. Of course, the net result was
F that he did not turn up on the day fixed. The tribunal having been fully informed about this, and I am reading from p 5 of the transcript of the judgment, said:

At the commencement of the proceedings we were informed by our clerk, Mr Whapham, that a communication had been received from the applicant stating that he was unable to attend the hearing as he had an interview for employment at that time. We were further informed that a lady from Blackfriars Settlement had telephoned to confirm the fact of the applicant's interview and to request that the matter be adjourned. The tribunal considered the request for an adjournment, but were unanimously of the opinion that the reason given was not sufficient either to justify an adjournment or to explain the applicant's failure to attend.

Again, it seems to me that that was a somewhat draconian view to take and it was not perhaps surprising, in those circumstances, that any court would strive if it could to find that that was an unreasonable exercise of a discretion. G

I am dealing with this because this case is unreported, and I know that there are those in the court who are very interested in this matter and will not of course have had an opportunity of understanding what it is about. The Lord Chief Justice went on to say this:

This is not an altogether easy case because the superior courts have always been hesitant to interfere with decisions of inferior tribunals relating to adjournments. That is because it is so often only the tribunal which is adjourning which really is possessed of all the circumstances and considerations to be applied, and to adjourn or not is a matter of discretion and H accordingly a matter which superior courts are loathe to interfere with when exercised by subordinate courts.

It is not so many years ago that Lord Goddard when Lord Chief Justice was saying that certiorari would never go merely to quash a decision of an inferior tribunal which had been taken after an adjournment had been refused. In fact he said this in the case of *In re Ekins* (1953) 117 JPJ 705: ". . . to make an order of certiorari because justices had refused to grant an adjournment would be to extend the ambit of the remedy beyond all authority and that the application must be refused".

Since Lord Goddard's day we have advanced somewhat on this point, and the court was referred today to *R v Thames Magistrates' Court, ex parte Polemis* [1974] 1 WLR 1371, where certiorari did go to set aside a decision of magistrates taken in breach of the rules of natural justice because taken at a J time when the defendant had been refused an adjournment necessary for him to present his case. The question really is whether in this instance we are to adopt the attitude which we adopted in *Polemis's* case or whether we distinguish this case and, if so, for what reason.

I think that it is still important to remember that superior courts must be hesitant before allowing certiorari to go where a subordinate court has reached a decision after the refusal of an adjournment and that action is complained of. I think that the somewhat special circumstances affecting adjournment decisions are something which I would not wish to destroy, but it is quite clear from *Polemis's* case that there is no absolute bar to the use of certiorari to quash a subordinate tribunal's decision in these circumstances.

What is the proper test to apply to decide whether certiorari should go? To begin with, it is perfectly clear, I think, that if one can say of the decision not to adjourn that it was a decision that no reasonable bench of magistrates or K other relevant authority could have arrived at, then there is no doubt that that is a case in which certiorari would go.

Mr Sedley

who was appearing for the applicant

says that the law has advanced a little further beyond that and that it is not necessary to apply the strict test to which I have just referred, and that the authorities as a whole now demonstrate that this court will allow certiorari to go in this type of case if it is clearly established that the decision is unjust, wrong and not to be sustained.

I think that that is the position since *Polemis*. I do not think it necessary strictly to show that no equivalent tribunal properly instructed could have L reached this decision. I think it suffices if this court can say with perfect confidence and conviction that this decision not to adjourn was wrong. It deprived the defendant of the hearing to which he was entitled, either under the rules of natural justice or under the statutory rules in the present instance.

I think if one applies that test to the present case the answer is that the application should have been adjourned. After all it was a very short period of notice which was given of the appointment. There was this important alternative engagement. If this man was genuinely seeking work, the chance of getting a job with EMI may have been the most important feature in his life, and one can well understand in those circumstances that he would not expect that the tribunal would require him to attend and possibly lose the chance of a job.

When one comes to ask oneself what it was about the case that caused the M tribunal to take what on the face of it was a very harsh decision, the reasons given I find unsatisfactory: indeed I find them virtually non-existent. All we were told, it will be remembered, was that the tribunal were unanimously of the opinion that the reason given was not sufficient either to justify an adjournment or to explain the applicant's failure to attend. That seems to me to be an appreciation of the situation which is quite unacceptable on the facts as can be placed before us, and I think this is one of those cases, albeit they may still be somewhat exceptional, where we ought, if confident that justice requires it, to be prepared to say that certiorari should go in order to restore the right to a hearing which in this case the applicant has been deprived of.

That case referred to the case of *R v Thames Magistrates' Court, ex parte Polemis*. That is an interesting case. Again, I do not want to

A go into it in too much detail, but as this is a matter of some importance going to the jurisdiction of this court, I am spending a little more time on it perhaps than I would otherwise. The short facts were that Mr Polemis was the master of a Greek vessel charged with discharging oil into navigable waters. He received a summons at 10.30 in the morning of the day on which the vessel was due to sail at 9pm. The summons was for hearing on that day at 2 o'clock. A considerable amount of activity took place by the solicitors instructed by the owners of the vessel, and at 2.30, when in fact the justices sat, they applied for an adjournment. There simply had not been any time to collect any of the sort of evidence necessary to refute the charge. The justices refused the application but stood the case over until 4
B o'clock. When the case came on, the defendant was unable to produce any evidence and was convicted.

The way Lord Widgery, who presided in that case (it was a Divisional Court case), put it was this, on p 1375:

> To start with, nothing is clearer today than that a breach of the rules of natural justice is said to occur if a party to proceedings, and more especially the defendant in a criminal case, is not given a reasonable chance to present his case. It is so elementary and so basic it hardly needs to be said. But of the versions of breach of the rules of natural justice with which in this court we are dealing constantly, perhaps the most common today is the allegation that the defence were prejudiced becase they were not given a fair and reasonable opportunity to present their case to the court, and of course the opportunity
> C to present a case to the court is not confined to being given an opportunity to stand up and say what you want to say; it necessarily extends to a reasonable opportunity to prepare your case before you are called upon to present it. A mere allocation of court's time is of no value if the party in question is deprived of the opportunity of getting his tackle in order and being able to present his case in the fullest sense.

He then referred to the well-known case of *Local Government Board v Arlidge* [1915] AC 120, and goes on:

> In this instance, on the brief and simple facts that I have related

and, parenthetically, I have tried to relate them briefly also

> can it be said that the applicant was given a reasonable opportunity to present his case? It seems to me to be totally unarguable that he was given such a
> D reasonable opportunity. He had no time to take samples, no time to see a report of the samples taken by the prosecution, no time to look for witnesses, no time to prepare any supporting evidence supportive to his own, and that too when he was a man with a very rudimentary knowledge of the English language in a country foreign to his own. When one just looks at those facts it seems to me to be a case in which any suggestion that he had a reasonable chance to prepare his defence is completely unarguable.

Later in a short passage on p 1377 C, Lord Widgery said:

> . . . I would hold that where the central allegation on which an order of certiorari is sought is that the defendant was not given a reasonable time to prepare his case, the mere fact that the matter became apparent as a result of a refusal of an adjournment does not prevent the court from treating the basic cause of complaint, namely, the failure to provide the defendant with ade-
> E quate time, as being a ground on which certiorari should go.

There is another unreported case called *R v Yorkshire Rent Assessment Tribunal, ex parte Lennard* before the Divisional Court consisting of Lord Widgery CJ, Eveleigh J (as he then was) and Slynn J on November 1 1976. I do not think that I need refer to the facts. There is a short passage in the judgment of Eveleigh J on p 3 which reads:

> I am greatly influenced by the fact that he

that is the applicant

> should, in the ordinary course of events, have been ready by December 12.
> F The tribunal were entitled to expect him to be ready to present his case on that day. His request, in effect, amounted to a request for a short adjournment and this was granted. In those circumstances I find it impossible to say that the tribunal came to a wrong decision and would dismiss this application.

Perhaps in explanation I should add that after having been granted a short adjournment he turned up on the adjourned hearing, applied for another adjournment and when that was not granted disappeared. The principle was followed and the Lord Chief Justice in that case referred back to what was said in *Bullen*.

It seems to me that in both those cases where the court allowed certiorari to go (in *Bullen* and in *Polemis*) and, I might also add, in *Hanson*, the ratio of the decision was that the refusal of an adjournment amounted to a denial of justice because the applicant in con- G
sequence was given no reasonable opportunity to present his case. In *Bullen* he had only 48 hours' notice and a very good reason for his application for an adjournment; in *Polemis* he had even less notice of the hearing and even better reason for asking for an adjournment. In both cases, therefore, the applicant was not in any way at fault. Looking at the tribunal in both cases, the court considered that they were at fault. In *Bullen* they really gave no reasons for not acceding to what the court considered a perfectly reasonable and justified request; in *Polemis*, as the passages later in the judgment from those I have read show, the court took the view that there were other matters which those justices should have taken into consideration to overcome the difficulty which was apparent to them that the ship was H
due to sail in a few hours. Similarly, when one looks at the case of *Hanson*, there was no fault on the part of the applicant there but the court felt that the tribunal had wholly failed to appreciate the applicant's situation and, indeed, the surrounding situation as well.

Of course, I am bound by those cases, and I think that the position is broadly this. While it is wrong to suggest that a court cannot go beyond the *Wednesbury* principle in considering cases of this kind, the only extension it will allow itself is when it can be said not only that refusal of an adjournment amounted to a denial of justice because the applicant was not given a reasonable opportunity to present his case, but also that looking at the background it would be fair to say that the applicant was not to blame and that the tribunal J
should have realised that the refusal of the adjournment would cause that type of injustice to which I have referred.

Against that background one has to consider what was before this rent assessment committee. They clearly considered the matter carefully. I am not going to read it but there is a whole foolscap page and rather more of closely typed material indicating what were the matters which were put in front of them and, quite clearly, they had taken them into account and properly taken them into account. They set out the matters urged by both Lady Burford and Mr Ballantine on behalf of the respective parties, and there is no suggestion from the applicants here that they missed out anything of importance which was in fact put before them by Lady Burford. It is not in particular K
suggested that Lady Burford's impression gained from her conversation with Sir Alan Dawtry was conveyed to the rent assessment committee.

The conclusion they came to was this, and I should read paragraph 6:

> After serious consideration of all the issues the committee announced that an adjournment was refused. It appears that the tenants had known since March 1981 that the landlords objected to the rent officer's determination and that therefore a committee hearing would be arranged in due course; and the panel committee of vice-presidents which had declined the request for adjournment on July 23 had reminded the tenants that the hearing would proceed as scheduled one month before, and left them two weeks to make alternative arrange- L
> ments for representation if they wished. The landlords' agents had justifiably gone ahead on the basis of that committee decision and the one new circumstance, the offer that any increase should be voluntarily back-dated, was, not unreasonably in this committee's view, unacceptable to the landlords at this late stage. Measuring the fairness of reaching a decision with as little delay as possible against the natural rights of the parties to have reasonable opportunity to prepare and present their case, it was decided to be on balance just to proceed; with the condition that copies of Mr Ballantine's case should immediately be made available to the tenants and that examination of this should not commence until 2 pm.

Paraphrasing the matter, it seems to me that the committee had before it two positions to consider. On the one hand there were the landlords maintaining first that they had at all times opposed any M
application and that the tenants knew this. Secondly, that they had in the circumstances justifiably gone ahead on the basis that the hearing would take place on August 6 and, third, that an adjournment would involve rewriting the landlords' case because the passage of two further months of inflation would make a significant difference.

On the other hand, there were the tenants who had been told on July 24 that an adjournment would not be granted in advance and that they must be prepared to present their case on August 6 if an application on that day to adjourn was refused, and that despite that warning had turned up unprepared. It seems to me that on those facts the committee were justified in taking the decision which they did,

A and whether one applies the principle of *Associated Provincial Picture Houses Ltd v Wednesbury Corporation* or the extension of it to be found in *Bullen* and *Polemis*, it is impossible, I think, to say that this committee acted wrongly.

I have much sympathy with the position of the tenants. I accept that they were genuinely under the impression that the application for adjournment on August 6 would be a mere formality, and I also accept that they were under a considerable disadvantage in not having the services of an eminent Silk who had agreed to present their case. Further, of course, they were at the disadvantage that they were not prepared to fight the case on that day, but having regard to the facts which I have set out — all too extensively, I fear — it seems to me that that was their fault as far as the tribunal knew. They had, as B the tribunal makes clear, had two weeks in which to arrange for alternative representation. For the reasons that Lady Burford mentioned they had not done so, but that they had those two weeks seems to me to be abundantly clear and was a matter which the tribunal was certainly entitled to take into account.

As I said earlier, this court operates a supervisory jurisdiction and not an appellate one. On the principles governing the exercise of that jurisdiction, I think this attack on the committee's decision not to adjourn must fail.

The application was dismissed with costs in favour of the rent assessment committee.

LANDLORD AND TENANT
RENT REVIEW

A **QUEEN'S BENCH DIVISION**

July 5 1981

(Before Mr Justice CANTLEY)

H WEST & SON LTD v BRECH AND OTHERS

Estates Gazette January 16 1982

(1981) 261 EG 156

Landlord and tenant — Rent review clause in lease — Landlords' notice requiring rent review to take place sent to tenants 18 months later than date prescribed by clause — Whether strict adherence to
B **the timetable should be regarded as of the essence of the contract — *United Scientific Holdings Ltd v Burnley Borough Council* discussed and decision in *Amherst v James Walker, Goldsmith & Silversmith Ltd* applied — Time not of the essence — Contrary submissions based on hardship to tenants and alleged unreasonable delay rejected — Landlords in present case had, despite tenants' assertion that the "trigger" notice was invalid, proceeded to invoke an arbitration clause in lease and to appoint an arbitrator, who made an award determining a rent for the second 7-years' term under the lease — It was conceded on behalf of the tenants that, if the notice was found to be valid, the rent for the second 7-years' term had been properly determined — Held that the rent had been properly determined in**
C **the sum fixed by the arbitrator**

The plaintiffs in this action were the landlords, H West & Son Ltd. The tenants, Michael Brech, Florence Brech and Edward Brech, were the defendants. The action concerned the interpretation and application of a rent review clause in a lease of a warehouse and land at East Street, Prittlewell, Southend-on-Sea, for a term of 21 years from February 17 1971.

M Nield (instructed by A G Freeman & Son, of Southend) appeared on behalf of the plaintiffs; A Nicholson (instructed by F T Fisher & Lang, of Southend) represented the defendants.

Giving judgment, CANTLEY J said: Under a lease dated February
D 17 1971 the defendants are tenants of the plaintiffs of land and buildings in Southend-on-Sea for a term of 21 years from February 17 1971. Clause 1 of the lease given to the defendants provides for a rent of £3,000 per annum during the first seven years of the term and "for the remainder of the said term the yearly rent calculated in accordance with the provisions of clause 8 hereof." Looked at strictly there is no clause 8 in this document, but it is quite obvious that by clause 8 in clause 1 is meant the clause which appears as clause 3(8) in the counterpart. What has happened has been a typist's or copying muddle in the preparation of the lease and counterpart resulting in a sort of cocktail of a document which happily is adequate for the purpose of deciding the rights and liabilities of the parties in this action.
E What I will call "clause 8" provides as follows:

> If the lessors shall not more than twelve months nor less than six months before the expiration of the seventh or fourteenth year of the said term give notice to the lessees requiring the rent to be revised then as and from the expiration of the said seventh or the fourteenth year the said yearly rent shall either be the rent hereinbefore reserved or such rent at which the demised premises might reasonably be expected to be let in the open market by a willing landlord by a lease for a term equal to the residue of the term then unexpired with vacant possession without a premium and disregarding, if applicable, those matters set out in section 34 of the Landlord and Tenant Act 1954 and subject to the provisions of this lease whichever shall be the F
> greater rent. And if the parties shall fail to agree such rent within three months after the service of such notice by the lessors on the lessees as aforesaid the amount of rent payable shall be referred to arbitration in accordance with the provisions for arbitration hereinafter contained. Provided that until such new rent shall have been determined the rent reserved in the period of seven years immediately before expired shall continue to be payable and any difference between that and the said new rent during such period as this proviso operates shall be added to and be payable with the next instalment of rent due after the new rent has been determined.

The provision for arbitration referred to in that clause is contained in what I will call clause 13 although, owing to the muddle which I have referred to already, it appears as a subclause of clause G
3. It reads as follows:

> Any dispute or difference arising between the lessors and the lessees for the settlement of which provision is hereinbefore made for arbitration shall be referred to the decision of a sole arbitrator or in default of agreement to two arbitrators, one to be appointed by each party subject to and in accordance with the Arbitration Act 1950 or any statutory modification or re-enactment thereof for the time being in force.

Notice in strict conformity with clause 8 should have been given by October 17 1977 for the purpose of increasing the rent for the ensuing seven years. Unfortunately the landlords overlooked the arrival of the relevant date. According to their letter of March 2
1979, this was owing to their having lodged the counterpart lease H
with their bankers and not having looked at it until the time fixed had passed. However, by that letter of March 2 1979 notice in accordance otherwise with clause 8 was given to the tenants. That notice, however, was 18 months later than the date prescribed as the last date by clause 8. Feeling that some explanation additional to the lodgment of the relevant document with their banker was due to the tenant the landlords obligingly included in their notice of March 2 1979 a reference to the decision in the House of Lords in *Cheapside Land Development Co v Messel's Service Co* [1978] AC 904. There appears then to have been some negotiation between valuers appointed by each party, but those negotiations led to no
agreement, and on July 10 1979 the defendants' solicitors replied to J
the plaintiffs' solicitors as follows:

> We have advised our clients, notwithstanding the two recent House of Lords decisions, that your clients are not now entitled to set in motion the machinery for calculation and determination of a new rent. We consider that there has been unreasonable delay on the part of your clients sufficient enough to preclude the review claim which has not been made according to the provisions of the lease.

The plaintiffs proceeded to invoke the arbitration clause. No sole arbitrator having been agreed, they appointed their arbitrator. They followed the procedure prescribed by the Arbitration Act
1950 for cases where one of the parties to an arbitration agreement K
refused to co-operate. The situation has been that the defendants have stood upon their assertion that the notice which was a condition precedent to the right of the plaintiffs to have a new rent assessed was an invalid notice. The arbitrator made an award which determined the yearly rent for the second seven years of the term at the sum of £7,250 per annum. It is now conceded that if the plaintiffs' notice of March 2 1979 was a valid notice the rent for the second seven years has been validly determined. The entire issue before me has been as to the validity of that notice.

A Mr Nicholson on behalf of the defendants has submitted that strict adherence to the specified timetable in clause 8 is of the essence of the contract, and as that timetable was not strictly adhered to the notice is invalid. He concedes that there is a presumption since the case of *United Scientific Holdings Ltd v Burnley Borough Council* and *Cheapside Land Development Co Ltd v Messel's Service Co*, to which I have already referred, that strict adherence to a timetable specified in a review clause of this nature is not of the essence of the contract. But he submits that for various reasons this particular review clause on its proper construction is of the essence of the contract. First of all he relies upon its wording. As was recognised by their Lordships in the case to which I have just
B referred, reference to the context and wording of the lease as a whole may clearly show that in any review clause the timetable is of the essence of the contract. Such a construction will naturally follow where the landlord's right to call for a review of rent can be linked with a corresponding right in the tenant to terminate the tenancy before the rent is raised. Similarly, of course, if a review clause does in terms specify that the timetable is of the essence of the contract full effect must be given to that agreement of the parties. Mr Nicholson says that although neither of those two features appears in this contract, none the less on the wording of clause 8 there is an imperative duty imposed upon the landlord to give notice within the specified period if he wishes to rely upon the right of review.

C I have considered what Mr Nicholson has said, but in the end it seems to me that on this point the review clause in question does not materially differ from other review clauses which were considered by the House of Lords in the case of *United Scientific Holdings Ltd v Burnley Borough Council* and held to be clauses where time was not of the essence of the contract. Mr Nicholson also relies in support of his submission on the fact that the sole initiative in reviewing the rent is given to the landlord and to the landlord only. This is a point which was argued before the House of Lords in the case to which I keep making reference. And in the view of Lord Diplock, Lord Simon and Lord Fraser this feature made no difference and the presumption in the absence of other circumstances will still be that
D time is not of the essence. It is true that on this point, which only arose under the terms of the lease in the case of *Cheapside Land Development Co v Messel's Service Co*, Lord Dilhorne took a view contrary to that of the noble Lords to whom I have already referred. Mr Nicholson therefore says that the observations of Lords Diplock, Simon and Fraser are obiter, and in applying what he presumed to call my better judgment I should take a view different from theirs. If it were necessary I would respectfully and without hesitation say that I am persuaded by the reasoning of Lords Diplock, Simon and Fraser and agree with their reasoning and their conclusion. But it is not necessary for me to deal with the point in that way, because their dicta have been cited and followed and
E applied by the Court of Appeal in the case of *Amherst v James Walker, Goldsmith & Silversmith Ltd* (1980) 254 EG 123. That decision at least is binding upon me even if contrary to the fact I did not agree with it.

Mr Nicholson also relies in construction of the review clause upon what he submits is the hardship which would fall upon the tenant if it were not strictly applied. Drawing on no doubt informed imagination he gave various illustrations of hardship which might fall upon a tenant if such a clause as this were not strictly applied. All his illustrations have occurred to some of the Law Lords in the case in [1978] AC 904, and it is sufficient for me to say that I agree with what Lord Diplock said about those suggested hardships.
F Prominent among the suggested calamities was the fact that a tenant whose dilatory landlord was able to enforce a large increase of rent after a lapse of time might be faced with a large sum of money which he might find it difficult to produce. It has already been pointed out by better men than I am that this hardship does not necessarily follow at all. In a lease such as this there is no question of there being a break clause and the tenant is obliged to remain the tenant unless he assigns the tenancy for the whole of the term provided. If he wishes to know what he may have to pay by way of increased rent at or after the first seven years, he can easily find out accurately enough by asking one of the many experienced valuers who are available for the purpose. If he is impatient to have the rent assessed, he has, as was pointed out by the House of Lords, G
the right after a reasonable time has expired to make time of the essence in this connection by requiring the landlord to serve his notice within a reasonable time thereafter. If he does not take that step the disadvantage is suffered primarily by the landlord who is not obtaining the rent to which he would otherwise be entitled, and the advantage is to the tenant who does not have to pay the increased rent in the meantime and can have the interest on his money. Of course there may be cases where a tenant could not afford the new rent and has no money on which he can enjoy interest. But happily I do know in this case that in July 1979 these tenants were able, if the landlord was willing to receive it, to find £60,000 for the freehold. I do not consider that there is in this case H
any feature of hardship which distinguishes it from the various circumstances visualised by their Lordships when they were considering the cases in the *United Scientific Holdings* case and the *Cheapside Land Development Co Ltd's* case.

In my view time was not of the essence in the contract in relation to clause 8.

Mr Nicholson has an alternative submission. He says if, as I have held, time was not of the essence of the contract in invoking clause 8, a landlord who wishes to invoke it should do so within a reasonable time and this landlord did not do it within a reasonable time. He had 18 months' delay, which is an unreasonable time. It certainly is quite a long time, although it should be judged not by J
itself but in relation to the period of seven years which is under review.

I am not satisfied that unreasonable delay of itself precludes a landlord from relying on a review clause. It is otherwise if such delay causes prejudice or hardship to the tenant, or if such delay is so inordinate as to lead reasonably to the inference that the right to review has been abandoned or that there is tacit agreement not to enforce it. So far as prejudice or hardship upon the tenant is concerned, it is sufficient to say that in this case none is alleged and none at all has been proved. There is therefore no question of my finding that by reason of this delay the landlord is estopped from enforcing the right to review which he would otherwise have had. I K
do not consider on the facts of this case that the delay was so extensive as reasonably to lead to an inference either that the right had been abandoned or that the parties had agreed not to apply it. Finally, judging the period of 18 months in relation to the period of seven years, I am not satisfied that it was so unreasonable as to preclude the landlord from relying upon the review clause if, contrary to my view, unreasonable delay of itself could so preclude him. In the case of *Accuba Ltd v Allied Shoe Repairs Ltd* [1975] 1 WLR 1559 there was a delay of 18 months and Goff J (as he then was) while describing it as a borderline case came to the conclusion that it was not too late. I have come to a similar conclusion.

For these reasons I find that the rent has been validly reviewed L
and determined at the rate of £7,250 per annum.

Following discussion the plaintiffs were awarded costs of the action and 12 per cent interest to run from the date of issue of the writ.

CHANCERY DIVISION

July 23 1981

(Before Mr Justice WARNER)

COVENTRY CITY COUNCIL v J HEPWORTH & SON LTD M

Estates Gazette February 13 1982

(1981) 261 EG 566

Landlord and tenant — Rent review clause in lease of commercial premises — Whether time of the essence of a stipulation setting a date for rent review notice — Review clause linked with clause giving lessee option to determine lease — Rent review notice not served before expiry of time-limit — Time assumed by lessors to be of the essence as these occurrences took place before decision of House of Lords in *United Scientific Holdings Ltd v Burnley Borough*

A *Council* — **Rent review notice served after House of Lords' decision — Held that time was of the essence because of the interrelation between the rent review clause and the option to determine — Judgment for lessees**

This was an originating summons by which the plaintiffs, Coventry City Council, sought to establish that a rent review notice served on the defendants, J Hepworth & Sons Ltd, after the expiration of the time-limit laid down in the rent review clause of a lease, was valid and effective, time not being of the essence of the stipulation. The lease in question demised to the defendants parts of a building in Coventry known as Broadgate House.

Anthony Cripps QC and R Denman (instructed by the chief B solicitor, Coventry City Council) appeared on behalf of the plaintiffs; Terence Cullen QC and A Rimmer (instructed by Biddle & Co) represented the defendants.

Giving judgment, WARNER J said: This is an originating summons raising questions as to the effect, in the events that have happened, of a rent review clause in a lease, that clause being linked to a clause giving the lessee an option to determine. The plaintiff is the Coventry City Council, which is, by virtue of the Local Government Act 1972, the successor to the Lord Mayor, Aldermen and Citizens of the City of Coventry. I shall call it "the corporation". The defendant is J Hepworth & Son Ltd, which I shall call "Hepworths". It is the parent company of a group of companies operating a substantial C chain of retail clothing shops throughout the country.

The lease was granted on August 10 1953 by the Lord Mayor, Aldermen and Citizens of the City of Coventry (therein called "the Corporation") to Hepworths (therein called "the Tenants"). It is a lease of parts of a building in Coventry known as Broadgate House, and it was granted for a term of a little over 42½ years. By clause 1 of the lease the term was expressed to begin on August 10 1953 and to expire on March 31 1996 "unless previously determined as hereinafter provided". By the same clause the rent was expressed to be one shilling a year for an initial period that does not now matter and, during the remainder of the term, £8,500 a year "subject to any revision in accordance with clause 4(3) hereof". Clause 4(3) D reads as follows:

> (a) The Tenants may determine the term hereby created on the Thirty-first day of March One thousand nine hundred and seventy-five provided that they shall have given notice in writing to the Corporation of their intention so to do on or before the Thirtieth day of September One thousand nine hundred and seventy-four and shall at all times up to the time of such determination pay the rent hereby reserved *PROVIDED* that on such determination this lease and everything herein contained shall cease and determine but without prejudice to the right of action of either party against the other in respect of any antecedent claim or breach of covenant
>
> (b) On or before the Thirty-first day of December One thousand nine hundred and seventy-three the Corporation may give to the Tenants notice E in writing of their desire to increase the rent hereby reserved in respect of the period from the First day of April One thousand nine hundred and seventy-five until the expiration of the term hereby created and if the Tenants and the Corporation shall thereafter agree upon the amount of such increased rent then the rent so agreed shall become payable hereunder in respect of the said period in substitution for the rent hereby reserved (but without prejudice to the provisions herein contained as to payment for services or any of the other terms and conditions herein contained)
>
> (c) If a notice having been served by the Corporation pursuant to Clause 4 (3) (b) aforesaid no agreement as therein mentioned shall have been reached by the Twenty-eighth day of February One thousand nine hundred and seventy-four then (without prejudice to the Tenants' right or power of determination contained in the foregoing paragraph (a) of this subclause) the question of the rent to be payable in respect of the period from the F First day of April One thousand nine hundred and seventy-five until the expiration of the said term shall be determined by a single arbitrator in case the parties can agree upon one within a period of one month next after the said Twenty-eighth day of February One thousand nine hundred and seventy-four otherwise to two arbitrators one to be appointed by each party on or before the Thirtieth day of April One thousand nine hundred and seventy-four and in either case in accordance with the Arbitration Act 1950 or any statutory enactment in that behalf for the time being in force and (unless the term hereby created shall be determined by the Tenants under the right or power contained in the said foregoing paragraph (a) of this subclause) the rent determined by such arbitration shall (subject as hereinafter mentioned) be substituted as aforesaid in respect of the said period from the First day of April One thousand nine hundred and seventy-five until the expiration of the said term notwithstanding the fact that the G result of such arbitration may not be known until after the said First day of April One thousand nine hundred and seventy-five *PROVIDED THAT* in the event of the rent so determined being less than the rent hereby reserved then such rent shall not be substituted as aforesaid and (unless the said term shall be determined by the Tenants as aforesaid) the rent hereby reserved shall continue to be payable for the remainder of the said term notwithstanding the result of such arbitration.

So that clause laid down an elaborate and precise timetable. On or before December 31 1973 the corporation might give to the tenants notice of its desire to increase the rent in respect of the period from April 1 1975 until the expiration of the term created by the lease, that is, roughly speaking, in respect of the second 21 years H of it. I shall call that a "rent review notice". If the corporation gave such a notice and no agreement had been reached by February 28 1974 (that is, within two months), there was to be a reference to arbitration. Clearly, what was envisaged was that, in the two months between December 31 1973 and February 28 1974, there would be negotiations between the parties, which would either produce agreement on a new rent or be abortive. Then, if the negotiations proved to be abortive, the parties were given one month in which to agree upon a single arbitrator. If they failed to do so, each must appoint an arbitrator. The last day for doing that was April 30 1974. The provisions of sections 7 and 8 of the Arbitration Act 1950 would then apply. On or before September 30 1974 the J tenants might give notice of their intention to determine the lease as from March 31 1975, that is the day before any increase of rent might take effect. At that time, that is on September 30 1974, assuming that the corporation had served a rent review notice, the tenants might not know precisely what the new rent was going to be. Clause 4(3)(c) expressly envisaged that the new rent to be determined by arbitration might not be known before April 1 1975. None the less, obviously, the tenants would, in the contemplation of clause 4(3), be in a good position to know, by September 30 1974, at least roughly what the new rent was likely to be. They would have been parties to the negotiations envisaged by the clause and parties also to any subsequent arbitration. By September 30 1974 any such K arbitration would be well under way. So the tenants would be aware of the upper and lower figures envisaged in the negotiations and in the subsequent arbitration.

There is before me a certain amount of evidence as to the circumstances in which the lease was entered into, but I do not think that that evidence throws any light on the construction of clause 4(3), though one aspect of it, which I shall mention later, is relied upon by Mr Cripps, who appears for the corporation, in that connection.

Owing to an oversight by officers of the corporation, the date, December 31 1973, by which a rent review notice should have been served was allowed to go by without such a notice being served. That oversight appears to have been realised by March 1974. At L that time, however, it was the generally accepted view of the law that time was of the essence of rent review clauses. So the officers of the corporation did not seek to do more than negotiate with Hepworths in an endeavour to "claw back" (as it was expressed to me) some of the loss occasioned to the corporation by the oversight. That loss was substantial. According to an affidavit of Mr Ingham, who is a Fellow of the Royal Institution of Chartered Surveyors and Hepworths' general manager for estates, the open market rent of the demised property as at April 1 1975 would have been of the order of £50,000 per annum. Mr Savage, who is also a Fellow of the Royal Institution of Chartered Surveyors and is the deputy director of the corporation's "Homes & Property Services", puts it at about M £80,000 per annum.

The negotiations between the parties do not appear to have led to anything of substance and in 1975 they fell into abeyance.

Then on March 23 1977 came the decisions of the House of Lords in *United Scientific Holdings Ltd v Burnley Borough Council* and *Cheapside Land Development Co Ltd v Messels Service Co* [1978] AC 904. These decisions were first reported in *The Times* on March 24 1977 and subsequently in the WLR and the All ER on April 9 and May 3 1977 respectively. As a result of them the corporation's officers reconsidered the possibility of serving a late rent review notice.

A Such a notice was, in fact, contained in a letter written by the corporation to Hepworths on August 4 1977. That letter is in these terms:

> Dear Sirs,
>
> *Re: 44-46 Broadgate, Coventry*
>
> I hereby give you notice of the Council's desire to increase the rent of the above premises in accordance with clause 4(3) of the lease dated August 10 1953, under which you hold the same.

Hepworths objected to that notice on the ground that it was out of time. Eventually, on October 24 1979, the present originating summons was issued.

B Two questions were argued before me. The first was whether I should hold, on authority of the two cases decided by the House of Lords on March 23 1977, that time was not of the essence of the stipulation in clause 4(3)(b) of the lease that a rent review notice should be given on or before December 31 1973. (I propose to refer for short to those two cases as "the *Cheapside* case", because the second of them was, on its facts, more like the present case than the first). The second question that was argued before me was whether, if time was not of the essence in that stipulation, I should none the less decline to relieve the corporation of the consequences of its having served the rent review notice late, on the ground of unreasonable delay on its part or on the ground of such delay C combined with prejudice to Hepworths. That second question can arise only, of course, if I decide the first question in favour of the corporation.

I will say at once that I have come to the conclusion that Hepworths are entitled to succeed on the first question, so the second question does not arise.

The general principle laid down by the House of Lords in the *Cheapside* case was stated by Lord Diplock in these terms (at p 930):

> So upon the question of principle which these two appeals were brought to settle, I would hold that in the absence of any contra-indications in the express words of the lease or in the interrelation of the rent review clause D itself and other clauses or in the surrounding circumstances the presumption is that the time-table specified in a rent review clause for completion of the various steps for determining the rent payable in respect of the period following the review date is not of the essence of the contract.

The important words there, for the purposes of this case, are:

> in the absence of any contra-indications . . . in the interrelation of the rent review clause itself and other clauses.

That is explained later in Lord Diplock's speech when he comes to deal with two earlier cases where rent review clauses were, as here, linked to clauses giving the tenant the right to determine the lease. The relevant passage is at pp 935 and 936 and it reads as follows:

E

> *Samuel Properties (Developments) Ltd v Hayek* [1972] 1 WLR 1296 may be regarded as the origin of the dichotomy between "option" on the one hand and "obligation" or "machinery" on the other: the word option having been used in the lease itself to describe the landlord's right to require the rent to be reviewed. It should be treated as overruled. There was a complication in that the rent review clause was associated with a break clause which gave to the tenant the right to surrender the residue of the term on any rent review day by giving prior notice. The time-table in the rent review clause for the determination of the new rent was obviously correlated with the time by which the tenant had to give notice of his intention to surrender, so as to enable him to make his decision whether or not to exercise that right in the knowledge of what the new rent would be if he continued in possession after F the review date. Had that been all, as it had been in the previous and rightly decided case of *C Richards & Son Ltd v Karenita Ltd* (1971) 221 EG 25, it would, I think, have been sufficient by necessary implication to make time of the essence of the rent review clause because of its inter-relation with the time by which notice was to be given under the break clause — a time which, for reasons I have given earlier, I consider to be of the essence of the contract.
>
> In *Samuel Properties (Developments) Ltd v Hayek*, however, the break clause itself contained a provision under which the period during which the tenant could exercise his right to surrender would be extended in the event of the reviewed rent not having been ascertained within the time stipulated in the rent review clause. So the implication that would otherwise have arisen from the association of the rent review clause with a break clause was negatived.

Lord Dilhorne, at p 940, expressed his agreement with what Lord G Diplock had said about those cases. Lord Simon of Glaisdale dealt with the matter at p 946. He said:

> However, where a rent review clause is associated with a true option (a "break" clause, for example), it is a strong indication that time is intended to be of the essence of the rent review clause — if not absolutely, at least to the extent that the tenant will reasonably expect to know what new rent he will have to pay before the time comes for him to elect whether to terminate or renew the tenancy (*cf Samuel Properties (Developments) Ltd v Hayek* [1972] 1 WLR 1296). That situation stands in significant contrast with those in the instant appeals.

Lord Salmon dealt with the *Hayek* case at pp 951 and 952. He said:

> In *Samuel Properties (Developments) Ltd v Hayek* the rent revision clause H which laid down the procedure for having the open market rent value ascertained at the end of the seventh and fourteenth years of the term and the rent then being raised to that level, was dressed up to look like an option. Indeed the word "option" appeared in the clause. But, for the reasons I have already stated, I do not think that it was a real option in the sense that any option to renew or determine a lease is an option. The clause required the landlord to give notice to the tenants six months prior to the expiry of the seventh year if he required the rent to be raised to the open market rental value. If within one month of the notice, the parties failed to agree the open market rental value this figure was to be determined by a valuer appointed by the President of the Royal Institution of Chartered Surveyors. But the date by which this determination was to be made was not specified. The landlord gave his notice about one month late. The Court of Appeal held that time was of the essence and that the landlord was precluded fron. J putting the rent revision clause into operation. Clause 5 of the lease so far as relevant gave the tenant a true option to determine the lease at the end of the seventh year of the term by giving the landlords at least one quarter's notice in writing. This break clause was obviously inserted to protect the tenant should he not wish to pay the increased rent during the next seven year period of the term.
>
> The proviso to the break clause strongly suggests however that the time provisions relating to rent revision were not of the essence of the contract. It reads:
>
> "Provided always that if one quarter before the expiration of the first seven . . . years of the term . . . the reviewed rent . . . shall not have been reviewed then the right of the lessee to terminate as herein provided shall be extended until the expiration of one month from the date of the notification K of the reviewed rent to the lessee."
>
> There is nothing in the proviso nor in any other part of the lease to suggest that the new rent may not be determined by the valuer and notification of this rent may not reach the lessee until after the expiration of the first seven year period. In my view *Samuel Properties (Developments) Ltd v Hayek* [1972] 1 WLR 1296 was wrongly decided and should be overruled.

It appears to me implicit in that passage that if the lease in the *Hayek* case had not contained the proviso extending the lessee's time for giving notice to terminate the lease, Lord Salmon, like Lord Diplock, would have held that the time provisions relating to rent revision in that case were of the essence of the contract.

Lastly, Lord Fraser dealt with the matter at pp 962 to 963. He said: L

> For these reasons I am of the opinion that the equitable rule against treating time as of the essence of a contract is applicable to rent review clauses unless there is some special reason for excluding its application to a particular clause. The rule would of course be excluded if the review clause expressly stated that time was to be of the essence. It would also be excluded if the context clearly indicated that that was the intention of the parties — as for instance where the tenant had a right to break the lease by notice given by a specified date which was later than the last date for serving the landlord's trigger notice. The tenant's notice to terminate the contract would be one where the time limit was mandatory, and the necessary implication is that the time limit for giving the landlords notice of review must also be mandatory. An example of such interlocked provisions is to be found in *C Richards & Son Ltd v Karenita Ltd* (1971) 221 EG 25 where the decision that time M was of the essence of the landlord's notice could be supported on this ground, although not, as I think, on the ground on which it was actually rested. The case of *Samuel Properties (Developments) Ltd v Hayek* [1972] 1 WLR 1296 is not in this class because, although there was a tenant's break clause, the time allowed to the tenant for giving notice was automatically extended until one month after the notification of the reviewed rent to the lessee.

My attention was also drawn to a recent, unreported, decision of the Court of Appeal in *Abdul Jabbar Al Saloom v Shirley James Travel Service Ltd* on May 14 1981* where, on the strength of what

*Since reported in full at (1981) 259 EG 420.

had been said in the House of Lords in the *Cheapside* case, particularly by Lord Fraser, it was held that time was of the essence of a rent review provision which was in the same clause as a tenant's option to break. The decision there was founded, however, on the wording of the relevant clause and I do not think that, for present purposes, it carries the matter further than the dicta in the House of Lords.

Mr Cripps invited me to hold that those dicta were inapplicable in the present case. He submitted that I must look at the circumstances of the case as a whole and that there were a number of features of it from which the inference should be drawn that the parties, when they negotiated the terms of the lease in 1953, cannot have intended the stipulation as to time in clause 4(3)(b) to be of the essence of the contract between them.

Mr Cripps pointed in particular to the following features of the lease:

(1) It was a lease for over 42 years, with only one break at the 21st year. This meant that the landlord's omission to serve a rent review notice in due time, if held to be fatal, could have catastrophic results for the landlord and confer a substantial windfall on the tenant.

(2) The parties envisaged that Hepworths would, from the start, underlet substantial parts of the demised property. That is shown by a provision in the lease itself (clause 2(12)) and also by the evidence. The case was not one, therefore, where the tenant was to occupy the whole property for the purposes of his own business, but one where he could cushion himself, to some extent, against the effect of a rise in his own rent by raising the rents of his underlessees.

(3) Clause 4(3) of the lease envisaged the lapse of a period of nine months between the date for service of the rent review notice (December 31 1973) and the date of service of any notice by the tenant determining the term (September 30 1974). That gave the tenant ample time, if he wished to, to serve on the landlord a notice making time of the essence of the contract.

(4) The provision in clause 4(3)(c) as to what was to happen if the result of the arbitration was not known by April 1 1975 showed that the parties envisaged that the tenant might not know the amount of the revised rent by September 30 1974.

Mr Cripps submitted that it was not in every case where a rent review clause was associated with a tenant's option to break that time was of the essence of the rent review clause. That is manifestly right, as the House of Lords' treatment of the *Hayek* case illustrates. I do not think, however, that the features of the present case to which Mr Cripps pointed are sufficient to negative the presumption that, where a rent review clause is linked to a tenant's option to break, time is of the essence of the rent review clause. The length of the term and the fact that the lease provided for only one break at 21 years are, I think, sufficiently explained by the circimstance that it was granted in 1953, at a time when inflation was at a much lower rate than now. The fact that substantial subletting by Hepworths was envisaged seems to me neutral. The period of nine months between December 31 1973 and September 30 1974 was, I think, intended, not to give the tenant an opportunity to serve a notice making time of the essence, but to enable any necessary arbitration to have been completed by September 30 1974 or at least to be sufficiently advanced by that date to give the tenant a good idea of what the revised rent was likely to be. Lastly, I do not think that it matters that there could be no certainty that the tenant would know the precise amount of the revised rent by that date. So far as the report of *C Richards & Son Ltd v Karenita Ltd* shows, there was no such certainty in that case either.

Mr Cullen, who appeared for Hepworths, reminded me of the paragraph in Lord Bridge's speech in *The Chikuma* [1981] 1 WLR 321 at p 322, where he said:

> It has often been pointed out that shipowners and charterers bargain at arm's length. Neither class has such a preponderance of bargaining power as to be in a position to oppress the other. They should be in a position to look after themselves by contracting only on terms which are acceptable to them. Where, as here, they embody in their contracts common form clauses, it is, to my mind, of overriding importance that their meaning and legal effect should be certain and well understood. The ideal at which the courts should aim, in construing such clauses, is to produce a result, such that in any given situation both parties seeking legal advice as to their rights and obligations can expect the same clear and confident answer from their advisers and neither will be tempted to embark on long and expensive litigation in the belief that victory depends on winning the sympathy of the court. This ideal may never be fully attainable, but we shall certainly never even approximate to it unless we strive to follow clear and consistent principles and steadfastly refuse to be blown off course by the supposed merits of individual cases.

That appears to me, if I may say so, entirely apposite in the present type of case also.

I accordingly hold that this summons must be dismissed.

CHANCERY DIVISION

November 6 1981

(Before Mr Justice Peter GIBSON)

LONDON & MANCHESTER ASSURANCE CO LTD v G A DUNN & CO

Estates Gazette April 10 1982

(1981) 262 EG 143

Landlord and tenant — Problem caused by oddly drafted rent review clause in lease which stated earliest date for setting in motion rent review machinery, but did not specify any latest date for doing so — First review period began on December 25 1977 — Notice by lessors specifying a figure of open market rental value for the review period had to be served not earlier than December 25 1976 — In fact such notice was not given until March 19 1980 — The first rent review period was due to expire on December 24 1982 — Lessees challenged the effectiveness of lessors' notice, contending that it should have been served not later than the expiry of the 12 months' period immediately preceding the rent review period, ie not later than December 24 1977 — Lessors sought declarations by originating summons that they were entitled to call for a rent review with effect from December 25 1977; that they were entitled to apply under the lease for an independent surveyor to determine the revised rent; and that such rent should be determined by reference to rental values current at the date of the award — Lessees contended that a notice in respect of rent for a review period must be served before the commencement of the period and that time was of the essence of that obligation — In the alternative they argued that the notice was invalid on four grounds, common law election, promissory estoppel, estoppel in pais or by representation, and unreasonable delay — These arguments were rejected by the judge, who held that they could not prevail against the express wording of the review clause — Interesting argument based on the agricultural holdings case of *Sclater v Horton* rejected — Declarations made in favour of the lessors as requested in the summons

In this originating summons the plaintiffs, London & Manchester Assurance Co Ltd, sought declarations favouring their construction of a rent review clause against the defendant lessees. The rent review clause in question was contained in a 21-year lease of ground-floor shops and basements at 54 and 56 Oxford Street and 57 and 58 Rathbone Place, London W1. The defendant lessees were the outfitters, G A Dunn & Co.

K Lewison (instructed by Wilde Sapte) appeared on behalf of the plaintiffs; Derek Wood QC and George A Ward (instructed by Wright, Johnson & Cheales) represented the defendants.

Giving judgment, PETER GIBSON J said: This case provides another illustration of the difficulties to which rent review provisions in a lease can give rise.

By a lease and deed of surrender made on February 4 1971 four ground-floor shops and the basements thereunder at the corner of Oxford Street and Rathbone Place in London, and known as 54 and 56 Oxford Street and 57 and 58 Rathbone Place, were let to the defendants, GA Dunn & Co, the well-known outfitters. The term of the lease was 21 years from December 25 1970. The rent was expressed to consist of two items, only the first of which is material.

A The relevant part of the reddendum in clause 1 reads as follows:

> . . . YIELDING AND PAYING therefor during the said term the said yearly and proportionately for any fraction of a year first the rents hereunder set out
> (a) for the first 4 years of the said term the yearly rent of £13,100
> (b) for the next 3 years of the said term the yearly rent of £14,600
> (c) During the next 5 years of the term commencing on December 25 1977 (hereinafter called "the first review period") the rent payable by the Lessee shall be whichever shall be the higher of £18,000 per annum and the open market value of the demised premises for the review period
> (d) During the next 5 years of the term and during the last 4 years of the term (hereinafter respectively known as "the second review period" and "the third review period") the rent payable by the Lessee shall be whichever shall B be the higher of the reviewed rent in the 12th and 17th year of the said term and the open market value of the demised premises for the second review period and the third review period respectively.
> Such reviewed rents to be determined in accordance with the provisions in that behalf contained in clause 5 hereof . . .

Clause 5 provides:

> PROVIDED THAT AND IT IS HEREBY AGREED as follows:
> (1) The expression the open market rental value as aforesaid means a sum in relation to the review period or second review period or third review period as the case may be determined in manner hereinafter provided as being at the time of such determination the annual rental value of the Demised Premises in the open market on a lease for a term of years certain equivalent C to the number of years then unexpired of the term granted by this Lease with vacant possession at the commencement of such term but upon the supposition (if not a fact) that the Lessee has complied with the obligations as to repair and decoration herein imposed on the Lessee such lease being on the same terms and conditions other than as to the amount of rent and the length of the term as are herein contained without the payment of any fine or premium and disregarding (if applicable) those matters set out in paragraphs (a) (b) and (c) of section 34 of the Landlord and Tenant Act 1954 as amended by section 1 of the Law of Property Act 1969 other than subsection (1) (d) thereof
> (2) The said open market rental value shall be determined as follows:
> (a) It shall be such sum as shall be specified in a notice in writing by the Lessor to the Lessee at any time not earlier than twelve months prior to the D expiration of the period of seven years in the case of the first review period immediately preceding the first review period or not earlier than 12 months prior to the expiration of the periods of 5 years in the case of the second review period and the third review period as the case may be or
> (b) As shall within three months after such notice be agreed between the parties in writing in substitution for the said sum or
> (c) It shall be determined at the election of the Lessee by counternotice in writing to the Lessor not later than three months after the Lessor's said notice (time to be of the essence hereof) by an independent surveyor appointed for that purpose by the parties jointly in writing or upon their failure to agree upon such appointment within one month after the date of the said counternotice then by an independent surveyor appointed for that purpose by the President for the time being of the Royal Institution of E Chartered Surveyors and every such determination shall be made in accordance (so far as not inconsistent herewith) with the provisions of the Arbitration Act 1950 or any statutory modification or re-enactment thereof for the time being in force and shall be subject to the further provisions of the next succeeding sub-clause hereof
> (3) In the event of the determination by such independent surveyor not having been made and communicated to both parties hereto prior to the commencement of the first review period or the second review period or the third review period as the case may be for any reason whatever then in respect of the period of time (hereinafter called "the said interval") beginning with the said commencement and ending on the quarter day immediately following the date on which such determination shall have been made and communicated as aforesaid the rent payable hereunder shall continue to be paid at the rate payable immediately prior thereto
> F PROVIDED that at the expiration of the said interval there shall be due as additional rent payable by the Lessee to the Lessor on demand a sum of money equal to the amount whereby the first reviewed rent or the second reviewed rent or the third reviewed rent respectively shall exceed the rent payable immediately prior thereto but duly apportioned in respect of the said interval.

The reference in the second line of clause 5(1) to "the review period" must of course be a reference to the first review period. The first review period was therefore to commence on December 25 1977 and to expire on December 24 1982, and the lessor could serve a notice under clause 5(2)(a) in respect of that period not earlier than December 25 1976.

G In June 1977 the reversion immediately expectant on the term granted by the lease became vested in the plaintiff, London & Manchester Assurance Co Ltd. Prior to that time no notice under clause 5(2)(a) had been served by the then lessor nor did the plaintiff thereafter until some 2¼ years into the first review period. The plaintiff demanded and received rent at the rate of £18,000 per annum for the first nine quarters of the first review period. On March 19 1980 the plaintiff's agents wrote to the defendants saying that they were giving notice pursuant to clause 5(2)(a) that the plaintiff required the rent to be reviewed and suggesting £140,000 as the open market rental. By letter of June 6 1980 the defendants' solicitors informed the plaintiff's agents that the defendants regarded the notice as invalid, but without prejudice to that asser- H tion they preserved their position by giving the counternotice under clause 5(2)(c), thus satisfying the strict condition as to time in that clause. No independent surveyor has been appointed.

On April 14 1981 the plaintiff took out the originating summons herein by which it seeks three declarations:

> 1. A declaration that upon the true construction of the reddendum in clause 5 of the above-mentioned lease and in the events which have happened the plaintiffs are entitled to call for a review of the rent payable under the above-mentioned lease with effect from December 25 1977.
> 2. A declaration that upon the true construction of the reddendum in clause 5 of the above-mentioned lease and in the events which have happened the plaintiffs are entitled to apply to the President for the time being of the J Royal Institution of Chartered Surveyors for the appointment of an independent surveyor as arbitrator to determine the amount of the reviewed rent to be paid by the defendant as aforesaid.
> 3. A declaration that upon the true construction of the reddendum in clause 5 of the above-mentioned lease and in the events which have happened the arbitrator should determine the amount of the reviewed rent by reference to rental values current at the time of his award.

There are two main issues between the parties. First, on the true construction of the lease is the lessor obliged to serve his notice under clause 5(2)(a) in respect of rent for a review period before the commencement of that period and, if so, is time of the essence of that obligation? If the answer be "Yes", as Mr Wood appearing for K the defendants submits, then the plaintiff was too late with its notice of March 18 1980 and the second issue does not arise. If the answer be "No", as Mr Lewison for the plaintiff submits, then the second issue arises: that is to say, in the events that happened is the plaintiff's notice invalid? Mr Wood, in submitting that the notice is invalid, puts this part of his case on four grounds which may be labelled as follows:

(1) common law election
(2) promissory estoppel
(3) estoppel in pais
(4) unreasonable delay

L I start with the first issue, the question of construction. Both sides agree that the draftsmanship of the rent review provisions leaves much to be desired. A comparison of those provisions in this lease with those in the lease the subject of the decision of Goff J in *Accuba Ltd v Allied Shoe Repairs Ltd* [1975] 1 WLR 1559 shows that in all probability a common precedent was used and adapted in each case; and it appears from the *Accuba* case (at p 1560) that that precedent is to be found in the 1966 edition of the *Encyclopaedia of Forms and Precedents*. The lease before me differs from that in the *Accuba* case in at least two material respects. First, whereas the lessor's notice in the lease before me is to be served "at any time not earlier than 12 months before the first review period", in the *Accuba* lease it was to be served at any time before the end of a M period ending a year before the review period. Second, there is an inevitable rent increase in the first review period in the lease before me subject only to the possible temporary effect of clause 5(3), whereas in the *Accuba* case the rent for the single review period was the higher of the previous fixed rent and the rent as determined in accordance with provisions the same as in clause 5(3).

The rent review provisions contain one remarkable feature. On the literal wording of clause 5 there is no fixed date such as the start of the review period for the date at which the open market rental value is to be determined. Mr Lewison (by way of primary submission) and Mr Wood are at one in saying that that rental value falls to

A be determined as at the time of the determination. Accordingly it is common ground that whether the determination of the rental value occurs before or after the review period has commenced, the rental value at the start of the review period is not the relevant value. As the determination by the independent surveyor pursuant to clause 5(2) is to be made in accordance with the Arbitration Act, it is clear that there could be considerable delays after the lessee had elected for a determination by an independent surveyor before the determination was made, for example, if there was a case stated for the decision of the High Court. When one bears in mind that the determination is to be of the rental value of the premises in the open market on a lease for a term of years certain equivalent to the B number of years then (ie at the date of determination) unexpired of the term and that those provisions are applicable in respect of all the review periods including the third which comprises the last four years of the term of the lease, it is readily apparent that the result of having the rental value determined at the date of determination could be very strange indeed. It may be that it was in the light of such obvious difficulties that in the *Accuba* case it was apparently not in dispute that the determination was to be of the rental value at the commencement of the review period (see the report at p 1565 C-E and p 1566 C-D). In the present case Mr Lewison made an alternative submission that as the rental value was, in the words of clause 1, the open market value of the demised premises for the C review period, on a determination after the start of the review period values over the whole period should be looked at. But this gives little or no effect to the express provisions of clause 5(1) that the open market value is to be the sum determined as being at the time of such determination the annual rental value, and I can see no escape from giving literal effect to those words, however curious a result they could produce.

Mr Wood submits that a term is to be implied in clause 5(2)(a) that the notice under that clause should be given not later than the expiry of the 12 months period immediately preceding the rent review period. He accepts that in the absence of any contra-indication time is not of the essence of a timetable for completion of D the various steps for reviewing rent under a lease (*United Scientific Holdings Ltd v Burnley Borough Council* [1978] AC 904). But he submits there is a contra-indication in this lease. He relies on the provision for the determination of the rental value as at the time of determination both for implying the term in clause 5(2)(a) and as the contra-indication making time of the essence of the implied time-limit. Further he relies on the decision of the Court of Appeal in *Sclater v Horton* [1954] 2 QB 1 as authority for implying a mandatory timetable in the present case.

Mr Wood's submissions are nothing if not bold. The lease expressly states the earliest date for the service of a notice under clause 5(2)(a) and in marked contrast it does not state the latest date for E such service; moreover the absence of a latest date is emphasised by the words "at any time". A further striking contrast is provided by the express time-limit for the lessee's notice of election, which was also expressly made subject to a provision that time was to be of the essence. Mr Wood accepts, as he must in the light of the *United Scientific* case, that the fact that the lessor alone had the right to initiate a rent review would not make time of the essence of a time-limit for initiating the rent review. Still less would that fact justify implying a time-limit not expressed. The only provision of the lease on which Mr Wood relies is the provision for determining the rent at the time of determination. He says that because of the possible injustice to either lessor or lessee if the rental value at the F date of determination were much lower or higher than it was at the start of the review period, the lease must be read in such a way that that determination will take place about the time of the start of the review period. I accept that it would have been entirely fair and reasonable for the parties to the lease to have agreed that the lessor's notice should be served by the start of the review period. But it is not permissible for the court to imply a term merely because it is fair and reasonable. Necessity is the appropriate test (see *Liverpool City Council v Irwin* [1977] AC 239). I do not see how in this lease it could be said that the suggested time-limit was a necessary implication. The lease expressly contemplates that the determination may take place after the start of the review period, G and Mr Wood accepts that even with his implied term there could be a long delay thereafter before the determination. Further, Mr Wood's submission must proceed from the premise that the parties intended that the rent for a review period should be the rental value about the commencement of that period. But the very terms of the provision on which Mr Wood relies, providing as it does for the rental value to be determined as at the date of determination, whenever that determination might be, appears to me to be inconsistent with that premise. Then there is the insuperable difficulty of reconciling the words "at any time" with the suggested implied time limit. In my judgment, therefore, as a matter of construction of the language of the lease, Mr Wood's submission cannot succeed.

H Does *Sclater v Horton* compel a different conclusion? That was a decision on the construction of section 8 of the Agricultural Holdings Act 1948 concerning rent reviews in agricultural tenancies. In that case the court was prepared to imply a term that a particular date referred to in the section had to precede another date referred to in the section, that being, as Sir Raymond Evershed MR put it at p 10, "the necessary and inevitable implication of the language which Parliament has used". Mr Wood submitted that because the section contained terms to be incorporated into tenancy agreements, it was in his words *in pari materia*. I cannot accept that. What Parliament may be presumed to have intended as a fair term in agricultural tenancies generally may be wholly different from what J parties to a particular business tenancy may be presumed to have intended. More important, there is no similarity between the language of section 8 and the language of the lease which I have to construe and the machinery for rent review laid down by the section is very different from that laid down by the lease. The reasoning which led the members of the Court of Appeal to their conclusion was largely based on the particular wording of the section. Mr Wood placed reliance on a passage in the judgment of Sir Raymond Evershed at p 11 where, after referring to the contrary argument as leading to the consequence that rents would be fixed by reference to a date when circumstances would have been quite different from those existing when the rent became payable, he said: "If a rent is to be K fixed, the normal and fair date at which it should be fixed is the date when it begins to operate". But that was said in the context of statutory provisions which did not contain anything corresponding to the provisions in clause 5 expressly recognising as they do that the determination might take place after the date when the rent was to begin to operate. *Sclater v Horton* seems to me to be plainly distinguishable and I cannot find in it any principle applicable to the present case. I should also record that Mr Lewison reserves the right to say elsewhere that *Sclater v Horton* was wrongly decided as, he says, the decision proceeded on the erroneous footing that rent had to be ascertained by the time it becomes due.

In the result I must reject Mr Wood's submission on the first L issue. In so doing I do not overlook that there are difficulties if clause 5(2)(a) is construed literally as enabling the lessor to serve his notice "at any time" after the earliest date specified without any limit in time, and indeed Mr Lewison was prepared to concede that the lessor could not serve a notice under clause 5(2)(a) initiating the rent review procedure for a review period at a time when the rent for an earlier period had not yet been determined. He was also prepared to accept that a notice for the first review period could not be served later than December 24 1981, the day before the earliest date when a notice for the second review period could be served. It is unnecessary for me to pronounce on the correctness of these concessions. Suffice it to say that, in my judgment, on the true M construction of the lease the lessor was not obliged to serve his notice before the commencement of the first review period or before the date when the plaintiff in fact served its notice. In reaching this conclusion I am comforted by the thought that this is at least consistent with the decision of Goff J in the *Accuba* case. In that case it was held that time was not of the essence in respect of the express time-limit for the service of the lessor's notice. It would be surprising if the lessor in the present case, where there was no time-limit expressed, was to be subject to a more stringent condition as to time. Of course the argument that Mr Wood has advanced in the present case on the significance of the provisions for determining rent at the date of determination does not appear to

A have been advanced to Goff J, and Mr Wood accordingly submits that the *Accuba* case was wrongly decided. While the reasoning of Goff J, based as it was on the then supposed dichotomy between option clauses and machinery clauses was incorrect, Lord Diplock in the *United Scientific* case (at p 936) expressly stated that the decision was right, and I respectfully agree.

I turn now to the second issue, and to Mr Wood's four grounds of attack on the validity of the notice in the light of the events that happened.

(1) Common Law Election:

The relevant rule is stated in *Spencer Bower and Turner's Estop-* B *pel by Representation,* 3rd ed at p 324 thus:

> When A in his dealing with B, being at liberty to adopt either of two mutually exclusive steps, proceedings, courses of action, or attitudes, in relation to B, elects to take or adopt one of them and to reject the other, or to "waive" his right in respect thereof, and A's declaration of such election or "waiver" by words, conduct or inaction, influences B to alter his position to his detriment, A is estopped, as against B, from thereafter resorting to the course of action which he has thus intimated his intention of relinquishing, dispensing with or "waiving".

The defendants' argument runs thus. The plaintiff on December 25 1977 had the choice either of demanding rent at the rate of C £18,000 per annum or of initiating a rent review, as it alone could do, by serving a notice under clause 5(2)(a). If the plaintiff had elected for the latter, the rent payable by reason of clause 5(3) was £14,600 per annum until the rent was determined, that being the rent previously payable. The plaintiff elected to demand rent at the rate of £18,000 per annum and the defendants were thereby influenced to pay rent at that rate. In so doing they lost the benefit of having the use of the difference between rent at the rate of £18,000 per annum and rent at the rate of £14,600 until the determination.

The crucial first step in the argument is that the plaintiff at the start of the first review period was put to the choice of electing between two mutually exclusive courses of action. It would appear D that implicit in this is the suggestion that the lessor had to initiate the rent review procedure if at all not later than the first demand for rent for the first review period, although Mr Wood did not in fact rely on clause 5(3) on the first issue. I must examine the rent review provisions in further detail to see whether the plaintiff was in fact faced with such a choice.

The relevant part of the reddendum required the defendants to pay whichever should be the higher of £18,000 or the open market value as determined in accordance with clause 5. The reddendum was not worded as a requirement to pay either £18,000 or at the election of the lessor the open market value as so determined. Thus the rent payable according to the reddendum could not be less than E £18,000 but could be higher if so determined under clause 5. Clause 5 also contained directory language for the determination of the open market value. The machinery laid down for that determination was, first, that the lessor should specify the rental value in a notice; second, if that value was not accepted by the lessee, the lessee within three months of the notice could either agree a different value with the lessor or elect for the determination of the value by an independent surveyor. The final part of clause 5(2) states "every such determination" (ie every determination by the independent surveyor) "shall be subject to the further provisions of the next succeeding sub-clause thereof". So clause 5(3) is expressed to have effect in respect of the determination by the independent surveyor F under clause 5(2)(c), but not in respect of a determination pursuant to clause 5(2)(a) and (b). Clause 5(3) contains a condition precedent to its operation, that is to say, "In the event of the determination by such independent surveyor not being made and communicated to both parties thereto prior to the commencement of the relevant review period". If that condition is satisfied, then in the period defined as "the said interval" beginning with the commencement of the review period and ending on the quarter day following the date on which "such determination" shall have been made and communicated to both parties, the rent is to continue to be paid at the rate payable immediately prior to that period. The reference to such determination can only be a reference to the determination by the independent surveyor. Accordingly, unless

there is such a determination there is no period marked out for the G payment of the rent at the previous rate. It follows that this provision cannot apply at a time when there might be no such determination, for example when the rent review procedures have not yet been initiated at all or when they have been initiated but have not reached the stage of the lessee's election pursuant to clause 5(2)(c) for determination by an independent surveyor. It can apply only when the lessee has elected but the independent surveyor has not yet made his determination and the review period has commenced. If that is right there can be no question of the plaintiff being put to its election with its first rent demand for the first review period or having elected by demanding rent at the rate of £18,000. £18,000 was the correct rate unless and until the rent review procedures had H been initiated and had proceeded as far as the lessee's election under clause 5(2)(c).

In so construing clause 5(3) I am aware that there are difficulties attendant thereto. One is that because clause 5(3) once operated specifies the rent for the whole of the said interval and that rent in respect of the first review period is £14,600, the defendants will (as Mr Lewison accepts) be entitled to a rebate. In contrast to the provision for the payment by the lessee of an additional sum when the rent is determined by the independent surveyor, there is no express provision for the payment by the lessor of any rebate. The possibility of any rebate could of course arise only in respect of the first review period. There is no difficulty in principle in a rate of rent J being applied retrospectively to an earlier date from which the rent was payable (see *C H Bailey Ltd v Memorial Enterprises Ltd* [1974] 1 WLR 728 as approved in the *United Scientific* case), and I do not think that this difficulty compels a different construction of clause 5(3). A second difficulty lies in the use of the word "continue" in that clause. Once rent has been paid at the rate of £18,000 per annum it cannot be said that the rent shall continue to be paid at the rate of £14,600. Again this difficulty arises only in respect of the first review period and I do not think that this provision should be construed differently in relation to the different review periods. In the context of a provision that the rent to be payable in the said interval is to be the same as the rent payable in the period immedi- K ately prior to the said interval, the word "continue" is not inappropriate. Again I do not regard this difficulty as sufficient to outweigh the other considerations pointing towards the construction I favour.

I would add that in any event I am not persuaded that the mere service by the plaintiff of a demand for rent at the rate of £18,000 per annum and the payment by the defendants in accordance with that demand is a sufficient foundation of fact for the defendants to succeed under the doctrine of common law election. There must be an unequivocal demonstration by the party alleged to have made an election of a choice of one alternative and the rejection or waiver of the other alternative. In the context of a provision for a minimum L rent of £18,000 and the possibility of a higher rent being determined once the lessor initiates the rent review procedure, some clearer indication of an election would, in my view, be necessary before the court would hold that the lessor had given up the right to a rent review. Certainly there is no evidence that the defendants thought that the plaintiff had deliberately chosen to abandon that right when demanding rent at the rate of £18,000 per annum. While a plaintiff will not be permitted to blow "hot and cold", he is entitled to blow "hotter", such as when he exercises one right to receive a payment and later exercises another right not inconsistent therewith to receive more (see *Lissenden v CAV Bosch Ltd* [1940] AC 412 at 429). In my judgment this is a case of the plaintiff legitimately blowing hotter. M

Thus despite the ingenuity of the argument and the skill with which it was presented, I cannot accept Mr Wood's submission under this head.

(2) Promissory Estoppel:

The second way in which Mr Wood puts his case on the second issue is on the basis of promissory estoppel. He says that the demand by the plaintiff for rent at the rate of £18,000 was a representation that the plaintiff would not initiate the rent review procedure under clause 5 and that in reliance thereon the defendants paid rent at that rate instead of at the rate of £14,600 per

A annum and thereby the defendants lost the benefit of the use of £3,400 per annum.

I take as a statement of the conditions needed to establish promissory estoppel the following passage from the recent judgment of Oliver LJ in *James v Heim Gallery Ltd* (1981) 41 P & CR 269 at 280 [256 EG 819]:

> In order to found a promissory estoppel, there has first to be found some clear and unequivocal representation, either by words or conduct, that the party claimed to be estopped will not rely upon his strict contractual rights. Secondly, the representation must be made with the intention, or at least the knowledge, that it is to be acted upon by the other party by altering his legal position, and thirdly he must so alter his legal position in reliance upon the
> B representation in such a way that it would be inequitable, or unfair, to permit the party claimed to be estopped from departing from the representation.

If I am right on the question of construction, the argument based on promissory estoppel seems to me to be bound to fail. The plaintiff could not be said to be making any such representation as is alleged as the rent demanded was the correct rent payable and the defendants' right to pay rent at the rate of £14,600 had not arisen in the absence of the initiation of the rent review procedure and the lessor's election. If I am wrong on that construction and the rent was payable only at the rate of £14,600, I am still unable to spell out of the bare demands for rent at the minimum rate payable under the
C reddendum any clear or unequivocal promise by the plaintiff not to serve a notice under clause 5(2)(a). Nor is there any evidence that in complying with the rent demand the defendants relied on any such representation. There is nothing before me to indicate that the defendants thought that the plaintiff was making any promise of any kind (compare *Kammins Ballroom Ltd v Zenith Investments* [1971] AC 850 at 994 per Lord Diplock).

Accordingly, I must reject Mr Wood's submissions under this head, too.

(3) Estoppel in pais:

The third way in which Mr Wood puts his case is on the basis of an
D estoppel in pais. For this, too, he must show a clear and unequivocal representation, but of a fact rather than of a promise as to future conduct, which was intended to induce the defendants to alter their position and on which the defendants relied in altering their position to their detriment. Again Mr Wood relies on the rent demands as being the representation. He says that thereby the plaintiff represented that the rent currently due was rent at the rate of £18,000 and not at the rate of £14,600, the latter being the appropriate rate if the plaintiff intended to preserve the right to initiate the rent review procedure; and that in reliance on that representation the defendants acted to their detriment by paying rent at the higher rate and thereby losing the benefit of the use for the time
E being of the difference between the two rates.

Again this argument cannot run if I am right on the question of the construction of clause 5(3). If I am wrong on that question, I am still unable to spell out of the bare rent demands for the minimum rent provided for by a reddendum an abandonment of the right to serve a notice at any one time initiating a rent review, and again I see no evidence that it was so understood by the defendants. In any event even if the rent demands constituted representations such as to estop the plaintiff from claiming that the rent for the nine quarters was other than £18,000, or was other than at the rate of £18,000 per annum, I do not see that it would prevent the plaintiff from having the rent reviewed in respect of the remainder of the
F first review period.

This attack, therefore, also fails.

(4) Unreasonable delay:

Mr Wood's fourth attack is on the basis of unreasonable delay. He says that there is an implication of law that a notice initiating a rent review must be served within a reasonable time from the date from which rent is to run and that this is a doctrine independent of estoppel. He says that the delay of 2¼ years after the start of the review period before the notice of March 19 1980 was served was unreasonable in the context of a five-year review period and in particular because of two factors. One is the provision for determining the rental value as at the date of determination. The other is the G
payment of rent at the rate of £18,000 for nine quarters instead of at the rate of £14,600. Mr Wood further says that it is unnecessary to show that detriment resulted from the delay, but that if it is necessary he points again to the defendants not having the use of £3,400 per annum.

The authorities on which Mr Wood relied were all cases where the rent review provisions specified a date by which the lessor had to take action though time was not made of the essence of that provision. Accordingly, in those cases it was meaningful to talk of delay occurring after that contractual time-limit had passed (see the *Accuba* case at p 1565 and *Telegraph Properties (Securities) Ltd v Courtaulds Ltd* (1981) 257 EG 1153). In the latter case Foster J H
referred to the dictum of Lord Salmon in the *United Scientific* case at p 951: "Any unreasonable delay caused by the landlords and which is to the tenants' prejudice would prevent the rent being reviewed after the review date". That remark was made in the context of a timetable (in respect of which time was not of the essence) being laid down by a lease. I note that in *Amherst v James Walker Goldsmith and Silversmiths Ltd* (1980) 254 EG 123 at p 125 Sir David Cairns expressed "no more than a tentative opinion" that a notice after a contractual time-limit had passed might well have to be within a reasonable time; while in *James v Heim Gallery Ltd (supra)* at p 278, when argument had been addressed on the basis of abandonment, Buckley LJ was similarly tentative: "If a J
party who unreasonably delays in asserting some right can on that ground, distinct from promissory estoppel, be held to have abandoned that right so as to be barred from asserting it, this is not, in my judgment, such a case". Let me assume, without deciding, that there is a principle that unreasonable delay after a contractual time-limit has passed prevents the party delaying from doing what he had agreed to do within that time-limit, even though time was not made of the essence thereof. It nevertheless does not follow that such a principle has any application to the present case where so far from there being a time-limit specified for the service of the lessor's notice under clause 5(2)(a), the wording of the clause is that the lessor may serve such a notice "at any time". I have been referred to K
no authority that supports the principle addressed to me by Mr Wood that the notice must be served within a reasonable time from the date from which rent is to run. In the circumstances I accept Mr Lewison's submission that it is not meaningful to describe the interval of time between the start of the review period and the service of the notice as a delay, still less that it should be categorised as unreasonable. Accordingly this argument, too, on the part of the defendants fails.

In the result I shall make the declarations as requested in the originating summons herein.

L

CHANCERY DIVISION

October 12 1981

(Before John MOWBRAY QC, sitting as a deputy judge of the division)

AMHERST v JAMES WALKER (GOLDSMITH & SILVERSMITH) LTD

Estates Gazette May 1 1982

(1981) 262 EG 442 M

Rent review clause in lease — Further proceedings following decision of Court of Appeal that time was not of the essence for the purpose of a "trigger" notice — Court of Appeal had indicated that, although time was not of the essence, the notice might none the less be invalid on account of delay — The present proceedings instituted by the landlord required a determination of this point — Review of case law by deputy judge — In order to establish that the notice was invalid by reason of delay it was necessary to show either that the delay was both unreasonable and caused prejudice or hardship to the tenants or that it was so great as to be evidence of abandonment — The tenants in the present case had not shown

prejudice or hardship; in fact they had benefited by the delay before the fixing of the new rent — As to abandonment, the circumstances did not show an intention to abandon — Although the due date for the "trigger" notice was December 25 1974, any delay before the House of Lords decision on March 23 1977 in the *United Scientific Holdings* case had to be discounted in view of the uncertainty of the law — Subsequent delay was something under 14½ months, a period during which advice had to be sought and consultations had to take place as to the effects of the House of Lords decision — Delay was not long enough to be evidence of an intention to abandon landlord's rights — Declaration in favour of landlord.

This was an originating summons in which the plaintiff was the landlord, William John Amherst, and the defendants the tenants, James Walker (Goldsmith & Silversmith) Ltd. The proceedings concerned the lease of business premises at 449 High Road, Wembley. They followed upon a previous originating summons between the same parties which came before Judge Mervyn Davies (as he then was), sitting as a deputy High Court judge, and subsequently before the Court of Appeal. The Court of Appeal (1980) 254 EG 123 upheld the judge's decision that time was not of the essence for the purpose of the landlord's "trigger" notice in the review clause of the lease. The present originating summons sought a declaration that the landlord was entitled to receive, as from June 24 1975, rent at a rate to be determined by arbitration, notwithstanding that the notice was not given until May 10 1979.

John Hamilton (instructed by Macdonald, Stacey & Co) appeared on behalf of the plaintiff; R J Moshi (instructed by Bulcraig & Davis) represented the defendants.

Giving judgment, JOHN MOWBRAY QC said: The plaintiff in this originating summons is the landlord, and the defendant company is the tenant by assignment of business premises in Wembley, under a lease for 28 years from June 24 1961. The rent for the first 14 years was £2,500. Thereafter, during the 14 years from June 24 1975, the rent was to be the market rent, to be assessed as follows:

(a) Such assessment shall be made in the first instance by the lessor and submitted to the lessee for approval in writing on or before the 25th day of December 1974;
(b) In the event of the parties hereto failing to reach such agreement as aforesaid on or before the date appointed (in respect of which time is to be deemed to be of the essence of the contract) then the yearly rent for the second period shall be fixed or assessed by an independent surveyor appointed for that purpose by the parties hereto, or failing agreement as to such appointment by the 25th day of January 1975 (time in this respect to be deemed to be of the essence of the contract) then by an independent surveyor appointed for that purpose by the President for the time being of the Royal Institution of Chartered Surveyors. The assessment fixed by the independent surveyor shall be communicated to the lessor and lessee in writing and immediately upon such communication the yearly rent so assessed as a reasonable rent for the second period or £2,500 per annum, whichever shall be the higher, shall be the yearly rent payable for the second period under the terms hereof.

On a previous originating summons between the same parties (1978 A No 3087) the Court of Appeal has upheld a decision by Judge Mervyn Davies, sitting as an additional judge of this court, that time for giving the landlord's assessment notice under paragraph (a) of that proviso was not of the essence. The Court of Appeal indicated that the assessment notice might none the less be invalid on account of delay, but that if it was not invalid then either party could go straight to the president of the RICS and ask for the appointment of an independent surveyor.

I have to decide whether the delay has been such as, in the relevant circumstances of the case, to invalidate the assessment notice, which was given on May 10 1979. Mr Moshi for the defendant company concedes that this question is not *res judicata*. He contended that on the previous originating summons it was decided that the giving of the landlord's assessment notice under paragraph (a) was a condition precedent to the rent review. Mr Hamilton for the plaintiff, rightly in my view, did not contend the contrary.

The circumstances of the case are these. The landlord's family solicitors allowed the due date, December 25 1974, to pass without giving the landlord's assessment notice under paragraph (a) of the proviso. On January 25 1975 they wrote a letter to the tenant company, which was not an assessment notice within that paragraph either, inviting the tenant to agree on the appointment of an arbitrator. The tenant would not co-operate in that, and on February 10 1975 the landlords asked the president of the RICS to appoint an arbitrator. The tenant objected and the president, in a letter from the then RICS administrative secretary dated March 5 1975, declined to act but indicated that if a court ruling were obtained setting aside the time-limit he would be pleased to make an appointment. In those days there was a body of first instance and Court of Appeal authority distinguishing between "trigger" or option provisions in rent review clauses on the one hand and mere "machinery" provisions on the other and holding that time was of the essence of trigger provisions. Paragraph (a) of the proviso here was a trigger provision for the purpose of those authorities, or at least could very well have been considered a trigger provision at the time. The landlord took no steps towards obtaining a court order nor did he communicate with the tenant for some time.

On March 23 1977 the House of Lords allowed landlords' appeals in *United Scientific Holdings Ltd v Burnley Borough Council* and *Cheapside Land Development Co Ltd v Messels Service Co* (later reported in [1978] AC 904). I shall call the decision in both those cases "the House of Lords decision". The House of Lords decided, contrary to the existing authorities in the lower courts, that in the absence of a contrary indication time was not of the essence of trigger provisions in rent review clauses. I am told that this was reported in *The Times* newspaper on March 24 1977 and it was subsequently reported in different law reports. It reached the *Weekly Law Reports* in April. On June 2 1978 the landlord's new solicitors wrote to the tenant's solicitors asking them, in the light of the House of Lords decision, to withdraw their objection to the appointment of an arbitrator. This they declined to do, and on October 12 1978 the landlord issued the previous originating summons (1978 A No 3087) that I have referred to. No assessment notice had been given, and none had been given on May 9 1979 when the originating summons came on for hearing before Judge Mervyn Davies in this court. Mr Hamilton for the landlord nobly shoulders the responsibility for this. He took the view that he could go straight to arbitration without such a notice. When it appeared that the learned judge took a different view, an assessment notice was given. It was given on May 10 1979, as I have said. That was before judgment, but after the originating summons, and the learned judge said, "It seems on the facts of this case as now before me that the landlord has delayed unreasonably in not serving the subclause (a) notice. It should have been served at least before the commencement of these proceedings."

Because of the way the hearing had gone, the learned judge declined to decide whether that notice was with or without effect, and dismissed the rest of the originating summons, with costs, but he none the less declared that time for giving the notice was not of the essence. That was the order that was approved by the Court of Appeal, which gave judgment on January 18 1980. It was enough to persuade the president of the RICS to appoint an arbitrator, which he did on March 20 1980. The landlord and tenant are discussing the amount of the market rent and the arbitrator has not yet made an award.

On October 1 1980 the landlord issued the present originating summons, seeking a declaration that he is entitled to receive, as from June 24 1975, rent at a rate to be determined by arbitration, notwithstanding that his assessment notice was not given until May 10 1979, and other relief.

Mr Moshi for the defendant, in a persuasively moderate argument, attacked the May 10 1979 assessment notice in two ways. First, he argued that the landlord must be taken to have abandoned his right to review the rent, after the RICS letter of March 5 1975 declining to appoint an arbitrator, and before the landlord's solicitors' letter of June 2 1978 reviving the claim. Secondly, he argued that the landlord's delay between December 25 1974, the due date, and May 10 1979, when he finally gave his assessment notice, was unreasonable and that such unreasonable delay of itself invalidated the notice. I use the words "unreasonable delay" here to mean delay which is beyond reason, but not so long as to indicate any abandonment of his rights by the party guilty of it. For

A completeness I should add that the defendant did not rely on any point of estoppel.

It is convenient to deal first with Mr Moshi's second line of attack. I adopt the finding of Judge Mervyn Davies that the landlord's delay in giving his assessment notice was unreasonable in the sense I have mentioned. If it adds anything, having considered the circumstances drawn to my attention in argument I respectfully agree with that finding. More important, Sir David Cairns, towards the end of his leading judgment in the Court of Appeal on the previous originating summons, expressly endorsed the reasoning of the learned judge with the rest of his reasons. But in my judgment even unreasonable delay by the landlord in giving **B** his notice does not invalidate the notice, unless the delay prejudiced the tenant or caused him hardship. There are four reasons.

The first is the analogy of contracts for the sale of land. There, where time for completion is not of the essence, even unreasonable delay in completion by one party does not excuse the other from his contractual duty to complete: it merely entitles him to serve peremptory notice on the party in default to complete within a further reasonable time.

Secondly, Lord Salmon said in the *United Scientific* case at p 951(E) and again at p 953(G), and p 956 (above C) that if a landlord is guilty of unreasonable delay *which prejudices the tenant* the landlord forfeits his rights. Mr Moshi submitted that this is **C** *obiter* and does not mean that prejudice to the tenant is a vital element. But it must be, or Lord Salmon would not have proceeded on p 956 to consider whether there was any prejudice in the case he was considering. He found, at p 956(C-D) that in that case (as in this and others) the tenants were benefited by the delay, because they had the use of the extra rent interest-free until the new rent was fixed. (Oliver L J added at the end of his judgment in *James v Heim Gallery (London) Ltd* (1980) 256 ESTATES GAZETTE 819 at p 826 that when the tenant does pay the difference it is in depreciated currency.) Furthermore, the sentence in Lord Salmon's speech between G and H on p 953 of the report could not be explained in the way that Mr Moshi suggested.

D Thirdly, Slade J in the *obiter* remark numbered (3) towards the end of his unreported judgment of May 4 1978 in *Kirkland Properties Ltd v GRA Developments Ltd* made it clear that he thought prejudice to the tenant was a necessary element.

Fourthly, Judge Mervyn Davies in the present originating summons seems to have thought so, since he dismissed most of the summons on the tenant's application partly on the ground that the tenant had not had an opportunity to prove prejudice, and Sir David Cairns in the Court of Appeal endorsed this and all his reasoning.

The defendant argued here that, if detriment or hardship is required, the effect is to punish a tenant who is foresighted and pru- **E** dent enough to set aside the extra rent and favour a feckless tenant who does not. But if there is no detriment, there is no punishment. The true explanation is that equity intervenes only where it is necessary to protect the tenant from some actual misfortune induced by the landlord's delay.

In *Telegraph Properties (Securities) Ltd v Courtaulds Ltd* (1981) 257 ESTATES GAZETTE 1153, where there was apparently total silence about the review from the landlord between the due date for notice on October 1 1972 and the notice of October 11 1978, Foster J said (at p 1154) "In my judgment the plaintiff has been guilty of such a delay as to make it unreasonable for him to call on the defendant for a rent review and to do so would be of necessity un- **F** fair on the defendant". As even quite a considerable delay normally benefits the tenant, I think Foster J must have been relying on something stronger which would make it unfair for the landlord to change tack. I think that must be treated as a case where there was not only unreasonable delay but such delay as indicated to the tenant that the landlord had abandoned his right to a review. That would make it unfair on the tenant to treat a subsequent trigger notice as valid. The learned judge did not merely say that the *delay* was unreasonable; he said it was such a delay as to make it unreasonable and of necessity unfair for the landlord to call for a review. I think only delay amounting to abandonment could have that as a necessary effect where there was no specific detriment to the tenant. Moreover, Foster J's reference, just after the passage I **G** quoted, to the *Heim Gallery* case reads to me as though he was considering the abandonment aspect of that case.

I conclude then that even unreasonable delay by a landlord in serving his trigger notice, if it was not so long as to amount to evidence of abandonment, would not invalidate the notice, unless it also prejudiced the tenant or caused him hardship. The tenant had an opportunity to prove detriment on this originating summons, but no such evidence is before me. In my judgment the tenant was benefited in the way I have mentioned. Therefore this first line of attack fails and I turn to the question of abandonment.

In *James v Heim Gallery (London) Ltd* Buckley LJ left open the question whether long delay could lose the landlord the right to **H** serve his notice on the ground of abandonment, but I think that a landlord's delay can disentitle him to serve a trigger notice if it has been, not only unreasonable in the sense I have mentioned, but so long and inexplicable as to amount to sufficient evidence that he has abandoned his right to a review. Again there are four reasons and again I start with the analogy of a contract for the sale of land where time is not of the essence.

Delay in completion in such a case may be so extended as to indicate abandonment of the contract by the party guilty of it. If that was the vendor, the purchaser could recover his deposit at common law and equity would not interfere. This was not a mere question of laches barring equitable relief, but a case of abandonment of **J** contractual rights.

Secondly, Slade J said in the *obiter* remark numbered (5) in the unreported judgment in the *Kirkland Properties* case I have referred to that the right to enforce a rent review could be lost, even in the absence of hardship to the tenant, by delay in enforcing it so great as to amount to evidence of abandonment. He cited some sale of land cases.

Thirdly, I think that, as I said, the decision of Foster J in *Telegraph Properties (Securities) Ltd v Courtaulds Ltd* ought to be treated as based on abandonment.

Fourthly, there is the judgment of Goff J (as he then was) in *Accuba Ltd v Allied Shoe Repairs Ltd* [1975] 1 WLR 1559. That case **K** was decided before the House of Lords decision, but Goff J treated the relevant provision as one of which time was not of the essence; so, as Mr Moshi rightly said, the persuasive force of this judgment is not substantially diminished by the House of Lords decision. Buckley L J treated this decision as based on abandonment in the *Heim Gallery* case, and I think I should do so too. Goff J used the word "unreasonable" to describe the necessary delay, but I think he must have been using the word to mean something more than I am — something amounting to abandonment. Buckley L J in the *Heim Gallery* case also used the word in that strong sense.

Before turning to the facts I should add that, though I think the analogy of contracts for the sale of land is persuasive, the cir- **L** cumstances are different. In delaying completion a party to such a contract is breaking his contractual duties, but a landlord is under no contractual duty to review his rent. Also, with a rent review there is a continuing relationship anyhow and the review usually benefits the landlord alone. So I consider that a shorter delay will indicate abandonment of such a contract than would be needed to show abandonment of a right to a review.

Then was the landlord's delay here enough to indicate abandonment? Delay before March 23 1977, the date of the House of Lords decision, is irrelevant for this purpose except perhaps as the historical background of the delay. That was the reason why the Court of Appeal in the *Heim Gallery* case held the delayed notice in **M** that case to be valid. The landlord is not to be assumed to have intended to abandon a right that no one knew he had got. If Foster J in the *Telegraph* case regarded the period of delay before the House of Lords decision as more than historical background, I do not think that can stand with the Court of Appeal decision. If we discount the period before the House of Lords decision in the present case, that leaves at most the period between March 23 1977 and the landlord's new solicitors' letter of June 2 1978, that is, under 14½ months.

The state of the authorities is this. Foster J said in the *Telegraph* case that a year and a half's delay after the House of Lords deci-

A sion was too long, but it followed a period of 4½ years when the landlord took no steps at all, and Foster J seems to have taken that into account. The previous delay here was only from about March 1975, just over two years, if delay before the House of Lords decision is relevant at all. In the *Accuba* case Goff J held valid a notice given after 18 months' delay in the context of a seven-year review period, as opposed to the 14-year period in the present case. He said it was on the borderline. The *Accuba* case is not mentioned in the report of the *Telegraph* case, though it was mentioned in the *Heim Gallery* case, which was cited in the *Telegraph* case.

In my view quite a long time should be allowed after such a B distinct change in the prevailing view of the law as was brought about by the House of Lords decision here, before further delay should be taken to indicate abandonment. The landlord's advisers need to learn of the decision, turn their minds to its application to the lease in question, advise their client, and put things in train for renewing the claim to review the rent. There may also be other possible claims, for instance against an agent or solicitor who allows the due date for a notice to slip by, which will need to be considered and probably discussed with those responsible, and even their insurers, before the landlord's solicitors can be sure precisely what approach should be made to the tenant.

Mr Moshi criticised the landlord's evidence in the present case C for not dealing with any of these points, or even making a blanket statement that the landlord had not abandoned his rights, though abandonment was argued in the tenant's evidence. I do not consider this criticism to be justified. The tenant's evidence contained no new or contradictory facts, and so no reply was called for. Like Goff J in his case I think the delay here is on the borderline, but on balance I do not consider that it is quite long enough by itself to show that the landlord intended to abandon his valuable right to a reviewed rent for the whole 14 years. So I am prepared to make a declaration favourable to the landlord and will hear counsel on the appropriate wording.

The judge awarded costs against the defendant company.

D

CHANCERY DIVISION

July 28 1981

(Before Judge Mervyn DAVIES QC, sitting as a judge of the High Court)

ESSO PETROLEUM CO LTD v ANTHONY GIBBS FINANCIAL SERVICES LTD

Estates Gazette May 15 1982

E **(1981) 262 EG 661**

Landlord and tenant — Rent review clause in underlease of commercial premises — Unusual clause providing for a two-year period during which a notice calling for a rent review could be served by the landlords on the tenants, but with a provision that if the landlords failed to give notice during that period they had the right "at any time thereafter" to review the rent on serving a month's notice — Landlords failed to serve a notice during the two-year period, but did serve a notice under the latter provision — Held, however, that, although no notice was served during the two-year period, a review had taken place as a result of informal discussions F ending in agreement between the agents of the parties, who thereby waived the provisions requiring formal notice — Such a review precluded a subsequent review pursuant to the notice under the "at any time thereafter" provision — Judgment given in favour of plaintiff tenants on their main submission; their alternative argument based on estoppel rejected

In this action the plaintiff company, Esso Petroleum Co Ltd, sought declarations to the effect that under the terms of an underlease, of which the interest of the underlessor was now vested in the defendants, Anthony Gibbs Financial Services Ltd, and in the events which had happened, the defendants were not entitled to claim a further increase of rent until 1984.

Michael Essayan QC and K M J Lewison (instructed by Durrant G Piesse) appeared on behalf of the plaintiffs; Derek Wood QC and J C Harper (instructed by Linklaters & Paines) represented the defendants.

Giving judgment, JUDGE MERVYN DAVIES QC said: This action concerns a rent review clause in an underlease dated January 29 1976. At all material times the plaintiff company, "Esso", has been the tenant of the underlease and the defendant company, "Gibbs", has been the landlord.

On October 9 1979, Gibbs gave Esso notice, under the rent review clause in the underlease, as a first step towards seeking an increase in rent. Gibbs say that the notice is a valid notice; Esso say it is not. In consequence Esso, on January 11 1980, issued a writ H claiming declarations to the effect that no increase in rent is permissible until 1984. Esso's claims are founded not only on their construction of the provisions of the underlease but also upon allegations that Gibbs are estopped from seeking any rent review until 1984.

I proceed to set out the facts. One starts with an underlease dated May 11 1973 between Capital & Counties Property Co Ltd (landlord) of the first part, Anthony Gibbs (Personal Financial Planning) Ltd (tenant) of the second part, and Anthony Gibbs & Son Ltd (surety) of the third part. During the hearing it was convenient to refer to this underlease as the "headlease", and I so refer to it now. The headlease demises parts of the basement, ground, J first, second, third, fourth, fifth and sixth floors of the building, comprising 7-10 Hanover Square and 17-19 Princes Street, W1. The parts in question are shown red on a plan annexed to the underlease. Such parts comprise — for present purposes, as I understand — the second, third and fourth floors of 7-10 Hanover Square. The term of the headlease is 25 years, from September 29 1972, so the term will run out on September 29 1997. The rent was £193,750, payable quarterly. There is also an annual service charge. Clause 3 of the headlease is a rent review clause. It provides that on July 6 1978, 1984, 1988 and 1992, the rent "shall be reviewed", and that from and after each review rent shall be such sum as shall be agreed to be the fair rack rental market value for a term K equivalent to the then unexpired residue of the term, without taking account of any goodwill of the tenant's business or of certain tenant's improvements. If the landlord and the tenant do not agree on a new rent "by the relevant date of review, then and in any such case, the question shall, as soon as practicable, be referred by either party" to an agreed nominee or to an expert nominated by the president of the Royal Institution of Chartered Surveyors.

Legal & General Assurance Society Ltd replaced Capital & Counties Property Co Ltd as the landlords of the headlease. As I understand, the tenant under the headlease intended an office user of the demised premises, either by itself or by one of its associated companies. Any such user was brief because in January 1976 there L were executed two underleases comprising the whole of the premises demised by the headlease. By an underlease dated January 14 1976, made between Anthony Gibbs (Personal Financial Planning) Ltd of the one part — that is "Gibbs PFP" — and Philips Electronic and Associated Industries Ltd — "Philips" of the other part — Gibbs PFP demised to Philips what was, in effect, the second floor of the building comprised in the headlease — that is of 7-10 Hanover Square — for a term beginning on May 1 1975 and ending on September 27 1997. Thus Philips were to hold for the rest of the term of the headlease less two days. The rent was £77,500 and, as well, a service charge was payable. Clause 6 of the underlease is a rent review clause. It is in many respects the same as M the rent review clause in the underlease I am about to mention.

It is common ground that by May 1978 the defendant Gibbs company was in the place of Gibbs PFP as the landlord of the underlease and that the plaintiff Esso was the tenant in place of Philips. I was not referred to the documents effecting these changes of ownership.

By an underlease dated January 29 1976 and made between Gibbs PFP of the one part, and the defendant Esso company of the other part, Gibbs PFP demised to Esso what was, in effect, the third and fourth floors of 7-10 Hanover Square for a term from March 25 1975 until July 6 1988. So the term was for about 12 years, and, therefore, less than the term of the (first) underlease dated January 14 1976. The rent was £197,225, with a service charge. This rent for the third and fourth floors was a little out of line with the rent for the second floor already mentioned. There were, I was told, good reasons for that, but I do not see that they bear on the issues before me.

This second underlease contains a rent review clause, which I must set out in full. The rent review clause is part of clause 5 in the underlease. Clause 5 begins with the words "PROVIDED ALWAYS and IT IS HEREBY AGREED and DECLARED that:" There follow nine subclauses not concerned with rent review and then there is subclause (x) in these terms:

> (x) (a) The Landlord shall by notice in writing given to the Tenant at any time during the review period (as hereinafter defined) have the right to call for a review of the yearly rent payable to the landlord in respect of the demised premises and if upon review it shall be found that the commercial yearly rent (as hereinafter defined) of the demised premises at the review date (as hereinafter defined) shall be greater in amount than the principal rent hereinbefore reserved and for the time being payable then as from the review date in question the amount of the principal rent payable to the Landlord in respect of the demised premises shall be increased in manner following and thenceforth the said rent shall be paid by the Tenant at such increased rate to the Landlord;
>
> (b) For the purpose of this subclause the following expressions bear the meanings assigned to them as follows:
>
> "the review period" means in the first instance the period commencing Twelve months immediately preceding July 6 1978 and Twelve months ending immediately following July 6 1978 and in the second instance the period commencing Twelve months immediately preceding July 6 1984 and Twelve months immediately following July 6 1984;
>
> "the review date" means in the first instance July 6 1978 and in the second instance July 6 1984;
>
> "the commercial yearly rent" means the best yearly rent at which the demised premises assuming the due performance and observance of the covenants on the part of the Tenant and conditions contained in these presents might be expected to be let at the review date by a willing Landlord in the open market with vacant possession and without premium or any other consideration than that evidenced by execution of a lease thereof to a willing Tenant for a term equal to the residue then unexpired of the term hereby granted by a lease in the same terms in all other respects as these presents there being disregarded:
>
> (i) any effect on rent of the fact that the Tenant or any other person may have been in occupation of the demised premises; and
>
> (ii) any goodwill attached to the demised premises by reason of the user of the demised premises by the Tenant or any other person or the purposes for which the demised premises may have been used;
>
> (iii) any effect on rent of any improvement carried out by the Tenant otherwise than in pursuance of an obligation under this Underlease.
>
> (c) With effect from the review date the amount of the principal rent payable under this Underlease shall be increased to the amount of the commercial yearly rent PROVIDED ALWAYS and IT IS HEREBY AGREED that the yearly rent payable hereunder from and after the review date during the residue of the term hereby granted shall at no time be less in amount than the yearly rent for the time being subject to review;
>
> (d) Such review as aforesaid shall first be made by the Landlord and the Tenant or their respective Surveyors in collaboration but if no agreement as to the amount of the increase (if any) to be made in the said yearly rent shall have been reached between the parties within two months after the date of the Landlord's notice calling for such review the question whether there shall be an increase in the said yearly rent and if so what will be the amount of the increased yearly rent shall be referred at the request of either party to the decision of a sole arbitrator who shall be appointed (in default of agreement between the parties) on the request of the Landlord by the President for the time being of the Royal Institution of Chartered Surveyors and the costs of the reference to him shall be in the award of such arbitrator and who shall proceed in accordance with the Arbitration Act 1950 or any statutory modification or re-enactment thereof for the time being in force;
>
> (e) Should the amount of the increase (if any) to be made in the said yearly rent not be agreed or decided before the review date in question the Tenant shall pending such agreement or decision continue to pay rent at the rate for the time being applicable and subject to review and on such agreement or decision being reached or made any necessary adjustment shall forthwith be made retrospectively in favour of the Landlord from such review date and the difference between the amount of rent actually received and the amount so due to the Landlord in respect of the period down to the date of payment of such difference to the Landlord shall be paid by the Tenant to the Landlord on the due date for payment of rent next following the demand made;
>
> (f) Notwithstanding subclause (a) of this Clause if the Landlord fails to give notice calling for a review of the yearly rent payable hereunder during the review period the Landlord shall at any time thereafter have the right to review the said rent upon giving to the Tenant not less than one month's notice in writing of its intention so to do and if the Landlord shall give to the Tenant a notice as aforesaid then from and after the date therein specified (which shall be a date not earlier than one month from the date of service of the said notice) the yearly rent then payable shall be increased to an amount which shall represent the commercial yearly rent of the demised premises at the date specified in the said notice and the amount thereof shall be agreed between the Landlord and the Tenant or determined in accordance with the provisions of subclause (d) of this Clause;
>
> (g) Should any such review at any time be restrained or restricted by any Act of Parliament instrument regulation order or public or local policy then upon the same being raised, relaxed or modified and on every such occasion the Landlord shall be at liberty to call for an intermediate review in manner hereinbefore provided mutatis mutandis by not less than three months' notice in writing to the Tenant for which purpose the date of service of such notice shall constitute the commencement of the review period and the date of expiration thereof shall constitute the review date provided however that the principal rent shall not be reviewed more than twice during the term hereby created.

The words I have underlined in 5 (x) (a) do not appear in the underlease, but counsel before me agreed that the words underlined ought to be inserted into subclause (a).

By mid-1978, so I was told, the defendant Gibbs company had replaced Gibbs PFP as the landlord of this second underlease. Accordingly, the position in 1978, for practical purposes, was:

(1) Legal & General were the head landlords of the second, third and fourth floors of 7-10 Hanover Square; the tenants were the defendant Gibbs company.

(2) There was an underlease of the second floor; Gibbs were the landlords and Esso were the tenants.

(3) There was an underlease of the third and fourth floors; Gibbs were the landlords and Esso were the tenants.

It will be appreciated that July 6 1978 was a "review date" for the headlease rent and as well for the rent payable under the underleases. But in the case of the underleases it was open to the landlord to serve a review notice not only 12 months before July 6 1978 but also 12 months after that date. (See in the case of the underlease dated January 29 1976 clause 5 (x) (a) read with the definition of "the review period".) The increase in the headlease rent was agreed following a notice dated February 16 1978 served by Legal & General on the defendants and said to be served pursuant to clause 3 of the headlease, followed by conversations between representatives for Legal & General and Mr David Rose, a chartered surveyor and a senior partner in the firm of John D Wood, acting for the defendant company. In a letter dated August 8 1978 Legal & General confirmed a new rent of £290,000. This sum appears to have been orally agreed on July 26 1978 during a telephone conversation with Mr Rose and the surveyor acting for Legal & General.

As well as negotiating with Legal & General Mr Rose was also negotiating with Esso respecting the rents of the underleases. Initially Mr Rose was in contact with Mr D G Dawson, a chartered surveyor who works for Esso. They came to no agreement and Mr D J W Larkin, a chartered surveyor of the firm of Jones Lang Wootton, on July 25 1978 took over negotiations on Esso's behalf. So Mr Rose and Mr Larkin had under consideration the rents of the underleases, but such consideration was not pursuant to any notice served by Gibbs pursuant to 5 (x) (a) of the underlease dated

A January 29 1976, or any notice served pursuant to the corresponding provision of the other underlease. No such notices had been served. Mr Rose realised that Gibbs had at least until July 6 1979 in which to serve such notices. Mr Larkin was very quickly off the mark. He and Mr Rose spoke on the telephone on July 25 or 26 1978. The surveyors agreed together a new rent for the second floor, that is the premises of the underlease dated January 14 1976 in the sum of £110,000 a year. They were also in agreement in respect of the third and fourth floors comprised in the underlease, dated January 29 1976. Their view was that no increase in rent was appropriate in that the rent currently payable, £197,225, was in excess of the current market rent. Mr Rose reported to Gibbs as
B follows in a letter dated July 26 1978:

Further to my letter of yesterday's date, Jones, Lang, Wootton have spoken with Esso and confirmed the review rent of £110,000 per annum, subject to contract and without prejudice.

The solicitors acting for Esso are Durrant Piesse (Michael Scott). Formal notice should now be served on Esso in accordance with the provisions in the underlease, dated January 14 1976 between yourselves and Philips. Perhaps this could be done by your solicitors to Esso's solicitors. You will see that the present rent reserved in this underlease is £77,500 a year.

In respect of the third and fourth floors, this is covered by an underlease dated January 29 1976 between yourselves and Esso at a reserved market rent of £197,225. This rent is in excess of today's market rent and therefore in excess of the sum agreed to be paid to Legal & General at the review date.
C I assume there is no point therefore in serving formal notice on Esso in respect of these two floors.

On the same day, Mr Rose wrote to Mr Larkin:

Subject to lease and without prejudice I confirm that on behalf of our respective clients, we have agreed a review rent of £110,000 a year in accordance with the provisions of the underlease dated January 14 1976 between Gibbs and Philips, of which your clients took an assignment during the past year ...

Mr Larkin reported to Mr Dawson of Esso as follows on July 28 1978:

Further to my letter of July 25 I had discussions with Mr D M Rose of John D Wood acting for your landlords at the above.
D I confirm our telephone conversation in that Mr Rose and I have agreed subject to lease and approval of your company and without prejudice, that the revised rent for the second floor and car parking accommodation etc will be £110,000 per annum exclusive with effect from July 6 1978.

Based on a floor area of about 8,000 sq ft and 32 car parking spaces, this equates to approximately £11.50 per sq ft on the office accommodation, £500 per space for the car parking and about £3 per sq ft for the storage accommodation in the basement. You will appreciate that the short time we have been involved in these negotiations has precluded us from carrying out our normal investigations to check and agree the floor areas and indeed confirm the number of car parking spaces, but I understand that you have satisfied yourself on these aspects.

On the same day Mr Larkin wrote to Mr Rose:

E Thank you for your letter of July 26 confirming our recent telephone conversations. You will appreciate the figure we have agreed is subject to our client's approval, as well as being subject to lease and being without prejudice and I am, therefore, seeking the appropriate instructions from our clients.

As appears above, Mr Larkin in his letter made no reference to any agreement respecting the underlease dated January 29 1976 or as to the third and fourth floors. However, in his report, within the Esso Group, Mr Dawson, in a letter dated August 8 1978 wrote as follows:

Further to my letter of June 22 Jones, Lang, Wootton have now carried out
F negotiations on our behalf with the agents to the landlords and have recommended the following:

That the rent of the second floor and the basement be agreed in the sum of £110,000 per annum with effect from the date of the rent review of July 6 1978. This is based on a figure of approximately £11.50 per sq ft on the office accommodation, £500 per space for the car parking and about £3 per sq ft for the storage accommodation in the basement. This is an increase from the present level of £77,500 per annum.

That the rent of the third and fourth floors remains at the existing level of £197,225 per annum, as the current rental value is below the existing rental value, and the rent review provisions in the lease are upward only.

No further steps were then taken respecting the underlease dated January 29 1976, but the matter of the increased rent payable under the underlease dated January 14 1976 of the second floor G was finalised by a memorandum on the underlease. No difficulty has ensued respecting the underlease dated January 14 1976. Difficulties arise in respect of the underlease dated January 29 1976 of the third and fourth floors.

Following these discussions in July 1978 about the rents of the second, third and fourth floors, there were in the following November other discussions. These other discussions arose when Mr Rose, on November 3 1978, wrote to Mr Larkin informing him that Gibbs were considering selling their interest under the headlease and enquiring whether Esso were interested in stepping into Gibbs' shoes. It is said by Esso that at this time Gibbs represented to Esso that the rent under the underlease dated H January 29 1976 must remain unchanged until 1984 and that Gibbs are now estopped from saying otherwise. I shall set out the situation concerning this alleged estoppel more fully later. I mention the matter at this stage for the sake of setting out the material events in the order of time.

Esso did not acquire the Gibbs interest. In 1979 Esso decided to sell the underleases and they entered into negotiations with Trust House Forte. Messrs Paisner, the solicitors for Trust House Forte, asked some awkward questions. On August 20 1979 their letter to Durrant Piesse, the solicitors for Esso, included the following:

(a) It would appear that the rent under the lease of the third and fourth floors has not been reviewed. Is there any evidence as to whether the J landlord has deliberately waived an increase or whether it has merely overlooked it?

(b) Has any notice been given under clause 5 (x) (f) or is the vendor aware of any intention by the landlord to serve such notice?

At that time it was too late for Gibbs to serve any notice under 5 (x) (a), but the possibility of a notice under 5 (x) (f) was brought plainly into the light. In the event, Gibbs did serve a 5 (x) (f) notice. It is the notice dated October 9 1979 I have already mentioned. It is in these terms:

On behalf of our clients, Anthony Gibbs Financial Services Ltd, your landlord under the above underlease, we hereby give you notice pursuant to clause 5 (x) (f) of the above underlease of our client's intention to review the K yearly rent payable under the above underlease and that from and after November 16 1979 the yearly rent payable shall be increased to an amount which will represent the commercial yearly rent of the demised premises at November 16 1979. Our clients or their agents will be in touch with you shortly to discuss the matter and we hope that agreement can be reached.

The "above underlease" there mentioned is the underlease dated January 29 1976. The underlease in question has in fact been assigned to Trust House Forte, but on terms which did not give rise to any need to join Trust House Forte in this action.

I now turn to the pleadings. It is convenient first to mention the Gibbs counterclaim. It is for a declaration that Gibbs have validly exercised their power to compel a review of the rent reserved by the L underlease dated January 29 1976 in accordance with 5 (x) (f) and that the rent payable from November 16 1979, the date mentioned in the notice set out above, is the commercial yearly rent of the third and fourth floors at November 16 1979, agreed or determined in accordance with clause 5 (x).

That was the Gibbs' attitude when Esso issued their writ on January 11 1980. I need not mention the statement of claims, save to say that it enabled Mr Essayan, for Esso, to make the following four submissions:

(1) On the true construction of clause 5 (x) of the underlease dated January 29 1976 and in the events which have happened there was an effective review of the rent of the third and fourth floors, for the purposes of M clause 5 (x) in July 1978; that review precludes Gibbs from seeking to review the rent again before July 6 1984.

(2) If there was no such effective review in July 1978, this was because the defendants deliberately abstained from carrying out such review, and the defendants could not thereafter avail themselves of 5 (x) (f) which only applies when the landlord has by inadvertence failed to give notice under 5 (x) (a).

(3) In any event Gibbs are estopped from seeking an increase in rent before July 6 1984, by reason of their having made a representation to Esso that the rent could not be increased before that date. The representation was said to be a representation of fact as regards a private right between Esso and Gibbs.

A (4) If wrong on (1) to (3) above, so that Gibbs was entitled to serve a 5 (x) (f) notice, then the relevant date for the fixing of rent is July 6 1978, and not November 16 1979, such rent to be payable from November 16 1979.

I take the first submission. Clause 5 (x) is a rent review clause. It is not nowadays to be construed in the strict fashion that prevailed before the case of *United Scientific Holdings Ltd* v *Burnley Borough Council* [1978] AC 904. It is to be construed evenhandedly as being an integral part of a bargain made between landlord and tenant at a time of rising rents with the tenant conceding a review clause as the price of obtaining assured possession for a period of years. A reading of clause 5 (x) as a whole brings out the following B points.

Gibbs has the right to call for a rent review in 1978 and 1984. The right is exercised by notice given in writing (see 5 (x) (a)). It is envisaged that the notice is followed by a "review" (see 5 (x) (d)). This review may involve an arbitration. The review leads to a finding (see 5 (x) (a)) where there are the words, "If upon review it shall be found ...." The notice in question may be given in respect of a 1978 review, at any time during the period July 6 1977 to July 6 1979.

Going on to 5 (x) (f), that provides that if the landlord fails to give notice during that two-year period the landlord has at any time thereafter the right to review the rent. Gibbs gave no notice pur- C suant to 5 (x) (a) during the two-year period. But they did give, pursuant to 5 (x) (f), the notice dated October 9 1979 that is now challenged. So, on the words of the clause alone, the validity of the notice seems sound enough.

But Mr. Essayan has drawn attention to the events of July 1978 and the accord then reached between the parties by their duly authorised agents, that is Mr Rose and Mr Larkin. Counsel for Gibbs accepted:

(1) That Mr Rose and Mr Larkin were, in 1978, appointed to negotiate the rents for all the three floors comprised in both the underleases.

(2) That they agreed that the market rent for office space (on the second, third and fourth floors of the building) was £11.50 per sq ft D overall.

(3) That they agreed that that market rent was less than the "passing" rent for the third and fourth floors; so the "passing" rent at £197,225 reserved by the underlease dated January 29 1976 was in July 1978 greater than the "market" rent.

(4) They further agreed that there could at that time be no increase in the rent payable in respect of the third and fourth floors. In these circumstances, that there was a review of the rent payable under the first underlease dated January 14 1976 of the second floor is beyond question. The market rent of the second floor was agreed at £110,000, a sum in excess of the passing rent. So the solicitors acting for Gibbs, on August 11 1978, wrote to Esso's E solicitors as follows:

> We have been instructed by our clients to serve on your clients a notice in accordance with the provisions of the said underlease, that is the underlease dated January 14, calling for a review of the yearly rent payable to the landlord in respect of the premises demised by the said underlease in the sum of £110,000. [Then a little later on], In fact, we understand that the reviewed rent at £110,000 has already been agreed between our respective clients subject to contract. We therefore enclose a draft rent review memorandum in duplicate for your approval, and look forward to hearing from you as soon as possible.

This letter resulted in a memorandum of increased rent being en- F dorsed, as I have said already, on the underlease dated January 14 1976. But since Mr Rose and Mr Larkin agreed that no increase was appropriate in respect of the third and fourth floors it is not surprising that no steps were taken to record by endorsement on the underlease dated January 29 1976 information to the effect that the rent had been reviewed in 1978, and that it had then been found that no increase in rent was appropriate.

Mr Essayan contended that just as there had been a 1978 review, within 5 (x) (a), as respects the second floor, so there had been a 1978 review as respects the third and fourth floors, so that 5 (x) (f) had not been usable on October 9 1979. Mr Wood, for Gibbs, contended that there had been no review as respects the third and fourth floors. There had been, he said, informal discussions only, G resulting in an agreement between the parties that no increase was then appropriate. In those circumstances, the landlord was entitled to reserve his rights under 5 (x) (a) until July 6 1979 and thereafter to make use of 5 (x) (f). The landlord, he said, had a choice. I am satisfied that Mr Essayan is right in his contention. The Rose/Larkin discussions of July 1978 about the third and fourth floors resulted in their agreeing a nil increase. That means they reviewed the rent. There was a review within the ordinary meaning of that word.

There having been a review, I examine the whole wording of 5 (x) to see whether in consequence of the review, Gibbs were precluded thereby from serving a 5 (x) (f) notice. It was argued that since the H review was unaccompanied by a 5 (x) (a) notice, or any other formality referable to 5 (x), the landlord was able, after July 6 1979, to give a 5 (x) (f) notice. I do not accept that argument. It would mean that the landlord would get a 1978 review and then another review in 1979. 5 (x) (f) contemplates a review being carried out pursuant thereto only if a review period passes without there having been a review pursuant to 5 (x) (a). Here there was a review pursuant to 5 (x) (a). Clause 5 (x) (f) is a fall-back provision that cannot be operated if a review has already been conducted in respect of the preceding review period. It may be said that on the strict reading of 5 (x) (f) Gibbs were entitled to serve a 5 (x) (f) notice. The relevant words in (f) are: J

> If the landlord fails to give notice calling for a review of the yearly rent payable hereunder during the review period the landlord shall at any time thereafter have the right to review the said rent upon giving notice.

It is true that Gibbs failed to give notice within the review period. But the correspondence and conversations between the parties' representatives up to and after July 26 1978 show that Gibbs failed to give notice under 5 (x) (a) because the circumstances in which Mr Rose, for Gibbs, conducted his discussions, first with Mr Dawson and then with Mr Larkin, made entirely unnecessary the serving of any formal 5 (x) (a) notice. I am satisfied that the parties proceeded with their review of the rent of the third and fourth floors on the K basis that they had clause 5 (x) in mind and on the basis that:

(1) prior to their agreeing on a nil increase, no notice was necessary, notwithstanding the opening words of 5 (x) (a), and

(2) that no notice was necessary after so agreeing. See in this connection Mr Rose's letter dated July 26 1978, where he "assumes there is no point therefore in serving formal notice on Esso in respect of these two floors".

In short, in conducting their review of the rent of the third and fourth floors, the parties implicitly waived the provisions respecting notice that was intended to initiate the review. Accordingly, I accept Mr Essayan's first submission, that is, that there was a rent review in July 1978 and that such review precludes Gibbs from L seeking a review pursuant to their notice dated October 9 1979. That means that the plaintiffs succeed in the action.

However, it is right that I should deal with Mr Essayan's other submissions. I have set out the second submission above. The essential element in this submission was that 5 (x) (f) is available to a landlord only if the landlord has by inadvertence failed to give a 5 (x) (a) notice. So, (f) ought to be read as if it opened with the words, "Notwithstanding subclause (a) of this clause if the landlord inadvertently fails to give notice." Reading 5 (x) (f) in the context of 5 (x) as a whole I am unable to regard the word "inadvertently" as being implicitly included in (f) in the way I have quoted. Let us suppose a situation in 1978 in which Mr Rose, M without communicating with any Esso representative at all, comes to the conclusion that the passing rent exceeds the market rent. Let us suppose that Mr Rose then considers the whole of 5 (x). He reports to Gibbs and says there is no point in giving a 5 (x) (a) notice because the rent review clause is upwards only; that it would be better to await a rise in rent, so that a rent review notice be deferred because a 5 (x) (a) notice is possible up to July 6 and a 5 (x) (f) notice is possible thereafter. Assuming further (1) that Esso are then told that no 5 (x) (a) notice will be given for the time being, and (2) that in October 1979 a 5 (x) (f) notice is given. In these circumstances the landlord would have failed to give notice within the

A review period, but he would not have inadvertently failed. On the other hand, I cannot see why a 5 (x) (f) notice given in those circumstances should not be good. Consequently, as I have said, I am unable to read (f) as applying only to an inadvertent failure.

It will be convenient next to deal with the fourth submission. This involves a pure point of construction. Mr Essayan drew attention to the fact that the phrase, "commercial yearly rent" used in 5 (x) (f) is a phrase defined in 5 (x) (b). That definition ties the rent to the review date, which, for the purposes of a 1978 review, is July 6 1978. So, it is argued, the phrase "at the date specified in the said notice" ought to be moved from its place in (f). It ought to be inserted after the word "increased". Mr Wood accepted that the
B definition of the phrase "commercial yearly rent" gave rise to difficulty, but his remedy for the difficulty was to say that the strict meaning of "commercial yearly rent" was obviously used inaptly in (f) and that the phrase must plainly be construed as the commercial yearly rent as at the date of the (f) notice and not as at the review date. I content myself by saying that I prefer Mr Wood's construction. Accordingly, I reject Mr Essayan's fourth submission.

I now turn to the third submission on the topic of estoppel. No promissory estoppel was relied on. There was said to be an estoppel in pais. I was referred to part of the judgment of Dixon J in *Grundt* v *Great Boulder Property Gold Mines Ltd* (1937) 59 CLR 641, as
C set out on pp 110 and 111 *Spencer Bower on Estoppel by Representation* (3rd ed). The estoppel is pleaded in paragraphs 7 and 8 of the statement of claim. The allegation is that on or about November 3 1978 Mr Rose, on behalf of Gibbs, offered to Esso an assignment of the Gibbs reversion expectant on the Esso underlease of the third and fourth floors; and it is said that when the offer was made Gibbs represented to Esso that the underlease rent was not capable of being increased with effect from any date earlier than July 6 1984. Then it is said — paragraph 8 — that Esso acted in reliance upon the representation and acted to its detriment so that had it accepted the offer the underlease would have merged in the headlease with the effect that Esso would thereafter have held on the terms of
D the headlease, the headlease containing no provision for any rent increase before July 1984. So, it is said, Gibbs are estopped from contending that the underlease rent was capable of increase before July 6 1984.

On November 3 1978, Mr Rose wrote to Mr Larkin asking if Esso would be interested in acquiring the Gibbs reversion on the Esso underleases. If Esso did so they would, of course, become the direct tenants of Legal & General. On November 6 1978 Mr Rose followed up this letter with a second letter. In this he said he had made "a few notes which might be helpful to you and your clients in considering our suggestion." On the same day Mr Rose sent another copy of the notes to Mr Dawson, who, as I have said, is a
E chartered surveyor employed by Esso. The notes consist of some very bare details of the headlease and the two underleases. As to the third and fourth floors the notes say: *"Anthony Gibbs to Esso.* The third and fourth floors approximately 16,000 sq ft. Lease expires 6-7-88. Rent £197,225. Rent review July 6 1984." So, it is said, Gibbs represented to Esso that the rent of the relevant underlease was not capable of being increased until 1984, and detriment ensued as already mentioned.

I now mention it was common ground that in November 1978, both sides — that is Esso and Gibbs — genuinely supposed that the underlease rent would not be reviewed until 1984. This belief stemmed either from the fact that the underlease was not read with such
F care as to appreciate the importance of (f) or from the fact that all parties supposed — as I have held — that a rent review had taken place in July 1978. However, there was this common belief. Mr Essayan said that the fact that there was this common belief did not mean that Esso could not raise any estoppel. In that, no doubt, he is right. See *Amalgamated Investment & Property Co Ltd* v *Texas Commerce International Bank Ltd* [1981] 2 WLR 554, where Robert Goff J said at p 571 C that:

> There may be cases where a representee has proceeded initially on the basis of a belief derived from some other source independent of the representor, but his belief has subsequently been confirmed by the encouragement or representation of the representor.

See also *Taylor Fashions Ltd* v *Liverpool Victoria Trustees Co Ltd* G
[1981] 2 WLR 576 at p 596 F. But it is necessary to have regard to some other considerations that Robert Goff J had in mind. The words I have quoted are part of the following extract from his judgment:

> First, the case advanced before me by the bank is not one of simple acquiescence by the plaintiffs in the mistaken belief of the bank; it is founded upon active encouragement by the plaintiffs or representations by the plaintiffs — encouragement and representations which I have already found were in fact given and made by the plaintiffs to the bank. Now, in my judgment, where an estoppel is alleged to be founded on encouragement or representation, it can only be unconscionable for the encourager or representor to enforce his strict legal rights if the other party's conduct has H
> been influenced by the encouragement or representation.
>
> Second, it is, in my judgment, no bar to any conclusion that the other party's conduct was so influenced, that his conduct did not derive its origin only from the encouragement or representation of the first party. There may be cases where the representee has proceeded initially on the basis of a belief derived from some other source independent of the representor, but his belief has subsequently been confirmed by the encouragement or representation of the representor. In such a case, the question is not whether the representee acted, or desisted from acting, solely in reliance on the encouragement or representation of the other party; the question is rather whether his conduct was so "influenced" by the encouragement or representation (I take the word "influenced" from the judgment of Bowen L J in *Edgington Fitzmaurice* (1885) 29 ChD 459, 481) that it would be unconscionable for the representor thereafter to enforce his strict legal rights. J
> Such a conclusion appears to be consistent with the decision of Oliver J in *Taylor Fashions Ltd* v *Liverpool Victoria Trustees Co Ltd; Old & Campbell Ltd* v *Liverpool Victoria Friendly Society* (Note) [1981] 2 WLR 576. The point can also be illustrated by a hypothetical example. Let it be supposed that A and B are neighbours, and that A proposes to build a wall on what is in fact B's land, though both parties mistakenly believe it to be A's. A invites B's co-operation in the building of the wall, for example, by providing a means of access over his land for the purposes of building work or by supporting an application for planning permission. B co-operates as requested, and the wall is built at A's expense. In such circumstances it may well be unconscionable for B thereafter to assert his strict legal rights.

K
That shows that I must consider whether Esso's conduct in the matter of the offer for sale of the Gibbs interest was so influenced by the representation of "no review until 1984" contained in the Rose notes that it would be unconscionable for Gibbs now to be allowed to operate a 5 (x) (f) review. I have considered the evidence in the case and I cannot find any sufficient suggestion than anyone on the Esso side was influenced at all by anything contained in those Rose notes. Thus Mr Dawson said in chief that the information in the notes accorded with his understanding of the position. In cross-examination he said that when he was shown the Rose notes they did not tell him anything that "he did not believe to be the case". There is no question in my mind that Mr Dawson was wholly uninfluenced by the notes. The same remark may be said of L
Mr Larkin. The Rose notes were without any influence in the consideration which Esso gave to the decision to buy or not to buy. See in this connection the remarks of Oliver J in *Taylor Fashions* v *Liverpool Victoria Trustees Co Ltd* at pp 596 F to 597 D. In chief, Mr Dawson said that if he had known of the possibility that the rent of the underlease could be increased at the behest of the landlord before 1984 then there would have been an advantage in purchasing the Gibbs interest so as to remove that possibility, and in cross-examination he went so far as to say, as I understand, that had he known of the possibility I have alluded to it would have been "commercially imperative to buy". But I do not see that remarks of that kind in any way indicate that the Rose notes played M
any part at all in encouraging or confirming the Esso side in its belief that the rent was unchangeable up to 1984. It may be said that the Rose notes confirmed the Esso side in its belief. That is not to influence the Esso side, at least not so to influence the Esso side that in the circumstances of this case it would, in the words of Robert Goff J "be unconscionable for the representor" — Gibbs — "thereafter to enforce his strict legal rights".

Another difficulty for the plaintiff in respect of the estoppel claim lies in the fact that the detriment alleged in paragraph 8 of the statement of claim has not been established in evidence with what I regard as sufficient certainty. Accordingly, if it were

A necessary, I should therefore hold that Gibbs is not estopped from contending that the rent is capable of increase before July 6 1984. As it is, I declare as in paragraphs 1 and 2 of the statement of claim. The counterclaim is dismissed.

The judge granted declarations in favour of the plaintiffs, the effect being that the defendants were precluded from requiring the plaintiffs or their successors in title to pay an increased rent before July 6 1984. The defendants' counterclaim was dismissed with costs and the defendants were ordered to pay the costs of the action.

B

CHANCERY DIVISION

November 24 1981

(Before Mr Justice PETER GIBSON)

LAING INVESTMENT CO LTD v G A DUNN & CO

Estates Gazette May 29 1982

(1981) 262 EG 879

Landlord and tenant — Rent review clause in lease — Landlords' claim to be entitled to operate rent review machinery despite failure C to carry out certain stages of the procedure — Review clause provided, first, for the parties to agree the rent by a fixed date; secondly, on failure to agree rent, for a surveyor to be appointed by agreement by another fixed date; thirdly, if surveyor was not so appointed by agreement, for him to be appointed by the president of the RICS — The first two stages had fixed time-limits, time being expressly made of the essence of the contract — The third stage, involving an application to the president for appointment of a surveyor to determine the rent, had no specified time-limit, nor was a time specified for a determination of the rent by the appointed surveyor — Landlords, without going through the procedures contemplated for the first two stages, claimed to initiate the third D stage by applying to the president to appoint a surveyor — Tenants submitted that neither party was entitled to disregard the first two stages, with their strict time-limits, and to initiate the procedure at the third stage — A number of recent rent review cases considered by the judge — Held, applying the principles of *Amhert v James Walker Goldsmith & Silversmith Ltd*, that landlords should be given the declarations they sought — They were entitled to apply, and require the tenant to concur in applying, to the president of the RICS to appoint a surveyor to determine the rent under the review clause

This was an originating summons by the plaintiffs, Laing Investment Co Ltd, landlords of shop premises at 52 King William E Street, Blackburn, seeking declarations as to their entitlement to operate rent review provisions in a lease for a term from January 1 1968 to March 19 2002. The defendants were the tenants, G A Dunn & Co.

L C Bromley QC and Christopher Priday (instructed by George Carter & Co) appeared on behalf of the plaintiffs; Michael Essayan QC and George A Ward (instructed by Wright, Johnson & Cheales) represented the defendants.

Giving judgment, PETER GIBSON J said: This originating summons raises a short but difficult question of construction of the rent review provisions of a lease. By a subunderlease made February 18 1971 between the plaintiff, Laing Investment Co Ltd ("Laing") F and the defendant G A Dunn & Co ("Dunn"), Laing let to Dunn shop premises at 52 King William Street, Blackburn, for a term from January 1 1968 to March 19 2002. In clause 1 the rent was expressed to be £2,750 per annum but this was made subject to a proviso in the following form:

PROVIDED ALWAYS AND IT IS HEREBY AGREED that the yearly rent payable by the Tenant during the period of seven years next following the nineteenth day of March One thousand nine hundred and eighty-one (hereinafter called "the first review period") and during the last fourteen years of the term hereby granted (hereinafter called "the residue") shall be the sum of Two thousand seven hundred and fifty pounds aforesaid or such sum whichever shall be the higher as shall be assessed as the fair market rent for the demised premises for the first review period or for the residue respectively or (if higher) the previous period such assessment to be made in the following manner that is to say G either

(a) such assessment as shall be agreed between the parties hereto in writing before the nineteenth day of July One thousand nine hundred and eighty in respect of the first review period or before the nineteenth day of July One thousand nine hundred and eighty-seven in respect of the residue

(b) in the event of the parties failing to reach such agreement as aforesaid on or before the dates appointed (in respect of which time is to be deemed to be of the essence of the contract) then the fair market rent for the first review period and for the residue respectively shall be fixed or assessed by an independent valuation surveyor appointed for that purpose by the parties hereto or failing agreement as to such appointment in respect of the first review period by the nineteenth day of September One thousand nine hundred and eighty and in respect of the residue by the nineteenth day of September One thousand nine H hundred and eighty-seven (time in this respect to be deemed to be of the essence of the contract) then by an independent valuation surveyor appointed for that purpose by the President for the time being of the Royal Institution of Chartered Surveyors. The assessment fixed by the independent valuation surveyor shall be communicated to the parties hereto in writing and immediately upon such communication the rent so assessed as the fair market rent for the first review period or for the residue as the case may be or Two thousand seven hundred and fifty pounds or the rent fixed for the first review period whichever shall be the higher shall be the respective rents payable for the first review period or for the residue under the terms hereof And for the purpose of assessing the fair market rent for the demised premises the assessment shall be based on the fair market rent obtainable for a lease of the demised premises subject to the same covenants as are herein contained

(c) the right of re-entry hereinafter reserved shall be exercisable by the J Landlord as well in the case of the non-payment of the rent payable under paragraph (b) of this Clause as in the case of non-payment of the rent originally reserved by the Lease or agreed between the parties hereto under paragraph (a) of this Clause and also in the event of the Tenant refusing to refer the question of the rent payable hereunder in accordance with the said paragraph (b)

(d) the Tenant as beneficial owner hereby declares that the demised premises shall stand charged with the payment of any rent payable under paragraph (a) or (b) of this Clause at the times and in the manner provided for the payment of the original rent hereby reserved and this remedy shall be in addition to and not in derogation of either other remedy in respect of such rent

(e) the fees payable to the independent surveyor hereinbefore mentioned in respect of the assessment to be made hereunder shall be borne by the parties K hereto in equal shares.

There was no attempt by Laing or Dunn to invoke the rent review procedure until December 17 1980, when Laing wrote informing Dunn that Laing had instructed a firm of chartered surveyors to negotiate and agree a revised rent due on March 19 1981. Dunn replied that it was too late to agree a rent in view of the provisions of proviso (a) to the reddendum that the rent had to be agreed before July 19 1980 and that time was of the essence in respect thereof. Subsequently, on December 24 1980, Laing by letter to Dunn claimed that as there was no time-limit stipulated in the lease in respect of a reference to the president of the Royal Institution of Chartered Surveyors it was still open to either party to apply to the L president. Dunn denied that claim and accordingly on May 14 1981 the originating summons that is before me was taken out by Laing. By it Laing seeks a declaration that on the true construction of clause 1 Laing is entitled to apply, alternatively to require Dunn to concur in applying, to the president to appoint an independent valuation surveyor to assess the fair market rent for the period of seven years from March 19 1981, and it also claims a consequential declaration relating to that rent as so determined.

At first blush the broad effect of the provisions for assessing the rent for the first review period might seem straightforward enough. First, there is to be an opportunity for the parties themselves to agree the rent provided that they do so in writing by July 19 1980. Second, M if the parties fail to agree this, then the rent is to be determined by an independent valuation surveyor. Such surveyor is to be agreed between the parties by September 19 1980, but, if not so agreed, the surveyor is to be such person as is nominated by the president of the Royal Institution of Chartered Surveyors. Effectively, therefore, there are three successive stages provided for the assessment of the rent, the first two stages having strict time-limits and those times must pass before the third stage can be commenced. But on a closer acquaintance with the provisions, and when one attempts to marry the facts of the present case with the language of the lease, the difficulties become apparent. Is it permissible for either party at any time after September 19 1980 to initiate the third stage procedure without having attempted to agree the rent or the surveyor within the strict time-limits laid down? What is the effect of the words making time of the essence in respect of the first two stages? I must revert to these questions later, but first let me set out certain matters relating to the provisions in question in respect of which Mr Bromley, appearing for Laing, and Mr Essayan, appearing for Dunn, are not in dispute.

First, the rent review procedures can be initiated by either Laing or Dunn. While the practical reality must be that Laing, as landlord, would be most likely to wish to initiate a rent review which can only raise the rent upwards, Mr Essayan accepted that there were circumstances (such as if an assignment were contemplated) when the tenant might wish to take steps to have the rent for the first review period ascertained. Second, there is no starting date for the rent review procedures to be initiated nor any particular form (such as a lessor's notice) for such initiation laid down expressly in the lease. Third, the phrases "in the event of the parties failing to reach such agreement" and "failing agreement" mean no more than "if there is no such agreement". At one point it seemed at least possible that the main battle in this case would be waged in respect of those words, and that Dunn might seek to resurrect the unsuccessful argument of the tenant in *Re Essoldo (Bingo) Ltd's underlease, Essoldo Ltd v Elcresta Ltd* (1971) 23 P& CR 1. In the *Essoldo* case the rent review clause contained a provision that the rent should be agreed before a specified date, and in default of such agreement the rent should be determined by a surveyor to be appointed by the president of the Royal Institution of Chartered Surveyors. The specified date having passed without any attempt at agreeing the rent, it was argued on behalf of the tenant that the rent could only be determined by the surveyor when an attempt had been made to reach agreement before the specified date. That argument was rejected by the then Vice-Chancellor, Sir John Pennycuick, who saw no reason to write in any words to qualify what he considered to be the plain meaning of "in default of agreement". Mr Essayan accepted that there were insuperable difficulties in writing in such words to the phrases that I have referred to. Fourth, whereas time is made of the essence in respect of the times specified for the first two stages, no time is specified in respect of the third stage, the application to the president for the appointment of the independent surveyor, or for the determination by the surveyor of the rent.

Mr Bromley accepts that there can be no application to the president made before September 19 1980. The determination itself may not be concluded before the first review period has begun, but in that event any rent so assessed by the surveyor would be applied retrospectively to the start of the first review period. In the absence of a time-limit for the third stage time was not made of the essence in respect thereof.

Fifth, the application to the president would have to be a joint application by both the tenant and the landlord. This is implicit in the forfeiture provision in proviso (c) with its reference to a forfeiture arising if the Tenant refused to refer the question of a rent payable under the lease in accordance with proviso (d).

Mr Bromley's argument, in its essentials, relies on the literal language of the lease. There was no agreement in writing or at all before July 19 1980; there was no appointment of an agreed surveyor by September 19 1980. Accordingly, the express condition precedent to having a surveyor appointed by the president being satisfied and no time-limit being prescribed for such appointment or for the assessment by such surveyor, it is open, he submits, to either party to obtain such appointments and the concurrence therein of the other party.

The significance of the provisions as to time being of the essence is explained by Mr Bromley as being intended to prevent either party from delaying the fixing of the rent by the independent surveyor. Each party, he says, would know that it could not extend negotiations for an agreed rent beyond July 19 1980 and it could not extend negotiations for an agreed surveyor beyond September 19 1980.

Mr Essayan, on the other hand, contends that the words in the proviso that the rent should be assessed "in the following manner", coupled with the three successive stages and the provision that time should be of the essence in the first two stages have the effect that neither party is entitled to ignore the first two stages and its strict time-limits by initiating the rent review procedure only at the third stage after, and possibly long after, September 19 1980. Moreover, he relies on the forfeiture provision in proviso (c) which might be invoked in the event of the tenant failing to concur in an application to the president. He says that if the landlord were free to apply at any time to the president, the lessee could be prejudiced by the effect of that forfeiture provision, and he says it would be very surprising in those circumstances if in fact it were open to the landlord to initiate the rent review procedures at the third stage at any time. Mr Essayan further points out that the provisions as to time being of the essence really achieve no practical effect if they were intended to do what Mr Bromley contends. He relies in particular on the remarks of Walton J in *C Bradley & Sons Ltd v Telefusion Ltd* (1981) 259 EG 337. In that case Walton J was concerned with a stipulation that time should be of the essence, and he had to determine whether that applied back to a provision that the parties should agree the rent by a certain date, or forward to another provision that an arbitrator should determine the rent by a particular date. At p 338 he said this:

> That, however, by itself, of course, does not solve the question whether one goes backwards or forwards. But I think one may be able to get some assistance from this general consideration, namely: What on earth would it mean to make time of the essence of the provision that the amount had to be agreed between the landlord and the tenant by a date not later than six months prior to the commencement of the second rent period? It seems to me that it would be meaningless to make time of the essence of that provision, and for this very simple reason, that the landlord and the tenant might one day later than the six months agree upon the rent which was properly payable, and if they wanted so to agree, why on earth should they not? They could do so, notwithstanding any prior agreements that they had made. So that it seems to me that to make time of the essence of that particular provision would be totally nonsensical. Of course, on a clause differently framed one might come to the conclusion nevertheless that time had been made of the essence of that provision so as to ensure that upon that breaking down, as it were, the landlord was then free to appoint the arbitrator. But I find nothing in this clause which would tie the landlord down to any particular time as to when the arbitrator is in fact appointed. Even if time were not of the essence of the six months' period (for the reasons which I have already given) it seems to me that the tenant could not possibly complain if, when that six months' period had expired, the landlord applied to the president for the appointment of an arbitrator. It would be quite idle for the tenant to say "Oh, but time is not of the essence of that six months' period, therefore you must wait before applying to the president of the Royal Institution of Chartered Surveyors." The doctrine that time is not of the essence does not work that way round at all. If somebody wishes to stick to the time-limit and it is within his power to make the next move, of course then he is entitled to.

Mr Bromley sought to rely on the sentence in that passage where the learned judge said that on a differently framed clause one might come to the conclusion that time had been made of the essence so as to ensure that the landlord was free to appoint the arbitrator. But Walton J is demonstrating the oddity of a provision making time of the essence in respect of something to be mutually agreed by a particular date such as in the first two stages of the present case, and also he is pointing out that, even if time is not made of the essence in respect of an earlier stage but a time-limit is specified, the party wishing to get on with a later stage is in fact free to do so without hindrance from the party who might otherwise wish to delay. I do not think that passage really assists Mr Bromley.

Mr Bromley accepted, I think, that the provisions as to time being of the essence were, on his construction, strictly unnecessary, but he said that they had the commercial merit of making the position clear to the parties that there could be no delays beyond the stipulated dates. It is, of course, plain that where time is made of the essence the courts must give effect to such a provision (see *United Scientific Holdings Ltd v Burnley Borough Council* [1978] AC 904, at p 923G per Lord Diplock). Accordingly, when a procedure is laid down providing successive stages in respect of each of which time is made of the essence, the court will give effect to that procedure (see *Weller v Akehurst* [1981] 3 All ER 411).

Mr Essayan submitted that the only way to make sense of the provisions was to imply a term that either party wishing to have the rent assessed in accordance with the rent review provisions must serve a notice to that effect on the other party before July 19 1980 and

A he said that there was further to be implied the term that time was of the essence in respect thereof. On this point he placed some, but cautious, reliance on the approach of the Court of Appeal in *Stylo Shoes Ltd v Weatherall Bond St W1 Ltd* (1974) 237 EG 343. In that case the court was concerned with a rent review clause under which the landlords were entitled to require the rent to be reviewed, and the rent was to be of such amount as might be agreed in writing, or, in the absence of agreement, as determined by an arbitrator to be nominated by the president of the Royal Institution of Chartered Surveyors on the application of the landlords made within a specified period ending three months before a specified date. Lord Salmon, at p 345, explaining the effect of the clause, said:

B Secondly, if the landlords require a revision of the rent, although the clause does not expressly oblige them to give any notice, it is to be implied that they can show that they require the rent to be revised only by giving the tenants notice of that fact.

The Court of Appeal held that time was of the essence of the time-limit for the application by the landlords for an arbitrator, and the notice that was implied was one that had to be served before the expiration of the three months' period. The actual decision in the *Stylo* case is no longer good law, as Lord Salmon himself said in the *United Scientific Holdings* case at p 953F. Mr Bromley rightly pointed out that it was a simple matter in that case to imply a term as
C to notice from the express provision in the relevant lease entitling the landlord to require a revision. The limited assistance I get from that case is merely that it is possible in appropriate circumstances in a rent review clause to imply a term as to giving notice. In the light of the actual decision in the *United Scientific Holdings* case it is very doubtful whether the court should have implied that time was of the essence of a time-limit in respect of such a notice.

It seems to me that I am faced with the invidious choice in the present case of having to prefer either a literal construction, with the consequence that provisions as to time being of the essence serve little practical purpose and are strictly unnecessary, or a construction which involves writing words into a lease. As I have already stated,
D the court refused to do this in the *Essoldo* case, but the wording of the lease in that case differs from the present; in particular that case lacked the feature of the present case that time was made of the essence in respect of no less than two stages of the rent review procedure, and in that case what was sought to be implied differed from what is now suggested by Mr Essayan. It is, of course, trite law that the court should be slow to imply a term and should only do so if it considers it necessary in a business sense to give efficacy to what parties have agreed, and it is not enough that the court should think that the implied term was a reasonable one (see *Liverpool City Council v Irwin* [1977] AC 239).

At the conclusion of the arguments of counsel when this matter
E first came on for hearing, it appeared to me that in the absence of compelling authority I should accept Mr Essayan's argument that a term was to be implied in order to give proper effect to what the parties had agreed. However, a few days later in another case involving a rent review clause* my attention was drawn to the decision of the Court of Appeal in *Amherst v James Walker Goldsmith & Silversmith Ltd* (1980) 254 EG 123, an authority which had not been cited to me previously. I gave the parties an opportunity to address me on the relevance (if any) of that decision to the present case and I have heard further argument from both Mr Bromley and Mr Essayan.

The *Amherst* case also concerned the construction of a rent review
F clause. The lease was for a term of 28 years. The rent for the first 14 years was fixed at £2,500 per annum, but for the second period of 14 years beginning on June 24 1975 it was provided that the annual rent should be either £2,500 or such higher yearly rent as should be ascertained in accordance with a proviso which provided for an assessment to be made in the following manner, that is to say:

(a) Such assessment shall be made in the first instance by the lessor and submitted to the lessee for approval in writing on or before the twenty-fifth day of December one thousand nine hundred and seventy-four; (b) In the event of the parties hereto failing to reach such agreement as aforesaid on or before the date appointed (in respect of which time is to be deemed to be of the G
essence of the contract) then the yearly rent for the second period shall be fixed or assessed by an independent surveyor appointed for that purpose by the parties hereto failing agreement as to such appointment by the twenty-fifth day of January one thousand nine hundred and seventy-five (time in this respect to be deemed to be of the essence of the contract) then by an independent surveyor appointed for that purpose by the President for the time being of the Royal Institution of Chartered Surveyors.

The second sentence in proviso (b) was similar to the second sentence in proviso (b) of the lease before me. The report in the *Amherst* case does not show what was proviso (c) in that clause. So it is not clear whether it contained a forfeiture provision as in proviso (c) in the present case. But proviso (d) was similar to proviso (e) in the present H
case.

As is apparent, the rent review provisions in the *Amherst* case differed from those in the present case in that there was an express provision that the rent review was to be initiated by the lessor submitting an assessment to the lessee and that that submission was to be by a particular date which the court held was the same date as "the date appointed", that is to say December 25 1974. The lessor on June 25 1975 wrote to the lessee asking for agreement as to the surveyor but indicating that he was applying to the president for the nomination of a surveyor and the lessor also suggested negotiating the rent. The question in that case was whether time was of the essence of the time-limit for the submission by the lessee of his J
assessment.

The provisions of that lease resembled those of the present case in that (1) the assessment provisions are introduced by the prefatory words "in the following manner" as with the present lease, (2) a three-stage procedure is laid down, in respect of two of which stages time is expressly made of the essence, and in the last stage of which no time-limit is prescribed. Again these exactly match the present case. Again time is made of the essence of two matters to be agreed between the parties by a particular time.

Judge Mervyn Davies QC (sitting as a deputy High Court judge) held that time was not of the essence of the time-limit for the submission of the lessor's notice. On appeal the Court of Appeal K
agreed with him. Sir David Cairns, giving the leading judgment, first found significance in the fact that time had been made of the essence of some provisions but not of others. Second he rejected as inconsistent with the speeches of the majority of the House of Lords in the *United Scientific Holdings* case a submission by the tenants advanced by their counsel, Mr Moshi, that the fact that the trigger notice could be operated by the landlord alone made time of the essence of the time-limit in respect thereof. He then continued thus, at p 125:

So far as other matters are concerned, I think that the most effective part of Mr Moshi's argument before us was to this effect, that, unless we hold that L
the taking of the first step was one which had to be taken by a specified date, the lessee was put in difficulty in relation to the later steps, because there was a fixed date which could not be departed from, because it was provided to be of the essence of the contract, up to which he was entitled to agree either as to the amount put forward by the landlord for the rent or as to the surveyor who was to be appointed to fix that rent. The difficulty of that argument appears to me to be this, that, putting it at its highest, assuming that it could be said that the period for the "trigger" notice was subject to an absolute limit of December 25, there is certainly nothing in the agreement to say that it must be given before December 25. If the date is to be of the essence of the contract at all, it is on or before December 25; it is quite impossible to put it at a reasonable time before December 25. So that, in effect, even if it were regarded as being of the essence of the contract, the position is that, if the notice were given by M
the landlord at the last moment, one would immediately be into what may be called the second stage.

It seems to me that the true meaning of this lease is that, if the notice is not given by December 25, it can still be given at a later date. It may well be that the later date would have to be within a reasonable time, but as to that I express no more than a tentative opinion. But if notice is given thereafter and if it is not given so late that it can be said it was not given within a reasonable time and that that makes it invalid, then the position would be that, indeed, in effect, the second and third stages are concertinaed and we come straight on to the position where either party can go straight to the president of the RICS and ask for the appointment of an independent surveyor.

That I think is the meaning and effect of the agreement between the parties . . . .

---

* *London & Manchester Assurance Co Ltd v G A Dunn & Co* (1981) 262 EG 143.

A Eveleigh LJ agreed, saying that he found some difficulty in seeing that in a case like the one before him a trigger notice had any real part to play, certainly once the date specified in the lease had passed. Megaw LJ also agreed. He said, at p 129:

> I should have expected that, if the intention had been that for which Mr Moshi contends, the latest date by which the lessor was required to give his assessment would have been expressly stated and not left to be, by implication, a reasonable time before December 25; and the clause would, by its express terms, have made it clear that, if that action was not taken within that time, the lessor would lose his rights.

Thus in the *Amherst* case, although the parties had expressly agreed that the initiating step should be taken by December 25 1974
B and all that was being asked of the court was to imply a term that time should be of the essence in respect of that time-limit, the court refused to do so. In the present case, as I have said, Mr Essayan is asking the court to imply not only that the initiating step should be taken by July 19 1980 but also that time is of the essence thereof.

Mr Essayan distinguishes the *Amherst* case by saying that it was simply an application of the principle of *expressio unius exclusio alterius*. In effect, he says, the lease in the *Amherst* case spelt out that there was to be a trigger provision exercisable by the lessor at any time and he says that is very different from a case such as the present where no trigger notice at all is provided. He says that the justification for implying a term for an initiating step, that is to say
C that otherwise the stipulations as to time would be pointless and the forfeiture provision could cause severe hardship on the tenant, is also the justification for implying a term that time should be of the essence thereof. The two go together. He cited *Beer v Bowden* [1981] 1 WLR 522 as an illustration of the court's approach to implying terms in a rent review clause. In that case the relevant provision of the lease was "such rent as shall thereupon be agreed between the landlords and the tenant". The court implied the word "fair" between the words "such" and "rent".

As Buckley LJ put it at p 528 D:

> If some such implication is not made, it seems to me that this would be a completely inoperative rent review provision, because it is not to be expected
> D that the tenant would agree to an increase in the rent if the rent to be agreed was absolutely at large. Clearly the parties contemplated that at the end of five years some adjustment might be necessary to make the position with regard to the rent a fair one, and the rent review provision with which we are concerned was inserted in the lease to enable such an adjustment to be made. The suggestion that upon the true construction of the clause it provides that the rent shall continue to be at the rate of £1,250 a year unless the parties otherwise agree would, in my opinion, render the provision entirely inoperative, because, as I say, one could not expect the tenant voluntarily to agree to pay a higher rent.

I do not see how in the present case one could say that the rent review provisions would be rendered completely inoperative if Mr Essayan's terms were not implied. The rent review provisions would of course work if the express time-limits were adhered to and if they E
were not there is the fail-safe procedure for determining a rent, that is to say either party could go to the president for the nomination of a surveyor to assess the rent, and the landlord can compel the tenant's concurrence in such a step. In my judgment, therefore, *Beer v Bowden* does not provide any principle that is applicable in the present case.

It seems to me that it would be a very surprising consequence if in a lease containing such similarities of language to the *Amherst* lease the lessor would find itself in a worse position than the lessor in the *Amherst* case who failed to comply with an express provision as to time, whereas the lessor in the present case was under no express obligation to take an initiating step. Like Megaw LJ in the *Amherst* F
case, I would have expected express provisions if the initiation of the rent review procedure by July 19 1980 was intended to be a matter of crucial importance, so as to make it clear that the lessor or lessee wanting a review would lose his rights if no step were taken by that date. Again as in the *Amherst* case the contrast between what is sought to be implied and the express provisions making time of the essence is very striking. Further there is the same difficulty in the present case in implying that the initiating step should be made by the date by which agreement had to be reached between the parties as to rent as there was in the *Amherst* case in respect of the trigger notice. To serve a trigger notice or to take an initiating step on or shortly before the date in question as a practical matter makes it impossible G
to reach agreement by that date.

Of course the Court of Appeal did not have the benefit of the arguments that Mr Essayan has put with such skill and lucidity to me, but that court does not appear to have been troubled by the fact that time-limits in respect of which the parties had laid such heavy emphasis by making time of the essence thereof could in effect be ignored by the concertinaing of the successive stages. In the light of the approach of the Court of Appeal and consistently with the *Amherst* case it seems to me that I would not be justified in holding that necessity requires the implication of a term.

In the result I shall make the declarations requested in the originating summons in the following form: H

1. A declaration that upon a true construction of clause 1 of the subunderlease the plaintiff is now entitled to apply and to require the defendant to concur in applying to the president of the Royal Institution of Chartered Surveyors to appoint an independent valuation surveyor to assess the fair market rent for the premises thereby demised for the period of seven years following March 19 1981.
2. A declaration that the rent so assessed by the said surveyor or £2,750, whichever shall be the higher, shall be the yearly rent payable by the defendant for the said period.

The plaintiffs were awarded costs.

MISREPRESENTATION

A **CHANCERY DIVISION**

October 8 1981

(Before Mr Justice DILLON)

WALKER v BOYLE

Estates Gazette March 20 1982

(1981) 261 EG 1090

Claim by purchaser for return of deposit paid under contract for sale of a property consisting of a house and some acres of land — Allegation of misrepresentation by vendor in reply to preliminary
B **inquiry — Inquiry related to existence of disputes as to boundaries and other matters — The answer referred to some difficulties which the vendor had had with a particular neighbour but failed to mention a boundary dispute with another neighbour relating to the position of a fence — Held that the purchaser was entitled to the return of his deposit with interest — Disclaimer in "small print" in preliminary inquiries form ineffective — Condition 17 in National Conditions of Sale (19th ed) not a protection to the vendor in the circumstances of the case — Effect of section 3 of Misrepresentation Act 1967 as amended by the Unfair Contract Terms Act 1977 discussed — Condition 17 did not satisfy the requirement of reasonableness in section 3, as amended, in the present case — Judge's**
C **comment that the survival through the ages of what became condition 17 did not entitle it to the "automatic accolade of fairness and reasonableness"**

These were in fact consolidated cross-actions, in the first of which Mr Walker, the purchaser, was the plaintiff and Mrs Boyle, the vendor, the defendant; in the second action the roles were reversed. The question in both actions was as to whether a deposit paid by Mr Walker should be returned to him or forfeited to Mrs Boyle. The question arose under a contract for the sale of a property called Stall House, near Pulborough, Sussex. The amount of the deposit was £10,500.

R J Furber (instructed by The Simkins Partnership) appeared on
D behalf of Mr Walker; R W Seymour (instructed by Streeter, Marshall & Wilberforce Jackson, of Croydon) represented Mrs Boyle.

Giving judgment, DILLON J said: I have before me two actions which have been consolidated. In the first action Mr Walker is the plaintiff and Mrs Boyle the defendant. In the second action, Mrs Boyle is the plaintiff and Mr Walker the defendant. The substantive question which is raised in each action is whether the deposit of £10,500 which Mr Walker paid under a contract of October 1 1979 for his purchase of a property called Stall House, near Pulborough, in Sussex, ought to be returned to Mr Walker or is forfeit to Mrs Boyle.

The claim by Mr Walker for the return of the deposit is based on
E an allegation that he was induced to enter into the contract by a misrepresentation in a reply to a preliminary inquiry before contract. There are four main issues. Firstly, was there a misrepresentation made on behalf of Mrs Boyle, the vendor? Secondly, if so, was that misrepresentation one of the factors which induced Mr Walker to enter into the contract? Thirdly, if the answers to questions (1) and (2) are in the affirmative, is Mr Walker none the less precluded by a general condition in the contract, namely, condition 17 of the National Conditions of Sale 19th ed, from claiming rescission on account of the misrepresentation? Fourthly, and alternatively, even
if condition 17 would otherwise be applicable, is Mrs Boyle pre- F
cluded from relying on it by section 3 of the Misrepresentation Act 1967 as amended by the Unfair Contract Terms Act 1977?

I should set out some of the facts which are not in dispute. Mrs Boyle purchased Stall House, which is a house, apparently, with four bedrooms and a couple of bathrooms and 4 or 5 acres of land including a pond or small lake near the northern boundary, in 1967, and it was the home of Mrs Boyle and her husband, Dr Boyle, although it was bought by Mrs Boyle alone.

In 1979 Mrs Boyle's marriage to Dr Boyle collapsed and on March 30 he left Stall House. He had in fact been living in London midweek for some time before. Mrs Boyle therefore decided to
purchase another property and sell Stall House. She contracted to G
buy the other property in reliance on arrangements for a bridging loan from her bank, and she put Stall House on the market, but initially there was no purchaser. In August, however, or thereabouts, a possible purchaser, a Mr Hunter, agreed to buy Stall House subject to contract, and on August 15 1979 Mr Hunter's solicitors submitted the usual formal inquiries before contract to Mrs Boyle's solicitors, Streeter, Howe & Wilberforce Jackson, of Croydon. Those inquiries were replied to by Mrs Boyle's solicitors on August 24 1979, and it is one of those replies which contains what is alleged to be the misrepresentation. In fact, Mr Hunter decided not to proceed with the purchase of Stall House. I do not
know why, and anyhow the reason does not matter, and Mr Walker H
came forward as the prospective purchaser and agreed through Mrs Boyle's estate agents, subject to contract, to buy Stall House for £105,000.

It was in September 1979 that this agreement, subject to contract, was reached, and indeed Mr Walker saw round Stall House. Forthwith Mrs Boyle's solicitor, Mr David Moore of Streeter, Howe, wrote on September 21 1979, to Mr Walker's solicitor, Mr Rutteman of The Simkins Partnership, enclosing a draft contract for approval. Mrs Boyle was at that time anxious for a quick contract and a quick completion because of the purchase of her other property. Mr Walker was under no pressure. He had the money available and did not have to rely on selling other property.
He liked Stall House very much, so he was keen to buy. J

The letter from Mrs Boyle's solicitor of September 21 1979, which I have mentioned, besides enclosing the draft contract for approval, also enclosed, with a view to saving time, a photocopy of Streeter, Howe's replies to Mr Hunter's solicitors' pre-contract inquiries. It enclosed a number of other documents which do not matter. The relevant pre-contract inquiry is no 2 which is headed "Disputes". The inquiry is as follows:

> Is the vendor aware of any disputes regarding the boundaries, easements, covenants or other matters relating to the property or its use?

The answer given to Mr Hunter's solicitors and now submitted to K
Mr Walker's solicitor, was:

> Not to the vendor's knowledge, but see points 3 and 4 in the letter of Messrs Anderson, Longmore and Higham enclosed herewith.

Anderson, Longmore & Higham had written a letter, which is the letter referred to, dated June 22 1979 to Mrs Boyle's estate agents. They had written this letter on behalf of a Mr Allfrey who lived in a nearby house called The Gables and was the owner of land which adjoined the Stall House property on the south and east sides, and also was the owner of the lane by which access to Stall House was obtained. The letter drew attention to the following points:

> (1) Access to Stall House is over a farm road belonging to our client; (2) the water supply to Stall House is by a spur from a private water pipe belonging to our client and is separately metered; (3) the boundary fence of Stall House on the eastward side belongs to the property, having been planted by Sir Howard Roberts and at the present time has grown out into our client's field to an appreciable extent and interferes with cultivation of the field. Our client has on a number of occasions requested your client to trim the hedge but it has not so far been done; (4) your client has erected a wicket gate in the eastward boundary fence of Stall House so as to give access to our client's fields without our client's knowledge or permission. This gate should be closed up and removed and proposing purchasers informed that there is no right of access to or of way over our client's land; (5) your client has formed two flowerbeds on either side of the front gate to his property at the edge of and encroaching upon our client's farm road at the entrance to the farm yard. We understand that these beds are now edged with masonry, that these beds constitute a trespass on our client's land should be pointed out to proposing purchasers. Our client requires these beds to be removed and the roadway reinstated as the beds are an interference with farm traffic.

I accept the evidence of Mr Walker and Mr Rutteman, that Mr Rutteman went carefully through these five points in the letter with Mr Walker. Mr Walker was concerned to see exactly what was involved. As to point (1) he was satisfied that there was a right of way over the farm road referred to. As to point (5) he was satisfied that the flowerbeds concerned had already been removed as was stated in the letter of September 21 1979. As to the letter from Messrs Andersons as a whole he took the view that it was a case of a somewhat pedantic neighbour who had learned that Stall House was on the market and was concerned to put his position on the record for the avoidance of doubts, and after consideration he decided that these were all matters he could live with and which would cause him no trouble at all. As to point (4) he appreciated that there was no right of way from the eastern boundary and he took the opportunity, as I find, on a visit to Stall House after he had considered the letter of September 21 1979 and the letter from Messrs Andersons enclosed and the replies to the preliminary questions, of having a look at the wicket gate referred to to be sure precisely what was involved.

Contracts were therefore exchanged on October 1 1979 for Mr Walker's purchase of Stall House from Mrs Boyle at the price of £105,000. The deposit of £10,500 was duly paid to Mrs Boyle's solicitors as stakeholders. The contract provided for a completion date on or before November 30 1979 and was on the usual form incorporating the general conditions known as the National Conditions of Sale, 19th ed. These included condition (13)

> That the vendor shall not be bound to show any title to boundaries, fences, ditches, hedges or walls,

and condition (17), which is important for present purposes and which I should read. It is headed "Immaterial Errors" and reads as follows:

> (1) Without prejudice to any express right of either party or to any right of the purchaser in reliance on section 24 of the Law of Property Act 1969 . . .

and I interpolate that that section has no relevance to the present case

> . . . to rescind the contract before completion, no error, misstatement or omission in any preliminary answer concerning the property or in the sale plan or the special conditions shall annul the sale, nor save where the error, misstatement or omission is in a written answer and relates to a matter materially affecting the description or value of the property, shall any damages be payable or compensation allowed by either party in respect thereof .
>
> (2) In this condition a preliminary answer means and includes any statement made by or on behalf of the vendor to the purchaser or his agents or advisers, whether in answer to formal preliminary enquiries or otherwise before the purchaser entered into the contract.

As I have already said, the important answer was the answer to preliminary inquiry no 2. That referred to the letter of Messrs Andersons on behalf of Mr Allfrey, but the trouble which has given rise to this action has nothing to do with Mr Allfrey or his land, or with the points (1) to (5) made in Messrs Andersons' letter. It is a trouble in relation to the northern boundary of Stall House. As I have mentioned, there is in the Stall House property a small lake, and beyond the lake there is a house and garden known as Staalcot which belongs to a lady who at times has been Miss Bilbe-Smith and at other times Mrs Earle. I shall for convenience refer to her as Mrs Earle. She is in fact a niece of Mr Allfrey and the granddaughter of a former owner of Stall House, but that does not matter.

The dispute concerns the boundary in the vicinity of the edge of the lake between Mrs Earle's property and Stall House. When Mrs Boyle purchased Stall House in 1967 she and her husband decided to erect a substantial fence along the boundary between Stall House and Mrs Earle's property. They were, of course, entitled to do that. Their reason was partly that Mrs Boyle owned a couple of Great Danes — she wanted to keep them in her property and prevent other dogs or animals coming in — and partly that Dr Boyle was minded to get a couple of swans to put on the lake.

However, a dispute arose in 1967 over where the boundary should run. Solicitors were involved and there was correspondence between solicitors, and ultimately there was a meeting on the site in the early autumn of 1967. After that meeting Dr Boyle got their gardener, Mr Cox, to put up a fence under Dr Boyle's supervision, but on October 16 Mrs Earle's solicitors, who happened to be Anderson, Longmore & Higham, wrote to Dr Boyle's solicitors, Peachey & Co, as follows:

> We refer to your letter of October 5. Our client has telephoned us today telling us that your client, Dr Boyle, has in her absence put up a fence which for a certain length is on her property and not on Dr Boyle's, and in connection with this work has pulled up and otherwise damaged various sections of a hedge which is her property. She also tells us that Dr Boyle and his agent had at various moments trespassed on her land. We are taking further instructions and shall be in touch with you again, but meanwhile on behalf of our client we protest most strongly at the action taken.

A "without prejudice" letter was written to Dr Boyle's solicitors at the same time offering a plan to help explain the situation. These were acknowledged by Dr Boyle's solicitors, but in fact no further action was taken by or on behalf of Mrs Earle at that stage in relation to the boundary or the fence. I accept Mr Anderson's evidence that in fact from 1967 until 1979 Mrs Earle was away in the United States for the greater part of the time, and only made occasional short visits to Staalcot.

There were between 1967 and 1979 various rows from time to time between Mrs Earle and Mrs Boyle about other minor matters. They were rows in the course of which tempers seem to have flared. If, however, the preliminary inquiry as to boundary disputes had fallen to be answered late in 1978, it could fairly have been said in my judgment by and on behalf of Mrs Boyle that there was no boundary dispute with Mrs Earle.

However, on February 22 1979 Messrs Andersons wrote a further letter on Mrs Earle's behalf to Dr Boyle's solicitors, Peachey & Co. In this they said:

> I refer to our correspondence with you in October 1967 and now write to inform you that having cleared some overgrowth along the bank of the lake at the bottom of her garden, our client has uncovered a strong growth of alders, willows and brambles having their roots in her bank and growing up on both sides of the fence. It is her intention to trim back all these saplings and small trees and ultimately to plant a screen of selected evergreen and coniferous trees along the top of the bank. You will recall that in our letter to you of October 16 1967 we objected that your client had erected part of the fence on our client's property by placing it on our client's side of a line of stakes which our client had erected at the northwest corner of the lake along the water's edge which forms the boundary. Our client pointed this out to your client at the time and again in the spring of last year, and we are now instructed to repeat our client's objection to the incorrect position of the wire fence at this point.
>
> In conclusion, we are to reiterate that our client sees no need for the fence, and would prefer to see it removed to give a more natural appearance to the landscape.

There is no evidence before me, despite the reference in the letter, to indicate that Mrs Earle did point out the incorrect position of the fence to Dr Boyle in the spring of 1978. Accordingly, I ignore that reference in the letter.

A Messrs Andersons sent Messrs Peachey a further letter on March 28 1979 objecting to the condition of two large trees growing on Mrs Boyle's land. It was said that these trees were dangerous. Nothing turns in this action on the condition of those trees. The letter is, however, part of the history of the matter in relation to Mrs Earle. I have no doubt that Dr Boyle received copies of both these letters from Messrs Peachey immediately after Messrs Peachey received them, and I have no doubt that Dr Boyle told Mrs Boyle the substance, as he understood it, of the two letters.

When in August 1979 Mr Hunter's solicitors submitted their preliminary inquiries to Mrs Boyle's solicitors, Mrs Boyle's solicitor, Mr David Lloyd, sent Mrs Boyle a copy of the form of B inquiries and asked her to fill in the answers. Mrs Boyle, throughout her married life, had left all business matters to Dr Boyle. She is a lady who is somewhat vague on points of detail, and had not been used to dealing with any business matters, and accordingly, she passed the form of inquiries to Dr Boyle for him to complete on her behalf. He dealt with the inquiry no 2 by simply inserting the one word "No", and sent that back to Messrs Streeter, Howe. They converted the answer into that which appears in the reply sent out, which I have already read, because they were aware, as Dr Boyle probably was not, of Messrs Andersons' letter of June 22 1979, written on behalf of Mr Allfrey, and they thought it prudent, as it undoubtedly was, to draw any purchaser's attention to that letter.

C Dr Boyle is a consultant physician at the Middlesex Hospital. He also had, and indeed he may still have, a Harley Street practice. He is a highly intelligent man, firm and decisive in handling matters, and of a dominant character, but I think he is prone to see only his own side of any question that he has to consider. I do not doubt that he was upset at the collapse of his marriage, but I do not think he is the sort of man who would be prevented by his own personal distress from handling business matters competently and firmly. I have no doubt that he had no intention to deceive or to suppress matters which he felt that he ought to disclose, but I think his attitude was, as he said in evidence, that he did not mention the dispute with Mrs Earle when he prepared the reply to the prelimi- D nary inquiry because to his mind there was no dispute. It had been resolved in 1967. His view of the letters from Messrs Andersons of October 16 1967 and February 22 1979 was that they were quite incorrect in saying that the fence was not in the correct position. The letters were misconceived and so there was no dispute.

It seems to me, with all respect to Dr Boyle, that that approach was misconceived. He ought to have disclosed the claim made by Mrs Earle's solicitors in the letter of February 22 1979, even if he also indicated that he thought that the claim was totally misconceived, because it was for the purchaser, whoever he might be, to satisfy himself as to the rights or wrongs of the claim over the boundary and to decide for himself whether he would be willing to E proceed in the face of a claim, the answer to which depended on matters which were not within the purchaser's own knowledge, but on which he would have to rely for evidence on Dr Boyle or others.

I take the view, therefore, that the answer to the preliminary inquiry no 2 involved a misrepresentation in that there was to the knowledge of Mrs Boyle and Dr Boyle on whom Mrs Boyle relied, a dispute over the position of the fence which had been put on record again by Messrs Andersons in the letter of February 22 1979.

My attention has been drawn to certain small print which appears above the replies on the form of preliminary inquiries. This small print reads as follows:

F These replies on behalf of the vendor are believed to be correct, but accuracy is not guaranteed and they do not obviate the need to make appropriate searches, inquiries and inspections.

This may be a warning to a purchaser's solicitors to do what any experienced solicitor would know was his duty and make appropriate searches, inquiries and inspections, but it cannot prevent the answers that are given from being representations of fact. The object of a purchaser in making preliminary inquiries is to obtain information on which the purchaser can decide whether to proceed to exchange contracts. Any vendor will know that that is the object of asking the questions, and will know that the answers are likely to be relied upon. A person who makes a representation of fact cannot G negative the representation by words such as those in the small print which I have quoted. I accordingly ignore those words.

I should now deal with the way the matter developed after the preliminary inquiries were replied to in August 1979. I have already explained that Mr Hunter dropped out; Mr Walker came on the scene, saw the property in September and the letter of September 21 was sent with the enclosures I have mentioned. At some stage Mrs Earle, or more likely agents on her behalf, removed a length of the fence which had stood since 1967 to which objection had been taken in the letter of February 22 1979. The length removed seems to have been some 20 or 30 yards, and she had the bank dug away so as to extend the lake to where the fence had H formerly stood.

Mr Cox, who was the Boyles' gardener and who worked at Stall House in 1979 for 5½ days per week, has said that this happened before September 20, which was the date on which Mrs Boyle moved out of Stall House, and that he, Mr Cox, discussed the removal of the fence with Mr Walker in September, that is to say, before Mr Walker exchanged contracts. If that is correct, Mr Walker knew all about the removal of the fence before contracts were exchanged, knew perfectly well, despite the reply to the preliminary inquiry, that there was a boundary dispute with Mrs Earle, but none the less decided to proceed. If that were right, Mr Walker can have no claim whatsoever to the return of his deposit on J the grounds of misrepresentation.

Mr Walker's evidence is that he had two or possibly three discussions with Mr Cox. One was a discussion in September before contracts were exchanged of a somewhat general nature. The other was a discussion on October 13 about taking Mr Cox into his own employment if he completed the purchase, and Mr Walker says that it was on that occasion, and not before, that he saw that the fence had been removed and discussed with Mr Cox what had happened, and was told that it had been removed by Mrs Earle.

Mrs Boyle, as I have said, is vague on dates, but her recollection is that she was told of the removal of the fence after she had left Stall House, that is, after September 20. What is clear beyond doubt is K that on October 18 Mr Walker's solicitors wrote to Mrs Boyle's solicitors raising a query in relation to the northern boundary and referring in that letter to the removal of the fence. They ask to know what the correct boundary is and what problems there have been in the past regarding the boundary.

It seems to me that all the probabilities are that Mr Cox is wrong in his placing of his discussion with Mr Walker, and Mr Walker is right. I find, as a fact, therefore, that the fence had not been removed by the time of Mr Walker's visits to Stall House before contracts were exchanged. It was removed during the period between the last of those visits in September and October 13. Mr Walker had no knowledge of the removal of the fence or of any L problem in relation to the fence or the boundary with Mrs Earle before contracts were exchanged. The fence was removed after Mrs Boyle had moved from Stall House. There is no evidence that its removal was before October 1 rather than during the period from October 1 to 12 and there is no evidence that the removal of this length of fence was made known to Mrs Boyle before contracts were exchanged.

Accordingly, the fact of the removal of the fence cannot be relied on as imposing any obligation on Mrs Boyle to correct the answer to preliminary inquiry no 2 before contracts were exchanged.

Mr Walker's attitude after he had learned in October of the removal of the fence was that he wanted to know what it was all M about. In a letter of November 6 1979 his solicitors asked Messrs Streeter, Howe to confirm that there had been no previous disputes regarding the boundary, and matters proceeded as far as Mr Walker was concerned towards completion in the usual way, subject to the qualification that he needed to be satisfied on this outstanding matter. In a postscript to a letter to Streeter, Howe of November 14 1979 his solicitors said that Mr Walker could complete on Friday November 23 provided the boundary question was satisfactorily resolved.

On November 10 Streeter, Howe acknowledged that letter, and they sent a copy of a letter of October 30 which they had received from Messrs Andersons as solicitors to Mrs Earle, together with copies of letters supplied by Messrs Andersons and a copy of a letter to themselves from their own client, Mrs Boyle. The letter from Messrs Andersons is somewhat difficult to follow because it says at one stage, "Our client's fence has always been in the correct position". It is plain that Messrs Andersons are persisting in the claim which Mrs Earle had made through them earlier. They enclosed copies of the letters of February 22 and March 28 1979 and these were thus sent on to Mr Walker's solicitors. The letter from Mrs Boyle herself to Mr David Moore, a copy of which was also sent on, says this:

> I herewith enclose your Ordnance Survey plan marking the boundaries of Stall House. Mrs Earle has been nothing but a trouble-maker ever since we have been there because for one thing she said it spoilt the country to have fences. What she really wants is to have use and pleasure of the lake. She has left some of the fence which should really be on the top of the bank but to keep her happy we put it almost in the water. She has just had it hacked and smashed down and should be made to replace it at once at her expense. She said her solicitor told her to do it.

Seeing a letter describing Mrs Earle as nothing but a trouble-maker naturally did not reassure Mr Walker. Moreover, the way Mrs Boyle has happened to phrase her letter suggests that Mrs Earle had been nothing but a trouble-maker over the fence and the lake. I think Mrs Boyle did not really do herself justice. She was writing at a time of considerable worry, both owing to the breakdown of her marriage and because she had the commitment to bridging finance for her new house, and it appeared now that after Mr Hunter had fallen out a difficulty was arising, thanks to Mrs Earle, with Mr Walker. The facts were that there had been incidents and some acrimony between Mrs Earle and Mrs Boyle, but not over the fence or the boundary by the lake.

Mr Walker took, on advice, the sensible course of suggesting that a meeting should be arranged to find out what the issue really was and to see if that issue could be resolved. The meeting was set up as a matter of urgency by Mrs Boyle's solicitor. It took place on November 27 at the site. It was attended by Mrs Boyle, Mr David Moore, her solicitor, Mr Walker, Mr Rutteman, his solicitor, Mrs Earle, who had arrived back in this country only a day or so before, and Mr Anderson, her solicitor, and by a surveyor from Mrs Boyle's estate agents to help if he could. Mrs Boyle's son may have been present for part of the time, but took no part in the meeting.

The first task was to find out what the dispute was really about. That had its complications. Mr Anderson has a point that the boundary on the deeds was fixed by reference to the boundary of the lake, but that the lake was subsequently extended, and so the boundary should be by the old lake boundary and not the current lake boundary. Dr Boyle, however, has told me that the boundary of the lake varied from time to time, because the lake was fed by a stream which collected drainage water from surrounding lands, and so at times there was more water in the lake than at other times and the boundary moved up the bank.

Unfortunately, the meeting failed to resolve the dispute or reach any conclusion at all, and it ended, as I find, in acrimony and bad temper on the part of Mrs Boyle and Mrs Earle, and a threat by Mr Anderson to have Mrs Boyle removed from the land. Mr Walker's position was that he was not concerned to take any part in the meeting but he wanted the dispute resolved between the two owners, Mrs Boyle and Mrs Earle. The actual line of the boundary did not matter very greatly to him, but he could not accept a position where there was still a continuing dispute with all the trouble that might involve for him as a successor coming into an acrimonious situation.

Accordingly, after the meeting, having discussed the matter with his solicitor, Mr Rutteman, he claimed the return of his deposit, rescinding the contract on the ground of misrepresentation, and the writ in the first action was issued.

On the same day on which that writ was issued Mrs Boyle's solicitors served a notice to complete on Mr Walker under condition 22 of the National Conditions of Sale. Mr Walker failed to comply with that notice, and Mrs Boyle's solicitors, treating his failure to comply as a breach of the contract, forfeited his deposit and issued the writ in the second action. The position as to this notice to complete is, as I see it, quite simply this. If Mr Walker was entitled to rescind the contract for misrepresentation then the notice to complete cannot have any effect. If, however, Mr Walker was not entitled to rescind the contract then the notice to complete was a valid notice with which Mr Walker failed to comply.

The subsequent history of the matter is that the property was sold in the spring of 1980 to another purchaser at a price in excess of £100,000, taking account of payment for chattels, but slightly lower than the price Mr Walker had agreed to pay. The facts of the dispute with Mrs Earle were fully disclosed in answers to preliminary inquiries to the other purchaser, but he, a Mr Smith, was not troubled and proceeded to completion.

Now, I have already said that there was in my judgment a misrepresentation and a failure to disclose the dispute with Mrs Earle in the answer to preliminary inquiry no 2. Had that dispute been disclosed before contracts, I have no doubt, and I find as a fact, that Mr Walker would have wanted to know all about it, and would have wanted the dispute resolved before contracts were exchanged. I do not regard that as an unreasonable attitude on his part. He could live with the boundary when settled, but was not prepared to accept hassle and trouble from a continuing dispute. Had there been disclosure, therefore, I think the meeting to resolve the situation would have been set up before contracts were exchanged, but I have no reason to suppose that a meeting at a slightly earlier date would have been more successful or less acrimonious than that which took place on November 27 and I find as a fact that had there been such a meeting before contracts were exchanged, Mr Walker would not have exchanged contracts.

In view of this and in view of the care which Mr Walker took over the matters in relation to Mr Allfrey which were disclosed in the reply to preliminary inquiry no 2, that there was not to the vendor's knowledge any boundary disputes except as indicated in Messrs Andersons' letter, I find that that reply was a factor among many which induced Mr Walker to enter into the contract, and the representation in the reply was a material representation.

I turn now to condition 17 and the argument that condition 17 precluded Mr Walker from claiming rescission or other relief in respect of the misrepresentation, which, as I find, was made. The gravamen of condition 17 is that:

> No error, misstatement or omission in any preliminary answer concerning the property shall annul the sale nor save where the error, misstatement or omission is in a written answer and relates to a matter materially affecting the description or value of the property, shall any damages be payable or compensation be allowed in respect thereof.

I am not concerned with the part of the condition which relates to damages or compensation, but it is strange that the condition should purport to exclude compensation for any oral misstatement, however grave. On its literal reading the words "No error, misstatement or omission in any preliminary answer shall annul the sale", would preclude a purchaser from obtaining rescission, even though the misstatement or omission was deliberately fraudulent on behalf of the vendor. Plainly, that cannot have been the intention of the condition, and no one has suggested that it would preclude relief for fraud.

Mr Furber, for Mr Walker, has submitted that the condition should be taken as inapplicable where the error, misstatement or omission was in respect of facts which the vendor knew or ought to have known. Mr Seymour, for Mrs Boyle, has said that it is to be qualified as inapplicable to matters of title because the vendor has an overriding duty to deduce a good title to the property and to disclose defects in his or her title.

These interpretations are derived from certain authorities. The authorities make it quite clear that the clause is not, in a court of equity, given literal force. I do not adopt quite either of the formulations by counsel of the limitations which the courts have put on the clause. The most helpful authority, I think, in many ways is *Nottingham Patent Brick & Tile Co v Butler* (1885) 15 QBD 261, which was decided at first instance by Wills J. In that case certain restrictive covenants affecting the title had not been disclosed, but it was said, firstly, that the purchaser would not be bound by the covenants and secondly, whether bound or not, he was bound to accept the title because of two conditions, no 4 and no 12 in the

A conditions of sale in the contract. No 4 provided that the property was sold subject to all matters, rights of way, tenants' rights and other easements and other matters or things affecting the property, whether disclosed at the time of the sale or not, and condition 12 provided that the property was believed to be and was to be taken to be correctly described, and any incorrect statement, error or omission found in the particulars or the special conditions was not to annul the sale or entitle the purchaser to be discharged from his purchase, nor was the vendor or purchaser to claim to be allowed any compensation in respect thereof. Wills J dealt with the matter at p 271 of his judgment. He said:

> The fourth condition provides that the property is sold subject to any matter B or thing affecting the same whether disclosed at the time of sale or not. Such a condition, however, does not relieve the vendor from the necessity of disclosing any incumbrance or liability of which he is aware but simply protects him if it should afterwards turn out that the property is subject to some burden or right in favour of a third person of which he is unaware. (*Dart on Vendors and Purchasers* 5th ed p 156). It would be nothing short of a direct encouragement to fraud if a vendor were at liberty by a condition of this kind to sell to a purchaser as an absolute and unburdened freehold, a property which he knew to be subject to liabilities which would materially reduce its market value. In the present instance the vendor knew of some of the restrictions and had the means of knowledge of all of them, and he cannot escape from the necessity of fairly disclosing them by omitting to make himself acquainted with his deeds or by forgetting their contents. In C honesty and in law alike he was bound to give the purchaser full and fair information what it was that he had for sale and was inviting him to buy, and having failed to do so he cannot insist upon the bargain procured by the suppression of material matters affecting the nature of the subject of a sale.

Then at the foot of the page Wills J goes on:

> The 12th condition provides that any error or omission in the particulars is not to annul the sale nor to entitle the purchaser to compensation. It is, however, settled law that such a condition will not protect the vendor where the misdescription being of such a nature as but for the condition would avoid the contract is due to the wilful reticence or negligence of the vendor.

Wills J's decision was affirmed by the Court of Appeal. The de- D cision of the Court of Appeal is reported in (1886) 16 QBD 778. In the course of his judgment Lord Esher, in referring to Wills J's decision, says, starting at p 786:

> But supposing he was right in the view which he took of that and that the case stands in the same position as if no statement had been made, then it is said that by reason of the conditions of sale and of the provisions of the Conveyancing Act, although there is a defect in the title, yet it can be forced upon the purchasers. It seems to me an astounding proposition that a vendor can by means of such conditions of sale force upon a purchaser a defective title, even though he (the vendor) knew of the defect. Now, that Butler did not know of the defect seems to have been assumed by Wills J. If you treat him as dealing himself with the purchasers he knew of the defect, and if the conversation between his solicitor and the plaintiffs' solicitor at which he E was present amounted to nothing at all, then it comes to this, that he, knowing the defect, by his agent put forth the condition of sale. If that is so the case is absolutely within the authority of *Heywood v Mallalieu* (1883) 25 ChD 357 as it seems to me. It is impossible for a vendor knowing of a defect in his title either by himself or his agent to put forward conditions of sale which are to force upon a purchaser a bad title of which he knew, but which he did not disclose. A court of equity would not be of much use if it could not meet such a case as that, and it seems to me clear that a court of equity would never have enforced a contract under such circumstances. Does the Conveyancing Act make any difference? I entirely agree that, whatever construction you put upon it, it cannot have that effect. It cannot cure that defect in the title of the defendant. If that is true there was a defect in the defendant's title, and he was insisting that the court should force on the F purchaser from him a bad title. The court will not do that. He cannot have specific performance of the contract when he has a bad title which was within his knowledge, and it seems to be that he cannot keep the deposit.

I have been referred to the passage in *Emmet on Title,* 17th ed pp 131 to 133 where the vendor's duty to show good title is expounded, and *Nottingham Patent Brick & Tile Co v Butler* (1886) 16 QBD 778 and *Heywood v Mallalieu* (1883) 25 ChD 357 are both cited as good authority. The passage in *Emmet* proceeds immediately at p 133 to consider the effect of an express representation as to title. The passage says:

> An express representation by a vendor in regard to his title to the land relieves the purchaser from an investigation of the facts of the case as he is G entitled to rely on such statement and it is no defence to say that he, the purchaser, had the means of discovery and might with reasonable care have discovered that the statement was untrue.
>
> Thus in *Hunt (Charles) Ltd v Palmer* [1931] 2 Ch 287 the particulars of sale of two leasehold shops described them as valuable business premises. In fact, the lease contained a covenant restricting the lessee from carrying on any trade or business other than that of a ladies' outfitter, fancy draper or the manufacture of ladies' clothing. The purchaser wanted the premises for use as a dairy. The conditions contained a condition as to the lease being open for inspection before the sale and that the purchaser should be deemed to have notice of the covenants in it. It was held by Clauson J that as the vendors had described the premises as valuable business premises, the purchaser was induced by that description to buy them and he was entitled H to require that the premises should fairly answer that description. The fact that the purchaser must be deemed to have had notice of the covenant was not relevant. The action for specific performance against the purchaser was therefore dismissed.

In fact in the report of *Hunt (Charles) Ltd v Palmer* it appears that the conditions of sale, which were the National Conditions of Sale of the time, contained not only the condition that the purchaser should be deemed to have bought with full notice of the contents of the lease, but also a condition no 10 that no error, misstatement or omission in the particulars, sale plan or condition should annul the sale, nor should any compensation be allowed either by the vendor or the purchaser in respect thereof. The argument by the plaintiffs J in that case to enforce the contract relied, as the report shows, on both the conditions which I have mentioned, and that argument was rejected by Clauson J, who held that non-disclosure overrode both conditions.

The Supplement to Emmet refers in relation to the passage on p 133 to a decision of my own in *Laurence v Lexcourt Holdings Ltd* [1978] 1 WLR 1128 where I applied Clauson J's decision in *Hunt (Charles) Ltd v Palmer* [1931] 2 Ch 287. I there held that it made no difference whether the misrepresentation involving a misdescription of the property was actually a term of the contract or was made in the course of the negotiations before the contract and I referred to a passage in a judgment of Tindal CJ in *Flight v Booth* (1834) 1 K Bing NC 370, which had been cited in a case referred to in Clauson J's decision. Tindal CJ said:

> We think it is at all events a safe rule to adopt that where the misdescription, although not proceeding from fraud, is in a material and substantial point so far affecting the subject matter of the contract that it may reasonably be supposed that but for such misdescription the purchaser might never have entered into the contract at all, in such case the contract is avoided altogether and the purchaser is not bound to resort to the clause of compensation.

It seems to me that the equitable barrier to specific performance extends not merely to matters of title where the vendor has failed to L disclose defects known to him in his own title, but also to misrepresentation where the vendor has, albeit innocently, misdescribed the property or made some other misrepresentation about the property, when the true facts were within his own knowledge. A trifling misrepresentation where the truth would have had no effect on the purchaser and the purchaser would have none the less entered into the contract, rests in a different category, because there the contract has not been induced by the misrepresentation, but here, as I find, the purchaser would reasonably have refused to contract unless the boundary dispute it disclosed to him had first been resolved.

Therefore, it seems to me that on equitable principles and consistently with the authorities I have mentioned, and consistently also M with the fairly recent decision of Walton J in *Faruqi v English Real Estates Ltd* [1979] 1 WLR 963, the vendor, Mrs Boyle, is not entitled in equity to rely on condition 17 in the circumstances of this case.

The alternative view is that whatever the position may previously have been in equity, the extent to which an exclusion clause can be valid is now governed by section 3 of the Misrepresentation Act 1967. As amended by the Unfair Contract Terms Act 1977, this section now reads as follows:

> If a contract contains a term which would exclude or restrict (a) any liability to which a party to a contract may be subject by reason of any misrepresentation made by him before the contract was made or, (b) any remedy

> available to any other party to the contract by reason of such a misrepresentation, that term shall be of no effect except in so far as it satisfies the requirements of reasonableness as stated in section 11 (1) of the Unfair Contract Terms Act 1977, and it is for those claiming that the term satisfies that requirement to show that it does.

If condition 17 has any validity or relevance in the circumstances of this case, it is a term which would exclude liabilities to which Mrs Boyle would be subject by reason of misrepresentation, and so section 3 is applicable. The requirement of reasonableness in section 11 is that the term shall have been a fair and reasonable one to be included, having regard to the circumstances which were or ought reasonably to have been known to or in the contemplation of the parties when the contract was made. I do not regard condition 17 as satisfying that requirement in the circumstances of this case. Another way of putting it is that Mrs Boyle has not shown that it does satisfy that requirement.

It has been submitted by Mr Seymour that as there were solicitors acting for both parties, it would be a very strong thing to say that any term of the contract which resulted is not a fair and reasonable one in the circumstances. That argument would have great force, no doubt, if the solicitors had specifically directed their minds to the problem and had evolved the clause which was under attack. In fact, however, neither solicitor directed his mind to condition 17, and they have both told me, and they are men of not inconsiderable experience as conveyancing solicitors, that they have never come across a case where any question under condition 17 has arisen. It was submitted that it was the duty of the purchaser's solicitor to advise his client, Mr Walker, of the implications of condition 17 and of the other terms of the contract, which Mr Walker was going to enter into, and he must be taken to have discharged that duty and satisfied himself and Mr Walker that the terms were reasonable. It is, of course, the duty of a solicitor to advise his client about any abnormal or unusual term in a contract, but I think it is perfectly normal and proper for a solicitor to use standard forms of conditions of sale such as the National Conditions of Sale. I do not think he is called on to go through the small print of those somewhat lengthy conditions with a toothcomb every time he is advising a purchaser to draw the purchaser's attention to every problem which on a careful reading of the conditions might in some circumstance or other conceivably arise. I cannot believe that purchasers of house property throughout the land would be overjoyed at having such lengthy explanations of the National Conditions of Sale ritually foisted upon them.

It has also been submitted by Mr Seymour that the court should be very slow to hold that a common form clause like condition 17 is not fair and reasonable. Of course it is true that there are common form clauses which have been evolved by negotiation between trade associations, associations of merchants or associations of growers or trade unions or other such bodies concerned to protect the rights of their members, which can be regarded as representing what consensus in the trade regards as fair and reasonable. Again, the National Conditions of Sale are not the product of negotiation between such bodies and it is plain from the conditions I have cited in the *Nottingham Tile Co* case and Clauson J's case of *Hunt (Charles) Ltd v Palmer* that what now appears in condition 17 has come down through the ages despite very drastic limitations imposed on it by the courts. I do not think it can be said that its precarious survival until 1977 entitles it to the automatic accolade of fairness and reasonableness.

It follows that Mr Walker succeeds in my judgment in his claim to the return of his deposit. The solicitors for Mrs Boyle in fact put the amount of the deposit which they had previously held as stakeholders into a deposit account and gave up their position as stakeholders in February 1980. Mr Walker's entitlement to the deposit carries with it, in my judgment, entitlement to the interest earned by the deposit since that date.

In addition, Mr Walker claims damages for the loss of interest from the time when he paid the deposit to Mrs Boyle's solicitors until they put it into an interest-earning deposit account. That claim is put as a claim for damages for misrepresentation under the Misrepresentation Act 1967. Before that Act, as the misrepresentation was, as I have found, innocent, no damages could have been recovered. Section 2(1) of the 1967 Act provides:

> Where a person has entered into a contract after a misrepresentation has been made to him by another party thereto and as a result thereof he has suffered loss, then if the person making the misrepresentation would be liable to damages in respect thereof had the misrepresentation been made fraudulently, that person shall be so liable notwithstanding that the misrepresentation was not made fraudulently, unless he proves that he had reasonable ground to believe and did believe up to the time the contract was made that the facts represented were true.

In the present case the representation was not made fraudulently. If it had been made fraudulently Mr Walker would have had a claim for damages, but I take the view that Dr Boyle did not in the circumstances have a reasonable ground to believe, although he did believe, that the answer he prepared and which was adopted by Mrs Boyle's solicitors was true. Therefore, I think Mr Walker is entitled to an inquiry as to the damage he has suffered by loss of interest over the period from his payment of the deposit to Mrs Boyle's solicitors until January 23 1980 when the deposit ceased to be held by them as stakeholders and earned the interest to which he is entitled as I have already held.

An order was made for the return of the deposit to the purchaser, with interest from the date when the deposit ceased to be held by the vendor's solicitors as stakeholders, and an inquiry as to the damage suffered by the purchaser by loss of interest on the deposit for the period from payment of it to the vendor's solicitors to the date when it ceased to be held by them as stakeholders; vendor to pay the purchaser's costs of the action.

NEGLIGENCE

A **QUEEN'S BENCH DIVISION**

November 13 1981

(Before Mr Justice RUSSELL)

LONDON AND SOUTH OF ENGLAND BUILDING SOCIETY v STONE

Estates Gazette February 6 1982

(1981) 261 EG 463

Action by building society against a chartered surveyor for negligence and breach of contract in respect of a mortgage valuation — Questions of liability and measure of damages — At date of valuation house suffered from serious structural defects — Settlement causing fractures — Visible cracks indicative of settlement should have been observed — Valuation report to building society stated that house was free from defects, stable and not needing repairs — House purchased for £14,850 and on basis of report society advanced £11,880 — Building deteriorated so that society's security, it was claimed, became worthless — Society, however, decided to carry out repairs and its expenditure, including an amount for the purchasers' alternative accommodation during works, eventually reached a total of £29,478 — Surveyor's liability found to be established, the main question being the correct measure of damages — Cost of repairs not correct measure — Guidance from decision of Devlin J in *Eagle Star Insurance Co Ltd v Gale & Power* — Correct measure the difference between the amount of the loan regarded as fully secured and the true value of the security, but in assessing the society's loss on the mortgage the value of the purchasers' personal covenants had to be taken into account — The security having been found to be worthless, the damages, apart from a small item which had been agreed, were held to be the total amount advanced, namely, £11,880, less a deduction of £3,000, mainly in respect of the personal covenants — A payment of £30,000, described as "ex gratia", made to the society by the insurance company with which the house had been insured, did not have to be brought into account

In this action the plaintiffs, London and South of England Building Society, claimed damages against the defendant, Barrie Stone, a chartered surveyor, for alleged breach of contract and negligence in relation to a building society valuation of a house called Lane End, Kingsdown Grove, Kingsdown, Corsham, Wiltshire. The action did not involve the purchasers as parties.

P A Twigg (instructed by Lawrance, Messer & Co) appeared on behalf of the plaintiffs; J C N Slater (instructed by Barlow, Lyde & Gilbert) represented the defendant.

Giving judgment, RUSSELL J said: As the title of this action indicates, the plaintiff is a building society. The defendant, Mr Barrie Stone, is a chartered surveyor. The nature of the claim is that the plaintiffs seek damages from the defendant alleging professional negligence and breach of contract arising out of a report given by the defendant to the plaintiffs in consequence upon his carrying out a survey on a dwelling-house. Mr Stone advised that the property was an appropriate one on which the plaintiff society could advance £11,880 to a prospective purchaser, the dwelling-house serving as a suitable security for such an advance. It is alleged that the defendant failed to detect that the dwelling house was suffering from serious structural defects at the time of his survey. In consequence it is said that the security was valueless.

I propose to deal with the issue of liability first: liability being denied. It would be convenient if I set out in summary form the salient facts of the matter, so far as possible in some sort of chronological order.

Lane End, Kingsdown Grove, Kingsdown, in Corsham, Wiltshire, is a semi-detached house built in about 1961. It was built on sloping ground on the site of an old quarry, the land sloping steeply from the front of the house to the rear. In the early part of 1976 it was owned by a Mr Williams. The house was insured with the Norwich Union Insurance Co.

The first event of any significance occurred in February 1976. Mr Williams made a claim upon his house insurance policy alleging that damage had occurred to the house as a result of subsidence. A Mr J R Probyn, a chartered surveyor and loss adjuster of vast experience, and a partner in the firm of Cunningham, Hart & Co, was instructed by the Norwich Union to investigate the claim. His report is to be found on the first page of a bundle of agreed documents (marked "A") to which I shall be referring from time to time during the course of this judgment. The material part of that report, at p A2, reads as follows:

> The damage extends to fractures in the outer wall in the region of the rear quoin with associated fractures on the rear wall, internal fractures affecting wall surfaces and ceilings in the lounge and kitchen on the ground floor, and a bedroom and bathroom on the first floor. In addition the small single storey extension (see photographs), which includes a wc and fuel store, has rotated slightly due to movement of the floor slab and this structure is badly distorted.

Later the report said:

> . . . once a structure has fractured it will be sensitive to thermal ground movements which would not otherwise have caused any visible movement in the fabric of the building.

At much the same time as that report was prepared, on February 19 1976, Mr Probyn took two photographs of the dwelling-house. They are agreed and have been lettered "A" and "B" respectively, on document P1. Photograph B shows the movement, to which reference is made in the report, on the north-east corner of the dwelling-house. Although no reference is made to this in the report, Mr Probyn told me when he gave evidence that at his inspection he detected that the concrete blocks below the line of the crack to be seen in photograph B were standing, he estimated, about one-eighth of an inch proud of the concrete blocks above the crack.

Mr Williams, the owner, had done some remedial work. As can be seen from the photographs he had done some pointing on the north-eastern corner. When Mr Probyn gave evidence before me he expressed the view that the indications of settlement seen in photograph B could not have been disguised to the extent that they were not readily observable by the following May of the same year, 1976. Indeed, the effect of Mr Probyn's evidence was that that crack in particular would be detected, or should have been detected, at any stage thereafter by the most cursory of professional examinations. The Norwich Union repudiated Mr Williams' claim or, if that did not occur, it was certainly not pursued. The view was taken at that time that the settlement had begun long before the insurance company came on risk.

In March 1976 a Mr Robinson comes on the scene. At that time he was unmarried, but he intended to be married to a lady who, indeed, ultimately became Mrs Robinson. The two of them were attracted to this dwelling-house and Mr Robinson told me that upon his original visits in March 1976 he did not notice anything really untoward with the property. He and Mrs Robinson (as I shall call her) decided to make an application for a mortgage from the plaintiffs, having agreed with Mr Williams to purchase the house for £14,850. The application by the Robinsons to the plaintiffs was for an advance of £12,800 repayable over 25 years. At that time Mr Robinson was earning £3,250, in connection with a business with which he was involved, and Mrs Robinson, who was a schoolteacher, was earning £4,000 per annum.

Upon receipt of the application for a loan the plaintiffs instructed the defendant, who was a surveyor of considerable experience in the area, to give them a report and a valuation of the property. Having visited the house and spent, as he told me, about an hour looking over it, Mr Stone, the defendant, prepared a report dated May 24 1976 (which is to be found at pp A27-A30 inclusive). In paragraph 1 of the report are to be found the valuation details and the extent of the loan required together with the agreed purchase price. Paragraph 4 deals with the construction of the dwelling-house and is headed "(State condition if other than satisfactory)". Mr Stone simply indicated in that paragraph that the walls and chimneys were constructed of concrete blocks, the roof of concrete tiles. So far as the outbuildings and boundaries are concerned, and this plainly embraced the outhouse to which I have referred, again the valuer was required to state its condition if other than satisfactory. Nothing was said untoward about the outbuildings in that paragraph. Paragraph 10 referred to repairs. A schedule of essential repairs or condition to be observed by the applicant was requested. The answer from Mr Stone was that none were considered necessary. Finally, under the heading "General observations", at paragraph 13, the answer to the question: "Do any circumstances exist which could endanger the stability of the property?" was "None observed". Paragraph 13(2): "Do you recommend the property as a suitable security for the advance and term requested?" Answer: "Yes". Then, so far as relevant, the following comment appears in Mr Stone's report:

> This is a well-built and maintained property on a small private residential development in an attractive position with fine views.

The valuation he gave for mortgage purposes was £14,850, and at the conclusion of the report Mr Stone certified as follows:

> I certify that the property described in this report has been inspected by me, that I have valued the property and prepared this report in accordance with the provisions of section 25 of the Building Societies Act 1962 (or any provisions re-enacting the same for the time being in force), and that I am not disqualified from reporting on the property by virtue of the said provisions.

The plaintiff society relied upon that report.

In the event, £11,880 was advanced to the Robinsons repayable over a period of 25 years, at the rate of £121 per month. That advance represented, in accordance with the policy of the plaintiff society, 80 per cent of the total purchase price. The defendant was paid £33.80 for his report.

The legal charge on the property, the formal mortgage deed, is to be found at p A43 in the bundle. It is, I imagine, in common form, but it is worth recording some of the provisions of the deed. The borrower covenanted with the society to pay the monthly instalments to which I have referred. The borrower, as beneficial owner, charged by way of legal mortgage all the property described in the deed, namely, this dwelling-house. There were various covenants into which the borrower entered. Among them, it is to be observed that the borrower covenanted at all times to keep any buildings on the property in good and tenantable repair to the satisfaction of the society. By condition 5(c) it was provided that the society would, from time to time, during the continuance of the security, advance to the borrower such further sum or sums as the society thought fit. Further advances were to be deemed to be advanced on the security of the property and the repayment thereof, with interest, was secured by the charge contained in the deed. Condition 5(i) reads as follows:

> . . . all money paid by the Society in completing repairing amending altering improving or insuring the property, or in making any payments for outgoings in relation thereto or for the protection or improvement of the security hereby created shall carry interest at the rate of interest herein provided for and shall be repayable by the borrower on demand and in the meantime shall be a charge on the property.

Condition 8 referred to the insurance of the property in the joint names of the society and the borrowers. The date of that deed coincided with the date of the conveyance, viz September 23 1976. The property was insured in the name of the plaintiff society and the Robinsons with the Sun Alliance Insurance Co. The Robinsons, the purchasers, did not have their own independent survey.

I heard Mr Robinson give evidence. He told me that when he was moving into this property in October 1976 he noticed, for the first time, pointing on the north-eastern corner, and at that date he detected that the brickwork was protruding slightly. As time went on he found that various doors in the house were difficult to open and close. The decorations he found to be fairly good. Within a matter of two or three weeks he discovered that the kitchen door into the hall would not shut. He thought, as a result of an examination, that this was because of a rise of about an eighth of an inch on top of the door at one corner. Conditions continued to deteriorate. At about Christmastime 1976 Mr Robinson told me that he detected the bathroom floor sloping and there was also, shortly afterwards, a slope which he noticed on the kitchen floor. During 1977 further defects were noticed by Mr Robinson so that eventually, in September 1977, Mr Robinson called in a Mr B C Hampton, a consulting engineer.

Mr Hampton examined this dwelling-house and his report upon it is to be found at p A52 and is dated September 16 1977. He made these observations:

> The retaining walls adjacent to the path leading down to the front door are cracked and show signs of movement. All the external walls and some internal walls of the house are cracked through movement of the foundations. The corner of the house overlooking the valley has sunk appreciably. The foundations on the side of the house facing the valley consist of block work piers with thin concrete lintels between them supporting the house. The piers and lintels show signs of movement.

In addition, on September 28 1977, Mr S J Westwood, a chartered surveyor instructed by the Sun Alliance took further photographs which are exhibited on document P1. Later, on February 2 1978, in the presence of the defendant further photographs were taken of the north-east corner. They show the same crack as was detected by Mr Probyn in February 1976 and when the defendant was asked by Mr Westwood whether he, Mr Stone, had detected the crack at the time of his examination, Mr Stone frankly conceded that he had not done so. Mr Westwood took the view, when he gave evidence before me, that making all allowances for the natural deterioration of this house and for the fact that remedial works may have been carried out, and certainly were carried out, by Mr Williams prior to the survey of Mr Stone, nevertheless, that crack, depicted in his photographs and also in Mr Probyn's photographs, should have been obvious to any reasonably competent surveyor carrying out an inspection in May 1976.

Another expert, called on behalf of the plaintiffs, was a Mr T R Gratton. He was a very experienced chartered surveyor, the principal of his own firm in Bath where he had been in practice for a number of years. For my assistance his report, prepared for the purposes of this case on October 7 1981, and marked P4, was placed before me. In that report, supplemented by his evidence, Mr Gratton told me that in July 1977 he had had occasion, on behalf of another building society, to inspect the adjoining property, known as Squirrel's Dray. At that time, in July 1977, he noticed evidence of minor settlement in that property, a number of the external blocks having moved, joints having widened and mortar having fallen away. He glanced across at the rear wall of the adjacent house, Lane End, and his evidence was that at that time that property showed similar signs of settlement though on a more extensive scale than Squirrel's Dray. Mr Gratton, corroborating the

A evidence of the other experts, took the view that the conditions depicted in photograph B must have been perpetuated until the date of Mr Stone's examination and he likewise expressed the professional opinion that a reasonably competent surveyor should have detected signs of settlement in this house when carrying out a survey on behalf of the building society in May 1976.

Mr Gratton prepared, for the assistance of those instructing him, what he considered should have been the report prepared by Mr Stone. I do not attach a great deal of weight to that part of Mr Gratton's evidence. It is to be observed that the report made no mention of the outhouse referred to by Mr Probyn and it is right at this stage to say further that the complaints levelled against Mr B Stone in the pleadings did not involve the outhouse at all but were principally concerned with the cracks in the structure of the dwelling-house itself.

Mr Stone gave evidence before me. If I may say so I found him a completely frank and, in my judgment, honest witness. He displayed a professional approach to the allegations that were being made against him and he emerged from the witness box, if I may say so, with credit. He told me that, as he had told Mr Westwood, he did not see anything untoward in this property when he surveyed it. He spent an hour looking over it. He took the view, reluctantly, that it may well have been that between Mr Probyn's examination in February 1976 and his own examination in May further remedial C works had been carried out by Mr Williams which successfully disguised the crack to be seen in photograph B. It was further suggested that the reality of this case was that the really serious settlement occurred after Mr Stone carried out his survey in May 1976. Of that there is no doubt, in the sense that conditions undoubtedly deteriorated substantially long after Mr Stone's examination and, indeed, after the occupation of the dwelling-house by the Robinsons.

The case, as it progressed, emerged against Mr Stone in this way. It was said, in the forefront of the complaints, that he should have detected the crack to be seen in the photographs, making all allowances for the remedial work that Mr Williams attempted and mak- D ing all allowances for the weathering-in of mortar with the passage of time. It is said that he should have detected the rotation in the outhouse; a defect that could not in any circumstances have been disguised. It is said that there should have been a detection of sticking doors and cracks in the internal decorations.

Having regard to Mr Williams' undoubted determination to better his property, for whatever reason, I am not satisfied that when Mr Stone inspected there was evidence within the house to indicate that it was settling. I reject the allegations relating to the sticking doors and the internal decorations. But, so far as the pointing is concerned, with the crack in the structure itself at the E north-east corner, and so far as the rotation of the outhouse is concerned, I am driven to the conclusion, on the totality of the evidence, that those defects were there to be seen at the time of Mr Stone's survey when, for whatever reason, he failed to detect them. I wish to make it clear that I am satisfied that this was an isolated lapse on the part of a gentleman of considerable expertise and experience. But I am driven to the conclusion, from the evidence that I have heard and the documents with which I have been provided, that upon the occasion of his survey, and I bear in mind of course the nature and extent of it — it was not a full structural survey — nevertheless, Mr Stone fell from his own normally high standards and did not on this occasion exercise the degree of care F and skill to be expected of a reasonably competent surveyor carrying out the type of examination that he was required to carry out on behalf of the plaintiffs. Accordingly, in my judgment, liability is plainly established.

I turn now to what I consider to be a much more difficult part of this case. What is the proper measure of damages to which the plaintiff society is entitled?

In order to consider this it is necessary to return to the history of the matter. What happened after Mr Hampton's inspection in September 1977? The Sun Alliance, it seems, were not prepared to help at that stage and on February 9 1978 the defendant was told that the plaintiffs intended to institute proceedings against him. On February 13 1978 the consulting engineer, Mr Hampton, inspected G again and he found, as was to be anticipated, some deterioration in the property. Mr Stone was put on notice that remedial work was contemplated. Mr Hampton obtained quotations for the work that he considered necessary and in particular he approached Pynfords Design Ltd, an organisation well known in the field of underpinning foundations. Their estimate, or quotation, arrived in June 1978. It is dated June 20 and is comprised in documents A95-A111 inclusive. At p A96 Pynfords gave this quotation:

> To strengthening the foundations by under-pinning to a maximum depth of 3.300 metres under all the internal and external walls of the property, excluding the party wall, using Pynford beam and base under-pinning method. For the lump sum of £7,462.00 plus VAT. In addition a pro- H visional sum of £1,500 should be allowed for providing [equipment] to service the site . . . via the difficult access conditions.

In round figures that quotation amounted to approximately £10,000, but, as I have already observed, it excluded any work on the party wall between Lane End and Squirrel's Dray; it also included a variation clause. That quotation was ultimately accepted by the plaintiffs and Mr Hampton told the society that in his judgment the work on the superstructure would probably involve an additional £5,000.

At this time (and I am speaking now of the period about August/September 1978) the attitude of the plaintiff society J toward their borrowers, the Robinsons, was beginning to emerge and I heard from the managing director of the plaintiff society, Mr A G Stoughton-Harris. The society, he told me, was hoping at this time to expand in the area. He felt that Mr Robinson was entirely innocent and in a sense a victim of the society's own agent. There were various options open to the building society. The Robinsons could not be expected to pay for the cost of the work in cash. If they left the premises and defaulted in their payments under the terms of the deed the society could foreclose. Alternatively, the moneys to be expended by the society upon repairs could be added to the mortgage debt and an adjustment of repayment made. In the further alternative, so said Mr Stoughton-Harris, some compromise K might be reached with the Robinsons.

These matters were canvassed in a letter from the plaintiffs' solicitors, dated August 11 1978, at p A123. The solicitors were acting both for the building society and for the Robinsons. In advising the building society, the letter reads, in part, as follows:

> I would be grateful if as a matter of urgency you could set out in a letter your position in this matter so that our client can plan for the future. It may be that you would consider it as a public relations matter to agree to bear the cost in whole or in part yourselves or you may perhaps feel that you could make some form of interest free loan to Mr and Mrs Robinson to be repaid over some considerable length of time. Alternatively you might feel that the sum should be added to the mortgage advance and that the mortgage L repayments should be increased in the future. Our client is as I understand it in a somewhat difficult financial position in that he would certainly not be in a position to incur a bill of £10,000 which he had to pay in say a year's time.

No decision as to the future, *vis-à-vis* the Robinsons, was taken at that time, and matters in that regard were left in abeyance by Mr Stoughton-Harris (as his letter of August 21 1978, at p A124 and 125, so indicates). What he did do, however, was to authorise, in the name of the plaintiff society, the work quoted by Pynfords. That work began in October 1978.

To cut a long story short, the Robinsons had to be rehoused for a time and the cost of the work escalated. There were further quotations from Pynfords for what is termed "crack bonding". It was M decided ultimately that the party wall, in addition to the three outer walls, would have to be under-pinned. This involved an additional £4,000 or £5,000. The work on the superstructure, it was found, could be done only on a day-work basis and the effect of that was that Mr Hampton's initial estimate of £5,000 was woefully too small. In the event it is common ground that no less than £29,478, including £12,153 for the superstructure, was expended by the plaintiff society upon this house, including an item relating to the costs of the Robinsons' alternative accommodation. It is that sum which Mr Twigg, on behalf of the plaintiffs, contends is the proper measure of damages.

To complete the factual history. On June 1 1979, by which date the work had been completed and the Robinsons were back in occupation, Mr Stoughton-Harris wrote to them. His letter is at p A192 and in it the managing director informed the borrowers that the society did not propose to seek an additional penny piece from them. The Robinsons continued to pay their monthly instalments of £121 until about August 1981, when they sold Lane End for £26,500. The mortgage was discharged and, taking into account their legitimate charges, the plaintiff society was repaid £11,914; in other words the whole of their advance. I should add, however, that it is common ground that there had been no improvement in the house by virtue of the repairs in the sense that it was merely restored to the condition described in the defendant's report.

One final matter to conclude the history in this part of the judgment. Among the agreed documents, at p A196, there is a memorandum which indicates that the plaintiff society received a payment of £30,000 from the Sun Alliance Co. Mr Stoughton-Harris described this as an ex-gratia payment. The matter was not investigated in depth during the course of the evidence, though it was introduced by Mr Slater. There are various possibilities which spring to mind, but I should make it clear that this payment will play no part whatever in my findings. I do not think that the plaintiffs have to bring it into account in any way in assessing the loss that they have sustained.

How then is the case put on each side? I am greatly indebted to both counsel not only for their respective, clear and skilful submissions but for their researches. Between them they cited some 15 or 16 authorities, and I hope I will be forgiven if I do not refer to each and every one of them. Some were of greater assistance than others. What plainly emerges, in my judgment, is that while the dicta in some of the cases assist me in the principles which I should apply, the precise formula very much depends upon the individual facts of the individual case.

First, I remind myself that this is a claim by a mortgagee. It is not a claim by the owner of property who has been misled by a negligent valuation. In the very nature of things therefore the interest of the mortgagee in the property is different to that of an owner-occupier. The mortgagee's interest in the property is that it serves as the security for his loan. It does not, and cannot, extend beyond that. It must, in my judgment, be equated with the extent of the loan. That may be, in an individual case, a substantial proportion of the total value of the property at any given time; it may be but a small fraction of the total value. The owner is in an entirely different position. His interest, subject of course to the mortgage, is in the whole of the property. That interest may subsist a lifetime; the mortgagee's interest will subsist only for so long as he is owed money on the security.

Mr Slater, on behalf of the defendant, contends therefore that when a mortgagee is misled by a valuation the maximum loss he can suffer is that he is left with an unsecured loan and what the court has to do is to make a valuation of that loss — the difference between a secured and an unsecured loan. That assessment has to be made at the time the security disappears as it were; in the instant case when the loan was made in 1976 when £11,880 was advanced or, perhaps, in September 1977 when the damage really became very apparent and substantial repairs were needed. In monetary terms the two dates do not really matter, for very little had been paid off the capital sum even by 1981 when the house was resold.

With one exception, Mr Slater points out, none of the authorities supports the suggestion that repair costs to real property, where that property is charged as a security, is a proper measure of loss to a mortgagee. That one exception is *Rust v Victoria Graving Dock Co* (1887) 36 Ch D 113. I have read the report. In my view that case had its own very special facts. It was not a case involving a negligent valuation; it was a case of direct physical damage to the secured property by the direct tortious activity of the defendant in flooding land, and the secured property consisted of houses in the course of erection. I do not think that the special facts of that case enable me to read into it any proposition of general application to claims by mortgagees whose security is diminished in value by a negligent surveyor's report.

Much nearer the facts of the instant case is the decision of Devlin J (as he then was) in *Eagle Star Insurance Co Ltd v Gale & Power* (1955) 166 EG 37. I read from the report as follows:

> This was a claim by the Eagle Star Insurance Co Ltd, . . . against Gale & Power, a firm of chartered surveyors . . . for damages for breach of contract for the valuation of a house.

On behalf of the plaintiffs it was said that

> in January 1952 defendants agreed to make an inspection and report on a house at Highfield Drive, Ewell, Surrey, of which plaintiffs were intending to become mortgagees. In their report, the defendants said the property was worth £3,350. Relying on defendants' report and valuation, plaintiffs granted a loan of £3,015 to a Mr Curtis, the owner and occupier of the property, with a mortgage on the house as security for the loan . . . defendants in their report stated that the house did not need any immediate repairs and was readily saleable at their valuation. They failed to notice that settlement or subsidence had occurred; they failed to see fractures in the external walls, movements in the internal walls and an extensive bulge in the south-east wall. In plaintiffs' opinion, the property was not worth more than £1,500 . . . defendants admitted the breach, [but submitted] that plaintiffs were not entitled to any damages because the mortgage of the house was only a collateral security . . . it was agreed that in January, 1952, it would have cost £950 to put the house in good repair.

After listening to submissions by counsel, his Lordship said:

> . . . that the only way of assessing damages was to calculate what plaintiffs would have got had they realised their securities. The fact that they had not done this could not influence his judgment. The two main items of security were the house, which his Lordship estimated as being worth £1,600, and Mr Curtis's personal covenant to repay £1,500 out of his retirement gratuity in August 1956. Together these amounted to £3,100 without counting the other personal covenants.

The learned judge said this:

> I cannot, therefore, say that by any method of calculation, plaintiffs have lost any of the money they advanced. I cannot accept defendants' submission that this is a case for nominal damages, and I think £100 would indemnify plaintiffs against the possibilities of their not being able to recover their money from Mr Curtis. Plaintiffs are also entitled to £52 10s, the cost of the surveys and reports made on the house subsequently.

There was judgment for £152 10s.

Devlin J there plainly rejected the plaintiffs' claim that the measure of damages was the cost of restoring the house, viz £950. He clearly took into account the mortgagor's personal covenant to pay and assessed the chances of that not being honoured by the mortgagor in the particular circumstances at seven or eight per cent of the total value.

Mr Twigg, on behalf of the plaintiffs, recognised that that case is against him and he urged me not to follow it. On the basis that I do follow it Mr Slater argued that I should discount from the total loan of £11,880 such a sum as I think should be subtracted to take account of the various powers which the plaintiff society had under the mortgage deed. I have already referred to them. There were the personal covenants of the Robinsons to continue paying the advance, whether in occupation or not, for the next 25 years from the date of the mortgage deed. If they wanted to remain in occupation there was the power to require them to pay for repairs by increased instalments. All these factors, it is said, I should take into account and in due course I shall return to them.

Further, argued Mr Slater, should I be tempted to examine the course adopted by the plaintiffs in repairing the property as a proper measure of damage the most cursory examination discloses the unreasonableness of that course. At no stage did the plaintiffs' interest in Lane End exceed £12,000 in round figures. By September 1978 the estimated repair cost was of the order of £15,000 and with the passage of time that figure escalated until it became more than 2½ times the value of the plaintiffs' interest in the property. At least some of that escalation, said Mr Slater, should have been foreseen. Mr Hampton knew at an early stage that there was a crack in the party wall and yet Pynfords were not instructed to quote for it in the first instance. When it became apparent, as it should have been apparent from the outset, that the party wall had to be under-pinned there was a substantial increase in the costs above the estimated £15,000. By November 1978 the estimated

A costs of Pynfords' work alone had risen by some £13,000. In these circumstances, said Mr Slater, the plaintiffs had no right to proceed with repairs and no right to hand the bill to the defendant who, in being negligent, could never have foreseen or contemplated that as a result of the defect in his report he would be faced with a repair account that was over twice the figure at which he valued the whole of the property. Whether the concept is properly described as causation or remoteness of damage and whether this was a tort or a breach of contract, or both, the cost of repairs, argued Mr Slater, is irrecoverable.

What does Mr Twigg say, on behalf of the plaintiffs? Quite the contrary. He begins with first principles. He cited *Dodd Properties* B *Ltd v Canterbury City Council* [1980] 1 WLR 433. The argument goes as follows. The plaintiff is entitled to receive as damages such a sum of money as will place him in as good a position as he would have been if no negligence or breach of contract had occurred. But that principle, it is conceded, must be qualified in all cases by a duty on the part of the plaintiff to act reasonably in mitigating the loss. What is reasonable is always a question of fact and not law and, in appropriate cases, provided that the plaintiff has acted reasonably, he may recover the losses incurred even though the resulting damages are, in the event, greater than the loss incurred. In the instant case, submits Mr Twigg, the measure of damages is the cost of making good the house so as to provide for the plaintiffs a proper C security for their advance. Dependent upon the extent of the advance the cost may not necessarily be the cost of putting the house in the condition described by the valuer, but where, as in this case, the house was valueless and the advance represented 80 per cent of the true value, the work done was necessary and reasonable to restore the security to a condition where it represented a proper security for the advance.

Assuming for the moment that this is the right approach, I must examine the course adopted by the plaintiff society. I must bear in mind at all stages that their advance was less than £12,000. Was it commercial good sense, I ask myself, to embark upon these repairs, thereafter forgoing any claim against the Robinsons on their per- D sonal covenants? Mr Twigg emphasises that the probable course of events, if the repairs had not been undertaken, would have been that the Robinsons would have been forced to leave the house in a dangerous condition; they would have been probably unable, and perhaps unwilling, to maintain their covenant to pay £121 per month. The plaintiffs were entitled to take into account that the Robinsons not only wished to remain in the property but were the innocent victims of a mistake made by a valuer instructed by the plaintiffs who felt under a moral obligation to indemnify their purchasers. There was, it was urged, a special relationship between a building society and a borrower and, as Mr Stoughton-Harris pointed out, the commercial reputation of his society was a relevant E consideration. Although the society had the contractual power to seek additional payments from the Robinsons they were under no duty to enforce it.

I have carefully considered all these factors but, at the end of the day, I am not persuaded that it was reasonable of the society to embark upon the course that they did, having regard to the extent of their interest in the property and what was from the outset plainly going to be a substantial expenditure which rose to comparatively astronomical proportions. In moral terms, while the conduct of the society was admirable in their dealings with the Robinsons, I must bear in mind that this was also a commercial transaction. To use the words of Morris LJ in *Philips v Ward* [1956] 1 WLR 471, at p 475, a F case involving a claim at the instance of the purchaser against a valuer who undervalues a property, the damages to be assessed are, and I quote, "such as could fairly and reasonably be considered as resulting naturally from the failure of the defendant to report as he should have done". The Court of Appeal in that case rejected the cost of repairs to bring the property into the condition certified by the valuer as the proper measure of damages but awarded the difference between the surveyor's valuation and the true value. If that be the true test for a purchaser's claim I cannot believe that a mortgagee is in a better position. I do not think that the cost of these repairs, unmitigated by any contribution from the purchaser, passes the test adumbrated by Morris LJ to which I have referred. Accordingly, I reject the cost of repairs as being the proper basis of the G award. In any event, as I have already indicated, I do not regard the course taken by the plaintiff society as being a reasonable one to mitigate their loss.

Mr Twigg, in the course of his closing submissions, acknowledged that if I were against him upon his primary submissions as to the measure of damages then I should adopt the basis of assessment used by Devlin J in the *Eagle Star* case. Indeed, that is how, by amendment, the claim is put in the statement of claim by way of alternative.

Should I make any discount from the figure of £11,880 claimed in the amendment? At one stage Mr Slater argued that because the H society had received this sum upon the resale of the property there had in fact been no loss as a result of the defendant's negligence, but that submission was not pursued. I have to look at the situation at the date of the report. The plaintiffs were left without any security in the house but the society had the continuing personal covenants of the Robinsons. I bear in mind those factors to which I have already alluded. Despite the difficulties that would have been encountered by the Robinsons in the event of their having to leave I think that their personal covenants must be something more than an illusory factor. As time went on, their income rose (they are now earning something like £15,000 per annum; their precise financial position was not investigated in any great depth). But doing the best J I can on the evidence before me I do not think that it would have been commercial good sense on the part of the society simply to have waived all their rights under the mortgage deed. It would probably have been both honourable and commercial good sense for the society not to have pursued the Robinsons to the *full* extent of their contractual obligations. I do not think to have done so would have been a necessary or reasonable mitigation of the loss and I do not think that the plaintiffs were under any duty to do so.

Accordingly, to take account of the Robinsons' covenants and the other facts to which I have referred, I shall make a deduction which reduces the claim from £11,880 to £8,880; to this must be added, by agreement, the reasonable costs incurred by the society K in investigating the matter. That sum I understand to be £253.13. Accordingly, there will be judgment for the plaintiffs in the sum of £9,133.13.

The plaintiffs were awarded the costs of the action and a sum for interest from January 19 1979, the date after which the plaintiffs became out of pocket in paying for repairs.

QUEEN'S BENCH DIVISION

July 28 1981 L

(Before Mr Justice BALCOMBE)

CRUMP AND ANOTHER v TORFAEN URBAN DISTRICT COUNCIL

Estates Gazette February 20 1982

(1981) 261 EG 678

Action against local authority for negligence and breach of statutory duty of the *Anns v Merton Borough Council* type — Decision on preliminary issue of application of Limitation Act 1939 — Severe damage to plaintiffs' house through vertical settlement of non-suspended floor slab — Settlement due to unsuitable filling below slab — Question as to when plaintiffs' cause of action accrued — Test is that cause of action accrues when damage due to negligent act is suffered by a plaintiff and this cannot be before damage is first detected or could by the exercise of reasonable skill or diligence have been detected — Onus on defendants, who raised issue of limitation, to show that plaintiffs' cause of action accrued more than six years before issue of writ — Held that defendants failed to discharge onus — Action not statute-barred — Trial of substantive issues to proceed M

This was a preliminary issue as to the application of the Limitation Act 1939 to the claim of the plaintiffs, Mr and Mrs Crump,

A against the defendants, Torfaen Borough Council (as successors to Pontypool Urban District Council) for alleged negligence and breach of statutory duty in the inspection and supervision of the building of a bungalow purchased by the plaintiffs. The bungalow, 1 St David's Close, was erected at Penygarn, Pontypool, Gwent, by Rural Homes (Cwmbran) Ltd, against whom the plaintiffs had obtained an award of £1,101.74 in an arbitration.

D G Morgan (instructed by Everett & Tomlin, of Pontypool) appeared on behalf of the plaintiffs; N Dennys (instructed by Dolmans of Cardiff) represented the defendant council.

Giving judgment, BALCOMBE J said: On November 7 1969 the architect for Rural Homes (Cwmbran) Ltd ("the builders") gave to
B the Pontypool Urban District Council a notice of intended building operations in respect of a detached bungalow and garage to be erected at Penygarn, Pontypool, Gwent, in accordance with plans which accompanied the notice. On November 26 1969 the council gave notice that they had passed these plans. Thereafter the builders proceeded with the construction of the bungalow and in September 1970 Mr and Mrs Crump, the plaintiffs in this action, bought the bungalow (then and now known as 1 St David's Close, Penygarn) from the builders. Mr and Mrs Crump entered into occupation of the bungalow in October 1970.

During the summer of 1971 Mr and Mrs Crump observed weeds growing from the junction between the skirting board and the floor
C in both of the bedrooms of the bungalow. They reported this to the builders, and a Mr Peterson, the managing director, came along and on two occasions effected remedial treatment by snapping off the weeds, which had grown to a considerable height, at floor level. When this failed to alleviate the problem, the Crumps consulted their solicitors who, on August 19 1971, wrote a letter of complaint to the builders' solicitors. This letter contains the following phrase:

> It would appear that your clients omitted to take some very fundamental steps in dealing with the oversite preparation of this property and the laying of the concrete floors and we would appreciate your clients' explanation and proposals for putting this matter right under the NHBRC certificate.

D Mr Dennys, for the defendant, relied on this letter as an indication that even at this early stage the Crumps realised that there was a serious defect in the structure of the bungalow. In my judgment, this is to read too much into the letter. It was no more than a solicitors' letter putting the clients' complaint in forceful terms.

However, the Crumps were also experiencing water coming under the front door of the bungalow and at the suggestion of Mr Parker, Mrs Crump's son who lived nearby, they instructed a surveyor, M K McKenzie of Pontypool. Mr McKenzie reported by letter dated September 13 1971. After stating that in bedrooms 1 and 2 he had found weed growth emerging from the junction between the floor tiling and the skirting he continued:

E

> The floor slab is concrete laid on polythene sheet. In this type of construction it is necessary to remove all vegetable soil over the area covered by the slab and where deep-rooted weeds exist they should be chemically treated to prevent further growth. In my view either the vegetable soil was not removed or no chemical weed killer was used. There would appear to be some shrinkage in the concrete slab as otherwise the junction between the edge of the slab and the outer perimeter walls would be too tight to permit the emergence of any plant. It is obvious that if a plant can grow through the crack, moisture in the subsoil will also be able to penetrate. It is imperative that the edge of the slab must be sealed and this may mean treating the whole perimeter of the slab. The seal will have to be capable of withstanding differential movement in the adjoining materials and must, of course, be
> F waterproof and inorganic. There is a further danger that, if the vegetable soil has been left under the slab, in the course of time, as vegetable and organic matter decays, the bulk will reduce and remove support to the underside of the concrete floor slab. This could produce cracks in the floor and in the partition walls built off the slab. I have also looked at the entrance lobby where the quarry tiles have been laid with a slight fall back to the wood door sill. This is causing dampness to rise in the adjoining walls. The hardword sill will require to be raised and the quarry tiles taken out and re-laid to fall to the front edge. A fuller and more comprehensive report can be prepared in this matter if circumstances so require.

A copy of this report was sent by the Crumps' solicitors to the builders' solicitors, and the builders then treated the weeds with a chemical weed-killer and filled in the gap between the skirting

G board and the floor with some form of mastic. However, matters did not improve. Not only did the weeds recur but further defects became apparent. (1) The Crumps became aware of draughts in the lounge and bedrooms. The draughts in the lounge caused them sufficient discomfort to require them to sit with their feet up. (2) Some minor cracks appeared in the walls of the bungalow. (3) A more serious crack developed in the east elevation wall between the main building and the garage. The Crumps reported these cracks to the builders, who came and filled them in. They assured the Crumps that there was nothing to worry about. (4) There was damp in the kitchen under the window. The builders told the Crumps that this was due to condensation. (5) Cracks appeared in the concrete path
H surrounding the bungalow and weeds grew in the cracks. The builders treated these weeds at the same time as they treated the weeds in the bungalow.

Throughout this period the builders constantly assured the Crumps that there was nothing wrong and that the matters of which they complained were only to be expected in a new building.

When they moved into the bungalow the Crumps had had wall-to-wall carpet fitted in the hall and the lounge. In late 1976 or early 1977 the carpets were taken up for redecoration and it then became apparent that a noticeable gap had developed between the floor slab and the skirting boards. The Crumps then again instructed their solicitors.

J In March 1977 the Crumps' solicitors gave notice to the builders that they proposed to refer their clients' complaints to the National House Building Council, and this they duly did. A representative of that council attended at the property on June 10 1977 and his report is dated June 16 1977. The relevant parts of the report are as follows:

> The main problem on this bungalow is subsidence of the infill under the solid floors. . . . The garage has been built on a concrete slab and this also has subsided causing the garage to slightly tilt and the brickwork bonded in the main bungalow to crack. . . . The problem is the builder's continuing liability, and I suggest he is instructed to investigate the complaints and put right the defects *which I consider to be major structural ones* [my emphasis]

K The builders did not take any immediate action in response to the National House Building Council's report, but in September 1977 they proposed further remedial works, and in particular the laying of a coat of screed on top of the floor slab so as to fill up the gap between the floor and the skirting board. However, these proposals were considered wholly inadequate by the National House Building Council and, indeed, the subcontractor sent by the builders to carry out this work refused to do so, giving as his reason that the problem was too serious to be treated in this manner. Thereupon, the Crumps instructed David Fowler, a consulting engineer, who made his first visual examination of the property on September 29 1977.
L He observed that movement had taken place in the floor slab, in addition to major structural cracking in the walls to the garage, and because of this it was decided to undertake an excavation through the floor slab to discover the nature of the sub-slab filling. One hole was bored through the floor slab of the lounge and another was excavated adjacent to the side wall of the garage. These were inspected, and soil samples recorded, by a soil expert on October 24 and 25 1977. This investigation showed that beneath the concrete floor slab 150 millimetres thick was a polythene membrane. Thereunder was a medium compact fill of clay, sand and broken stone 160 millimetres thick. Thereunder was a loose fill of sand and clay topsoil, with occasional dirty bricks, fragments of concrete, sand-
M stone, slate, glass and various artefacts, such as a mineral water bottle a car fan belt, chain sprockets and a steering wheel hub. This layer was 1,350 millimetres thick, was inadequately compacted and contained organic matter.

Mr Fowler's report was dated December 8 1977. After listing the structural damage he had found in the property, he gave his observations on the cause of the damage:

> The weed growth previously referred to and the slab settlement clearly relate to the use of vegetable soil as filling below the non-suspended ground floor slab. The present investigation confirms the use of totally unsuitable material as fill below the floor slab in the location examined. It is the writer's opinion that the slab will probably continue to settle if the

> A existing fill is relied on for support. It is possible that vertical settlement in the future could be as much again as had already been recorded. I have no doubt that the reason for the settlement in the floor slab and subsequent distress that is likely to occur to the internal partition walls is due in total to the (unsuitable) quality of filling material below the slab. The material that has been used could in no way be said to be normal accepted building practice, and the use of organic filling material, such as that which has been found, would be bound to cause damage in the long term.

He went on to propose remedial measures.

The Crumps pursued their claim against the builders, and eventually this claim went to arbitration. In due course, by an award
B dated May 18 1978, the arbitrator awarded the Crumps the sum of £1,101.74 and some, though not all, of their costs.

On April 12 1979 the Crumps issued their writ in this action claiming damages against the defendants, Torfaen Borough Council (as successors to the Pontypool Urban District Council), alleging negligence and breach of statutory duty on the part of the council in the inspection and supervision of the building of the bungalow. This is a type of action with which the courts have become increasingly familiar in recent years, the leading case being *Anns v Merton Borough Council* [1978] AC 728 in the House of Lords. The statement of claim was duly served on April 18 1979 and on June 29 1979 a defence was served, alleging (*inter alia*) the plaintiffs' cause
C of action arose more than six years before the issue of the writ and was barred by the provisions of the Limitation Act 1939. At the beginning of the hearing both parties asked me to deal with this issue (of limitation) as a preliminary point, and to this request I acceded. Accordingly, I heard evidence and submissions directed to this issue only. If I were satisfied that the point was a good one I would, of course, dismiss the action.

Before turning to consider the relevant law, I should mention that Mr Crump died in July 1980 and accordingly this action is being carried on by Mrs Crump alone. For some three years before his death Mr Crump had suffered from arthritis, but there was no evidence to indicate that his health was in any way affected by the condition of the bungalow.

D The tort of negligence is traditionally described as damage, which is not too remote, caused by a breach of a duty of care owed by the defendant to the plaintiff: see *Clark and Lindsell on Torts*, 14th ed, para 859. Thus it is apparent that there is no cause of action unless and until damage has been caused. For some time in cases of this type it was uncertain whether the damage was caused, and the cause of action arose, when the building was constructed with inadequate foundations, ie at the same time as the breach of the duty of care: see *Dutton v Bognor Regis Urban District Council* [1972] 1 QB 373 at p 396; *Higgins v Arfon Borough Council* [1975] 1 WLR 524. However, this uncertainty was finally resolved by the House of Lords in *Anns v Merton Borough Council*, when it was held, follow-
E ing the Court of Appeal in *Sparham-Souter v Town and Country Developments (Essex) Ltd* [1976] QB 858, that physical damage to the house was a requisite of the cause of action. Further, it is not any physical damage, however trivial, that gives rise to a cause of action. Because of the nature of the duty of care owed by a local authority exercising duties imposed, and powers conferred, by the Public Health Act 1936 and building bye-laws, the damage must be of such a kind as to endanger the health or safety of the occupants of the building. Lord Wilberforce, with whose speech Lords Diplock, Simon of Glaisdale and Russell of Killowen agreed, put the matter as follows at p 760 "A":

> F When does the cause of action arise? We can leave aside cases of personal injury or damage to other property as presenting no difficulty. It is only the damage for the house which requires consideration. In my respectful opinion the Court of Appeal was right when, in *Sparham-Souter v Town and Country Developments (Essex) Ltd* [1976] QB 858, it abjured the view that the cause of action arose immediately upon delivery, ie conveyance of the defective house. It can only arise when the state of the building is such that there is present or imminent danger to the health or safety of persons occupying it. We are not concerned at this stage with any issue relating to remedial action nor are we called upon to decide what the measure of the damages should be; such questions, possibly very difficult in some cases, will be for the court to decide. It is sufficient to say that a cause of action arises at the point I have indicated.

Lord Salmon, at p 770, is to the like effect. I need cite only the G
following passage at p 770 E:

> The plaintiff, Mrs O'Shea, however, acquired her maisonette on December 12 1962. The writ was issued on February 22 1972. If it could be proved that the building suffered damage prior to February 1966 which endangered the safety of its occupants or visitors, Mrs O'Shea's claim would be statute-barred.

I was at one stage of the hearing troubled by the words "present or imminent danger" in the passage cited from the speech of Lord Wilberforce. Taken literally, those words might suggest that the building must be on the point of collapse before the cause of action
arises. It would be an absurd situation if the owner of a house, H
knowing that if no remedial measures were taken the house would ultimately collapse, had to defer those measures until the last possible moment, for fear of losing his right of action against a negligent local authority. I have come to the conclusion that these words need not be interpreted so as to lead to this result. The words of judges, however eminent, do not fall to be construed in the same way as the language of a statute. They must be construed in the context of the facts of the case before them; and in *Anns v Merton Borough Council* it seems that when the damage to the building was first discovered it constituted a present or imminent danger to the safety of the occupants: see the facts as set out at p 733 E to G. Further, earlier in his speech Lord Wilberforce had expressed the
matter in terms which I find easier to apply to the facts of such a case J
as the present, when he said at P 759 G:

> for the whole purpose of the byelaws in requiring foundations to be of a certain standard is to prevent damage arising from weakness of the foundations which is *certain* [my emphasis] to endanger the health or safety of occupants.

In my judgment, therefore, once it is established that the physical damage to the building is of such a nature that it is certain to endanger the safety of the occupants, the cause of action is constituted, even though it may be that the building is not on the point of collapse.

I now turn to the other question which was debated before me: K
How far is the plaintiff's knowledge of the damage relevant to the accrual of the cause of action? In his speech in *Anns* case Lord Wilberforce does not consider this point. Lord Salmon, however, does so in a passage immediately following that previously cited, where he says:

> It seems to me, however, that since in fact no damage manifested itself until February 1970 it may be very difficult to prove that damage had in fact occurred four years previously. In the unlikely event of the defendants overcoming this difficulty, the fact that the damage went undetected for four years would not prevent the statute running from the date when the damage first occurred.

This passage suggests that Lord Salmon considered that knowledge L
is irrelevant.

However, in *Sparham-Souter's* case two of the three members of the Court of Appeal approved a test involving the knowledge of the plaintiff. Lord Denning MR said at p 868:

> When building work is badly done – and covered up – the cause of action does not accrue, and time does not begin to run, until such time as the plaintiff discovers that it has done damage, or ought, with reasonable diligence, to have discovered it.

To the like effect is Geoffrey Lane LJ (as he then was) at p 881 B: M

> The cause of action accrues when the damage caused by the negligent act is suffered by the plaintiff and that cannot be before that damage is first detected, or could by the exercise of reasonable skill or diligence have been detected.

Since in *Anns* case Lord Wilberforce (and by inference those other learned Lords who agreed with him) did not in any way qualify their acceptance of the decision in *Sparham-Souter*, I am content to accept the test proposed by Geoffrey Lane LJ as a correct statement of the law. This whole question, and in particular how far the extent to which the House of Lords in *Anns* case approved the decision in *Sparham-Souter*, has been considered by the Law Reform Commit-

A tee, but unfortunately their consultative document on latent damage arrived after this judgment had been prepared.

I must apply the principles laid down in *Anns* and *Sparham-Souter* to the facts of the present case, bearing in mind that the onus lies on the defendants, who have raised the issue of limitation, to satisfy me that the plaintiffs' cause of action accrued more than six years before the writ was issued; that is, before April 12 1973. It seems to me that I have to answer the following questions: (1) What was the nature of the physical damage to the bungalow which constituted a certain danger to the health or safety of the occupants? (2) When did that damage first occur? (3) When was that damage first detected? (4) When, by the exercise of reasonable skill B or diligence, might it first have been detected?

(1) The nature of the damage. On this issue I heard the evidence of Mr Fowler (to whose report of December 1977 I have already referred) on behalf of the plaintiff and of Jeffrey Digweed, another consulting engineer, on behalf of the defendants. From their evidence I am satisfied that the damage to the bungalow which constituted a certain danger to the safety of the occupants was the vertical settlement of the non-suspended floor slab. The fact that this settlement was caused by the unsuitable nature, and the depth, of the organic soil used for filling below the floor slab does not mean that the cause of action was constituted once this type of fill was C used, even though the nature of the fill meant that settlement was inevitable. To accept the submission of Mr Dennys for the defendants that the cause of action arose once there was actual or presumed knowledge of the facts — namely, that there was here a non-suspended floor slab built on a defective fill which was bound in due course to cause a settlement which would endanger the safety of the occupants — would in my judgment involve going back to the test, discredited in *Anns* case, that damage is suffered from the moment you have a building with defective foundations, qualified only by the additional factor that knowledge of the existence of the defective foundations is a necessary ingredient to constitute the cause of action. Accordingly I reject this submission. I also reject D Mr Denny's further submission that the presence of weeds and dampness constituted a danger to the health of the occupants of the bungalow. This submission was wholly unsupported by any evidence.

(2) The occurrence of the damage. Mr Fowler and Mr Digweed were agreed that the symptoms that were observed and reported on by Mr McKenzie in 1971 might have been attributable to horizontal movement of the floor slab caused by normal shrinkage in the new building. Such symptoms might also have been attributable to vertical movement of the floor slab, caused by a settlement, and the fact that the damp-proof polythene membrane had been punctured (as evidenced by the penetration of the weeds into the bungalow) E was some indication that there had by then been vertical movement as well as horizontal movement, although the membrane might have been punctured by a careless builders' labourer. The evidence of both experts, assisted by a considerable degree of hindsight, was that it was probable that by late summer of 1971 some settlement of the floor had started, and I am satisfied on the balance of probabilities that the settlement of the floor slab had occurred before April 12 1973.

(3) The detection of the damage. It is clear that the damage — that is, the settlement of the floor slab — was not in fact detected until 1977.

F (4) When might the damage have first been detected? Mr Dennys inevitably relies strongly on Mr McKenzie's report of September 1971 and in particular that passage where he points out the danger that reduction of the organic matter under the floor could remove support to the underside of the floor slab, producing cracks in the floor and in the partition walls built off the slab, as in fact happened. Nevertheless, what Mr McKenzie in fact observed — the weeds emerging from the junction between the floor and the skirting and the dampness in the entrance lobby, although not the major cracks which appeared later — were only symptoms and not the disease itself, to borrow the felicitous analogy used by Judge Edgar Fay QC in his unreported judgment in *Eames London Estates Ltd v North Hertfordshire District Council*, October 31 1980,* of which I have G been supplied with a transcript: see in particular pp 28 to 30. Clearly Mr McKenzie did not in 1971 positively identify the cause of the symptoms which he observed, since if he had he would not have expressed that part of his report last quoted in a subjunctive sense, nor would he have advised a fuller and more comprehensive report only "if circumstances so require". It was not unreasonable for Mr and Mrs Crump, in the light of this report, and in view of the continued assurances they were receiving from the builders, not then to incur the considerable further expense and disruption which a more detailed investigation would have required.

In my judgment neither in 1971 nor at any time before April 12 1973 did the Crumps fail to exercise reasonable skill or diligence in H seeking to discover the cause of the defects in the bungalow, which, as I have already said, were but symptoms of the underlying disease, the settlement of the floor slab. Accordingly, the defendants have failed to satisfy me that the action is statute-barred and it may continue.

The plaintiffs were awarded the costs of the hearing and the issue of limitation.

COURT OF APPEAL J

February 3 1982

(Before Lord Justice ORMROD, Lord Justice DUNN and Sir Sebag SHAW)

PIRELLI GENERAL CABLE WORKS LTD v OSCAR FABER & PARTNERS

Estates Gazette May 29 1982

(1982) 262 EG 879

Action for negligence in relation to building work — Period of limitation — Present obscurity of the law as to when cause of action accrues — Whether at time when defective work is carried out, or K **when it becomes visible, or when the plaintiff taking reasonable care should have discovered it — Negligence in supervision of the construction of a chimney — Writ issued in 1978 but damage had become visible in 1970 — Official referee gave judgment against defendants, consulting engineers, but question arose as to whether plaintiffs' claim was statute-barred — Court of Appeal, considering themselves bound by *Sparham-Souter v Town and County Developments (Essex) Ltd*, dismissed defendants' appeal, but drew attention to the lack of clarity of the law and the public importance of the issue — *Sparham-Souter* case difficult to reconcile with House of Lords decision in *Cartledge v E Jopling & Sons Ltd* — *Anns v Merton London Borough Council* leaves matter in some doubt —** L **Present case "ideal" for leave to appeal to House of Lords — Leave given without further discussion of the issues**

Appellants, Oscar Faber & Partners, appealed against a decision of Judge Sir William Stabb as official referee in an action brought against them for negligence by the respondents, Pirelli General Cable Works Ltd. The action arose out of supervision by the appellants of the construction of a chimney and the judge had given judgment against the appellants, the defendants in the action.

D G Wright QC and J B Storey (instructed by Beale & Co) appeared on behalf of the appellants; Patrick Garland QC and John Dyson (instructed by Herbert Oppenheimer, Nathan & Vandyk) represented the respondents. M

Giving judgment, ORMROD LJ said: This is an appeal by a firm of consulting engineers who were responsible for building a chimney for the plaintiffs and, in the course of an action heard before Sir William Stabb as official referee in July 1980, judgment was given against the defendants, the consulting engineers, on the basis of negligence in the supervision of the construction of the particular chimney.

The question arose before the learned official referee as to whether or not the plaintiffs' claim in this case was statute-barred. To put it in

*Now reported in full at 259 EG 491: see at p 498.

A a sentence, the writ in the action was issued on October 7 1978. It is apparently clear that the damage to the chimney had become visible, from the top at any rate, by April 1970, which would be before the beginning of the limitation period, which would be October 1972. The case, therefore, raises in the clearest form the question as to whether in an action for negligence the cause of action accrues at the time when the defective work is done, or when it becomes visible, or only when the plaintiffs, taking reasonable care, have been able to discover it.

Mr Wright, who appeared for the appellants, concedes that this court is bound by its own decision in *Sparham-Souter v Town and County Developments (Essex) Ltd* [1976] 1 QB 858, but he submits B that that decision cannot be reconciled with *Cartledge v E Jopling & Sons Ltd*, reported at [1963] AC 758, and he goes on to say that a subsequent case in the House of Lords, *Anns v Merton London Borough Council* [1978] AC 728, leaves the law in an obscure state, so that it is not at all clear from speeches in that case whether their Lordships were approving the decision in *Sparham-Souter* of this court or not. While conceding that this court is bound to dismiss the appeal, Mr Wright asks for leave to appeal to the House of Lords.

It is plain that this raises an issue of great public importance, and it is equally plain that the law is far from clear. This is an ideal case in which this court should give leave to appeal to the House of Lords. Mr Garland, on behalf of the respondents, concedes that that is the C position.

It is only necessary to mention that the course of the proceedings in this case has been a little unusual because, at the conclusion of the case before the official referee, an application was made for what I might call, for short, a leap-frog certificate to the House of Lords. A petition was duly prepared to the House of Lords by the proposed appellants for the purpose of applying for leave to appeal to the House in accordance with the practice. That petition for leave was not proceeded with because the present appellants, Oscar Faber & Partners, had, on consideration, wished to consider carefully whether they ought not to challenge the learned official referee's finding of negligence against them. That would have meant an D appeal to this court and so the petition was not proceeded with and instead a notice of appeal to this court was served. Hence the parties are here.

Second thoughts have prevailed and it is now accepted by the appellants, for the purpose of this case, that the official referee's finding will not be challenged. They accept all his findings of fact and the dispute, therefore, is confined to the point of law which has been outlined. In those circumstances I would dismiss this appeal and give leave to appeal to the House of Lords.

DUNN LJ: I agree.

SIR SEBAG SHAW: I agree.

There was no order as to costs.

E

QUEEN'S BENCH DIVISION

October 21 1981

(Before Mr Justice DRAKE)

LYONS AND ANOTHER v F W BOOTH (CONTRACTORS) LTD AND MAIDSTONE BOROUGH COUNCIL

Estates Gazette June 5 1982

(1981) 262 EG 981

F **Action by house-owners against building contractors and local authority arising out of severe structural defects due mainly to settlement—Extensive cracking and movement of floor slabs away from skirting—As building contractors had gone into liquidation the claim for damages on the ground of negligence was pursued against local authority—Authority alleged to be negligent in the consideration and passing of plans and in failure of proper inspection to ensure compliance with building by-laws—Parallel action (not the subject of this report) by present plaintiffs and other owners against insurance company claiming that damage fell within scope of owners' policies—Claim against local authority based on negligence held to be established—*Anns v Merton London Borough* G *Council*—Authority's inspector negligent in failing to see that a soil survey was carried out, in not carrying out an inspection of the underfill on which the ground-floor slabs were laid, and in failing to see evidence that the subsoil was unsuitable as a load-bearing medium—underpinning of main foundations necessary as it was not clear that settlement had ceased—Ground-floor slabs had to be suspended so as to be supported by the strip foundations—Owners entitled to damages including a sum in respect of the diminution of the value of the property as a result of the bad reputation which it had acquired—Issues as to quantum of damages referred to an Official Referee—Judge's evaluation of evidence should be of special interest to building surveyors**

H The plaintiffs in these proceedings were Mr and Mrs Lyons, leasehold owners of a maisonette, 70 Roseholme, Maidstone, but the case was a test case of interest also to other long leaseholders in the same building. The maisonettes were built in about 1963 and the leases were for 999 years. The first defendants, F W Booth (Contractors) Ltd, had gone into liquidation and the claim against them was not pursued. This report is concerned with the claim against the second defendants, Maidstone Borough Council. The plaintiffs had bought the leasehold interest in no 70 in 1972 and began to see cracks appearing in 1973.

M Myers QC and C Aldous (instructed by Gulland & Gulland, of Maidstone) appeared on behalf of the plaintiffs; J A T Loyd QC and J C A W Gibson (instructed by Badhams) represented the defendants.

Giving judgment, DRAKE J said: This action, brought by Mr and Mrs Lyons against the Maidstone Borough Council, and the associated action brought by Mr and Mrs Lyons and a number of neighbouring house-owners against the Legal & General Assurance Co Ltd is concerned with a block of eight maisonettes which were built in Maidstone in about 1963. Mr and Mrs Lyons and the other plaintiffs hold their respective properties on 999-year leases. For the purpose of these actions I shall refer to them simply as owners.

In the early days no one noticed anything wrong with the building; but at some time in or about 1973 and thereafter serious things were seen to be happening. Numerous cracks appeared both outside and K inside a number of the maisonettes, some of them small but some of them quite large. The concrete floors in a number of the downstairs rooms sunk so that in some cases they parted from the skirting boards at the bottom of the internal wall. At one end of the block the outside concrete steps to a doorway sunk, so did a nearby concrete path; and some retaining walls, one at the rear of the block next to the forecourt to the garages, and two others in front of the building, developed severe cracks and were apparently being pushed over by the ground they were intended to retain. The total cost which will be involved in making good the entire block might possibly, on one estimate, be in the order of £100,000.

These major defects became obvious for all to see, and they were L and are so serious that building societies were not prepared to grant mortgages on the maisonettes, so that the owners found themselves virtually unable to sell their properties. Mr and Mrs Lyons who had bought the leasehold interest in their maisonette in 1972 tried to sell it when Mrs Lyons was about to have her first child in 1973; but they could not find any buyer who would complete a purchase — so they are still there, now with two growing boys, in 1981.

It is obvious, from what I have said, that something went gravely wrong with the construction of this building. But to make matters worse for the owners, the building company who built the houses and who are the first defendants in the first action have gone into liquidation. Furthermore, the insurers, with whom the house-M owners all insured their property, have denied that the damage falls within the scope of the policies.

Faced with this situation the owners have brought the two actions now before me. In the first action Mr and Mrs Lyons claimed damages for negligence in the building of their property by the builders; but, as I have said, the building company have gone into liquidation, so the claim against them was not pursued. They also claim damages against the Maidstone Borough Council as the local authority, alleging negligence in the consideration and passing of the plans for the building and also in failing to inspect the site and building so as to ensure that the building complied with the local by-laws.

A Although Mr and Mrs Lyons of 70 Roseholme, Maidstone, are the sole plaintiffs in this action I have been told that it is brought as a test case in respect of the owners of five other properties, namely nos 58, 60, 62, 64 and 66. A seventh owner (of no 56) wishes to claim but has not yet done so: the eighth property (no 68) has for some years been empty and no claim is made in respect of it.

In the second action the six owners sue the insurance company, seeking a declaration that the damage to the buildings falls within the scope of their respective insurance policies which, so far as is relevant to the claims, are in identical terms. The defendants in each action deny liability.

The evidence in the two actions was, by consent, heard during one B hearing. It has also been agreed that having heard full argument in the first action I should first give judgment in that action, which I am now doing, so as to deal with all matters other than the quantum of damages relating to the cost of remedying the defects in the building and expenses which will be incurred by the occupiers moving to alternative accommodation while the remedial works are carried out. The parties involved in the second action concerning the insurance policies will then have a short time (until tomorrow morning) to consider the judgment, which will be accepted as dealing with all factual disputes. Argument will then be heard and I shall in due course give judgment in that action.

C *Description of the building*

This is best given by reference to the numerous plans and the agreed experts' report used in evidence, so I need not lengthen the judgment by referring to it in detail.

It comprises eight maisonettes within a two-storey building, four maisonettes being on the ground floor with the other four above them but having a staircase down to a small entrance hall and door at ground level. The building, referred to on some of the original plans as "Block E" — which is a description I shall continue to use — is in three parts joined together, being stepped in two places so as to allow for the slope averaging 6 degrees from north to south on which it was built. In the section on the highest part of the slope are two D maisonettes (no 56 on the ground floor and no 58 above): in the middle section are nos 60 and 64 (on the ground) and 62 and 66 (above) and in the lowest section at the southern end is no 68 (empty for some years on the ground-floor level) and no 70, the home of Mr and Mrs Lyons above.

Block E formed part of one substantial estate development by the same builders comprising a number of detached and semi-detached houses, maisonettes and flats all surrounding a new road called Roseholme. The whole site is on ground sloping from north to south down towards the River Medway; and in days gone by part of the site, including the area on which Block E has been built, was in use as watercress beds, as appears from the 1908 Ordnance Survey map. E This was no doubt because of the presence of spring water which, at least on some occasions, probably emerges from the sloping ground on or just below part of the site.

Block E itself is a brick-built structure on strip concrete foundations. As can be seen from the photographs, two types of facing brick were used, namely calcium silicate bricks on the two end sections and clay bricks on the longer middle section, which is partly weather-boarded in the front but wholly faced with clay bricks at the rear.

The calcium silicate brickwork on each of the end sectors is keyed into the clay brickwork in the middle section, so that although the building is stepped to allow for the slope it is one unit and not three F separate units. Stresses developing in any section are therefore capable of being transmitted to another section.

The floors of the maisonettes are constructed of four-inch thick concrete slabs. At the first-floor level the weight of the slabs bears on those walls which themselves are founded on the concrete strip foundations to the building. But at ground-floor level the concrete floor slabs do not bear on the strip foundations; they are independent of the main walls of the building and bear down directly on 4 in of hardcore resting on some fill material which in turn bears on the ground underneath. The fill underneath the ground-floor slabs has undoubtedly been one of the causes of damage to these properties. It is up to 2 m in thickness, but instead of having been compacted before the floor slabs were laid, it was loose. With fill of G this thickness it was essential to have a granular or similar material of even size so that it could satisfactorily be compacted; but in fact the fill was found on inspection to be loose, comprising a clayey silt containing numerous fragments of stone of varying thickness and in one place even an old metal bath. Such fill could not easily be compacted and what happened was that it settled, ie compacted gradually over the 17 years or so which have elapsed since the floor slabs were laid. This led to a void ranging between 20 and 140 mm (ie nearly 1 in up to nearly 5 in) below the concrete slabs, which, being unsupported, have fallen and moved, so that in some places they have come away from the skirting board. The concrete floor slabs on the ground floor are themselves the foundation for one internal H partition wall in each ground-floor maisonette, and the movement of the floor slabs has consequently caused movement of and damage to these internal walls.

Damage has also occurred to three retaining walls which were built at about the time the maisonettes were built, probably immediately after the main parts of the block were finished but before the building was regarded as ready to be lived in. Two of these retaining walls are in front of the main building, each running out from the place where the sections of the block are stepped, ie at the point where the calcium silicate brickwork is keyed into the clay brick of the centre section. These walls serve to retain the soil in the small front gardens to the property to prevent it falling down the slope. J

The other retaining wall, clearly shown on plate 4 of the agreed bundle of photographs, is at the top end of Block E and was designed to retain the fill laid to the north of the block so as to provide the support for a large garage forecourt and a number of garages. This wall, which runs along the length of the north gable end of Block E, is about 2 m high and was built so as to be 4½ in from the gable wall to the block.

Each of these three retaining walls has been damaged by the pressure of the ground or material it was designed to retain. The wall retaining the garage forecourt has been forced over so that its top is now in contact with the gable wall of Block E (maisonette 56). The K front walls are badly cracked and unless put right will probably fall. In the case of the wall next to nos 60 and 62 the effect will be to block the entrance to these maisonettes.

The damage

When Mr and Mrs Lyons jointly bought 70 Roseholme in 1972 they did not have a private survey carried out, but Mr Lyons saw no signs of anything wrong with no 70 or anywhere else in the block. They obtained a mortgage from the Woolwich Equitable Building Society, and paid for the society to have a survey. Mr Lyons was not shown the result, but it must have been reasonably satisfactory because a mortgage was granted. L

In 1973 he saw cracks starting to appear in his own home and then elsewhere in the block, and they got bigger and more numerous. Mr Lyons repointed cracks under windows of the main room in front and the kitchen at the back, and part of his work can be seen in one or more of the many photographs, eg plate 10 of the main bundle. But more cracks appeared; and the concrete floor slab to his entrance hall moved and came away from the skirting leaving a gap of between 3 and 4 mm and it also came away from the stairs and left a gap of similar size, and at least three of the floor tiles cracked. Mr and Mrs Lyons wanted to move but it became impossible to sell the property. The defects in Block E had become well known to local estate agents, and building societies would not grant any mortgage upon any of the M maisonettes.

I need not catalogue the extent of the defects to Mr Lyons' property or elsewhere in Block E because they are set out in detail in the various reports and shown on the numerous photographs and plans.

Two experts gave evidence and submitted various reports, namely Dr A G Weeks for the plaintiffs and Mr L Langford for the Maidstone Borough Council. They also compiled an agreed report to which the Legal & General Assurance Society's expert, Mr A P Blackham, was a party, although he did not give evidence in the trial.

A The experts agree

(i) that some settlement has occurred to the strip foundations due to consolidation of the underlying materials;

(ii) that the depth of fill beneath the ground-floor slabs is excessive and has settled and no longer provides support to these slabs and to the internal partitions which are founded upon it; and

(iii) that the retaining walls are badly damaged.

Mr Langford and Dr Weeks also agree that some of the external cracking to the calcium silicate bricks is due to shrinkage of those bricks caused by moisture and temperature changes. Mr Langford thinks that much of the cracking in the clay bricks in the centre section is due to temperature changes and to stress transmitted from
B the shrinkage of the calcium bricks. He agrees that some damage in the clay bricks is due to settlement, as also is some damage in the calcium silicate bricks in the extreme south-west corner of the block, ie at nos 68 and 70, but his conclusion is that cracking throughout Block E is predominantly due to shrinkage and thermal movement. Dr Weeks thinks that the calcium silicate cracking is of comparatively minor importance and in his view the predominant cause of cracking to both types of brick is the settlement of the strip foundations. This is one of the major issues which I have to resolve.

These experts are also in disagreement as to the remedy required to deal with the settlement damage which has occurred. Mr Langford believes that the settlement is now virtually complete, so that the
C damage can be remedied simply by repairing the cracks and other damage which has occurred. Dr Weeks believes that the settlement is as yet only about 75% complete and that it is essential to underpin the strip foundations in order to avoid future damage by the remaining 25% settlement still to occur.

There is agreement that the ground-floor concrete slabs will have to be removed in order to remedy the existing and probable future damage; but there is disagreement as to the remedial measures required. Dr Weeks (supported, according to the agreed report, by Mr Blackham) thinks that the floor slabs should be suspended, ie spanned on to the external walls and supported by the newly underpinned strip foundations.

D Mr Langford is in favour of a new entirely independent floor slab resting on newly prepared underfill.

If the ground-floor slabs are to be supported by the strip foundations, he agrees that the increased pressure upon those foundations will mean that they must be underpinned. But as he does not regard underpinning as necessary to avoid future settlement damage to the external walls he thinks that the high cost of underpinning should not be incurred merely to deal with the floor slabs. His proposed remedy for the floor slabs is first to deal with the existing faulty underfill. This should either be removed and replaced with a new material which can be satisfactorily compacted, or treated with a chemical injection which will bind together the existing
E fill. A lightweight structure should then be built above the fill to support new timber flooring.

The experts agree on the damage which has occurred to the retaining walls due to pressure of the retained materials and they also agree that the remedy required is to rebuild these walls.

Involvement of Maidstone Borough Council

I next turn to consider the involvement of the borough council, who are and were in 1963 the authority responsible for granting planning approval under the Town and Country Planning Acts and also for making and ensuring compliance with building by-laws under the provision of the Public Health Act 1936. The nature and
F scope of the duties of the council under that Act were clearly indicated by the House of Lords in *Anns v Merton London Borough Council* [1978] AC 728 to which I must make further reference later.

In 1963 the council considered plans in respect of Block E. By that time they had certainly approved plans and inspected various stages of building work in respect of numerous other properties forming part of the fairly large estate development by the same contractors, F W Booth (Contractors) Ltd. Records exist and have been produced of the plans and inspection reports made in respect of some of the other properties. Unfortunately a large number of the council's records were lost or destroyed, possibly as a result of flooding of the borough engineer's office in 1968, so that some of the records, in particular the inspection reports relating to Block E, have not been G
produced. However, some of the approved plans have been produced and, as a result of the evidence, the council properly and realistically on the balance of probabilities have conceded some of the matters which were in issue as to the inspections carried out by one of their building inspectors. The history, as revealed or conceded as being probably true, is as follows:

(1) Plans for the maisonettes were drawn up by some unqualified person on behalf of the contractors and sent to the council in June 1963. I need refer only to three of those plans. One, no 15/37 dated February 1963, shows "Typical Elevations" and a section drawing shows "Concrete Foundations — (Depth to suit site conditions)".
Plan no 15/38A of the same date has notes, including H

6. Foundations. To approval according to site conditions.
17. Layout. Plans repeated to form terrace, stepped to suit site conditions.

Another plan, no 15/40A dated May 1963, is entitled "Plans, Elevations and Section" and also shows in section a foundation marked "Depth varies according to site conditions". Notes include note A, "Foundations — To approval according to site conditions". Each of these plans is stamped by the council as having been "Approved 26 June 1963".

(2) Plan no 15/56 dated September 2 1963 is of "Details of Maisonettes" and on the drawing "Typical Section" it shows
concrete foundations 9 in thick and 2 ft (or in the case of foundations J
to an internal wall 1 ft 6 in) wide. The depth is not marked, but scaling the plan shows the depth to be 2 ft 6 in. This plan is marked by the council "Approved 17 September 1963"; and the original, produced by the council, has written on it "detailed plan of maisonettes agreed, office copy".

(3) The council in 1963 had an establishment for four building inspectors whose duties included that of ensuring compliance with the council's building by-laws. The council operated a system of inspections at eight stages of the development of which stage 1 was of the undeveloped site before any building work, stage 2 was inspection of the foundation trenches before any foundations were
laid, stage 3 was of the foundation concrete before any brickwork, K
and stage 4 was of brickwork up to damp-proof course level with the foundations covered in and with concrete over the ground surface within the external walls. I am not concerned with the subsequent stages.

Reports have been produced of a number of inspections of buildings (detached and semi-detached houses and a flat) during 1963 — but none exist for any of the maisonettes in Block E. The reports produced show a virtually complete series of inspections at the eight stages. The council have realistically conceded the probability that inspections were carried out on the maisonettes in Block E.

As a matter of practice the building inspector would have had with L
him, at the time of his inspections, copies of the relevant approved plans.

It is a matter for the discretion of the council as to the manner in which they will carry out their duties to ensure compliance with the building by-laws. But as the plans marked "Approved in June 1963" showed "Foundations — To approval according to site conditions", it is conceded by the council that in this case they were under a duty to inspect the site, which they did.

One of the major issues in this case which I have to decide is what the building inspector saw or ought to have seen on his inspection. The plaintiffs say that the inspections were carried out negligently
with the result that the building by-laws were not complied with M
because the foundations required and constructed were not such as to ensure the stability of the building. The council deny this.

The inspection was probably carried out by the then senior building inspector, a Mr Mitchell. He remained with the council until April 1981, when he became mentally unwell and ceased to work, and was retired in September 1981. During the course of preparation of this action it was Mr Mitchell who provided all the available information to the council and its insurers' solicitors. However, nothing has been made available to indicate what he knew about the inspections beyond the answers to interrogatories sworn by Mr Osborne, now the assistant planning officer employed by the

A council. Mr Osborne said that information came from Mr Mitchell.

I therefore have to decide this major issue as to what the building inspector saw or ought to have seen on the basis of evidence given as to the sort of experience and qualifications held by building inspectors; what Mr Osborne told me of the way in which such inspections are made; and on the evidence of the two experts, Dr Weeks and Mr Langford. But before I deal with this I must first refer to the scope and nature of the duty owed by the council to the plaintiffs, and to the relevant by-laws.

Duty of the council

The plaintiffs allege that the council were negligent at two stages,
B ie:

(i) in approving the plans and
(ii) on the inspections of the site and works.

There is no issue between the parties as to the nature and scope of the duty of the council under either of these heads.

As regards the duty on inspection under the by-laws I have already referred to *Anns v Merton London Borough Council* which deals fully with this matter. That case was not concerned with any duty when considering and giving approval to plans under the provisions of the Town and Country Planning Acts, but Mr Loyd, for the council, concedes that in principle the duty should be the same as that
C arising on an inspection to ensure compliance with by-laws. I fully agree and hold that the nature and scope of the duty is indeed similar.

In the present case this is probably only of academic importance. Although the plans considered in June 1963 were marked "Approved", this was in fact only qualified approval, because the plans themselves stated that the foundations were subject to further approval according to site conditions; and the council concede that this meant that a site inspection was required. By September 1963, when the detailed plan (no 15/57) was marked "Approved", it seems probable that the necessary inspection had been carried out. If it was done negligently then the subsequent approval of the plan was done on the basis of a negligent inspection. If the inspection was carried
D out without negligence then I do not think the council could be held negligent in approving the plans on the basis of an inspection carried out properly. In either event I think the duty of the council towards the plaintiffs was the same.

I need refer only to two passages in the speech of Lord Wilberforce in *Anns v Merton London Borough Council.* At p 755D he said:

> Passing then to the duty as regards inspection, if made. On principle there must surely be a duty to exercise reasonable care. The standard of care must be related to the duty to be performed — namely to ensure compliance with the by-laws. It must be related to the fact that the person responsible for construction in accordance with the by-laws is the builder, and that the inspector's function is supervisory. It must be related to the fact that once the inspector has passed the foundations they will be covered up, with no
> E subsequent opportunity for inspection. But this duty, heavily operational though it may be, is still a duty arising under the statute. There may be a discretionary element in its exercise — discretionary as to the time and manner of inspection, and the techniques to be used. A plaintiff complaining of negligence must prove, the burden being on him, that action taken was not within the limits of a discretion bona fide exercised, before he can begin to rely upon a common law duty of care. But if he can do this, he should, in principle, be able to sue.

and at p 758F he said:

> *The nature of the duty.* This must be related closely to the purpose for which powers of inspection are granted, namely, to secure compliance with the by-laws. The duty is to take reasonable care, no more, no less, to secure that the
> F builder does not cover in foundations which do not comply with by-law requirements. The allegations in the statements of claim, in so far as they are based upon non-compliance with the plans, are misconceived.

I turn to the relevant by-laws of the Maidstone Borough Council, which are as follows:

Under Part III — Buildings

> *Foundations*
>
> 18(1) The foundations of every building shall be:— (a) so designed and constructed as to sustain the combined dead load of the building and imposed vertical and lateral loads and to transmit these loads to the ground in such a manner that the pressure on the ground shall not cause such settlement as may impair the stability of the building, or of adjoining works or structures.

I need not refer to 18(1)(b) or (c). G

> 19(1) The foundations for the load-bearing structure of a domestic building where constructed as strip foundations of plain concrete situated centrally under the walls or piers, shall be deemed to satisfy the requirements of subparagraph (a) of the last preceding by-law if —
>
> (a) there is no wide variation in the type of subsoil over the loaded area, and no weaker type of soil exists below that on which the foundations rest within such a depth as may impair the stability of the structure.

By-law 19(1)(b) is concerned with the width of the foundation which is required to be not less than that specified in a subjoined table. This table was referred to in evidence because the council say that the building inspector was not shown to have been negligent in seeing that by-law 19(1)(b) was complied with, which involved him H making an accurate assessment of the type of subsoil.

Although some of the other by-laws, including by-law 21, were referred to in argument I do not find it necessary to set them out in this judgment.

There is, however, a dispute as to whether or not the retaining walls are "buildings" within the meaning of that word as used in by-laws 18 and 19. I refer to by-law 1, which is the "interpretation" by-law, only to say that "building" is not among the words or phrases there defined. By-law 2(2), so far as I need refer to it, is as follows:

> 2(2) For the purpose of these by-laws any of the following operations shall be deemed to be the erection of a building:

and at (c) is: J

> the roofing over of any open space between walls or buildings.

The issues

I am now in the position to formulate the issues which I have to decide, which are as follows:

(1) As to the main foundations was the building inspector negligent when he inspected the site and/or building works? In particular should he have called for a soil survey, and if he had done so would that have resulted in foundations deeper than those in fact laid?
(2) Are the ground-floor slabs "foundations" within the meaning of the by-laws? If so then:
(3) Was the building inspector negligent in failing to ensure that these slabs were dealt with differently from the way they were in fact dealt with, ie should K he have required either:
(a) that they were supported by the "main" foundations to the external walls, or
(b) that the underfill to these slabs was properly compacted?
(4) Are the retaining walls "buildings" within the meaning of these by-laws?
(5) How much of the damage to Block E is due to the settlement, and how much due to shrinkage and/or thermal movements of the bricks?
(6) Is it now necessary as a remedial measure to underpin the main foundations?
(7) What is the proper remedy in order to deal with the ground-floor slabs?
(8) If the council are liable what sum should be awarded to the plaintiffs for:
(a) diminution in value of the property as a result of the bad reputation it has L acquired (a head of damage similar to that considered in *Payton v Brooks* [1974] RTR 169);
(b) diminution in the value of the property to the plaintiffs during the time they have lived in it and been affected by the damage. This includes the inconvenience caused by the defective state of the maisonettes and by having people come to inspect them and dig boreholes and trial pits nearby and the general worry and distress they have suffered?

It is agreed that at this stage I am not in a position to quantify any damages which might be recoverable under two other important heads, namely, first, for remedying the defects in the building and, secondly, for expenses which will be incurred in moving to alternative accommodation while the remedial works are done.

So I turn to deal with these issues. M

Was the building inspector negligent with regard to the main foundations?

The plaintiffs say he was negligent in failing to call for a soil survey. Dr Weeks gave detailed evidence as to the geology of the site. He is a very experienced and very highly qualified and skilled civil and geotechnical engineer who for many years has specialised in geotechnics. His evidence as to the geology of the site is not disputed. I need not deal with it at length. It is mainly contained in his reports and in the agreed report, although he explained and enlarged on these in his evidence with the help of drawings on a blackboard.

A Suffice it to say that much of the estate development is on a sandy subsoil known as the Hythe Beds, but in some parts, including the area of Block E, the Hythe Beds join up with another underlying subsoil known as Atherfield Clay. At this junction spring water emerges — but there is an issue between Dr Weeks and Mr Langford as to how easily this would have been noticeable to the building inspector in 1963.

The Hythe Beds and Atherfield Clay are found at varying depths up to about 6 m or less below the present ground level, and are covered with a layer about 2 to 3 m thick of material deposited in the last Ice Age known as "head". Above the head is a soft sandy and clayey material called "colluvium" varying between about 1 and 2 m B in thickness, and over this is a thin layer of the original topsoil, which includes decaying roots and other vegetable matter, and from there up to the existing surface the original ground has been covered by a fill — much of which was probably taken from the excavations for foundations and so on elsewhere on the estate.

The "colluvium", being a soft material, will consolidate, ie compact from the weight of any building if the foundations bear on it; and this is what has caused the settlement in Block E.

Two boreholes, dug in April 1980, clearly reveal the structure of the ground at each end of Block E; and those and a number of trial pits dug in June 1980 show that the strip foundations were constructed in, and generally in the upper part of, the colluvium, and C in one case (trial pit 80/3) in the topsoil over the colluvium. The results of these excavations are all contained in the agreed report.

Over the years the weight of the building transmitted through the foundations led to compaction of the soil underneath. This compaction was not uniform, as the compressibility of materials on the site increases as you go down the slope. Accordingly the building sank more at the lower end than at the top end — in the order of 4 in at the lower end compared with only 1 in at the top end. This resulted in damage to the block.

There is no doubt that any qualified engineer surveying this site and considering plans to build Block E down the slope, ie at right angles to the contours, would at once have appreciated the dangers D involved and would have called for a soil survey. Mr Langford agreed with Dr Weeks on this and agreed that, although not possessing Dr Weeks' geological expertise, he would certainly have carried out a soil survey and on what he found would have called for pile foundations for Block E.

But the issue is whether a building inspector ought to have called for such a survey or at least have seen sufficient to make him refer the matter to the borough engineer who, it is agreed, would have carried out such a survey.

Building inspectors are normally unqualified except by practical training and experience. Mr Mitchell, who was senior building inspector in 1963, was a former carpenter-joiner with another local E authority, but had become a building inspector, probably after some training, including short courses and practical instruction.

Dr Weeks says that with his local knowledge, with the availability of geological maps and old Ordnance Survey maps showing that there had been watercress beds on the site, with knowledge of certain codes of practice relating to foundations, but above all having regard to the signs of damp and probably of standing or running water a little way further down the slope, and because of the length of this terrace block running across the contour lines, down the slope, any building inspector ought to have asked for a soil investigation before approval was given to the plans. At the very least, says Dr Weeks, the building inspector should have recognised potential problems so as F to refer the matter to the engineering staff of the council.

Dr Weeks further thinks that a competent building inspector should have recognised the dangers of the site when inspecting the foundation trenches at stage 2 of the inspections. At that stage he could and should have seen that the soil was possibly unsuitable as a base for the foundation, and by carrying out a "heel test" (digging his heel into the soil) and using the yard-long metal probe he would have had with him, he should have been put on notice that a soil survey was required.

Mr Langford disagrees. He thinks it unlikely and beyond the normal practice for such an inspector to have consulted old maps or the geological survey or that he would be familiar with the particular codes of practice referred to by Dr Weeks. He says that no building G inspector would know of a material called "colluvium", which even he, Mr Langford, had never heard of until Dr Weeks referred to it, and that the colluvium which the building inspector saw would have been reasonably classified by him as head material which, in dry weather, would have seemed firm and suitable to take the foundations. He agrees that the spring water would have meant that there were damp patches on the ground with lush growth of vegetation coming from them, but thinks that there would have been no visible standing or running water. The water now visible on the surface he thinks comes mainly from the rainwater soakaways and surface water from Block E.

As to the inspection of the open trenches Mr Langford does not H think that the building inspector should have been put on notice that there was anything wrong with the ground as a base for the foundations, and does not think the heel test or probe would have revealed anything of significance.

That is a summary of the conflicting views of the experts.

In reaching my decision on this issue I have been careful to bear in mind that Dr Weeks is so highly skilled that he might expect too high a standard of less experienced people; secondly, that the building inspector was not a qualified man, save by practical training and experience; and thirdly, that the virgin site in 1963 may have looked very different from the photographs of the surrounding land in 1980-81 (in particular I have in mind the fact that rainwater coming J from Block E and its surrounds must surely have added to the dampness of the ground below the block); and fourthly, the burden of proving negligence rests firmly on the plaintiffs.

As to the geological and old Ordnance Survey maps, I doubt whether the building inspector consulted them nor do I think him negligent not to have done so. I am not satisfied that he would or should have been familiar with the codes of practice referred to by Dr Weeks — although he may have been, and he would certainly have read and been familiar with some textbook and practical written guidance about building work including foundations.

But I have reached the clear and firm conclusion on all this evidence that the building inspector *did* fall below the required K standards in failing to see that a soil survey was carried out — either by requesting one himself or by referring the matter to the engineering staff, which, on the evidence, would have led to the same result.

On the balance of probabilities I think that signs of boggy, marshy or wet ground *were* visible in 1963. The spring lines certainly result in water being present on or near to the surface, and the former existence of the water-cress beds show that probably the amount of fresh spring water emerging at this site was considerable.

The photographs show a lot of water present below the site in October 1981, and figure 1 to the agreed experts' report dated October 7 shows that water issues from the bottom of the bank just L below the site, leading to a "boggy area". I do not think, on the evidence, that this is entirely due to water drainage from Block E and its surrounds.

When it was put to Dr Weeks in cross-examination that a building inspector could not be expected to have known Block E was to be built on or near the spring lines he retorted that the inspector would have found himself up to his knees in soft boggy material and that it needs no expert to know what *that* means.

As to the lack of professional qualifications of an inspector, I bear that in mind. But the essential qualifications required of an inspector are a sufficient practical expertise to enable him to secure compliance with the by-laws; and I think that such practical experience and M training should have led this inspector, whoever he was, at least to have recognised that this building and site demanded that at very least he referred the matter to the engineering staff.

As to the subsequent inspection of the open foundation trenches the same findings apply, but with this addition. At that stage the inspector was able to see down some way into the colluvium and I think it likely that evidence of dampness and/or softness was available to him then; and that the use of the heel test and probe *would* have been likely also to indicate the need for a soil survey. My findings that the building inspector was negligent are reinforced by the fact that at one point the foundations rest on the topsoil. Any building inspector would at once have known that topsoil was not a proper base for the foundations. The fact that part of the foundations was built on the topsoil is an indication that the inspections were not carried out with the thoroughness required, ie to see that the byelaws were being complied with.

In deciding this issue I have had to be influenced greatly by the opinions expressed by the two experts as to what a building inspector would be likely to have seen in 1963 and what conclusion he ought to have drawn from the facts available to him. In this as in other matters the two experts both gave their evidence clearly and carefully. I preferred most of the views expressed by Dr Weeks, who, on this point, was very much more emphatic and certain than Mr Langford in the conclusion that he reached. I therefore decide this issue in favour of the plaintiffs. It is common ground, arising from Mr Langford's evidence, that had a soil survey been made it is likely that deeper, piled, foundations would have been required.

Are the ground-floor slabs foundations within the meaning of the by-laws?

Having regard to the evidence that they support an internal partition wall, Mr Lloyd conceded this point. Mr Langford, his expert, agreed that the ground-floor slabs are foundations. I was referred to a decision of Woolf J, in *Worlock v SAWS* (unreported as yet*, judgment given on June 19 1981), in which he held that such a floor slab which had been specially thickened in certain areas so as to take the weight of internal walls is a foundation for the purpose of building regulations similar to the by-laws with which I am concerned. Woolf J said (at p 14 of the transcript with which I have been supplied)†:

> I . . . regard a foundation as an object which is placed in position on or in the ground in the course of constructing a building . . . the function of which is to provide support for the building so that it in fact transmits load to the material beneath.

I agree with this definition, extending it so as to read, ". . .the function of which is to provide support for the building *or any part of the building,* so that it in fact transmits load to the material beneath". I see no need to limit the application of the definition, where it applies, to a floor slab specially thickened in certain areas.

Was the building inspector negligent with regard to the ground-floor slabs?

It seems unlikely that he carried out any inspection of the underfill on which these floor slabs were laid, because stage 3 of the council's system was carried out before the fill under the slabs was fully in position and stage 4 was carried out only after it had been covered. Since the fill was itself resting on an unsuitable base (ie the colluvium) it follows from my findings with respect to the main foundations that I think the building inspector was negligent with regard to the floor slabs.

But in my view he was negligent also in respect of the fill. In the decision of Woolf J, to which I have just referred, he held that not only the floor slab but also the hard core beneath it constituted a foundation; and Mr Myers for the plaintiffs argued that the definition should therefore be extended to cover the fill in the present case. I think this is right and that by-law 18 therefore requires that the fill should be so designed and constructed as to support the building and any part of it.

Furthermore, the unusual depth of underfloor fill to be used here was easily seen by reference to the plans and should have called for exceptional care by the building inspector to see that it consisted of suitable and/or suitably treated material. Since no inspection at all was carried out I am of the view that the council must be held to have been negligent by omission.

It is true that they have a discretion as to the manner in which they ensure that the building by-laws are complied with (see *Anns v Merton London Borough Council*). But I do not think that discretion was properly exercised by operating a system which virtually precluded inspection of the underfill to ground-floor slab foundations.

Alternatively, I would hold that in carrying out his inspections to the building the building inspector was negligent in the case of this particular building in not appreciating that the underfloor fill required special attention from him or, if he did appreciate that fact, in not doing anything to inspect the fill.

Are the retaining walls "buildings" within the meaning of the by-laws?

As a matter of fact I do not think that any of these walls form part of the main building, Block E. The two front walls are not keyed in to the main structure and were almost certainly built after it was completed but before any occupants moved in. The forecourt retaining wall was probably built at about the same time. Dr Weeks thought it possible that it rests on the main foundations of the building but this was no more than speculation, doubted by Mr Langford and I cannot accept it as a proved fact.

The only authority cited to me was the decision of Romer J in *Wood v Cooper* [1894] 3 Ch 671, in which he held that a covenant by a lessee not, without the consent of the lessor, to erect or build on the demised premises "any other building whatsoever" save for a stable and coach-house, was broken when he erected a trellis-work screen of wood above his boundary fence. He held that the trellis-work was an erection, and that had it been a wall the matter would not even have been open to doubt. I do not think that decision is of much assistance to me in the existing case, although it does somewhat favour the view that a wall may, for some purposes, be called a building.

Mr Osborne, for the Maidstone Borough Council, said that the council had never regarded any retaining wall as part of a building and never inspected them or the foundations to them. He referred to by-law 2(2), part of which I have already read.

I note that on none of the plans in respect of the premises is there any reference to any of these retaining walls. This fits in with Mr Osborne's evidence that the council did not regard such walls as structures which required planning or by-law approval, and indicates that the contractors, and whoever drew the plans and dealt with the council, also did not think that planning or by-law approval was required. But the opinion held by the council or contractors does not, of course, assist me in reaching my decision as a matter of law. I think that since in ordinary language a wall has to be built there is some ground for calling what is built a "building". Some dictionaries give such a definition, eg the *Shorter Oxford* (1944 edition) has one definition "Building . . . that which is built; a structure, edifice". But I am cautious to accept this. It is true that in normal language a wall is said to be "built"; but so is a road and I certainly do not consider that a road is a building within the by-laws.

Although I was not referred to this or any other dictionary definition I have looked at a number without, however, deriving much assistance. *Jowitt's Dictionary of English Law* (2nd ed, 1977) refers to a bewildering number of different meanings which can be given to the word in different circumstances. From that dictionary, and by reference to the case cited, I find that in *Moir v Williams* [1892] 1 QB 264 at p 270 Lord Esher considered that what is ordinarily called a building is an inclosure of brick or stonework covered in by a roof. In 1981 I do not think that what is ordinarily called a building need be made only of stone or brickwork, but subject to that it is the meaning which I think would ordinarily be given to the word today. This is close to the "deemed to be the erection of a building" definition given in by-law 2(2).

Looking at the matter largely as one of impression against the objects of these by-laws as a whole I have come to the conclusion that a wall which does not form a part of another building but stands independently is *not* a "building" within the scope of these by-laws.

In reaching that decision I have *not* been influenced by Mr Osborne's evidence that it would be wholly impracticable to require local authorities to inspect every garden or ornamental wall that is constructed. But I think his opinion on that matter is correct, and I am therefore not sorry to have reached the conclusion I have done.

On this issue, therefore, I find in favour of the defendants.

How much of the damage to Block E is due to settlement?

I have already referred to the different opinions held by the two experts, Mr Langford and Dr Weeks. On this issue I have had no

*Now reported fully at (1981) 260 EG 920.
†260 EG at p 927.

A difficulty in preferring the views of Dr Weeks.

When he first looked at Block E in July 1979 Mr Langford concluded that virtually all of the crack damage was due to drying shrinkage of the calcium silicate bricks. In a letter to the council's insurers, dated January 31 1979 he expressed this view and said that he saw no reason to suppose that the ground was unsuitable as a load-bearing medium.

Mr Langford is an experienced, skilled and careful engineering expert, but I think that on his initial inspection of the plans and building on that occasion he jumped too quickly to a faulty conclusion. Perhaps he was misled by some of the information with which he was then supplied, including his conversations with the B occupants of nos 58 and 62. Although he concluded that the calcium silicate bricks were the source of the main damage he did not visit any of the neighbouring properties, some of which were built of those bricks, to see if similar damage had occurred. Had he done so he would have seen nothing comparable to the cracking in this building. Subsequent further inspection with fuller information has led him to agree that some of the damage *is* due to settlement. The extent of the damage is now known to be greater than that known by him in January 1979. He remains of the view that shrinkage of calcium silicate bricks is responsible for the major part of the crack damage, including some of the damage to the clay bricks. He thinks that some of the other damage to the clay bricks is due to thermal expansion C and contraction.

In my judgment the body of evidence points strongly the other way. This block has undoubtedly settled quite considerably and on the balance of probabilities I think that the major part of the damage is due to this. The absence of any evidence of similar trouble with silicate bricks elsewhere on the estate and the accepted fact that part of the damage to the silicate bricks at the lower end of the block is due to settlement favour Dr Weeks' views. So too, in my judgment, does the degree and pattern of cracking in the clay bricks in the centre section. On this issue I have no hesitation in preferring the views of Dr Weeks where they conflict with those of Mr Langford, and I believe it likely that Mr Langford has, probably without realising it, D had his present views coloured by the hasty conclusions he came to on his initial brief inspection of the building.

I should add here that it was urged on me that because on that initial visit, Mr Langford, an expert, saw no reason to conclude that the subsoil was in any way unsuitable as a load-bearing medium, it would be wrong to hold that a much less qualified building inspector was negligent when he inspected the site and trenches in 1963. I reject that argument, firstly because I think that Mr Langford jumped to an admittedly false conclusion based on what he saw, being misled by looking at the calcium silicate cracking, and secondly because the circumstances were wholly different. The building inspector did not have the calcium silicate cracking to mislead him; he had the virgin E site, later the opened trenches, at his disposal for inspection.

There is no detailed evidence on which an accurate assessment can be made of the proportion of damage due to settlement and that due to the calcium silicate shrinkage. Dr Weeks believes that by far the larger proportion is due to settlement: Mr Langford puts it the other way round. Having accepted Dr Weeks' view, and doing the best I can on the evidence, I hold that the calcium silicate shrinkage has had only a small effect — none at all in the clay centre section and only a small part in the end sections. The lower section (including nos 68 and 70) has been badly affected by settlement. In relation to the highest section, ie the other end from nos 68 to 70, I hold that 90% of the damage is due to settlement. In the centre section I reject the F evidence that thermal contraction is of any significance and hold that 100% of the damage is due to settlement, and in the lowest section, including Mr and Mrs Lyons' maisonette, where it is agreed that there has been severe settlement, I hold that 95% of the damage is due to that cause.

Is underpinning necessary?

I have no hesitation in holding that it is.

Again this issue must largely depend on deciding between the opinions given by the two experts and I have preferred that of Dr Weeks.

The differences of opinion largely turn on whether or not the G settlement which has undoubtedly occurred so as to affect the building is by now virtually complete. Dr Weeks believes that there is still a substantial amount of settlement to come, part of which will be due to rising and falling levels of the underlying water table caused by the climate. Mr Langford believes that after 16 or 17 years the compaction of the ground under the strip foundations is now virtually complete and that any water tending to collect in the ground at the lower level of Block E will drain off through the colluvium. Mr Langford does agree that if there is any further significant settlement to come underpinning is required. He regards a further 15 mm of settlement as being sufficient to require this remedy.

In his report of July 1980 Dr Weeks had made some highly H technical calculations to show the amount of settlement which could be predicted from the facts now known about the building, its foundations and the ground. After some cross-examination on these calculations it was agreed that he should discuss the matter with Mr Langford to see if their differences could be resolved. This resulted in a measure of agreement set out in a short joint written report dated October 15. Although Dr Weeks had somewhat modified his calculations it did not at all affect his views that further significant settlement is likely to occur.

Mr Langford in reaching his conclusion that settlement is virtually complete took into account a "crack monitoring survey" carried out by him over the period of 26 weeks starting with March 12 1981. They J showed no indications of further settlement. His report (dated September 1981) concluded that the measurement period was too short to give conclusive indications, but that the lack of settlement over the period was *some* indication that settlement is complete. Dr Weeks considered that the period was too short to be of any practical use.

I bear in mind Mr Lyons' evidence, which I accept, that although Block E was built in about 1963-64 the first signs of cracking did not occur until about 1973 when cracks appeared internally and about 1974 externally. Since then they have got very much worse. Figure 3 of the experts' agreed report dated July 1980 shows only one internal crack in Mr Lyons' property; but he told me that there are now three K or four cracks under the first window and another crack under the second window shown in that diagram. He also referred to Figure 5 and said that there are now further cracks in the rear of his maisonette.

As I have rejected Mr Langford's opinion that the cracking is mainly due to calcium silicate brick shrinkage, it follows that in my view this increased cracking during 1980 is due to settlement. This goes some way to neutralise any conclusions to be drawn from the crack monitoring survey.

But having on this matter preferred the evidence of Dr Weeks I would, in any event, have concluded that underpinning of the strip foundations is required.

L However this issue is very much related to the seventh issue:

What is the proper remedy for dealing with the ground-floor slabs?

I have not the slightest hesitation on this issue in holding that they must be suspended so as to be supported by the strip foundations. And in that event Mr Langford completely agreed that underpinning of the strip foundations will be required.

Mr Langford's proposed alternative for dealing with the floor slabs was, in my judgment, wholly unacceptable. The alternative envisaged the possibility that the existing fill might have to be dealt with. If so, Mr Langford proposed that it should be removed and replaced with new fill. But when it was put to him in cross-M examination that the removal of the existing fill would set up severe stress which would cause further serious damage he seemed uncertain and suggested a possible alternative of binding the existing fill with some cementatious or chemical material. But he was unable to refute the suggestion that it would be necessary to use a chemical material which would involve as much expense as underpinning the strip foundations.

I do not think that Mr Langford's alternative for dealing with the floor slabs is in sufficiently definite form for me to find it as acceptable.

Furthermore his alternative is based on his opinion that settlement of the building is virtually complete — a view which I have rejected.

A The necessary remedies for dealing with the strip foundations and the floor slabs are interlinked but in my judgment, whether these issues be considered separately or together, the conclusion is the same — that underpinning is required.

Damages

I can deal briefly with the two heads on which I am asked to quantify damages.

Firstly, as to diminution in value of the property as a result of the bad reputation it has acquired, the council concede in principle that this is a properly recoverable head of damage, subject to proof that such bad reputation does exist.

B I am satisfied on the evidence of Mr Lyons that local estate agents and building societies are well aware of the history of Block E, and I think that such a reputation is bound to exist having regard to the history, nature and extent of the damage.

I have no expert guidance as to the amount by which the selling price of the plaintiffs' property will be lowered, nor have I been told of the existing market value of the property in repaired condition — although I am aware from documents before me of the price paid by Mr and Mrs Lyons in 1972.

I hold that the plaintiffs are entitled to damages under this head, but I am reluctant without further evidence to quantify it. Therefore I propose to leave this to be decided along with other outstanding C issues as to damage, by another tribunal.

The issues as to quantum of damages should in my view be referred for assessment to an Official Referee.

QUEEN'S BENCH DIVISION

January 14 1982

(Before Sir Douglas FRANK QC, sitting as a deputy judge of the division)

FISHER AND ANOTHER v KNOWLES

D Estates Gazette June 12 1982

(1982) 262 EG 1083

Action against surveyor for breach of contract or negligence — Instructions from clients were to carry out something more than a bare valuation but less than a full structural survey — Defendant surveyor had been instructed to make a valuation report for a building society for the purpose of a mortgage on a house, but was also given independent instructions by the intending purchasers, the plaintiffs, to make a survey as above mentioned — Judge's interpretation of defendant's instructions was to provide a report on the general state of the property, to draw attention to those matters E which might give rise to suspicion, such as springing floors or a musty smell which might require further investigation, and to report any matters which might cause the plaintiffs to withdraw or to bargain for a lower price — Report drew attention to one or two minor defects, but supported a purchase at the asking price of £14,900 — After some four years the plaintiffs complained of the condition of the property and brought the present action — Although the plaintiffs' claim particularised 14 alleged defects, the judge found that the defendant was liable only for failure to report on rot in window frames and defects in ceiling joists and door joinery — Damages assessed at £500, not justifying a High Court action and attracting costs on county court scale only

F This was an action by Mr and Mrs Fisher against Victor Rodney Knowles BSc FRICS, a chartered surveyor, alleging negligence or breach of duty in a report, made on the plaintiffs' instructions, in respect of a house at 6 Hawkwell Close, Hockley, Essex, which the plaintiffs were proposing to purchase and subsequently did purchase.

Richard Greening (instructed by Kingsford, Dorman, agents for Drysdales & Janes, of Southend-on-Sea) appeared on behalf of the plaintiffs; H James Latter (instructed by Kenneth Wood & Co) represented the defendant.

Giving judgment, SIR DOUGLAS FRANK said: In 1976 the plaintiffs were minded to buy a house at 6 Hawkwell Close, Hockley, in Essex. The asking or negotiated price for the house was £14,900. The G Cheltenham and Gloucester Building Society had been asked to make an advance of £12,500 towards the purchase price; they instructed the defendant to make a report and valuation for the mortgage and he did that.

The plaintiffs wanted a survey and they found out who was carrying out the survey for the building society and instructed the same surveyor, namely the defendant. The defendant carried out an inspection and, following that inspection, he made his report to the building society valuing the property at £14,900 and also made a report, which is entitled "Valuation", which he sent to Mr Fisher.

There has been some debate as to the nature of the terms of employment of the defendant and how he was instructed, but at the H end of the day I do not think that anything turns on those questions, because it seems that what the defendant was being instructed to do was something more than the bare valuation which he gave to the building society and something less than a full structural survey. Indeed, what he did was to produce a report which was more than a bare valuation and less than a structural survey. I think what he was by implication being asked to do, and what he in fact did, was to provide a report on the general state of the property and to draw attention to those matters which might give rise to suspicion, as, for example, springing floors or a musty smell, which would require further investigation in the nature of a structural survey. I think it was also implicit in his terms of employment that he should draw to J the plaintiffs' attention any matters which might cause the plaintiffs either to wish to withdraw from the sale or to bargain for a lower price. I think that impliedly that is how Mr Knowles approached the matter. He is a chartered surveyor of high qualification and of many years' experience, including valuable experience in the Valuation Office of the Inland Revenue, which is notorious for the care that it takes in carrying out inspections.

That report was submitted on April 5 1976 and one or two defects were referred to. For example, the eaves gutters were rusted and corroded in places, but in general Mr Knowles expressed the view that he saw no reason to assume that the property would not prove to be a satisfactory dwelling for many years to come, nor any reason K why the plaintiffs should not proceed with the purchase at the price at which the property was offered, namely, £14,900, which figure they considered to be realistic.

The plaintiffs bought the property and went to live in it. They say that there were a number of defects which the defendant should have noticed and which he should have drawn to the plaintiffs' attention. On May 9 1980, that is some four years later, they made their complaint against the defendant. They subsequently brought these proceedings for breach of contract and/or negligence.

Before I turn to the allegations made, it is proper that I should emphasise that the burden is on the plaintiffs to prove that on the balance of probabilities the defects are defects to which the L defendant should have drawn attention when the inspection was carried out. That burden seems to be even more heavy, having regard to the fact that there was a time lapse of four years between the inspection and the first making of the allegations, and also having regard to the fact that some of the evidence — for example, the window frames — had been destroyed.

The word "defects" is a word having a number of connotations. Defects may be something which is inherent in the type and age of a building.

For example, an Elizabethan cottage is likely to have the defect of having low ceilings, a defect both in the ordinary sense and in the statutory sense. To some people, though, that would be considered M to be an enhancement. There are defects of some kind in every home. There are creaking floorboards, gaps between the wainscot and the floor. There are cracks in ceilings. One would be very lucky indeed to buy even a new house, let alone a 25-year-old house, as this one was, without finding some cracks in the ceiling. It seems that in approaching this matter in the instant case I should have regard to those defects which were of such a nature as a prudent person would have put right and which would have been apparent to the defendant exercising the ordinary skill of an ordinary competent chartered surveyor, and taking into account the type and age of the property. The defects are particularised and there are 14 in number. Some of

A them have been made good, but others still have not been made good.

It seems that both from the defendant's own qualification, knowledge, and experience, and the way in which he gave his evidence, he is a man who does exercise the skill of a competent chartered surveyor. It is apparent from his report that he made a fairly full survey of this property, but of course even the most competent professional man can make mistakes. Did he in this case?

The most important defect relates to the window frames and timbers, which were taken out and renewed with a different style of window at a cost, having allowed for a different type of window, of £1,000. The defendant says that it was his practice, which is the
B common practice of any surveyor, to go round and test window frames, and indeed all timbers, with a penknife. Although his memory cannot go back for him to say, "Yes, I remember doing it in this case", he says, "That is what my practice is and I have no doubt that I would have done that as a matter of course." On the other hand, the plaintiff says, "Within a short time of moving in I found that I could plunge a screwdriver into the window frames and in not one but in many places". His wife gave corroborative evidence of that. That is something which occurred soon after the purchase and the window frames were renewed, although not until 1979 because the plaintiff says that he had no money to do it.

It seems, having regard to the plaintiff's own evidence, and he
C struck me as a witness of truth, to Mrs Fisher's corroborative evidence, and, the fact that the windows were renewed, that on the balance of probabilities at the time of inspection there was rot in these window frames which should have been noticed by the defendant, but which for some reason was not noticed.

I now turn to the other items and deal with one in particular relating to the ceiling joists under the bathroom floor. I think that, seeing that wc, it should have been fairly obvious to the defendant
D that work had been carried out not only in contravention of the building regulations but by some inexpert hand and that he should have been put on an inquiry and investigated further.

Apart from those two matters, however, I am not satisfied that there was any evidence of damp in the roof space at the time of the inspection by the defendant. I am not satisfied that any sagging or bowing in the floor was of such a kind as to be any reason for affecting the valuation or that it was such that one might not expect to find in a house of this kind at this price of that age. I do not think that it was incumbent upon the defendant to report in such detail as is envisaged by these particulars.

E I do not think that in general these were matters that affected the value, or in some cases, as I say, I am not satisfied that they existed. So I am left with the two matters, the joists and the window and door joinery.

I am satisfied that the correct approach in assessing damages is to take the difference in value. There may be exceptions to that general rule, but this case is not one. It seems that it is wholly appropriate in the instant case because window frames and external doors notoriously require periodic maintenance and painting and re-puttying and so on. The plaintiffs had the benefit of these frames, not only of having new for old and therefore a longer life, but also in postponing the costs of maintenance and, incidentally, of being able
F to take advantage of renewal by putting in a type of window frame which presumably made the property more attractive. Doing the best I can, I have come to the conclusion that the plaintiffs claim succeeds but to the extent that I assess damages at £500.

Judgment was given in favour of the plaintiffs for £500, with three years' interest at 12 per cent and with costs on old County Court Scale 3. An order was made for the legal aid taxation of the plaintiffs' costs.

RATING

A

COURT OF APPEAL

November 3 1981

(Before Lord DENNING MR, Lord Justice OLIVER and Lord Justice WATKINS)

WATNEYS OF LONDON LTD AND OTHERS v CITY OF LONDON CORPORATION

BURROWS (VO) v CITY OF LONDON CORPORATION AND OTHERS

Estates Gazette March 20 1982

B (1981) 261 EG 1089

Rating — Procedure of Lands Tribunal — Whether Lands Tribunal can be required to state a case for the decision of the Court of Appeal on an interlocutory matter — Interpretation of section 3(4) of Lands Tribunal Act 1949 — Meaning of "decision" — Refusal of tribunal in a rating appeal to order discovery of documents relating to takings of public houses in the City of London — Tribunal member's ruling that he had no power to state a case on this question — Following an application for judicial review, Judge Newey, sitting as a High Court judge, directed the tribunal to state a case, which was done — Held by Court of Appeal that the Lands C Tribunal cannot be required to state a case on an interlocutory point — Proviso to section 3(4) of the 1949 Act refers to final decisions only — Other machinery is available to deal with interlocutory matters, namely, application to a High Court judge under section 3(6)(c) of the Act and rule 38 of the Lands Tribunal Rules or an application to the Registrar of the Lands Tribunal under rules 40 and 45 — Appeals against Judge Newey's direction allowed

These were proceedings by way of appeal from a decision given by Judge Newey QC, sitting as a judge of the High Court, following leave given to apply for judicial review. Judge Newey ordered that the Lands Tribunal should state and sign a case for the decision of D the Court of Appeal on the ruling given by J H Emlyn Jones FRICS that the tribunal had no power to state a case on the tribunal's refusal to order discovery. In the first appeal the appellants were Watneys of London Ltd and three other brewery companies (Chef & Brewer Ltd, Truman Ltd and Goodhews Ltd) and the respondents were the City Corporation. In the second appeal the appellant was Richard Burrows (valuation officer) and the respondents were the City Corporation and the four brewery companies. The decision of the Lands Tribunal was reported at (1981) 258 EG 561.

In the first appeal David Widdicombe QC and Miss Susan Hamilton (instructed by C Cullum Smith and Crossman, Block & Keith) appeared on behalf of the appellants; Bernard Marder QC and Jeremy Sullivan (instructed by the Comptroller and City Solicitor) E represented the respondents. In the second appeal Alan Fletcher (instructed by the Solicitor of Inland Revenue) appeared on behalf of the appellant; Bernard Marder QC and Jeremy Sullivan (instructed by the Comptroller and City Solicitor) represented the first respondents (the City Corporation); David Widdicombe QC and Miss Susan Hamilton (instructed by C Cullum Smith and Crossman, Block & Keith) represented the remaining four respondents (the brewery companies).

Giving judgment, LORD DENNING MR said: There are three public houses in the City of London. One is the "Sir Christopher Wren" in Cathedral Place: another is "The Cock Tavern" in Fleet F Street: and the third is "The Magogs" in Russia Row. A question has arisen as to what the rateable value of those public houses should be. The valuation officer inserted his figures in the valuation list. But the Corporation of the City of London were dissatisfied with them. So the matter went on appeal to the Lands Tribunal.

In the course of the proceedings the Corporation of the City of London said that the valuation was too low. They submitted that the rateable value ought to be assessed on a new basis. In aid of this submission the Corporation of the City of London want to see the profit and loss accounts and other details of the business of these three public houses. They want discovery of the documents relating to the takings in those houses. G

Before the appeals were heard by the Lands Tribunal, there were pre-trial applications. No application was made for discovery then. But it was raised at the beginning of the appeals. Mr Emlyn Jones, the member of the Lands Tribunal who was hearing the appeals, refused to order discovery at that stage. He also expressed the opinion that he had no power to state a case and refused a stay of the proceedings.

The corporation then went straight to the Divisional Court and asked for leave to apply for judicial review. They asked for an order that the Lands Tribunal should state and sign a case for the decision of the Court of Appeal. They said that the Lands Tribunal were bound to state a case. H

His Honour Judge Newey accepted the corporation's submission and directed that a case should be stated by the Lands Tribunal on the matter. A case has now been stated. But there is a preliminary point to be decided. Was the judge right in ordering the Lands Tribunal to state a case? It is submitted to us that the Lands Tribunal cannot be required to state a case on an interlocutory point such as discovery of documents. It can only be required to state a case on a final award.

This point depends on the true interpretation of section 3(4) of the Lands Tribunal Act 1949, which provides:

> A decision of the Lands Tribunal shall be final:
> Provided that any person aggrieved by the decision as being erroneous in J point of law may, within such time as may be limited by rules of court, require the tribunal to state and sign a case for the decision of the court . . .

The Corporation of the City of London say that the word "decision" in that subsection includes not only a final decision but an interlocutory decision. But on the other side it is said: "No; it refers only to a final decision of the Lands Tribunal and not an interlocutory decision".

My own view is that the Lands Tribunal can state a case only on a final decision, not on an interlocutory matter. The principal reason is because there is ample other machinery available for interlocutory matters. K

One way is by applying to a High Court judge. This is provided for by section 3(6)(c) and rule 38 of the Lands Tribunal Rules. It brings in section 12(6) of the Arbitration Act 1950, which says:

> The High Court shall have . . . the same power of making orders in respect of . . . (b) discovery of documents and interrogatories . . . as it has for the purpose of and in relation to an action or matter in the High Court:
> Provided that nothing in this subsection shall be taken to prejudice any power which may be vested in an arbitrator or umpire of making orders with respect of any of the matters aforesaid.

It seems to me that the rules of the Lands Tribunal, by incorporating section 12(6) of the 1950 Act, show that the High Court has power to make an order for discovery of documents in relation to proceedings which are going on in the Lands Tribunal. But that power is without prejudice to any power which the Lands Tribunal itself may have of making orders.

Another way is by applying to the registrar of the Lands Tribunal. Rule 40 gives the registrar power to order discovery. It says:

> A party to proceedings shall deliver to the registrar on his request any document or other information which the Tribunal may require and which it is in the power of that party to deliver and shall afford to every other party to the proceedings an opportunity to inspect those documents (or copies of them) and to take copies:
>
> Provided that nothing in this rule shall be deemed to require any information to be disclosed contrary to the public interest.

This is followed by rule 45, which tells how interlocutory applications are to be made. It says:

> (1) Except where these Rules otherwise provide, an application for directions of an interlocutory nature in connection with any proceedings shall, unless otherwise ordered by the President, be made to the registrar . . .
>
> (7) The registrar may, and shall if so required by the applicant or by a party objecting to an application under this rule, refer the application to the President for decision.
>
> (8) A party aggrieved by a decision of the registrar on an application under this rule may appeal to the President by giving notice in writing to the registrar and to every other party within seven days after service on him of notice of the decision or within such further time as may be allowed by the registrar, but an appeal from a decision of the registrar shall not act as a stay of proceedings unless so ordered by the President.

Thus there are two ways available for getting discovery. One way is to make an application to the registrar and thence to the president. The other way is to apply to the High Court. It has an overriding power, if an application is made to it, to order discovery of documents.

So machinery is provided by the rules: first by applying to the tribunal: and then, if the application is refused, by applying to the High Court judge. That seems to me a good and effective machinery for obtaining discovery. It follows that section 3(4) need never be invoked for interlocutory applications. It should be limited to final decisions of the Lands Tribunal.

We were referred to one or two other matters in the case. I will put on one side the Tribunals and Inquiries Act 1971 because that has its own provisions in section 13 for dealing with decisions of tribunals including interlocutory orders. The Lands Tribunal and the valuation courts are specifically excluded from section 13.

Then we were referred to two cases. First, *Norwich Rating Authority v Norwich Assessment Committee* [1941] 2 KB 326, in which on different wording it was held that the decision could be the subject of an appeal by way of case stated. There were no rules such as we have here. Secondly, *Oswestry Corporation v Hudd (Valuation Officer)* [1966] 1 WLR 363, which is parallel to our present case. It was held in that case that "decision" applied only to a final decision.

I would go simply on the Lands Tribunal Act and the rules made thereunder. It seems to me that the rules make ample provision for justice to be done in interlocutory applications to the Lands Tribunal — or, alternatively, to the High Court: and section 3(4) can very properly be confined to final decisions.

For these reasons I am afraid I take a different view from Judge Newey. I would allow the appeals. There is no jurisdiction to state a case from the interlocutory order.

Agreeing, OLIVER LJ said: The question turns entirely upon the correct construction of the word "decision" in section 3(4) of the Lands Tribunal Act 1949. It simply provides: "A decision of the Lands Tribunal shall be final". It then goes on to provide for a person aggrieved a right to have a case stated on a point of law for the court — not just an option but a right. And subsection (11) makes it clear that the court referred to in subsection (4) is the Court of Appeal. So one has here for a start a provision which confers a right upon the litigant to have a point of law referred by way of case stated to the Court of Appeal. In my judgment, the natural construction of that would be, as a pure matter of commonsense, that it related only to the final decision of the court.

That seems to me to be borne out by a consideration of the Act as a whole. Section 1(3) sets out the various matters which are referred to the Lands Tribunal for decision — a series of what are described as "questions" or "disputes". Section 4(1) provides:

> His Majesty may by Order in Council direct that questions which are required or authorised by any Act . . . or instrument made under any such Act, to be determined by any statutory tribunal shall or may be determined instead by the Lands Tribunal if it appears to His Majesty — (a) that the questions are appropriate for the Lands Tribunal

and so on.

Subsection (4) of that section provides:

> Without prejudice to the generality of the last foregoing subsection, the provisions thereby authorised to be contained in an Order in Council shall include provisions for the following purposes, that is to say — (a) making decisions of the Lands Tribunal in the exercise of the jurisdiction transferred enforceable in the same way as those of the statutory tribunal from which the jurisdiction is transferred.

Clearly there, I think, "decisions" is used in the sense of a final decision. Then in subsection (6) of section 3 one finds again that the rules which may be made under the subsection may make provision as to the form in which the decision of the tribunal is to be given. It seems to be singularly inappropriate to interlocutory decisions on matters such as discovery and so forth.

It makes, as Mr Fletcher has said, perfectly good sense to give an absolute right to a litigant to have a case stated in the Court of Appeal in the case of a final decision; but it does not seem to me to make any sense at all to give a similar right — a right which is not conferred in any other court — to go straight to the Court of Appeal without leave in the case of an interlocutory decision. One would require some persuading that that is the effect of the section even if, contrary to the view I take, the word "decision" is not *prima facie* applicable to a final decision.

As Mr Fletcher has submitted, Parliament envisaged that the Lands Tribunal procedure should be in essence a modified form of the arbitration procedure which is provided by the Arbitration Act because subsection 6(c) of section 3 enables rules to be made and "apply in relation to the Tribunal any of the provisions of the Arbitration Acts, 1889 to 1934" — updating them to the relevant time for this appeal, the provisions of the 1950 Act.

It would, one would have thought, have been a sensible thing that there should be either in the tribunal or in the litigant before the tribunal the right, subject to the control of the High Court, to have a case stated in the same way as before an arbitrator. Rather curiously perhaps, the provisions for consultative cases have not been applied by the rules, which do not, in fact, incorporate in the provisions which are to be applied to the Lands Tribunal the provisions of section 21 of the Arbitration Act 1950. But, as Mr Fletcher submitted, it is not right to construe the Act — or what is the true meaning of the Act — by reference to the rules which have subsequently been made under it by the rule-making body. The fact is that the Act here contemplated a method which is that a perfectly sensible right of challenge should be conferred in the case of interlocutory decisions, and indeed the rule-making power has incorporated the current jurisdiction of the High Court under the Arbitration Act by applying to the Lands Tribunal the provisions of section 12(6) of that Act.

Construed in the way in which my Lord has construed the relevant section, that is, section 3(4), with which with respect I concur, it seems to me that one gets a perfectly sensible meaning out of the Act; and it would, in my judgment, be a misconstruction to treat the word "decision" in subsection (4) of section 3 as applying to the sort of interlocutory ruling which is in question in this case.

For those reasons, therefore, and for the reasons which my Lord has given, I too agree that the appeal should be allowed.

WATKINS LJ: I agree, and there is nothing I can usefully add.

A **CHANCERY DIVISION**

July 31 1981

(Before Mr Justice NOURSE)

MOBIL OIL CO LTD v RAWLINSON

Estates Gazette January 23 1982

(1981) 260 EG 261

**Legal mortgagee's right to possession — Restatement of the law by Nourse J, following *Birmingham Citizens Permanent Building Society* v *Caunt* — Passages in *Fisher and Lightwood* and *Halsbury's Laws*
B disapproved — Mortgagor in arrear with payments due under a legal charge, but seeking to counterclaim and set off against arrears certain liquidated and unliquidated sums — Master made an order for possession, but subject to a proviso that it was not to be enforced if the mortgagor should within a time-limit lodge a sum (representing a balance of indebtedness) into court — Master's order wrong — Even if the debtor's cross-claims exceeded the amount of the mortgage debt (as to which he cannot make a unilateral appropriation) the master should have made an unconditional order for possession, unless satisfied that there was a reasonable prospect of the mortgagor's paying the arrears in full, in which case an adjournment for a short time could have been granted — Practice
C as to adjournment of matter by master to judge after order has been perfected — Master's order varied by deletion of proviso**

Motion to set aside or vary master's order on summons by mortgagees for possession. The motion was necessary as the mortgagees had not asked the master to adjourn the matter to the judge before the order was perfected. The notice of motion sought the deletion of part of the master's order.

E A Davidson (instructed by Metson, Cross & Co) appeared on behalf of the plaintiffs; H A P Picarda (instructed by Pardoe, David & Shaw, of Bridgwater) represented the defendant.

Giving judgment, NOURSE J said: This case raises in new guise the old question whether, contract and statute apart, the court has any
D but the most limited power to deny a legal mortgagee possession of the mortgaged property.

The defendant, Mr K A F Rawlinson, is the owner of the equity of redemption in the Heathfield Garage, North Petherton, near Bridgwater in Somerset. That property is subject to a legal charge made on September 25 1975 between a company then controlled by Mr Rawlinson as mortgagor of the one part and the plaintiffs, Mobil Oil Co Ltd, as mortgagees of the other part. Mr Rawlinson took a conveyance of the equity of redemption in 1977 and he is now in possession of the property. It is unnecessary for me to go into the provisions of the legal charge in detail. It is enough to say that it is in a form commonly used in cases where the mortgagor owns a
E filling station and the mortgagee is a petrol company supplying petroleum products for sale on the premises. There were also supply and equipment loan agreements in force between Mobil and Mr Rawlinson. In addition to principal and interest the legal charge secured payment of all moneys for the time being due from and payable by Mr Rawlinson to Mobil under the supply and equipment loan agreements.

In circumstances into which I need not go Mr Rawlinson fell into arrears under the legal charge. They are agreed to have amounted to £26,936.89 as at April 30 1981. But he seeks to counterclaim and set off against those arrears certain other liquidated and unliquidated sums being expenses and losses incurred by him as a F
result of alleged breaches of contract by Mobil. I need not go into the nature of those cross-claims at this stage.

In due course Mobil issued an originating summons against Mr Rawlinson claiming various familiar forms of relief, including possession of the mortgaged property. That summons came before the master on May 14, when Mobil asked for an order for possession. The parties were not represented by counsel and the master evidently did not have the benefit of a full argument on the matter. In the event he made an order for possession, but subject to a proviso that it was not to be enforced if Mr Rawlinson should on or before May 28 (or subsequently within four days after service of the order) lodge the sum of £8,000 in court. It appears that that sum was the G
master's estimate of the difference between the admitted amount of the arrears under the legal charge and the aggregate amount of the expenses and losses which Mr Rawlinson seeks to counterclaim and set off against them. The aggregate amount of the liquidated cross-claims is £16,846 or something over £10,000 less than the agreed arrears. That means that the master credited Mr Rawlinson with something over £2,000 in respect of his cross-claims of unliquidated amount. The master then gave directions under Order 28, rule 8, for the further conduct of the proceedings, including an order that Mobil's affidavit in support of the summons should stand as the statement of claim and Mr Rawlinson's as the defence and counterclaim. Mobil did not ask the master to adjourn the summons to H
the judge and the order was perfected on May 22. The £8,000 was duly lodged in court within the stipulated period.

In due course Mobil's solicitors instructed counsel to settle a draft reply and defence to counterclaim. They had not realised that there might in law be ground for submitting that the master ought not to have included in his order any provision to the effect of the proviso which he did include. Accordingly, they did not instruct counsel to consider that point. And that was why no adjournment to the judge had been sought and the order had been perfected in the meantime. However, on July 7 counsel communicated with Mobil's solicitors and indicated that in his opinion there was ground for submitting that the proviso should not have been included in the order. J
Accordingly, on July 10 Mobil issued a notice of motion seeking in effect to delete the proviso from the master's order. That application came before me last week.

Contract and statute apart, a legal mortgagee has, as an incident of his estate in the land, an unqualified right to possession of the mortgaged property. He may, as Harman J put it, go into possession before the ink is dry on the mortgage and without default on the part of the mortgagor. If he does so, he is liable to account on the footing of wilful default and that will often deter him from taking possession himself. But during the present century it has become usual for mortgagees to seek an order for possession as a preliminary to realising the security by means of a sale with vacant pos- K
session out of court under the statutory power. In a normal case the court is not asked to grant any other relief.

Before 1936 a mortgagee who was only asking for possession had to commence his proceedings in the Queen's Bench Division. That was because possession alone could not be sought by summons in the Chancery Division. But in 1936, Order 55, rule 5A, of the then Rules of the Supreme Court was amended so as to make that possible. The Chancery judges of the time issued a practice direction which said, among other things, that when possession was sought and the defendant was in arrear with any instalments due

A under the mortgage or charge and the master was of the opinion that the defendant ought to be given an opportunity to pay off the arrears, the master might adjourn the summons on such terms as he thought fit. That direction caused confusion. It led to a general view among the Chancery masters, no doubt assisted by the benevolent attitude which the legislature had by then assumed towards tenants faced with eviction by their landlords, that they had a discretion to adjourn a legal mortgagee's application for possession, at any rate in instalment cases, against the wishes of the mortgagee in order to enable the mortgagor to catch up on instalment arrears; and (inferentially) a right if he did so to continue to deny the mortgagee possession notwithstanding that on the default the whole of the
B mortgage money had become and thereafter remained repayable. By the end of the 1950s it had become necessary for the Chancery judges of a later generation to reassert the legal mortgagee's right to possession. In the van of that movement was Harman J, although even he subscribed to the view that the practice direction had qualified the right in the case of an instalment mortgage; see *Fourmaids Ltd v Dudley Marshall (Properties) Ltd* [1957] Ch 317 at 321.

In 1961 the whole question was fully argued and considered in the case of *Birmingham Citizens Permanent Building Society v Caunt* [1962] Ch 883 and it is on Russell J's judgment in that case that the foregoing summary of the earlier history is based. I well
C remember that decision and the general view of the profession that it had settled once and for all the limited extent of the court's power to adjourn a legal mortgagee's application for possession. The rule in regard to instalment mortgages (and *a fortiori* in regard to ordinary mortgages) was stated by Russell J at the end of his judgment at p 912, in the following terms:

> Accordingly, in my judgment, where (as here) the legal mortgagee under an instalment mortgage under which by reason of default the whole money has become payable, is entitled to possession, the court has no jurisdiction to decline the order or to adjourn the hearing whether on terms of keeping up payments or paying arrears, if the mortgagee cannot be persuaded to agree to this course. To this the *sole exception* is that the application may be
> D adjourned for a short time to afford to the mortgagor a chance of paying off the mortgage in full or otherwise satisfying him; but this should not be done if there is no reasonable prospect of this occurring. When I say the sole exception, I do not, of course, intend to exclude adjournments which in the ordinary course of procedure may be desirable in circumstances such as temporary inability of a party to attend, and so forth.

The reason for the exception is that the court has never allowed a mortgagee to enforce his rights under the mortgage in the face of a concrete offer by the mortgagor to redeem.

Since then the court has twice been given additional powers of adjournment in cases where the mortgaged property consists of or includes a dwelling-house; see section 36 of the Administration of
E Justice Act 1970 and section 8 of the Administration of Justice Act 1973. But the general rule continues to apply to other types of property, for example commercial premises of the kind with which the present case is concerned. Moreover, in *Western Bank Ltd v Schindler* [1977] Ch 1 the Court of Appeal reaffirmed the *prima facie* unqualified right of a legal mortgagee to possession in unequivocal terms. And the judgment of Buckley LJ demonstrates that there may well be occasions in the conditions of modern society when a mortgagee will be wise to go into possession in spite of his liability to account on the footing of wilful default; see *ibid* at pp 9H-10B.

It is accepted that the legal charge in the present case in no way
F restricts Mobil's right to possession as a matter of contract. Apart from an argument based on the particular circumstance that no adjournment to the judge was sought before the order was perfected (I will deal with that point later), Mr Picarda, who appears for Mr Rawlinson, bases himself on the following passage in *Fisher and Lightwood's Law of Mortgage*, 9th ed at p 325 where, having referred to the general rule established in *Birmingham Citizens Permanent Building Society v Caunt*, the learned editor proceeds as follows:

> However, if there is a *bona fide* defence, such as that the plaintiff is a moneylender and the requirements of the Moneylenders Acts have not been satisfied, or any default has been waived or there be some estoppel which can be raised against the mortgagee or counterclaim or set-off arising G
> out of the mortgage, the master or registrar will adjourn the matter into court.

In reference to the possibility of a counterclaim or a set-off arising out of the mortgage there appears this footnote:

> Even an admitted liquidated claim which exceeds the mortgage does not *per se* discharge the mortgage and is no reason for adjourning the application: *Samuel Keller (Holdings) Ltd v Martins Bank Ltd* . . . See also *Inglis v Commonwealth Trading Bank of Australia* . . . But where the amount due under the mortgage is determined by some other transaction between the parties, eg where the mortgage is to secure the amount owing from time to time for goods supplied by the mortgagee to the mortgagor, the possession H
> claim may be adjourned, if the mortgagor has a cross-claim for goods supplied to the mortgagee.

Mr Picarda relies strongly on the proposition embodied in the second part of that footnote. He accepts that is is not supported by any authority, but he says that it is consistent with principle and ought to govern this case. The essence of Mr Picarda's argument is that the set-off claimed by Mr Rawlinson goes to the root of his secured floating indebtedness to Mobil, and enables him to claim that there is in truth no such debt in existence. It is on that ground that Mr Picarda seeks to distinguish *Samuel Keller (Holdings) Ltd v Martins Bank Ltd* [1971] 1 WLR 43 and *Inglis v Commonwealth* J
Trading Bank of Australia [1972] 126 CLR 161, which are the two cases referred to in the first part of the footnote.

I am prepared to assume in Mr Rawlinson's favour that the amount of his cross-claims exceeds the amount of the mortgage debt. I say at once that I regard that as an assumption of extremely doubtful validity — on the master's estimate there is a shortfall of about £8,000 — but I will make it none the less. However, I find it impossible to make any distinction between this case and *Samuel Keller (Holdings) Ltd v Martins Bank* [1971] 1 WLR 43. Megarry J's statement of the principle at pp 47H-48B, which was expressly approved by the Court of Appeal, is in entirely general terms. The principle is that a mortgagor cannot unilaterally appropriate the K
amount of a cross-claim, even if it is both liquidated and admitted (and *a fortiori* if it is unliquidated or not admitted), in discharge of the mortgage debt. On that footing the origin and nature of the cross-claim and its relationship to the mortgage debt are wholly irrelevant.

In the circumstances this case must be approached on the footing that when the matter came before the master there were substantial arrears outstanding. Consistently with the general rule established in *Birmingham Citizens Permanent Building Society v Caunt* he ought then to have made an unconditional order for possession, unless of course he was satisfied that there was a reasonable prospect of Mr Rawlinson's paying the arrears in full, not into court but L
to Mobil, or otherwise satisfying Mobil, in which case he should have adjourned the application for a short time.

It follows that I must disapprove the proposition embodied in the second part of the footnote in *Fisher and Lightwood* and also in one in almost identical terms in *Halsbury's Laws of England*, 4th ed vol 32, at para 834. I should add that even without the authority of *Samuel Keller (Holdings) Ltd v Martins Bank Ltd* (which was not concerned with a claim for possession) I would regard the proposition as being unsound. If the general rule is that the mortgagee is entitled to possession unless there is a reasonable prospect of the mortgagor's paying him off in full or otherwise satisfying him within a short time I do not see how payment into court, even in full, or M
anything else for that matter, can be a compliance with the rule. Any other view would only serve to perpetuate the former confusion between the unqualified right to possession and the state of the account between the mortgagor and the mortgagee.

I do not intend to suggest that there may not be other circumstances in which the court will refuse a mortgagee possession on terms that the mortgagor pays the full amount of the mortgage debt into court. A number of other possibilities are mentioned in the passage in *Fisher and Lightwood* which I have already read. Without attempting to decide any point which does not arise for decision in the present case, I will only say that it seems to me that that course could be adopted only in a case where there was a substantial

A question as to the existence or enforceability of the right to possession, for example where it was claimed that the mortgage was void for illegality or that the mortgagee was in some way estopped from asserting his right. It appears probable that *Lidco Investments Ltd v Hale* [1971] Estates Gazette Digest 669 is a case which falls within the former category.

That leaves only the question whether Mobil's position has been prejudiced by its not having asked for the matter to be adjourned by the master to the judge before the order was perfected. This has meant that it has had to come to the court by way of motion to set aside or vary the master's order. The practice of the court is stated by Templeman J in *Whiteoaks Clifton Property Services Ltd v*
B *Jackson* [1975] 1 WLR 658 at p 660H in the following terms:

> Of course the jurisdiction to set aside an order made by a master will be sparingly exercised where the aggrieved litigant does not ask for the matter to be adjourned to the judge before the order is perfected. The jurisdiction will be exercised even more sparingly if a considerable time elapses between the date of the order and the date of the motion to set aside.

That shows that the jurisdiction will be sparingly exercised where, as in this case, no adjournment to the judge is requested before the order is perfected. However, Mr Picarda very properly accepted that he would be in some difficulty in resisting this application if I came, as I have, to the conclusion that the master had no jurisdic-
C tion to impose the proviso. Further, Mobil did act with great expedition once the point had been drawn to its attention by counsel. In the circumstances, it seems to me that I should certainly accede to Mobil's application and that the proviso should be deleted.

The master's order for possession was varied by deletion of proviso that it should not be enforced if the defendant lodged a sum in court within a time-limit. A stay of 28 days was granted in view of prospect that the mortgage debt might be repaid in full.

D

COURT OF APPEAL

January 13 1982

(Before Lord Justice CUMMING-BRUCE, Lord Justice DONALDSON and Sir Sebag SHAW)

COHEN v NESSDALE LTD

Estates Gazette May 1 1982
(1982) 262 EG 437

"Subject to contract" condition — Case illustrating the effectiveness of this condition or qualification even where it leads to an unmeritorious result which has to be accepted with regret — Ap-
E peal from decision of Kilner Brown J — Appellant (plaintiff in court below) had sought specific performance of an alleged agreement to grant him a leasehold term of 99 years of a flat — In order to succeed he had to be able to put in evidence an alleged firm oral agreement arrived at in the course of a meeting — Before this meeting there had been prolonged correspondence some of which had been expressly stated to be subject to contract, but others had not — No reference made at the meeting to the discussions then taking place being subject to contract — Held by Court of Appeal, affirming decision of Kilner Brown J, that the subject to contract qualification covered the discussions at the meeting, so that the oral agreement then reached could not be the foundation for an ac-
F tion for specific performance — Negotiations had been interrupted but not abandoned — "Subject to contract" protection can be removed only if both parties expressly agree that it should be expunged or if such an agreement must necessarily be implied — Appeal dismissed with expressions of regret in view of unmeritorious conduct of respondents.

This was an appeal by the plaintiff, Ronald Cohen, from a decision of Kilner Brown J refusing an order of specific performance sought by the plaintiff against Nessdale Ltd, the present respondents, of an alleged agreement to grant the appellant a lease for 99 years of Flat 2 at 11 Rutland Gate, London SW7. The appellant was the statutory tenant of this flat.

R Bailey-King (instructed by Coode, Kingdon, Somper & Co) G
appeared on behalf of the appellant; David Neuberger (instructed by Harold Stern & Co) represented the respondents.

Giving judgment, CUMMING-BRUCE LJ said: This appeal raises first, and, on the view which I have formed, raises only, one short point. The facts have been set out in the judgment of the learned judge below, which judgment has been reported in [1981] 3 All ER 118, and in the circumstances it will not be necessary for me to elaborate very much upon the facts which I am content to take from the findings of the learned judge.

The background of the matter is that at all material times the plaintiff below — the appellant in this court — was statutory tenant of a flat in a building called 11 Rutland Gate, holding from the H
defendants below — the respondents in this court — a company called Nessdale Ltd. The history of the tenancy prior to the year 1976 included a protracted argument in which the tenant was alleging that the landlords had failed to observe their covenants for repair, which had led him to withhold his rent. The history includes the fact that the landlords, for practical purposes, consisted of two persons: someone called Sadleir and a person called Henderson, who had been charged by the company with responsibility for negotiations with tenants or prospective purchasers as to the terms of their leases or purchases.

The well-known firm of Goddard & Smith had instructions from the landlords to deal with tenants or prospective purchasers of J
leases on three matters, and three matters only: collection of rents; business arising out of the obligations of the landlords to observe their covenants; and thirdly, in connection with the negotiation of long leases, such work as was requisite for the purpose of ascertaining the detail of apportionment of service charges between the landlords on the one hand and their several tenants on the other.

The first and, on my view of the appeal, the only question that arises on this appeal is whether the learned judge was right in holding that the discussion and agreement entered into and arrived at between Nessdale Ltd on the one hand and Mr Cohen, the tenant, on the other hand, on November 18 1977, was, by necessary implication, proceeding under the same umbrella as that which had covered earlier negotiations in the year, whereby those negotiations K
were all subject to contract.

For the purpose of deciding whether the judge was right in his conclusion about that, it is necessary to look in a little detail at the history between March 15 1977 and May 10 1977, which represented the first stage of negotiations between the parties connected either with the negotiations of a long lease or, alternatively, terms for surrender by Mr Cohen of his interest in the premises. I say that was the first stage in the negotiations, but it is right to refer to the fact, though it is not of importance, that in 1971 there had been earlier discussions which had come to nothing.

So one comes to consider that history which conveniently can L
begin with a letter from Mr Henderson, on behalf of the landlords, to Mr Cohen, their tenant, dated March 15. I read it:

> I refer to our meeting last year and wonder whether you have further considered purchasing a leasehold interest of the above. We are prepared to sell a term of 99 years from 25/12/1968 for the sum of £20,000, subject to contract. The ground rent will be £50 pa for the first 33 years doubling every 33 years thereafter. There will be a service charge of £60 pa payable on account being 25% of the landlords' outgoings. Should this not be of interest to you we are prepared to pay you a sum of monies for the surrender of your tenancy providing this takes place in the relatively near future. I look forward to hearing from you in due course.

So there is a proposal in which the interest proposed to be transferred is a term of 99 years. The consideration is offered at M
£20,000 together with the stated ground rent and the proposed service charge, which, as the tenant later observed, called for clarification, because it is offered as "£60 pa payable on account being 25% of the landlords' outgoings." Therefore it did require some clarification in order to see the exact obligation that was being undertaken over the years.

On May 2 the tenant replied, thanking Mr Henderson for his letter. He said:

> I have considered what you say and would point out the following; I assume that the price you are quoting for approximately 90 years takes into account that I am a sitting tenant, but nevertheless I am advised that

A £20,000 is rather on the high side for this particular flat. However, I would be prepared to consider, subject to contract, a figure of £15,000 on the same conditions subject to clarification of the Service Charge which I don't, at the moment, quite follow. Alternatively, bearing in mind my own circumstances and what this would entail for me to vacate the property, I would be prepared to consider subject to agreement a payment of £12,000 to surrender my position here.

So there, at the beginning of that correspondence, you have both the landlords' representative, Mr Henderson, and the tenant both negotiating, expressly stating, that they are so doing, subject to contract.

Those negotiations continued. The landlords, on May 3, in a letter with a label at the bottom "SUBJECT TO CONTRACT AND B WITHOUT PREJUDICE", say:

Having given the matter some consideration, we would be prepared to proceed at a compromise figure of £18,500, subject to contract.

On May 5 Mr Cohen came back, in a letter addressed to Mr Glover of the landlords, saying:

I don't really feel that £18,500 is a compromise and therefore I think that the best course would be to proceed along the second alternative suggested in your letter of March 15 on which I have made an offer.

On the next day Mr Glover answered, and said:

If you do not feel that £18,500 . . . is a compromise figure, I do not think that it is very likely that you will consider the maximum which we are able C to pay by way of compensation for vacant possession to be a compromise, as we certainly cannot reach anywhere near your figure. I would, however, advise you that if you were to give vacant possession by the end of June, we would be able to pay to you compensation of £5,000 or if you were to complete the purchase of the long lease by that date, we would be prepared to make a small further reduction to £17,500, subject to contract.

That letter has "SUBJECT TO CONTRACT AND WITHOUT PREJUDICE" written at the bottom.

The next material matter is that Mr Cohen received notice of an application by the landlords to the rent registration office for an increase in rent. That he felt was an action really quite inconsistent with the kind of negotiations that were proceeding. He was evidently very angry. It may be — I do not know — that the ex- D planation was that Goddard & Smith, who were dealing with the rent matter, were a right hand not aware of what the left hand was doing. However, Mr Cohen wrote to Mr Glover on May 10:

Thank you for your letter of May 6. In the meantime I have received notice from the rent registration office upon your application for an increase in the rent. In these circumstances obviously there is nothing further we can do in the matter.

That is the letter which was relied upon below, and in this court, as bringing those negotiations subject to contract to a complete end, thus finally ending and interrupting that negotiation. It is against that background that the submission on behalf E of the appellant is that what happened afterwards in November was not in any sense a continuation of negotiations, but that it was the opening of a new chapter which was not affected by what had gone before.

It is necessary to note the further activities of the tenant in his correspondence with Goddard & Smith. Goddard & Smith were, of course, dealing with the application to the rent officer for an increase in the statutory rent, and they were also responsible for dealing on behalf of the landlords with the controversy about the landlords' alleged failure to comply with, and observe, their covenants to repair. So it was that, on July 7, Mr Cohen wrote to the senior partner in Goddard & Smith, and in that letter, in the se- F cond paragraph, he said:

I would venture to put you in the picture about this flat . . . I am holding over as a sitting tenant from an original lease which expired in 1969 and recently in May Messrs Nessdale, the landlords, wrote to me and asked me if I would like a new 99 year lease or, alternatively, would I like to be bought out by surrendering my interest. We started some correspondence on the matter which did not finalise because in the middle of the correspondence I had notice from the fair rent tribunal that an application had been made to increase the rent. I was quite flabbergasted to find myself firstly in negotiation with my landlord at his instigation and then he applies to have the rent increased, whereas one would have expected that since he was in negotiation with me he would have written about this and suggested something instead of carrying out the procedure in this extraordinary manner. As a result of this I terminated the negotiations.

Then he deals with his troubles about the failure of the landlord G to comply with his covenants and he goes on to say this:

I am trying to be reasonable about this matter and this is why I am writing to you because I am quite prepared now to have my premises done up completely if we can only get some sense out of Messrs Nessdale and conclude a lease at a proper price and not £20,000 being asked for a sitting tenant.

He says that the rent people have suggested that "the correct price ought to be £10,000." He then says:

Bearing in mind all the above . . . matters perhaps you could try and see whether we can do something constructive on the lines suggested . . . Thanking you very much for your assistance, I await hearing from you.

Mr Cohen was there hoping to use the good offices of Goddard & Smith, who, I may say, come very well out of this history, in H order to get the landlords to take what Mr Cohen thought to be a more realistic view of the appropriate consideration for the negotiation of a long lease. Goddard & Smith acknowledged his letter.

Then, on August 2, Mr Cohen wrote another letter to Goddard & Smith, in which he said this:

Now I have to do all these repairs and, as I have said before, I am quite prepared to have the porch roof repaired whilst my builders are here but it is going to be very expensive if I have to have them in twice in relation to having the flat done up. I must now ask the landlords to deal with this matter properly and I repeat that I am willing to pay £10,000 for the lease, which is the correct price, and then to have the flat done up and all the J repairs carried out. Can I now please have some action on this so that we can stop all this going backwards and forwards which is getting no one anywhere.

On August 22 Mr Cohen, again to Goddard & Smith, mentions the problem of the porch and says:

May I remind you that your clients came to me to either take a lease or to give up my tenancy and, at the same time, went to the fair rents tribunal to increase the rent . . . I want to get the builders in here in one go and I want to get this matter of the lease resolved, together with the question of the rent.

Goddard & Smith, on August 25, answered and said this:

We are forwarding a copy of your letter to our clients with whom you have been negotiating the terms for a new lease, in order that they may deal K direct. We will [get on with] remedial works . . . to the porch and balcony.

Goddard & Smith did that in their letter to Mr Sadleir of Nessdale, enclosing a copy of Mr Cohen's letter.

The next letter to which I may refer is a letter from Goddard & Smith to Mr Cohen dated October 25 1977. It refers to dealings with contractors and refers to a letter of October 12 from Mr Cohen addressed to Mr Ongley of Goddard & Smith. That letter is not in the bundle and we do not know what it said. I only refer to it because the last paragraph of that letter says this:

With regard to the other matters raised in your letter, we would respectfully point out that these are matters which are subject to discussion between L yourself and our clients, and we are therefore unable to comment.

Thus, the stance of Goddard & Smith throughout, which was perfectly correct, was that they were not going to come in to the negotiations for a new lease, which was why they had sent Mr Cohen's earlier letter about the new lease and resumption, or beginning, of negotiations straight on to Nessdale, Mr Sadleir.

So one comes to the crucial period of the history when at last Nessdale Ltd respond to the approaches of Mr Cohen made to them through the intermediary, Goddard & Smith. On November 2 1977 Mr Henderson wrote to Mr Cohen:

You may recall our earlier meeting concerning our proposals for the flat you occupy. Namely the possibility of you surrendering your tenancy in favour of a capital sum, or in purchasing a leasehold interest. I would have M thought it to be in your interest to take advantage of your position as a sitting tenant rather than continue occupying on a somewhat negative basis as at present. If you would care to communicate with me at the above I will be happy to discuss the matter further. Accordingly I look forward to hearing from you in due course.

That led to a vital meeting between Mr Henderson, for the landlords, and Mr Cohen to discuss the question of a new lease or surrender. About that meeting, the learned judge has made findings of fact. I quote from the judgment:

There is a conflict of evidence as to what was said at that meeting and I accept the plaintiff's evidence and reject that given by Mr Henderson. The plaintiff's evidence was to the effect that there was a general discussion trying to get the whole thing settled up. He acknowledged that rent for the

A year 1977 had not been paid because in his opinion he was entitled to withhold this rent because the landlords had defaulted on their obligations and that the negotiations for sale of the lease had broken down because the landlords were wholly unreasonable. Therefore taking all these matters into consideration a figure of £17,000 by way of premium was agreed. It was further agreed that the maintenance charge should be £100 a year being 25 per cent of the outgoings and that ground rent should be £50 a year payable as from January 1 next. The lease and contract were to be drawn up and completed by the end of the year. At the end of the discussion Mr Henderson said that he was so glad that everything was agreed and they shook hands on it. Nothing was said at any time about the decisions reached being subject to contract. I find as a fact that that is what happened. I do not believe Mr Henderson when he gave evidence to the effect that he made it plain that the whole discussion was subject to contract in the accepted sense. The contemporary and subsequent documents indicate that he
B regarded the agreement as final and binding and he gave a very significant answer when giving evidence as to his understanding of the words "subject to contract". He said and I quote from the transcript as well as relying on my note "I understand them to mean subject to the terms being reduced in an agreement which would be finalised by the solicitors or in writing at a later date.

The judge goes on:

In the afternoon of the same day, November 18, Mr Henderson sent a letter in the following terms. "Further to our discussion, I have pleasure in confirming the sale of the above to you for the sum of £17,000, subject to contract. This is for a term of 99 years from December 25 1968 with a service charge of £100 per annum payable on account being 25% of the landlords' outgoings. The ground rent will be £50 per annum for the first 33 years doubling every 33 thereafter. I have requested our solicitors Messrs Harold
C Stern & Co of 6 Holborn Viaduct . . . to submit a draft contract to Messrs Kood Kingdom Sompen (sic) attention Mr J M Davies for their approval."

In fact, as the judge finds:

Mr Henderson wrote, not on that day, but his letter was dated November 21. This began with the significant words "I have pleasure in advising that I have sold the above flat to the tenant, Mr R Cohen subject to contract."

In the letter of November 18 in which Mr Henderson began with the words, "Further to our discussion, I have pleasure in confirming the sale of the above to you for the sum of £17,000 subject to contract", he added at the bottom of the letter in capital letters the ditty "SUBJECT TO CONTRACT & WITHOUT PREJUDICE."

The judge, in his summary of findings, namely, no 11 in his
D judgment, summarised it in this way:

Later the same day, November 18, Mr Henderson confirmed the agreement in writing but introduced the words subject to contract in that confirmatory letter.

So the facts are that in the conversation between Mr Henderson for the landlords and Mr Cohen, on his own behalf as tenant, there was no express mention that the discussion or agreement was subject to contract. It is against that background that this court has to consider the correctness or otherwise of the judge's finding that, although there was no express reference to that agreement being subject to contract, there was an implication to that effect because
E that discussion was still under the umbrella, subject to contract, which had clearly been erected over the negotiations which had ended in May when Mr Cohen had called the negotiations off.

I do not think, for the purposes of the limited question with which I am dealing, that it is necessary for me to repeat or to recite any part of the judge's findings on the history after the letter of November 18, save to say that the effect of those findings was that Mr Sadleir, with great acuteness and indeed cunning, succeeded in keeping Mr Cohen under the illusion that Nessdale Ltd were intending to honour the agreement of November 18 entered into between Mr Henderson and Mr Cohen, and, by a clever strategem of carefully avoiding answering Mr Cohen's letters, they left him ex-
F pecting and assuming, quite wrongly, that Mr Sadleir's standards of commercial conduct were such that, though it was taking a long time, the agreement was clearly going to be honoured, though there seemed to be some delay about completing the documentation.

The judge's findings make it plain that any reliance of Mr Cohen upon Mr Sadleir's standards of commercial behaviour was wholly mistaken, as it is perfectly clear from the histories recited by the judge that this was a careful strategem executed, as I say, with great shrewdness, but Mr Cohen took far too long to awake to the vendor's methods of doing property business. In the event, finally on September 2, when I suppose Mr Sadleir decided that for his own reasons it was time to come out into the open, he called the whole deal off, and said that Nessdale Ltd were not bound by any

contract and therefore there was no point in proceeding any fur- G
ther. Not surprisingly, Mr Cohen and his solicitors went into action in order to see whether the law could give them a remedy in order to surmount this manifest history of trickery.

As it is a matter of real property law, in which the law has, necessarily, to be settled and has reached a high degree of certainty, Mr Cohen was, of course, in very considerable difficulty. Below and in this court, every argument that can possibly be put forward on Mr Cohen's behalf has been put forward. For myself, I take the same view as the learned judge: if one could help Mr Cohen by finding a legal remedy in order to frustrate the artifices of Mr Sadleir clearly one would be eager to do so, but real property law has to attain a certain certainty and the question is, whether, on the facts H
that I have summarised, it can be shown that the judge was wrong in the implication that he found.

The starting point, as Mr Bailey-King recognises, is really the statement of the law recited in this court in the case of *Sherbrooke v Dipple* (1981) 41 P&CR 173* and in particular the judgments in this court, reported on p 176. The Master of the Rolls said this:

There is this overwhelming point: Everything in the opening letter was "subject to contract". All the subsequent negotiations were subject to that overriding initial condition. We were referred by Mr Parish to a decision of Brightman J

as he then was

in 1972. It is *Tevanan v Norman Brett (Builders) Ltd* (1972) 223 EG 1945. J
Brightman J said: "parties could get rid of the qualification of 'subject to contract' only if they both expressly agreed that it should be expunged or if such an agreement was to be necessarily implied".

He went on to deal with the application of that principle to the facts of the case before him. Templeman LJ quoted the whole of the passage to which the Master of the Rolls had referred from the judgment of Brightman J, and I quote it:

when parties started their negotiations under the umbrella of the "subject to contract" formula, or some similar expression of intention, it was really hopeless for one side or the other to say that a contract came into existence because the parties became of one mind notwithstanding that no formal contracts had been exchanged. Where formal contracts were exchanged, it was true that the parties were inevitably of one mind at the moment before
the exchange was made. But they were only of one mind on the footing that K
all the terms and conditions of the sale and purchase had been settled between them, and even then the original intention still remained intact that there should be no formal contract in existence until the written contracts had been exchanged.

Brightman J

thought parties could get rid of the qualification of "subject to contract" only if they both expressly agreed that it should be expunged or if such an agreement was to be necessarily implied.

I do not think that any assistance is to be derived from looking at the facts in *Sherbrooke* or in the case of *Tevanan v Brett.* Brightman J neatly and accurately stated the question: in such a
situation, have the parties expressly agreed that subject to contract L
qualifications should be expunged or should it necessarily be implied? In this court counsel for the appellant has submitted — I hope that I do him justice when I try to summarise his submission — that what happened in this case was this: there was an initial period of negotiation which was clearly subject to contract, but that was terminated; as Mr Cohen himself said in his correspondence, he terminated that negotiation. I observe that it is perfectly plain, and Mr Cohen would accept, that he did not terminate that negotiation because of any difficulty in continuing the negotiation having regard to the terms that the parties were respectively putting forward for agreement. He terminated because he
was very angry at what he thought was really outrageous behaviour M
on the part of Nessdale: on the one hand trying to deal with him as a gentleman on the negotiation of the long lease, and then behind his back going along to the rent people to raise his rent, which he probably interpreted as an attempt to bring pressure on him to accede to the negotiating offer then being put forward by the landlord. Whether he was right about that is another matter, but that was the reason why he terminated the negotiations. Counsel submits that that very context explains the fact that when Mr Cohen terminated the negotiation he intended to terminate it. It

*Also reported at (1980) 255 EG 1203.

A was his state of wrath about his view of the conduct of the landlords that led him to break the whole thing off altogether, and, having broken it off altogether, what began to happen in November was the opening of a completely new chapter in the history and not a continuance of the earlier negotiation. I hope that summarises the way in which counsel put it in this court.

There are, to my mind, and I say it with regret, very great difficulties about accepting that approach. First, the parties to the transaction: on the one hand the landlords, who were manifestly experienced, and indeed fly, commercial men; secondly, the tenant himself, who described himself in his correspondence with a rather splendid though faintly obscure title, as a consultant in commerce B and the law. It was those two parties who in May had been carefully covering their offers, their negotiations, with the qualification "subject to contract". Indeed, Nessdale seem commonly not only to have put it in the text of the letter, but, in case the recipient cannot read, they also put it in capital letters at the bottom.

How far do the facts support the submission that the termination was such a final termination that it is a misunderstanding to think in terms of a resumption or continuance of the negotiations begun in May? The difficulty is this: although Mr Cohen after May did not, until November, communicate directly to Mr Sadleir or Mr Henderson, he did, in the correspondence to which I have referred, make a number of requests to the landlords' agents, albeit agents C with a limited scope of authority, to encourage the landlords to reopen the negotiations. I have referred to the letters and I have described how Goddard & Smith did pass correspondence on to Mr Sadleir. So, far from the truth being that Mr Cohen, like Achilles, got into his tent and stayed in it, he got into his tent, but then he put out feelers through Goddard & Smith with a view to resumption of negotiations.

It is against that background that one comes to the letter of November 2 written by Mr Henderson, with the label "WITHOUT PREJUDICE" at the bottom, when he used this expression:

> You may recall our earlier meeting concerning our proposals for the flat you occupy . . . If you would care to communicate with me at the above I D will be happy to discuss the matter further.

In that letter Mr Henderson is inviting further discussion on the matter which had been the subject of the landlords' earlier proposals. To my mind it is very difficult, against that history of the transactions between the parties, to take the view that, when Mr Henderson and Mr Cohen began their discussions on November 18, they were not, by so doing, resuming the interrupted negotiations which had been interrupted for what I would call a collateral reason, namely, the alleged misbehaviour of the landlords in going behind the tenant's back while they were negotiating and trying to get an uplift in rent. There is no question here of any express agreement to expunge the qualification involved in the words "subject E to contract."

Is it necessary, against this history, to imply that the parties on F November 18 intended that, though they had previously always been negotiating subject to contract, they were on that day intending to enter into a firm contract in which all the essential terms had been agreed, so the solicitors would do nothing except put on the dots and fill in the gaps which could readily be filled in? I do not think so.

These parties must have realised, both of them, that in connection with this kind of real property transaction the normal procedure that both parties would expect was a procedure by which there would be no concluded contract until solicitors had settled the terms of the contract, signatures had been obtained from the parties and contracts exchanged. I can see no reason to suppose G that, when Mr Cohen and Mr Henderson met and had a discussion in which they managed to reach agreement on November 18, it was a common intention that the parties would abandon the stance which they had hitherto taken up, a stance which may be described as the normal stance of experienced business people dealing with a property transaction, and had decided to take up a new position in which a firm and final contract without any qualification and without anything in writing was to be completed or made on November 18 in an oral discussion.

It would to my mind be perfectly reasonable to imply such a term, but the question for decision is not whether it is reasonable to imply the term but whether it is necessary to do so. Like the judge, H I find myself constrained, reluctantly to the view that, however badly these landlords, Nessdale Ltd, behaved — and, of course, they did behave badly — it is not possible to arrive at the conclusion that it is necessary to imply into the conversation of November 18 that the stance which had been a most important qualification accepted by both of them a few months before had been abandoned so that the qualification no longer applied. For those reasons, I would dismiss the appeal.

Agreeing, DONALDSON LJ said: I share my Lord's regret that this appeal must be dismissed, but dismissed it must be. The effect of the subject to contract qualification to the earlier negotiations was not in my judgment spent when agreement was reached between J the parties on November 18 1977. It follows that that agreement was, in law, provisional in nature and so unenforceable.

I, too, would dismiss the appeal.

Also agreeing, SIR SEBAG SHAW said: I reach that conclusion with regret, particularly as, upon the findings of fact of the learned judge, the conduct of this matter by those who acted for Nessdale — I mean of course Mr Sadleir and Mr Henderson — reflects little credit upon their integrity or upon their sense of business propriety. However, the matter is concluded upon the learned judge's findings, which are not subject to any challenge.

I would dismiss the appeal.

K The appeal was dismissed with costs. Leave to appeal to the House of Lords was refused.

TOWN AND COUNTRY PLANNING

A

QUEEN'S BENCH DIVISION

July 17 1981

(Before Sir Douglas FRANK QC, sitting as a deputy judge of the division)

CARPET DECOR (GUILDFORD) LTD v SECRETARY OF STATE FOR THE ENVIRONMENT AND ANOTHER

Estates Gazette January 9 1982

(1981) 261 EG 56

Town and Country Planning Act 1971 — Town and Country Planning (Use Classes) Order 1972 — Appeal against a decision of B inspector dismissing appeal against enforcement notice served by planning authority — Use of premises as wholesale warehouse or repository (Class X in schedule to order) — Use of vaults over a long period for storage of documents — Appellants used vaults for storage of carpets which were delivered to the site by manufacturers for storage pending distribution to appellants' shops — Planning authority served enforcement notice alleging that the use of storing carpets was without planning permission — On appeal inspector accepted that the use of the vaults for storage of carpets was a use as a repository within Class X and thus within the same class as the storage of documents — He considered, however, that the scale of operations, with the additional traffic, noise and movement, consti- C tuted a material change of use for which planning permission was required — Held that the inspector was in error — Once the inspector concluded that the change was one within Class X he should not have proceeded to consider whether the change was material — Use for another purpose within the class does not constitute development — Appeal allowed

The appellants, Carpet Decor (Guildford) Ltd, appealed against the decision of an inspector appointed by the Secretary of State dismissing an appeal against an enforcement notice served by Guildford Borough Council. The notice related to the use by the appellants of vaults under a site at Pewley Hill Fort, near the top of D Pewley Hill, south east of Guildford. The appellants had constructed a parking area and ramp on the site and used the vaults for the storage of carpets.

Michael Barnes QC (instructed by Janners) appeared on behalf of the appellants; Joseph Harper (instructed by the solicitor to Guildford Borough Council) represented the second respondents; the first respondent, the Secretary of State for the Environment, was not represented and took no part in the proceedings.

Giving judgment, SIR DOUGLAS FRANK QC said: The appellants apply to have quashed a decision of an inspector appointed by the first respondent whereby he dismissed an appeal against an enforcement notice served by the respondent council. The Sec- E retary of State being willing to submit to a judgment does not appear, but the council appear by Mr Harper to oppose the application.

The site of the land concerned lies half a mile south east of Guildford near the summit of Pewley Hill. It originally formed part of Pewley Hill Fort, constructed about 100 years ago, but generally is within a low-density residential area. Within the site are a parking area and a ramp, both of which have been constructed by the appellants. The ramp descends to an entrance to underground vaults, in which the appellants store carpets.

I turn now to the history of the use of the vaults. From 1938 to 1946 they were used by a building society for the storage of docu- F ments. From 1950 to 1978 they were used by the National Provincial Bank also for the storage of documents. In 1978 they were leased to the appellants, who constructed the parking area and ramp and then used the vaults for the storage of carpets. Deliveries to the site are made by manufacturers for storage in the vaults pending distribution to the appellants' shops.

On July 20 1979 the council served an enforcement notice on the appellants alleging that the use of the vaults for the storage of carpets and the construction of the ramp and parking area had been carried out without the grant of planning permission. The appellants appealed on grounds (a), (b), (f) and (g) of section 88 of the Town and Country Planning Act 1971. It is common ground G that if planning permission was not required for the change of use there was no planning objection to the parking area and ramp.

It is also common ground that the storage of documents by the bank and the building society came within Class X of the Schedule to the Town and Country Planning (Use Classes) Order 1972. That was considered by the inspector in his conclusions where he said:

> I agree with the view put on your clients' behalf that their use of the vaults as a repository also falls within the same Class, if only because the House of Lords decision removes the possibility of any doubt about it. It follows that, in so far as the application of the Use Class argument is concerned, there has been no development within the meaning of Class X of the 1972 Order. I am H not, however, convinced that the change from the storage of archives to the storage of carpets within that Class can be said to be not a material one in this case. The notice alleges the storage and distribution of carpets, and while it may be that the latter aspect is ancillary to the former, there can be little doubt from the evidence that as a matter of fact and degree the present scale of operations represents a marked and significant change in planning terms from the simple storage of papers. The amount of traffic now generated, the transfer operations between the vaults and the hardstanding areas and the degree of noise and movement to which these give rise cannot, in my opinion, be regarded as anything but totally different from the previous storage use. Therefore I conclude that the change of use is material and needed planning permission which has not been obtained. Consequently the appeal under ground (b) fails.

It seems to me that the paragraph I have quoted contains an J inherent misconception. Planning permission is required only for the carrying out of development, that is to say for certain operations and the making of any material change of use. Unless development has been carried out, planning permission is not required and, therefore, any exceptions to the definition of development can apply only to something which would otherwise constitute development. Section 22 of the Town and Country Planning Act 1971 defines "development" in the way I have summarised and then in subsection (2) sets out a number of operations or uses which are not to be taken to involve development and includes "(f) in the case of buildings or other land which are used for a purpose of any class specified in an order made by the Secretary of State under this K section, the use thereof for any other purpose of the same class." It follows from my reasoning that that exception, which is expressed in Class X of the Use Classes Order, can apply only to a change of use which is material. It further follows that once the inspector had concluded that the change was within Class X, he was wrong then to have gone on to consider whether the change was material.

Mr Harper for the council argued that the starting point was the planning permission granted in 1950. That permission was in the usual form granted with the printed words granting permission for

A the development described and subject to a printed condition that "no variations from the deposited plans and particulars will be permitted unless previously authorised by the Guildford Borough Council". The development as described in the application form was "as store for papers of National Provincial Bank Ltd and as residence for caretaker in employ of said Bank but for no other type of store or for any other person or corporation". The reference to the residence of a caretaker was to part of the application relating to a dwelling-house. In answer to another question requiring the applicants to "state whether the permission is desired for permanent development or use for a limited period and if the latter for what period", the applicants replied "during ownership by National
B Provincial Bank Ltd only".

Mr Harper argued that both the use and its duration were defined in the application and incorporated in the permission and therefore cut down the scope of the permission, thereby making it not open to the appellants to rely on the Use Classes Order. He referred to the *City of London Corporation v Secretary of State for the Environment* (1972) 23 P&CR 169, where it was held that a condition which prohibited the use of the premises within the same use class was *intra vires*. Alternatively, he argued that if the permission in this case did not, in its terms, contain a condition restricting the use and duration, then it was open to the court to imply one, and he cited *Kwik Save Discount Group Ltd v Secretary of State for Wales* [1981]
C JPL 198* He referred also to *Williamson and Stevens v Cambridge County Council* (1977) 34 P&CR 117 LT as authority for the proposition that words in a permission can have a function of significance. He also cited *Wilson v West Sussex County Council* [1963] 2 QB 764. He submitted that the planning permission was limited in time to the ownership by the National Provincial Bank.

Mr Barnes said that the true ratio of the *West Sussex* case is that when planning permission is given to change from use A to use B and, in fact, the change is from use A to use C, then however similar are B and C there is a breach because there is no planning permission, but if the use is changed from A to B and then a change from B to C, the question whether there is a breach depends on whether
D the change is development. That proposition, he said, was supported by the *Kwik Save* case and that if at Oldham the change from A to B had been more than minimal, then the limitation in the planning permission would not have prevented the operation of the General Development Order.

I think that this case turns on the proper construction of the planning permission. As a general principle, where a local planning authority intend to exclude the operation of the Use Classes Order or the General Development Order, they should say so by the imposition of a condition in unequivocal terms, for in the absence of such a condition it must be assumed that those orders will have effect by operation of law. As I have said, the parts of the planning
E permission relied on here are in common form and printed on the documents. Their purpose is to ensure that the operations are carried out in accordance with the deposited plan and that the premises shall be used for the purpose described. I do not read this document as doing more than that, and certainly not as excluding the operation of the Use Classes Order.

That is sufficient to dispose of the appeal and it is unnecessary for me to consider whether, had I decided otherwise, the enforcement notice was bad in form.

It follows that the appeal is allowed and it therefore is remitted for rehearing and redetermination in accordance with the opinion of the court that the change of use of which the enforcement notice
F complains was not a breach of planning control.

The appeal was allowed with costs against the second respondent (Guildford Borough Council).

---

*Reported in full at (1980) 257 EG 169.

QUEEN'S BENCH DIVISION G

July 31 1981

(Before Sir Douglas FRANK QC, sitting as a deputy judge of the division)

J A PYE (OXFORD) ESTATES LTD v SECRETARY OF STATE FOR THE ENVIRONMENT AND ANOTHER

Estates Gazette January 30 1982

(1981) 261 EG 368

Town and Country Planning Act 1971 — Appeal against decision of Secretary of State dismissing appeal against refusal by local planning authority of permission for housing development — The first ground on which the Secretary of State's decision was challenged, H **namely, that it was inconsistent with an earlier decision relating to a nearby site, was rejected by the judge, who considered the cases distinguishable — The second ground was that the Secretary of State had misinterpreted the planning authority's local planning policy — The inspector's report, which the Secretary of State had approved, had failed to indicate that the development site had been excepted by the authority from a restrictive policy towards residential development in the area — Hence the Secretary of State had not taken into consideration a matter which he ought to have considered and on the principle of *Ashbridge Investments Ltd v Minister of Housing and Local Government* the decision would have to be quashed** J

This was an application by J A Pye (Oxford) Estates Ltd to quash an order made by the Secretary of State for the Environment whereby he dismissed an appeal by the applicants against the refusal of Cherwell District Council to grant planning permission for the erection of houses on a site of 4.25 acres at Yarnton, near Oxford.

Alan de Piro QC (instructed by Darby & Son, of Oxford) appeared on behalf of the applicants; Simon Brown (instructed by the Treasury Solicitor) represented the first respondent, the Secretary of State; the second respondent, Cherwell District Council, was not represented and took no part in the proceedings. K

Giving judgment, SIR DOUGLAS FRANK QC said: This is an appeal against a decision of the Secretary of State dismissing an appeal by the appellants against a refusal of planning permission by the second respondents. The planning permission sought was for the erection of 28 houses on a site of 4.25 acres at Yarnton. There was a second application but that was for a smaller area within the same site and it will be unnecessary to refer to it again. Yarnton is a village about 1½ miles from the most northerly suburb of Oxford and adjoins the western side of the Woodstock Road (A34). It is triangular in shape, bounded on the south by Cassington Road and on the west by Rutten Lane, which has a junction with Woodstock Road. The appeal site lies just below the apex of the triangle L between those two roads and is bounded on the south for the most part by Aysgarth Road. The inspector recommended that the appeal should be dismissed on the general ground that he "found no reason to question the inclusion of the major part of the appeal sites in the proposed green belt, pending definition of its boundary in the forthcoming local plan and approval of the green belt proposals as a whole". He also said that the proposed development would extend the village built area northwards on to agricultural land, albeit low grade. While the flanking hedges would substantially screen them from the view from Rutten Lane and the A34, the proposed houses would, because of the slope of the appeal site and the relatively open nature of the sites' southern boundary, represent a major intrusion into the view from Aysgarth Road. M

In this court the decision was challenged on two grounds, namely:

(1) that the Secretary of State had ignored or failed to give any or due weight to one of his own previous decisions; and
(2) that he had wrongly ignored or misinterpreted a local planning policy.

The first ground relates to a site at the southern end of the village which I shall call the Wimpey land. That was the subject of an appeal inquiry held in January 1980 against a refusal of permission to erect 45 houses. On May 20 1980 the Secretary of State allowed the appeal. That was about two months before his decision in the instant case. There the inspector had concluded that the exclusion

A of the site from the green belt would have no significant effect but recommended that the appeal be dismissed on the ground that the future use of the site should be left for determination by the local plan which he thought should not be long delayed. The Secretary of State, however, concluded that there was no substantial objection to the development of the site on the grounds of amenity and that it would represent a logical rounding-off of the adjacent housing estates. Therefore he considered that the release of the site should not compromise the application of the structure plan and council policies aimed at restraining development in the area and, that being so, there was no justification for delaying a decision on the future use of the site until the local plan was available.

B Mr de Piro for the appellants said that each and every one of the points referred to by the Secretary of State applied to the Pye site. He said they were identical applications decided differently within two months. That, he said, was a material consideration which the Secretary of State was bound in law to take into account.

I cannot find that these were virtually identical applications; there were two cogent distinguishing features. In the Pye case the Secretary of State accepted the inspector's conclusion that he found no reason to question the inclusion of the major part of the appeal sites in the proposed green belt, pending definition of its boundary in the forthcoming local plan and approval of the green belt proposals as a whole. That contrasts with the inspector's conclusions in C the Wimpey case that the exclusion of the site from the green belt would not have any significant effect. Secondly, in the Pye case the Secretary of State agreed with the inspector's conclusion that the proposed houses on the Pye site, because of its slope and the relatively open nature of its southern boundary, would represent a major intrusion into the view from Aysgarth Road. That contrasts with the Secretary of State's opinion that there was no substantial objection to the development of the Wimpey site on grounds of amenity. It follows that in my judgment this ground of appeal has not been made out.

I turn to the second ground of appeal, which is that the inspector's conclusion proceeded on a false basis of fact relating to the council's D planning policies.

The first document is expressed to explain the development of the interim rural settlement policy proposed to be adopted by the council and a summary of the findings of a survey of the factors determining the development potential of individual villages. It was expressed to have been amended in step with the then emerging structure plan. The villages were dealt with in two appendices; the first setting out those not having any growth potential other than infilling and the others appearing to have potential growth and warranting detailed study. In the second appendix, which summarised the results of the investigation, one of the villages specified is Yarnton, where five sites with potential growth were considered E but only two were recommended, namely the Pye site and the Wimpey site. Of these it was said that "development of these sites, which are fairly well screened, would relate to the existing settlement and have well-defined boundaries".

The other document sets out the policies which the council has adopted for determining proposals to build dwellings in the rural area. It is stated that it is intended that the policies will, after approval of the structure plan and subject to such amendments as may be necessary, be adopted as a local plan under sections 11-15 of the Act of 1971 and, pending approval, the policies will be operated on an interim basis. It is stated that exceptions to the general policy (which is one of restraint) "will be considered for F proposals in villages with a primary school, a shop and a bus service every weekday and do not amount to infilling where sites are available free from objection on agricultural or other valid planning grounds but in no case will other than minor development be permitted". There is an appendix which lists the villages which have a primary school, shop and weekday bus service; Yarnton is included.

The first finding of fact made by the inspector was that in the policy to which I have just referred "the village of Yarnton is regarded as one where new residential development will generally be restricted to infilling within the existing village limits". That plainly was at the best misleading and at the worst wrong. It certainly could not have drawn the Secretary of State's attention to the fact that the Pye site was one at least regarded as likely to be shown G for residential development in the local plan. It seems to me, therefore, that the Secretary of State has failed to take into consideration that which he ought to have done and for that reason the decision should be quashed: see *Ashbridge Investments Ltd v Minister of Housing and Local Government* [1965] 1 WLR 1320 at p 1326. Mr Brown argued that it was clear that the inspector had taken account of and accurately stated the policy. For the reasons I have stated, I do not think that the policy was accurately and certainly not sufficiently stated and even if the inspector took it into account, he did not draw the attention of the Secretary of State to it.

The appeal was allowed with costs. H

COURT OF APPEAL

November 27 1981

(Before Lord DENNING MR, Lord Justice OLIVER and Lord Justice WATKINS)

JENNINGS MOTORS LTD v SECRETARY OF STATE FOR THE ENVIRONMENT

J

Estates Gazette March 13 1982
(1981) 261 EG 994

Town and Country Planning Act 1971 — Secretary of State's decision to uphold enforcement notice challenged — Discussion of theories of "new planning unit" and "new chapter in planning history" — Latter theory preferred by Master of the Rolls — Enforcement notice requiring discontinuance of use of a new building for the purpose of the repair, servicing and maintenance of motor vehicles — New building erected to replace old building on a small portion of a half-acre site which had existing use rights for a mixed use in connection with repair, servicing and maintenance of vehicles — Divisional Court in present case had upheld minister, K **regarding themselves as bound by *Aston v Secretary of State for the Environment* to decide that the erection of a new building ipso facto gave rise to a new planning unit and put an end to any existing use rights in respect of the land on which the new building was erected — Statements in *Petticoat Lane Rentals Ltd v Secretary of State for the Environment* and in the *Aston* case criticised as too wide — Appellants in present case entitled to benefit of existing use rights in respect of new building — Appeal allowed, reversing decision of Divisional Court — Matter remitted to Secretary of State for rehearing and determination in the light of the Court of Appeal's ruling — *Aston* decision criticised**

L

The appellants, Jennings Motors Ltd, appealed from an order of the Divisional Court (Donaldson LJ and Bristow J), dated April 22 1980 upholding a decision of the Secretary of State which in turn upheld an enforcement notice served by the local planning authority on the appellants. The notice was in respect of the erection of a new building on an industrial site at Dibden Purlieu, near the New Forest, and required the discontinuance of the use of the building for workshops for the repair, servicing and maintenance of motor vehicles.

Michael J Burrell (instructed by Malkin, Cullus & Sumption, agents for Lamport, Bassitt & Hiscock, of Southampton) appeared on behalf of the appellants; Simon Brown (instructed by the Treasury Solicitor) represented the respondent. M

Giving judgment, LORD DENNING MR said: The village has an attractive name, Dibden Purlieu. It goes back to the times of the Norman French. It is a mile or so inland from Southampton Water on the west side, near the New Forest. Some part of it has been developed in recent years for residential purposes. It is designated in the Hythe Town Map, Hampshire, for residential use. But, in the midst of this residential area there is an industrial site. It only takes up about half an acre all told. It has been used for the last 20 years for commercial purposes in connection with motor vehicles. It is a mixed use in connection with the repair, servicing and maintenance

A of vehicles. There is a garage workshop in one corner. It takes up one-twelfth of the site. There is a showroom and office in another. It takes up one-thirteenth of the site. No question arises about those. But then there is a new building. It occupies only about one-seventeenth of the site of half an acre. It is about 55 ft long and 23 ft wide and 10 ft high. The rest of the site is for access and parking motor vehicles.

It is that new building which is in question. Previously there was a garage workshop there. The occupiers had applied in 1975 for permission to pull it down and put up the new building. It was refused. They never got permission for it. But, in spite of the refusal of permission, the occupiers completed the building. The local B authority did not take enforcement proceedings in respect of the building because "it was considered that the new building was more satisfactory in appearance than those it had replaced". Nevertheless, the local authority resolved that enforcement proceedings should be taken to secure the discontinuance of the use.

Now the point of law that arises is this: The whole of the half-acre site is being used for a "mixed" use for which it has been used for the last 20 years, since 1962. The new building is being used for one of those mixed uses, namely, the repair and servicing of motor vehicles. It is not suggested that there has been a change in the "mix" of use on the site as a whole. Nor is any charge made in the enforcement notice that there has been a change owing to the C "intensification" of use. The enforcement notice related only to the new building (coloured pink on a plan). It was dated August 12 1976 and was in these terms:

> . . . It appears to the Council that after December 31 1963 there has been a breach of planning control in that the said land has been developed by the making of a material change in the use of the buildings coloured pink on the attached plan situate thereon to a use for the purpose of workshops for the repair, servicing and maintenance of motor vehicles without the grant of permission. . . . Now therefore . . . the Council hereby require you within two months . . . to discontinue the use of the buildings coloured pink on the attached plan for the purpose of workshops for the repair, servicing and maintenance of motor vehicles.

D The minister upheld the enforcement notice. In his decision letter of June 6 1978 he said:

> . . . In the light of the judgment of the *Petticoat Lane* case it is considered that when the new building was erected a new planning history commenced in respect of it, and the building on completion had no actual, established or permitted use . . . it is considered that this use (for the repairing, servicing and maintenance of motor vehicles) must involve a material change of use from no use, and there has therefore been a breach of planning control.

The Divisional Court upheld the minister. They did so reluctantly as they thought they were bound by the unreported case of *Aston* (*infra*). The occupiers appeal to this court. We have been referred to all the cases. They disclose two theories. The one is the theory of E the "new planning unit". The other is the theory of the "new chapter in planning history". I will consider each theory separately.

"The new planning unit"

According to this theory, when a man applies for permission to erect a new building, either where none existed before or to replace an old building, he creates a "new planning unit". He can use it for any purpose specified in the permission or, if no purpose is specified, for the purpose for which it was designed to be used; see section 33(2) of the 1971 Act, subject to any conditions contained in the permission. If he erects a new building without any permission at all, he starts with a nil use and must get permission for any F use. Once he erects that new building, he cannot fall back on previous existing use rights. This theory was stated by Widgery LJ in *Petticoat Lane Rentals Ltd v Secretary of State for the Environment* [1971] 1 WLR 1112. In that case the new building covered the whole site. Widgery LJ said:

> In my judgment one gets an entirely new planning unit created by the new building. The land as such is merged in that new building and a new planning unit with no planning history is achieved. That new planning unit, the new building, starts with a nil use . . .

In the later case of *Aston v Secretary of State for the Environment* (not reported, April 9 1973) Lord Widgery CJ (as he then was) applied it to a case where the new building covered only *part* of the G site — just about half the site. He said:

> where you have a new building erected, that part of the land which was absorbed in the new building and covered by the new building is merged in it; you start with a new planning unit which has no permitted planning uses except those derived from the planning permission, if any, and from section 33(2) of the Town and Country Planning Act 1971, which allows such a building in many instances to be used for the purpose for which it was designed.

That theory was accepted by Lord Fraser in *Newbury District Council v Secretary of State for the Environment* [1981] AC 578 in the House of Lords, when he said (at p 606): "The only circumstances in which existing use rights are lost by accepting and imple- H menting a later planning permission are when a new planning unit comes into existence."

This theory has been extended by some observations in the House of Lords to a case where a man applies to change the use of a building so as to make it available for occupation for several families. If he acts on the permission and makes the change — by putting in internal partitions and doors — then he creates a "new planning unit". He must abide by any conditions inserted in the permission. He cannot fall back on previous existing use rights: see the *Newbury* case *supra* [1981] AC at p 607 by Lord Fraser and at p 618B by Lord Scarman.

J

"A new chapter in planning history"

According to this theory, when a man applies for permission to erect or alter a building — or to make a change in the use of land — in such circumstances as to effect a radical alteration in the nature or use of the site — then it may be interpreted as the opening of a "new chapter in the planning history". If he then acts on the permission — and erects or alters the building or changes the use of the land — he must abide by the conditions on which the permission was given. He cannot afterwards revert to any previous existing use rights. This theory was stated clearly by Lord Parker in *Prossor v Minister of Housing and Local Government* (1968) 67 LGR 109. In that case a garage proprietor applied for planning permission to K erect a new building on *part* of the site to replace an existing repair shop. He was granted permission on the condition that no retail sales were to take place in the new building. The garage proprietor afterwards claimed that he had existing use rights for selling motor cars. Lord Parker CJ said (at p 113):

> . . . Assuming that there was at all material times prior to April 1964 an existing use right running on this land for the display and sale of motor cars, yet by adopting the permission granted in April 1964 the appellant's predecessor, as it seems to me, gave up any possible existing use rights in that regard which he may have had. The planning history of this site, as it were, seems to me to begin afresh on April 4 1964 with the grant of this permission, a permission which was taken up and used. L

This theory was restated by Lord Lane in the *Newbury* case [1981] AC at p 626 C-D:

> . . . The holder of planning permission will not be allowed to rely on any existing use rights if the effect of the permission when acted on has been to bring one phase of the planning history of the site to an end and to start a new one.

The difference in the two theories

In many cases the two theories give the same result. Thus in the *Newbury* case [1980] AC 578 there was no new building at all. The M two hangars remained the same throughout. So there was no "new planning unit". Equally the use of those hangars remained substantially the same throughout for storage purposes. So there was no "new chapter of planning history". Lord Lane said (at p 626F): "The change of use from repository to wholesale warehouse could not by any stretch of the imagination be said to have started a new planning history or created a new planning unit. Indeed no one has so contended."

But in some cases the two theories give different results. Thus, where an old building is pulled down and a new one put in its place, there is no "new planning unit". But the change of use may be so radical that the new use to which the building is put may open a

A "new chapter in planning history". That is what happened in *Prossor*'s case (1968) 67 LGR 109. The new use was for a repair shop and stores. The existing use (on which the occupier relied) was for the display of secondhand cars for sale. The repair shop was so radical a change that it opened a "new chapter in planning history".

In the *Aston* case, and in our present case, the two theories give different results. In each case there was a new building on part of the site, and thus a "new planning unit". But in neither case did the new use open a "new chapter in the planning history". I think that the *Aston* case was wrongly decided, so also the decision of the Divisional Court in this case, which followed it.

B *Result*

In the light of experience, I think we should discard the theory of the "new planning unit". In future it should no longer be thought that a new building creates a "new planning unit" which starts with a "nil use". Certainly not when it is just the replacement of an old building. The better theory is the opening of a "new chapter in planning history". This may take place when there is a radical change in the nature of the buildings on the site or the uses to which they are put — so radical that it can be looked upon as a fresh start altogether in the character of the site. If there is such a change and the occupier applies for permission and gets it subject to conditions — and acts upon that permission — he cannot afterwards C revert to any previous existing use rights.

Conclusion

Before us Mr Simon Brown pleaded for guidance. He told us that those in the ministry were much perplexed as to the right principle to adopt. He submitted that the right theory was the "new chapter in the planning history". I agree with him. Applied to this case, I think there was no change in the planning history at all. There is one whole site of half an acre with existing use rights. All that has been done is to erect a new building in place of an old one, on a little portion of the site. The occupiers are entitled to the use of those rights inside the new building. I would allow the appeal accordingly.

D Agreeing that the appeal should be allowed, OLIVER LJ said: The Divisional Court held that it was bound by the unreported decision in *Aston v Secretary of State for the Environment* to find that where, either with or without permission, a building has been erected upon land previously unbuilt on, that building is, from the time of its erection, a new planning unit, so that the building owner is unable, where the question in issue is whether or not there has been a material change of use, to pray in aid the pre-existing user of the land on which the building stands.

It is, I think, impossible to escape from the conclusion that that is what *Aston v Secretary of State* decided. The pith of the matter is E contained in the following passage from the judgment of Lord Widgery:

> . . . The principle which one derives from the authorities and applies to the present case is that where you have a new building erected, that part of the land which was absorbed in the new building and covered by the new building is merged in it; you start with a new planning unit which has no permitted planning uses except those derived from the planning permission, if any, and from section 33(2) of the Town and Country Planning Act 1971, which allows such a building in many instances to be used for the purpose for which it was designed.

The principle there expressed is enunciated as a universal one, and the short question raised by the present appeal is: Is it right? F Inevitably this involves some review of the relevant authorities from which the universal principle is said to stem.

The first of these is *Prossor v Minister of Housing and Local Government* (1968) 67 LGR 109. There had been for some years an established existing use as a petrol filling station and motor repair shop. A planning permission was granted for the erection of a new building and the replacement of an existing repair shop and stores. It is not clear how far the new building covered the whole of the site, but having regard to its size and shape and the nature of the user, I infer that it covered part only. It was, however, granted subject to a condition relating to the whole site that it should not be used for retail sales other than sales of spare parts and the rede- G velopment took place pursuant to that permission. An enforcement notice to prevent the user of the forecourt for the sale of secondhand cars was upheld by the Divisional Court and on one analysis of the decision (subsequently suggested in *Gray v Minister of Housing and Local Government* (1968) 68 LGR 15) the basis for this was, in effect, an estoppel arising from the acceptance of the permission with the condition attached and the subsequent implementation of it. If that were indeed the ground of the decision — and the judgment of Lord Parker CJ does not make it entirely clear that it was — it is now clear from *Newbury District Council v Secretary of State* [1981] AC 578, that it cannot be sustained as a correct decision on that ground. But the way in which it was put by Lord Parker was this: H

> By adopting the permission granted in April 1964, the appellant's predecessor, as it seems to me, gave up any possible existing use rights in that regard which he may have had. The planning history of this site, as it were, seems to me to begin afresh on April 4 1964 with the grant of this permission, a permission which was taken up and used. . . .

I draw attention to this because it is consistent with Mr Simon Brown's submission on the present appeal that what the court is concerned with is not so much "a new planning unit" (an expression which, he suggests, may be misleading) but simply with the question whether an event or a concatenation of events has taken place which can, as a matter of *fact*, be said to have opened a new chapter in the planning history. That is a question of intention and degree. J The expression "planning unit" (which nowhere appears in the legislation) is, however, one which is now hallowed by usage and is, I think, a convenient phrase for identifying, in cases where the question is whether there has been a material change of use, both the area whose planning history requires to be studied for that purpose and, in an appropriate case, the starting point of that history. The concept of the creation of a new "planning unit" by the opening of a new planning history, not so expressed in *Prossor*'s case, emerges in the judgments of the Divisional Court in *Petticoat Lane Rentals v Secretary of State for the Environment* [1971] 1 WLR 1112. The history of the land with which that case was concerned had started with the use of the land as an open site upon which a K market was held. A building was then erected over the whole site pursuant to a planning permission which designated the use as an office, warehousing, supermarket, car parking and loading area (to be used for market trading on Sundays). The question was whether the existing use of the site for weekday markets prior to the erection of the building survived that event. The Divisional Court held unanimously that it did not. There was some discussion in the judgment of Widgery LJ (as he then was) of the ratio of *Prossor*'s case, but he found it unnecessary to express any conclusion and was content to say that the principle clearly applied "where, as here, one has a clear area of land subsequently developed by the erection of a building over the whole of that land". It seems, however, that when L he spoke of "the principle of *Prossor*'s case" he must have had in mind the reference in the judgment of Lord Parker to the fresh beginning of the planning history, for he went on to say at p 1117:

> Where that happens . . . one gets in my judgment an entirely new planning unit created by the new building. The land as such is merged in that new building and a new planning unit with no planning history is achieved. That new planning unit, the new building, starts with a nil use, that is to say immediately after it was completed it was used for nothing, and thereafter any use to which it is put is a change of use, and if that use is not authorised by the planning permission, is a use which can be restrained by planning control.

M It is thus clear that what brought about the result described by Lord Widgery was not the acceptance of the planning permission, which was arguably inconsistent with the prior existing use, but the mere erection of the building, and it would follow that the same result must ensue even if there were, as in the instant case, no planning permission at all. Bridge J (as he then was) and Lord Parker CJ agreed, but reserved the question of what the result would be where (as here) the building was on part only of a site having a prior existing use as a whole. The concept here, therefore, is that of the creation of a new and different unit by the physical alteration of the land to an extent enabling it to be treated as if it

A were a new creation different and divorced from the land which was there before. Thus Bridge J referred to the "disappearance" of the land to which the existing user was attached.

A similar concept is to be found in *Leighton and Newman Car Sales Ltd v Secretary of State* (1976) 32 P & CR 1, where there had been a pre-existing garage business on a site forming part of the relevant land, a business which included the sale of cars. The site, together with two adjoining sites, was redeveloped entirely as a petrol filling station under a permission which contained a condition restricting the use of certain car-parking space to the parking of vehicles of the occupiers and users and prohibiting the sale or display for sale of vehicles on the forecourt. The Secretary of State's
B dismissal of the tenant's appeal against an enforcement notice was attacked (*inter alia*) on the ground that the inspector's report had not included a relevant matter, namely the existing use of the garage site before the redevelopment. This court upheld the dismissal by a Divisional Court of an appeal from the Secretary of State. In the judgment of the court, delivered by Browne LJ, there is this passage (at p 10): "Further, nos 271 to 275 were not merely a different 'planning unit' from the old no 271; they were a completely new and different physical unit to which we think the previous use of no 271 was irrelevant".

In my judgment this is the essence of the matter. Where there has been a total change in the physical nature of the premises it is easy
C to infer — indeed, the inference may be irresistible — that reliance upon any prior user is being abandoned and a new planning history is to begin. Such an inference may equally be drawn — and may equally be irresistible — where there is no change or a less radical change in the physical nature of the site but a change in what I may call its planning status which is inconsistent with the preservation of a prior existing use — for instance its subdivision into smaller units of occupation or its incorporation into a larger single unit.

Whether the alteration is of such a character as to produce this result is, I think, in every case a question of fact and degree.

How, then, does the matter stand when, as in the instant case, what has occurred is that there has been some physical alteration to
D part only of an occupation site, for instance by the erection on it of a new building, the alteration of an existing building, or an application for and grant of a planning permission subject to conditions inconsistent with the prior user? This may pose very difficult problems in the interpretation of the facts, particularly where, as here, there is a mixed site or where a particular use has been intensified as a result of the change. But these are problems only of fact and degree, not of principle, and they are not insoluble.

There are, so far as the industry of counsel has been able to discover, only two reported cases — the *Aston* decision is, as I have said, unreported — which relate to the problems created in the field of change of user by the erection of a new building on part only
E of the site. The first is the case of *Joyce Shopfitters Ltd v Secretary of State*, a decision of the Divisional Court, reported only in the notes of cases in [1976] JPL 236.* The report is a very sketchy one, but what seems to have occurred there was that a building on the site was used partly for the manufacture of fencing and garden furniture and partly for car repairs and car breaking. The remainder of the site, which included two cottages, had been used for purposes ancillary to those purposes. The site owners, without planning permission, built an extension to the main building and demolished the cottages. Two enforcement notices were served, one requiring the removal of the extension and one requiring the discontinuance of the user of the site for the purposes of a joinery and shopfitting
F business — a business which the Secretary of State found was not a change of use because it was merely a change from one industrial use to another within section 22(2)(f) of the Town and Country Planning Act 1971. He declined to grant permission for the retention of the extension to which the first enforcement notice related, but he upheld the second enforcement notice relating to change of use in respect of the site of the extension and (on the ground that there was a separate planning unit with a nil user) in respect of the site of the cottages. The Divisional Court remitted the matter for reconsideration on the basis that, so far as the site of the extension
G was concerned, there was an existing use which had not been extinguished by the erection of the extension, and so far as the site of the cottages was concerned the planning unit to be considered was the site as a whole.

It is not easy from the very brief report to get any very clear idea of the court's reasoning, but the case is consistent — indeed really only consistent — with the view that the court did not consider that the changes which had taken place in the site as a result of the extension and the demolition were of a sufficiently radical nature to justify the inference that in either case there was the abandonment of the existing industrial user and the commencement of a new planning history. The case is thus entirely consistent with what has
H been said above and inconsistent with the view that the mere erection of a building constitutes a new planning unit. The second case is that of *Hilliard v Secretary of State* (1978) 37 P & CR 129. Here an area of farmland had existing use rights for agricultural purposes with ancillary storage and wholesale distribution of both indigenous and non-indigenous produce. The owner applied for and was granted a planning permission to erect a building, the permission having attached to it a condition that it should only be used for the storage of agricultural produce and farm implements in conjunction with the use of the farm for agricultural purposes. The building subsequently came to be used for the wholesale distribution of fruit and vegetables and the local authority, instead of
J serving an enforcement notice to enforce the condition, served a notice requiring the discontinuance of material change of user on the footing that there had been an intensification of user amounting to an unauthorised change of user. This court, on appeal from the Divisional Court's upholding of the Secretary of State's decision to uphold the enforcement notice, reversed the decision and remitted the matter to the Secretary of State on the footing that the evidence had been directed entirely to the intensification of the user of the building when it should have been directed to the intensification of the user of the farm as a whole. The case is not, I think, of very much assistance in the context of the present inquiry, because, *ex concessis*, the planning unit there was the farm as a whole and it was to the
K whole of that unit (as is, indeed, the case here) that the enforcement notice was directed. The case therefore turned entirely upon whether the evidence could be said to have established a change of user by intensification of the farm as a whole. It has, however, this relevance, that, as Mr Burrell points out, a building erected upon part only of the land was not there treated as, *ipso facto*, creating a new planning unit, and indeed the decision of the court is only consistent with their view that it had not done so. Presumably it did not do so because the alteration of the unit was not such, as a matter of fact and degree, as to justify the inference that the existing user was being abandoned and a new and different planning unit created.

L Speaking for myself, I have some sympathy with Mr Brown's suggestion that "planning unit" has become perhaps a slightly confusing expression, combining, as it does, concepts both of geography and history. It is used in the temporal sense of a separate and distinct period of planning history with its own beginning and end relating to a given area of land, and it is used to distinguish and isolate the geographical area of land the history of which, for a given purpose, has to be studied. Nevertheless, it is, I think, a convenient phrase, and there is an extremely helpful general test contained in the judgment of Bridge LJ (as he then was) in *Burdle v Secretary of State for the Environment* [1972] 1 WLR 1207 at 1212-3.

M The historical content of the term is well brought out in the speeches of Lord Scarman and Lord Lane in *Newbury District Council v Secretary of State for the Environment* [1981] AC 578. Lord Scarman, at p 617, says:

> Clearly it will be much more difficult to establish the creation of a new planning unit or the beginning of a new chapter of planning history where the unnecessary permission which has been granted subject to conditions purports to authorise only a change of use. But such cases can exist . . . [and he gives an example]. There is [he continues] in such a case a wholly new departure, a new chapter of planning history.

Lord Lane, at p 626, is to the same effect. He said:

> We were asked by Mr Newey to say that the principle [ie the principle of the *Petticoat Lane* decision] can only apply where the permission

*In fact the decision, given on July 14 1975, is reported verbatim and in full at 237 EG 576.

A granted is to build or rebuild or the like and can never apply to cases where the permission is simply to change the use. I do not consider that any such limitation would be proper. It is not the reason for the break in planning history which is important. It is the existence of the break itself, whatever the reasons may have been. . . . In the present case there is no such break in the history. The change of use from repository to wholesale warehouse could not by any stretch of the imagination be said to have started a new planning history or created a new planning unit.

Coming back, then, to the *Aston* case, in my judgment, the principle there stated by Lord Widgery CJ is, if I may say so respectfully, too widely expressed. In my view the authorities show B not that a new building, per se, has to be equated with a new planning unit but that it is one of the factors — it may in many cases be a conclusive factor — to be taken into account in considering whether there has taken place in relation to the particular land under consideration a change of so radical a nature as to constitute a "break in the planning history" or a "new planning unit" (the expressions are used interchangeably). As Mr Burrell points out, the building in the *Aston* case was, as a matter of fact, pretty much on all fours with that in the *Petticoat Lane* case, for it replaced an earlier building, destroyed eight years before, which had covered rather less than half the available area, and it itself covered 90 per cent of the available area. There is, therefore, no ground for think- C ing that the case was wrongly decided on the facts. But the stated ground for the decision was, in my judgment, too widely expressed.

In the instant case, the local authority could perfectly well have exerted their planning control over the site by serving an enforcement notice for the removal of the building, and then, if it was the user and not the building that they objected to, granting a planning permission subject to a condition as to user. They could have done that at any time within four years from 1975, but they chose not to do so. In serving an enforcement notice to discontinue the use of the building they necessarily opened up a consideration of the question whether the circumstances, including the erection of the building without permission and any intensification of user, led to the infer- D ence of the creation of a new planning unit. It is not for this court now to decide that question, and I content myself with observing only that the buildings concerned were merely a replacement for already existing buildings on the site. I am, however, clearly of the view that in considering himself bound by the *Petticoat Lane* decision to hold that the erection of the new buildings per se constituted a new planning unit and that he was thus constrained to depart from his inspector's recommendation, the Secretary of State misdirected himself. This, I think, would clearly have been the view of Donaldson LJ if he had not felt himself bound by the *Aston* decision — a disability from which this court is relieved — and I would allow the appeal.

E It follows from what I have said above that in my view the difference between a new planning unit and a change in the planning history is largely a semantic one. The former expression is, as I have said, hallowed by long usage, and I for my part think it would be a pity to discard it so long as the concepts which it embraces (which include a change in the planning history) continue to be clearly appreciated.

WATKINS LJ, whose judgment was read by Oliver LJ, said: I have had the advantage of reading the judgment of Oliver LJ with which I entirely agree. The Secretary of State, we were told, seeks guidance upon the use of the expressions "planning unit" and "a change in planning history". It is my firm opinion that the use of the former F should be preserved and the guidance provided by Bridge LJ (as he then was) in *Burdle*'s case on its application, which obviously involves a study of the history of the land in question, followed.

The appeal was allowed with costs in the Court of Appeal and below. The Divisional Court's order was rescinded and the matter was remitted to the Secretary of State for rehearing and determination in the light of the court's opinion.

COURT OF APPEAL G

November 13 1981

(Before Lord DENNING MR, Lord Justice OLIVER and Lord Justice WATKINS)

WAIN AND OTHERS v SECRETARY OF STATE FOR THE ENVIRONMENT AND OTHERS

Estates Gazette April 24 1982
(1981) 262 EG 337

Town and Country Planning Act 1971, sections 180(1) and 183(3) — Purchase notice — Appeal from decision of Webster J who quashed decision of Secretary of State refusing to confirm a purchase notice served by owners of land on local authority — Land H subject to purchase notice consisted of about 37 acres, part of which could be used for livestock grazing and hay production and part was incapable of reasonably beneficial use — The owners, whose application for planning permission for residential use had been refused, served a purchase notice on the local authority covering the whole of the 37 acres — The authority were unwilling to comply with the notice, and the Secretary of State refused to confirm it, on the ground that it had not been shown that the whole of the land was incapable of reasonably beneficial use — Part of the land covered by the notice was capable of such use — Webster J construed the words in section 180(1)(a) of the 1971 Act, "the land has become incapable of reasonably beneficial use", to mean "the land taken as a whole", J not "each constituent part of the land", and quashed the Secretary of State's decision — Held by Court of Appeal that, reading section 180(1) and section 183(3) together, this construction was incorrect — If a part only of the land is incapable of reasonably beneficial use and the other part is capable of such use, the owner cannot insist on the local authority buying the whole; he can only insist on their buying the part which is incapable — In the present case the purchase notice related to the whole land, but part of it was capable of reasonably beneficial use — Appeal accordingly allowed

This was an appeal by the Secretary of State for the Environment from a decision of Webster J, reported at (1981) 259 EG 857, quashing a decision of the Secretary of State, who had refused to K confirm a purchase notice. The notice was served by Wigan Metropolitan Borough Council on Arthur Wain and others, owners of the land subject to the notice situated at Martland Mill, Wigan.

Simon Brown (instructed by the Treasury Solicitor) appeared on behalf of the appellant; N R Macleod QC and J Howell (instructed by Sidney Torrance & Co, agents for Ellison, Blank, Goldsmith & Co, of Manchester) represented the respondents, Arthur Wain and others, the owners of the land; Wigan Metropolitan Borough Council and Greater Manchester County Council, the second and third respondents, were not represented and took no part in the proceedings.

Giving judgment, LORD DENNING MR said: Under the Town and L Country Planning Act of 1971 there are provisions by which the owner of the land can compel the local council to buy his interest in the land. The land here is an area of 37 acres, about a couple of miles north of the town centre of Wigan. It is geographically and naturally divided into two halves. The southern half was in former time part of a farm called Martland Mill Farm. It is at a higher level than the northern half. The southern half is gently undulating, with several gullies running through it, dotted with thorn bushes and much of it covered with grass, some of it waist high. In one of the corners of it there was a Dutch barn and one or two buildings which are becoming dilapidated. The northern part of the site is at a lower level. It is covered with a mixture of grass, weed and reed, with M concrete protrusions marking a line, a reed-covered pond, and the like.

The planning inspector, after a full inquiry, placed those two halves in different categories, with regard to their reasonable beneficial use. The southern half, which is about 56 per cent of the whole, could be used for livestock grazing and hay production. It has not got enough buildings for winter shelter, but it certainly could be used quite usefully for grazing and hay in the spring and summer months. It could, with making good some of those buildings, without any extra planning permission, be made into reasonably beneficial use.

A The northern half is different. It would need a substantial initial outlay to reclaim it at all or to bring it into any form of agricultural use. The return even then would not be such as to warrant the outlay. So the northern half has become incapable of reasonably beneficial use.

That is the important point of this case. The southern half is capable of reasonably beneficial use, whereas the northern part is not.

The owners of the land some little time ago applied to the local planning authority for permission to develop the land. They applied for the whole of the area, the whole 37 acres, to be developed. One application was for residential development. The other was for
B making the roads and infrastructure for industrial development.

Those applications were made to the local authority. In each case the local authority turned them down. There was no need for housing development and the area was unsuitable for industrial development. That refusal opened the way for the owners of the land to apply under the statutory provisions to the local council to buy up the land.

The point is this. The owners of the land say that the local council are bound to purchase the whole of the 37 acres. If that be so, it might be the total price would be quite high, because the southern half is capable of beneficial use. But the council say that they can apply only for the northern half which is useless — the price of it
C would be very low.

The section on which the case depends is section 180(1):

> Where, on an application for planning permission to develop any land, permission is refused or is granted subject to conditions, then if any owner of the land claims — (a) that the land has become incapable of reasonably beneficial use in its existing state . . . he may serve . . . a notice

on the local council

> requiring that council to purchase his interest in the land. . . .

The local council say that that only applies when *all* the land has become incapable of reasonably beneficial use in its existing state.
D So section 180(1)(a) means that the *whole* of the land has become incapable of reasonably beneficial use. In this case they say that it is not. Only the northern part has become incapable of reasonably beneficial use in its existing state. The southern half has not. So on that ground the local council, and now the Secretary of State, says this compulsory purchase provision does not apply. The section only applies when all of the land is incapable of reasonably beneficial use.

That interpretation of this section is borne out by section 183(3) of the Act. It deals expressly with the position of part of the land as distinct from the whole of it. It says:

> If it appears to the Secretary of State that the land, or *any part* of the land,
> E could be rendered capable of reasonably beneficial use within a reasonable time *by the carrying out* of any other development for which planning permission ought to be granted, he may, in lieu of confirming the purchase notice, or in lieu of confirming it so far as it relates to *that part* of the land, as the case may be, direct that planning permission for that development shall be granted in the event of an application being made in that behalf.

That section seems to me to be directed at a case where part of the land is not now reasonably capable of beneficial use, but if planning permission were given it could be developed so as to be rendered capable of reasonably beneficial use. The subsection says that in such a case as that the Secretary of State can say: "Well, I will give planning permission for the part which can reasonably be
F developed, and I will confirm the purchase notice in regard to the other part which cannot in any event be reasonably developed."

Reading section 183(3) with section 180(1) it seems to me to fit everything together. If a *part* of the land only is *incapable* of reasonably beneficial use and the other part is *capable* of reasonably beneficial use as it is, the owner cannot insist on the council buying the whole. He can only insist on their buying the part which is incapable.

I need not pursue it further. It seems to me the judge was in error in the interpretation which he eventually put upon the section. The true interpretation is that the owner cannot claim the right to have the council purchase his land compulsorily, except when all the land has become incapable of reasonably beneficial use. If part of the G
land is capable of reasonably beneficial use, then he cannot insist on a compulsory purchase.

I would allow the appeal accordingly.

OLIVER LJ: For the reasons given by my lord, the Master of the Rolls, I agree that the appeal should be allowed.

WATKINS LJ: I also agree.

The appeal was allowed with costs in the Court of Appeal and below. The Secretary of State's order not confirming the purchase notice was restored.

H

COURT OF APPEAL

January 11 1982

(Before Lord Justice WALLER, Lord Justice DONALDSON and Sir David CAIRNS)

BACKER AND ANOTHER v SECRETARY OF STATE FOR THE ENVIRONMENT AND ANOTHER

Estates Gazette May 8 1982

(1982) 262 EG 549

Town and Country Planning Act 1971 — Enforcement notice — J
Appeal from decision of Divisional Court holding that an enforcement notice was time-barred under section 87(3) (c) of the 1971 Act — Permission granted for erection of a second storey of a building for use as a loft subject to a condition that the additional storey should not be used for any purpose other than storage connected with the residential use of the remainder of the premises — In fact the second storey was used as a separate unit of residential accommodation — Held by Court of Appeal, upholding decision of Divisional Court, that the enforcement notice was time-barred — The breach of planning control consisted of "the making without planning permission of a change of use . . . to use as a single dwelling-house", with the result that under section 87(3) (c) the enforcement K
notice could be served only within four years from the date of the breach, which period had expired before service — It was immaterial that the change of use was also a breach of condition — Appeal dismissed.

This was an appeal by the London Borough of Camden from a decision of the Queen's Bench Divisional Court (Ackner L J and Skinner J) on December 5 1980 holding that an enforcement notice served by the borough as planning authority and upheld by the Secretary of State for the Environment on appeal was time-barred. The proceedings related to a building in the grounds of 18 East Heath Road, Hampstead. The respondents to the present appeal L
were R G Backer, the freehold owner of the property, and A Aird, the tenant.

J Sullivan (instructed by S Nickson, London Borough of Camden) appeared on behalf of the London Borough of Camden; D Robins (instructed by Bennetts & Partners) represented the respondents. The Secretary of State for the Environment was not represented and took no part in the proceedings.

Giving judgment, WALLER LJ said: This is an appeal from a decision of the Divisional Court, when that court held that an enforcement notice under the Town and Country Planning Act 1971 was time-barred. The respondent is the freehold owner of 18 East Heath Road, Hampstead, in the London Borough of Camden, and M
that house has been converted into flats, although at one time it was a detached house in its own grounds. Some buildings have been erected in the grounds. This appeal relates to the upper half of one of the buildings.

The findings of fact by the minister's inspector were that an upper storey was added to the building in 1968-69; thereafter in June 1970 a planning application was submitted and in October 1970 planning permission was granted for the erection of a second storey for use as a loft. A condition of the permission was that the additional storey should not be used for any purpose other than storage connected with the residential use of the remainder of the premises.

A

The fact that it had been used for a separate unit of residential accommodation from March 1 1973 was also found by the inspector. He found that it had been continuously occupied since then by the same tenant.

The enforcement notice, after quoting the conditions, went on:

> The matter alleged to constitute a breach of planning control is that the said development has been carried out but the said condition has not been complied with, in that the said land is being used as an individual unit of residential accommodation.

The appellants appealed to the Secretary of State against that enforcement notice, and the inspector held that there was a breach of planning control. The Secretary of State, in considering the inspec-

B

tor's report, held that the appeal on the grounds of section 88 of the 1971 Act — namely that a four-year period had elapsed before the date of service — failed because it was a breach of a condition after 1963, when the earlier Act had abolished the universal benefit of the four-year delay.

Section 87 of the Town and Country Planning Act 1971, by subsection (1), says:

> Where it appears to the local planning authority that there has been a breach of planning control after the end of 1963, then, subject to any directions given by the Secretary of State and to the following provisions of this Section, the authority, if they consider it expedient to do so having regard to the provisions of the development plan and to any other material con-
>
> C
>
> siderations, may serve a notice under this section (in this Act referred to as an "enforcement notice") requiring the breach to be remedied.

Subsection (2) says:

> There is a breach of planning control if development has been carried out, whether before or after the commencement of this Act, without the grant of planning permission required in that behalf in accordance with Part III of the Act of 1962 or Part III of this Act, or if any conditions or limitations subject to which planning permission was granted have not been complied with.

Subsection (3) says:

> Where an enforcement notice relates to a breach of planning control consisting in — (a) the carrying out without planning permission of building, engineering, mining or other operations in, on, over or under land; or (b)
>
> D
>
> the failure to comply with any condition or limitation which relates to the carrying out of such operations and subject to which planning permission was granted for the development of that land; or (c) the making without planning permission of a change of use of any building to use as a single dwelling-house, it may be served only within the period of four years from the date of the breach.

By section 88 (1) (c):

> A person on whom an enforcement notice is served, or any other person having an interest in the land may, at any time within the period specified in the notice as the period at the end of which it is to take effect, appeal to the Secretary of State against the notice on any of the following grounds — ... (c) in the case of a notice which, by virtue of section 87(3) of this Act, may
>
> E
>
> be served only within the period of four years from the date of the breach of planning control to which the notice relates, that that period has elapsed at the date of service.

By subsection (3):

> Where an appeal is brought under this section, the enforcement notice shall be of no effect pending the final determination or the withdrawal of the appeal.

The Divisional Court allowed the appeal against the decision of the Secretary of State, holding that section 87(3) (c) was effective. Mr Sullivan has submitted to this court that the local authority had a choice. They could serve that enforcement notice either based on a breach of condition or based on a change of use contrary to the

F

planning permission. He accepted, in the course of argument, that had this notice been based on change in use to a single dwelling-house, he would have no answer. But he says that as the notice was on the basis of a breach of condition — and he drew attention to the contrast between the earlier parts of the same section, subsections (a) and (b), between carrying out without planning permission and a failure to comply with any condition — the planning authority (or the minister as the case might be) would be free to decide, because of a breach of condition, that the time bar did not take effect.

In my judgment the fact is that there has been a change of use of the building from use as a loft for storage to use as a single

G

dwelling-house. The permitted use was for storage, so therefore the case is exactly covered by the words of section 87 (3) (c), and the fact that the change of use is also a breach of condition does not override this.

I agree with Skinner J in his judgment in the Divisional Court that the contrast between the word "relates" which occurs in subsection (3) of section 87, where the "enforcement notice relates to a breach of planning control", and the word "specify" in subsection (6) of the same section which states that an "enforcement notice shall specify" gives added support to the view which I have just expressed.

Mr Sullivan has also drawn attention to an amendment made by the Local Government and Planning (Amendment) Act 1981 which

H

substitutes a new section 87 for the section which I have already read and which has, in the equivalent of subsection (3), an additional subsection dealing specifically with a breach of condition resulting in use of a building as a single dwelling-house.

But in my judgment, since I have clearly come to the conclusion that the facts here were completely covered by subsection (3) (c) of section 87, it is not necessary to consider the effect of subsequent legislation.

Accordingly, I would dismiss this appeal. I would only add that had it not been for the argument of Mr Sullivan in this appeal, I would have been content to accept the judgment of the Divisional Court.

J

Agreeing, DONALDSON LJ said: I, too, would dismiss this appeal. It seems to me that the facts fall plainly and completely within the terms of section 87(3) (c) of the 1971 Town and Country Planning Act. The enforcement notice related to a breach of planning control and that breach of planning control consisted in the making, without planning permission, of a change of use of the building to use as a single dwelling-house.

The only doubt which I entertain about this matter at all is one which has not been ventilated, namely, whether this is in fact and in law a breach of the condition. I mention that doubt merely in case it arises in another case. My doubt arises out of the form of the planning permission, which I have not seen in its original form

K

but which I take to be accurately recorded on the enforcement notice, in which it would appear that permission was given to develop the land by the erection of a second storey for use as a loft. There was then a condition that the additional storey could not be used for any purpose other than storage connected with the residential use of the premises.

It seems to me to be arguable that the condition applies and limits and restricts loft user, and that it is only if you are using the premises as a loft otherwise than in accordance with the condition that then it can be said to be a breach of the condition. If you are not using it as a loft at all, I would have thought it arguable that you simply were acting without planning permission.

L

But, as I say, apart from recording that doubt, which is immaterial to the result of this appeal, I merely say that this is plainly within the section and this appeal to my mind should fail.

Also agreeing, SIR DAVID CAIRNS said: I agree that this appeal should be dismissed. I entirely agree with the reasons for their conclusion given in the Divisional Court by the two judges of that court.

The fresh argument that was not available to the Divisional Court but which was advanced in this court was to the effect that the provisions of the Act of 1981 indicate an intention of Parliament to change the law and show that Parliament put the same construction upon this particular provision as Mr Sullivan invites this court to put.

M

I do not take the view that it was Parliament's intention in 1981 to change the law as laid down in the Act of 1971. The two judges in the court below, and both my brethren in this court, have found this to be a perfectly plain provision. It is apparent from the decision of the Secretary of State that he did in fact take a contrary view of what the section in the 1971 Act meant. In those circumstances I think it is not unnatural that Parliament should take the trouble, in the 1981 Act, to specify in terms simply to make it clear to everybody what the intention of the 1971 Act was.

The appeal was dismissed with costs.

A **QUEEN'S BENCH DIVISION**

December 7 1981

(Before Mr Justice GLIDEWELL)

NORTH SEA LAND EQUIPMENT v SECRETARY OF STATE FOR THE ENVIRONMENT AND ANOTHER

Estates Gazette May 15 1982

(1981) 262 EG 668

Town and Country Planning Act 1971, section 246 — Appeal from decision of Secretary of State for the Environment dismissing appeals against three enforcement notices — It was conceded during B the hearing that the appellants' criticism of one of the notices, which concerned the storage of containers, was valid and that this notice must be quashed — The other two notices required discontinuance of the use of a building and land comprising the planning unit for the purpose of repairing, as distinct from storing, containers — The Secretary of State had concluded that the applicants' use was for the storage and repair of containers and that this was materially different from the previous use, which was confined to the storage of timber — Appellants contended that the requirement of the enforcement notices to discontinue the use for repair of containers was excessive as it might be thought to prohibit repair work which was ancillary to the established use for storage — After a C review of case law the judge held that the enforcement notices were not open to this criticism — If the appellants could show that a certain amount of repair activity was a normal incident of the use of the site for the storage of containers they would not be in breach of the notices — It was not essential to include in the notices some qualifying or limiting expression *ex abundante cautela* as in *Monomart (Warehouses) Ltd v Secretary of State for the Environment* — Appeal against the two notices dismissed

This was an appeal by North Sea Land Equipment under section 246 of the Town and Country Planning Act 1971 against the dismissal by the Secretary of State of appeals against three enforcement notices served on the appellants by Thurrock Borough Coun-
D cil. The notices related to land of some 5 acres on the northern edge of an industrial estate in West Thurrock.

W Hicks (instructed by Merriman, White & Co) appeared on behalf of the appellants; Simon Brown (instructed by the Treasury Solicitor) represented the first respondent, the Secretary of State; the second respondents, Thurrock Borough Council, were not represented and took no part in the proceedings.

Giving judgment, GLIDEWELL J said: This is an appeal under section 246 of the Town and Country Planning Act 1971 against the decision of the Secretary of State dismissing appeals to him against three enforcement notices served by Thurrock Borough Council, the second respondents. The enforcement notices are
E dated April 12 1979. The Secretary of State's decision letter was dated October 15 1980. The three enforcement notices all related to parts of the same land, which is agreed, in total, to be the correct planning unit, which comprises an area of some five acres on the northern edge of an industrial estate lying north of London Road, West Thurrock. The site is mostly open in the sense that it is not built upon, but there is one large building standing on it, approximately in its centre.

The three enforcement notices, during the hearing of the appeal before the Secretary of State's inspector and before me, were called respectively A, B and C. Notice A related to the building only. That alleged a breach of planning control consisting of a material
F change in the use of the building to use for the purposes for the repair of containers, and required discontinuance of the use of the building for the repair of containers. Notice B related to land less the building, that is to say, the land less the site of the building. It related to an alleged breach of planning control consisting of use for the purpose of the repair of containers, and it required similarly discontinuance of that use. Finally, Notice C, which related also, like B, to the land less the building, alleged a breach consisting of a material change in the use of the land to use for the purposes of the storage of containers and required discontinuance of that use.

Of the facts found by the inspector the following, in my view, are material to the matters before me. The industrial complex, that is the industrial estate south of the site, was owned and operated G
for many years by a firm called Magnet Joinery, who imported timber and stored it for their local machine shops. They also stored timber on the appeal site for later onward transporting to their factories in York. The operation closed down in 1975. In 1950 or 1951 Magnet Joinery extended timber storage on to the appeal site, first on the western part which they owned, and later on to the whole of the site. The timber was stored approximately 16 ft high in lanes served by diesel trains and mobile cranes. It was found that that was the use of the land well before the beginning of 1964. The inspector having made that finding of fact said:

> Although the destination of the timber was either to other parts of the appeal site for making up into furniture or to another factory in Yorkshire the H
> primary purpose for its being on the site was to keep it until required. Therefore I do not consider that it could be successfully argued that this use could have been other than as a repository, notwithstanding the Council's argument that part of the site might be claimed to have enjoyed a Class IV use before that. In my view the storage use created a new planning unit and that was the unit which the appellants took over.

And then he said:

> The purpose for which the appellants took over the site was to store containers.

He then made findings about the number of containers, and in relation to repair, he said:

> Nor could there be a strong argument against the repairs aspect being J
> ancillary

that is to the storage;

> if it were only a minor operation. However, the facts are that 80% or so of the containers brought on to the site require some form of work doing to them before they can be stacked involving some 1,360 containers as a maximum figure. In my view this represents an equal, if not a major, part of the present use pattern and so the proper definition of the use should be a mixed use for storage and repair of containers over the whole planning unit notwithstanding that the separate elements operate at different levels on parts of the site. Therefore I would consider that the allegations in all the notices should more properly have been a change to a mixed use for the purposes of the storage and repair of containers. K

Finally, in relation to the building, he noted that on November 7 1978 a determination under section 53 was given that planning permission for the storage of containers in the building was not necessary.

The appellants took possession of the site in 1978. They then commenced to store containers on it and to do work of repair to many of them as it was needed, both in the building and in the open. The average number of containers on the site was said to be about 1,000, but, as I have already said, the inspector found that there could be up to 1,360. At the time of the inquiry there were approximately 800 on the site. L

The appeals against the enforcement notices were on several of the grounds set out in section 88(1) of the Town and Country Planning Act 1971, but so far as I am concerned the arguments before me related to a limited number of grounds. So far as Notice C is concerned, there were two, that is to say (b) that there had been no breach of planning control or (d) that if there had been, the development took place before the beginning of 1964. In the case of Notices A and B all that was argued was ground (f), that the steps required by the notice to be taken exceed what is necessary to remedy any breach of planning control.

I deal first with Notice C. Mr Hicks advanced three arguments, which I can describe shortly in these terms. First, that there had been no material change of use between the use of the land for the M
storage of timber and the use of the land for the storage of containers, and he did not refer me to, but he had referred the inspector to, the *East Barnet* case *(East Barnet Urban District Council v British Transport Commission* [1962] 2 QB 484). Secondly, he argued that the whole site was a repository within the meaning of the Town and Country Planning (Use Classes) Order. It did, when it was used for the storage of timber, fall within Class 10 of the Schedule to that order and thus no planning permission was needed because by definition there was no material change of use in continued use as a repository, albeit that what was being stored in the repository was now containers rather than timber. Thirdly and

A alternatively, he argued that: (a) the building has an established use for storage; (b) article 2(3) of the Town and Country Planning (Use Classes) Order 1972, reads, "References in this Order to a building may, except where otherwise provided, include references to land occupied therewith and used for the same purposes"; and (c) it followed that as the building clearly was in use as a repository within Class 10 the effect of article 2(3) was that the land which was occupied with the building could also be said to be within Class 10, not by the direct route of his first argument but by the indirect route of going via article 2(3), and thus again it followed that there was no material change of use.

That third argument is not to be found anywhere in the notice of
B motion despite the fact that the notice of motion contains a series of different points, not all of which were in fact argued before me. At the start of the hearing Mr Hicks sought leave to amend the notice of motion to add this as a ground and, without objection from Mr Simon Brown, I allowed the amendment.

I should say that the fact, as I know from an earlier application made in this case, that Mr Hicks himself was instructed only the day before the hearing started and the coincidence of the hearing starting with an application for an amendment is at any rate some *prima facie* evidence that the point is one in which Mr Hicks himself had a part. If it be so, it merely shows the benefit of having an additional mind bearing on a problem.

C But be that as it may, as soon as Mr Brown heard the argument developed on this third point he asked for some little time to consider it and take instructions, and 20 minutes or so later he conceded that the point was a good one and the enforcement notice C must therefore be quashed. It is therefore unnecessary for me to consider the first and second arguments in relation to that enforcement notice. The only materiality of the lateness of the amendment to the notice of motion is as to costs, to which I shall have to come in due course.

I come to Notices A and B. As I have said, they both deal with an alleged use for the purposes of the repair of containers. The Secretary of State's conclusion was that the applicants' use was,
D and is, a mixed use for the storage and repair of containers, and that it is materially different from the previous use which was confined to the storage of timber. That conclusion of the Secretary of State is on a matter of fact and degree, and there is no right to challenge it in this court. For that reason Mr Hicks did not attempt to challenge it, but what he did argue was that the requirement in the enforcement notices, which were confirmed by the Secretary of State, to discontinue the use of the land in the one case and the building in the other, for the purposes of the repair of containers exceeded what is necessary because it may be thought to prohibit an extent of repair work which is ancillary or incidental to the storage which is an established use.

E On this argument I was referred to three decisions of the Queen's Bench Divisional Court, namely, *Mansi v Elstree Rural District Council* (1964) 16 P & CR 153; *Newport v Secretary of State for the Environment* (1980) 40 P & CR 261; and a very recent decision in *Cord v Secretary of State for the Environment and Torbay Borough Council* [1981] JPL 40. In *Mansi* the minister found that there was an established use of part of the site for retail sales in addition to the overall established use of the site as a plant nursery. The enforcement notice required the discontinuance of all retail sales and the court held that went too far and should have been amended by the minister. Mr Brown submits, and I agree, that *Mansi* was a case of an established mixed use, that is to say a
F primary use as a nursery and a secondary limited use for retail sales, and is not a case of a single established use to which some other activity is said to be incidental or ancillary. So understood it becomes clear, as Mr Brown submitted, that the decision in that case is not relevant to the present case.

In *Newport* it was found as a fact that at a motor repair garage a small number of cars, not exceeding four at a time, was sold from time to time before 1964. The inspector described the uses as a composite use, which I apprehend to mean the same as what I mean by a mixed use. But the Secretary of State held that the sale of cars was an ancillary use to the principal use of the garage for the purpose of repairs. The court held, as in *Mansi*, that the Secretary of State should have amended the enforcement notice to G
allow car sales to a limited extent. In my view, despite the way the Secretary of State treated it, *Newport* can properly be regarded as another example of a composite or mixed use. Certainly this seems to have been the approach of Stephen Brown J who gave the second judgment in the Divisional Court, and if that is right then, like *Mansi*, it is not relevant to the present problem. But if it is wrong I must say that I regard *Newport* as a decision related only to its own special facts.

The two decisions of the Divisional Court which are, in my view, clearly relevant are the recent decisions in *Cord* and the earlier decision in *Monomart (Warehouses) Ltd v Secretary of State for the Environment* (1977) 34 P & CR 305. In *Monomart* planning per- H
mission had been given for the erection of a builders' merchants' warehouse. After the building was erected the appellants started to use it as what was called a "Do-it-yourself supermarket," which in more formal language means for retail sales. An enforcement notice was served requiring discontinuance of retail sales. On appeal the Secretary of State upheld the notice but he added the words in relation to the use to be discontinued "except as may be incidental to the use of the premises as a builders' merchants' warehouse." In his judgment, Lord Widgery CJ, with whom Parker and Peter Pain JJ agreed, said at p309:

> .....this was not a case which required amendment by the Secretary of State at all. I think that this notice on its construction and as it originally stood J
> was a notice complaining of the use of the premises for retail sale in excess of retail sales permissible as ancillary or incidental to the main use. I think that the initial requirement was similarly limited. I form that opinion because one must read the enforcement notice as a whole, and it is quite impossible to make sense out of this notice unless one assumes that the author in each case where he refers to the retail sale of goods is referring to the retail sale of goods in excess of the rights which may be incidental to the warehouse use. I do not think that the words need have been amended at all, but I certainly take the view that the Secretary of State is not to be criticised for having amended them, and he was perfectly entitled to take the view *ex abundante cautela* that this should be done.

In *Cord*, where the facts were in essence similar to though rather more complicated than *Monomart*, the Secretary of State had not K
added any words to an enforcement notice which required discontinuance of an unestablished composite use. The court held that he was not obliged to do so and dismissed the appeal.

Mr Hicks submits that unless some such words as were used in *Monomart* are added to the enforcement notice it will not be clear what his clients can or cannot legally do and thus they will be at risk if they are prosecuted in the magistrates' court. I remind myself that the appellants' right of appeal to this court is on a point of law only. On this issue the test is that derived from the judgment of Upjohn LJ, as he then was, in the well-known decision of *Miller-Mead v Minister of Housing and Local Government* [1963] 2 QB 196, and to quote his Lordship: "Does the notice tell the ap- L
pellants fairly what they have done wrong and what they must do to remedy it?" In my judgment these notices satisfy that test. They tell the appellants that they must discontinue the use of the land for the repair of containers. If the appellants can show that a certain amount of repair activity is a normal incident of the use of the site for the storage of containers they will not be in breach of the notices. If the Secretary of State thinks it desirable in a particular case to add words such as those used in *Monomart*, that is a matter for his discretion. In this case he was in no way wrong in law in not adding such words. The appeal against enforcement notices A and B therefore fails and it is dismissed.

I have already heard argument on the question of costs, an- M
ticipating that Mr Brown would be unable to be here. The position now is this, that I have found for the appellants on the enforcement notice which was the most important from their point of view and on which everybody accepted that the major part of the argument would have centred, and indeed did centre, as far as Mr Hicks' argument was concerned, but I have found for them on a ground which was adumbrated only at the last minute. Mr Brown made the point, which he was perfectly entitled to make, that had that ground been in the original notice of motion it must be taken as being very probable that the Secretary of State would have been advised to concede that argument before ever the matter came to

A court. It occurs to me that if that had happened it may well have been that the matter would not have come to court at all. Whether it would still have been desirable to argue Notices A and B I simply do not know. Be that as it may, on Notices A and B of course the appellants have failed. Both counsel asked that their respective clients should have costs, possibly limited to a proportion. One can go through a series of mathematical exercises to try to arrive at what is fair in this case, but I take the view that overall the major point is that very probably the Secretary of State would not have resisted the appeal if the amended point had been put in the notice of motion originally and I think it right that he should have some of his costs. The order that I have concluded is fair in that the ap-
B pellants should pay half the Secretary of State's costs.

Leave to appeal, if necessary, was given.

COURT OF APPEAL

March 24 1982

(Before Lord DENNING MR, Lord Justice EVELEIGH and Lord Justice WATKINS)

MALVERN HILLS DISTRICT COUNCIL v SECRETARY OF
C STATE FOR THE ENVIRONMENT AND ANOTHER

Estates Gazette June 19 1982

(1982) 262 EG 1190

Town and Country Planning Act 1971—Outline permission for housing development in the Malvern Hills followed by approval of reserved matters—By virtue of section 42(2)(b) of the Act development had to be begun not later than the expiration of two years from final approval of reserved matters—Under section 43(2) development was to be taken to be begun on the earliest date on which any "specified operation" began to be carried out—Relevant operation in present case was "any operation in the course of laying out or constructing a road or part of a road" (section
D **43(2)(d))—Question as to whether "pegging" of lines of part of new estate road, which was carried out before the expiry of the time-limit, was "an operation in the course of laying out" the road—Enforcement notice served by local authority on ground that development had not been begun in time—Sir Douglas Frank QC, sitting as a deputy judge of the Queen's Bench Division, held that the development had been begun in time—Court of Appeal divided—Lord Denning MR, in the minority, considered that the driving in of pegs was only a preliminary act, not a specified "operation"—Majority of court (Eveleigh and Watkins LJJ) held that the pegging constituted a "laying out" on the ground and was therefore an "operation" within section 43(2)(d)—Development**
E **consequently held by majority to have been begun in time—Appeal from decision of Sir Douglas Frank QC dismissed, but leave given to appeal to House of Lords**

This was an appeal from a decision of Sir Douglas Frank QC dismissing an appeal by Malvern Hills District Council against a decision of the Secretary of State for the Environment who had accepted the inspector's conclusion that the pegging out of part of the road constituted a "specific operation" which satisfied the time-limit for the beginning of development under section 43(2)(d) of the 1971 Act.

J Sullivan (instructed by Sharpe Pritchard & Co, agents for Paul
F Graham, Malvern Hills District Council) appeared on behalf of the appellants; Simon Brown (instructed by the Treasury Solicitor) represented the first respondent, the Secretary of State; A T Smith QC (instructed by Jacobs Bird & Co, of Birmingham) represented the second respondents, the developers, Robert Barnes & Co Ltd.

Giving judgment, LORD DENNING MR said: The Malvern Hills look down on a very English landscape to the west. In the valley there are charming old villages, dotted here and there among the woods. One of them is Cradley, which gives its name to the Cradley Brook running through it. It has quite enough houses for everyone there, but developers wish to build more. They plan to build a new housing estate of 25 houses. They are confident that they would be bought by
purchasers who would commute to work elsewhere in the cities and G
towns of the Midlands: or else by purchasers who would come to the village to retire. The parish council and all the people of the village are vigorously opposed to it. They fear that their village life will be replaced by a commuter society. An inspector at a local inquiry has found that the development would be socially unacceptable. Yet the minister has felt bound in law to let the developers go ahead and to build this housing estate. It all depends on the construction of a few words in the Town and Country Planning Act 1971.

The story starts on February 22 1974. The Hereford County Council granted outline planning permission for residential development. It was an area of 4½ acres in the old part of the village
close to the church and school. The outline permission was granted H
subject to conditions, and many matters were reserved for further consideration. These reserved matters were not approved for three years. Before these were approved a petition with 267 signatures opposed the development. But it was no good. On March 28 1977 the Malvern Hills District Council (who were now the planning authority) very reluctantly granted approval of the reserved matters.

The statutory provision

That date, March 28 1977, is crucial because section 42(2)(b) of the Town and Country Planning Act 1971 provides that:

> . . . the development to which the permission relates must be begun not later
> than whichever is the later of the following dates — J
> (i) the expiration of five years from the date of the grant of outline permission;

that date is here February 22 1974

or

> (ii) the expiration of two years from the final approval of the reserved matters or, in the case of approval on different dates, the final approval of the last such matter to be approved

that date is here March 28 1977.

So it was necessary for the development of these 4½ acres to be begun not later than March 28 1979. If the enactment stopped there it would
have been very difficult to say when the "development" was K
"begun". But Parliament endeavoured to remove this difficulty by enacting section 43. It says:

> 43(1) For the purposes of sections 41 and 42 of this Act, development shall be taken to be begun on the earliest date on which any specified operation comprised in the development begins to be carried out.
> (2) In subsection (1) of this section "specified operation" means any of the following, that is to say —
> (a) any work of construction in the course of the erection of a building;
> (b) the digging of a trench which is to contain the foundations, or part of the foundations, of a building;
> (c) the laying of any underground main or pipe to the foundations, or part of the foundations, of a building or to any such trench as is mentioned in
> the last preceding paragraph; L
> (d) any operation in the course of laying out or constructing a road or part of a road;
> (e) any change in the use of any land, where that change constitutes material development.

What was done before March 28 1979

In view of those statutory provisions, it is important to know exactly what was done before March 28 1979. Nothing was done by the original applicants. They sold the site of 4½ acres to a new firm of developers and gave them vacant possession on December 21 1978 — when there was only three months to go. The approved detailed plan showed 25 new houses to be built and an estate road to get to them.

During these three months the new developers made preparations M
for drainage work. They brought a lot of pipes and manhole rings on to the site. They deposited them in the centre of the site. A JCB excavator was sent to unload them and to excavate a drainage trench. But the JCB machine became bogged down and was removed from the site. No excavation was done there by this machine.

Also during those three months the new developers made preparations for the estate road. In January 1979 they put in pegs to show the centre line of the proposed new estate road, and also others to show the line of the pavements on each side. These pegs did not get very far. They only went the first 250 ft of the proposed estate road at intervals of about 50 ft. They stuck the pegs only about 6 to 10 in out

A of the ground. Later the developers put in pegs beyond 250 ft, but they did no more than put in pegs. They were held up by the weather.

No details of the layout of the road were submitted to or approved by the highway authority. And condition 16 of the application said in terms that:

> No work on the site of the development shall be commenced until these details have been approved by the highway authority and an agreement entered into.

I presume that the highway authority might have required the site of the proposed road to be altered and the pegs removed before work was commenced. The position of the pegs was only tentative or provisional.

B *Differences arise*

Early in April 1979 differences arose. The new developers said that they had begun the development by March 28 1979. The Malvern Council said they had not. At first it was thought that the differences could be resolved by an application under section 53 of the Act. But after about a year that was found to be inapplicable. So on May 9 1980 the council served an enforcement notice on the developers requiring them "to cease construction of the road and all building, engineering or other operations". On May 29 1980 the developers appealed to the minister against this enforcement notice. Thereafter, on June 13 1980, the council served a "stop notice" on the C developers under section 90 of the Act. This effectively stopped the developers from doing any more work on the land. But, if the enforcement order were afterwards quashed, the council would be liable to pay compensation for any loss suffered by the developers owing to the stop notice: see section 177 of the Act.

When was development begun?

That history shows that everything depends on the date when the development of the site was "begun" by the developers. If it was begun *before* March 28 1979 the enforcement notice must be quashed. The developers will be able to go on with the development of the housing estate. They will also be entitled to compensation for all loss suffered by them on account of the bad enforcement notice. D They put it at £500,000 payable by the Malvern Council. But, if the development was begun *after* March 28 1979 the developers will not be able to develop the housing estate, and will receive no compensation.

So, much depends on whether the development was begun by March 29 1979 or not. I have set out what was done. Now I turn to the law.

Section 43

In construing subparagraph 2(d), it is very right and proper to look at the subparagraphs with which it is associated. The Latin maxim is *noscitur a sociis*. Put into English it means that, in order to construe E a word or phrase, you can find out by looking at the companions with which it is associated. Looking at subparagraphs 2(a), (b), (c) and (e), it is significant that they all refer to the actual work that is being done. They exclude any acts which are only preparatory to it. Thus, in subparagraph 2(a) — about erecting a building — the development does not begin when you bring the bricks or the sand and cement on to the site, or put in pegs. It is only begun when you start to lay the concrete or the bricks. In 2(b) — digging the foundations — the development is not begun when you bring the JCB on to the site, or get the men there with their picks and shovels, or put in pegs. It is only begun when you actually start to dig the trench. And in 2(c) — laying the pipes — the development is not F begun when you bring the pipes on to the site. It is only begun when you start to lay a pipe in the trench. And in 2(e) — change in the use — when you have permission to use your front garden for selling goods by retail, the development does not begin when you buy the stall or the goods or bring them on to your land. It only begins when you actually start selling them. So also subparagraph 2(d) should be construed not to apply to preparatory acts, but only to the actual operation itself.

Note that in the last sentence I use quite naturally the words "the actual operation itself". That brings me straight to the word "operation" itself in subparagraph 2(d). To my mind the word "operation" in the singular is used in the same sense as in the many places in the statute where the word "operations" is used in the plural. That G must be so as a matter of statutory interpretation: for, as we all know, in a statute, "words in the singular shall include the plural, and words in the plural shall include the singular". The plural "operations" is used in many places in this statute, notably in section 22(1) where permission is required for "the carrying out of building, engineering, mining or other operations in, on, over or under land": and in section 290 when "engineering operations" includes the formation or laying out of means of access to highways. Also in section 45(4)(a). We considered its meaning in those sections in *Parkes v Secretary of State for the Environment* [1978] 1 WLR 1308, when I said that the word "operations" comprises activities which result in some physical alteration to the land, which has some degree H of permanence to the land itself. For instance, I do not think the putting in of pegs — without permission — would be a breach of planning control. It would require something more permanent. I think that the same meaning should be applied to section 43(2)(d) also: especially as it fits in so well with subparagrahs 2 (a), (b), (c) and (e).

Applied to this case

Applying this interpretation to the present case, it is plain — and it is accepted — that the bringing of the pipes and manhole rings on to the site was not a "specified operation" within subparagraph 2(c). The argument has been confined to the driving in of pegs into the J ground to show the proposed line of the road. I do not think these were a "specified operation" within subparagraph 2(d). They were tentative and provisional until approved by the highway authority. They were not permanent at all. They could be easily removed. They were preparatory acts — acts done in anticipation of "operations" which were to follow. They were not part of the "operation" of laying out the road.

Conclusion

This is a difficult point of construction on which different minds may well come to different conclusions. Everyone will have sympathy with the developers who were prevented by bad weather K conditions from beginning operations before March 28 1979. But equally everyone will have sympathy with the villagers at Cradley, who are anxious to preserve its character and safeguard it from intruders from outside. These sympathies are so evenly balanced that we must decide it simply by the true interpretation of the statute. I hold that the development was not before March 28 1979 and that thereby it is not now permitted. I would allow the appeal accordingly.

Disagreeing with the Master of the Rolls and holding that the appeal should be dismissed, EVELEIGH LJ said: I can find no suggestion in the inspector's report or in the minister's letter to indicate that the builders, the respondents, were not entitled to begin L work on constructing any estate roads. It is true that condition 16 of the approval of reserved matters dated March 28 1977 referred to details of the layout and alignment, width and levels of the proposed estate roads, and stated: "No work on the site of the development shall be commenced until details have been approved by the highway authority and an agreement entered into". The appellants have placed no reliance upon this condition before and there is no indication that the matter has been argued at any stage of these proceedings. The parties have accepted and the minister concluded that the relevant date by which development had to be begun was March 28 1979. Section 42(2) (b) (ii) provides that development must M be begun not later than "the expiration of two years from the final approval of the reserved matters or, in the case of approval on different dates, the final approval of the last such matter to be approved". If further approval were required in the present case, the effect would be to extend the time-limit beyond March 28 1979. I feel completely unable to place any reliance upon condition 16 in these circumstances and it would be clearly unfair to the builders to do so. I can only assume that there is some explanation of this apparent irregularity.

I take the following extract from the inspector's report:

> On January 8 1979 Mr Bevan, the managing director of the appellant company, and Mr French, examined the drawings of the proposed site works

A and inspected the land, which they found to be overgrown and marshy in places.

Starting work on January 11 1979 and continuing on January 12 and 16 Mr Bevan and Mr French, assisted by a labourer, cut lines to the boundary hedges and measured the position of the centre line of the first 250 ft of the estate road on the site, which they marked by putting in pegs at intervals of about 50 ft; they also put in "off set" pegs to mark the line of the pavements on each side at similar intervals.

The pegs used were about 1½ in or 2 in square by 2½ ft long, and pointed at one end and were driven into the ground to leave from 6 to 10 in above the ground; steel pins were used where it was necessary to place a peg into the metalled track on the site.

Mr Doyle, Mr Bevan's assistant, checked the position of the pegs marking B the first 250 ft of the estate road during February 1979; these pegs were sufficient for the driver of an excavating machine to follow when the construction of the road was begun, following the company's practice; the pegs were dug out with the topsoil on the road by a machine during the first part of April 1979, when the weather had improved sufficiently to allow work to proceed.

During the second part of January 1979 and before March 28 1979 Mr French continued to place pegs to mark the centre line of the estate road beyond the first 250 ft, and he also marked with pegs and hammer-head turning places on the estate roads; some of these pegs were plotted on plan E by the council's officers in April 1979.

On February 12 1979 because of bad weather the appellants decided to leave work on the site until further notice and they drove in the road marker pegs to just above ground level to prevent them being removed by C unauthorised persons.

Sections 41 and 42 of the Town and Country Planning Act 1971 relate to the duration of planning permission, and their effect as regards the present case is that development had to be begun not later than March 28 1979. Section 43 (1) and (2) provides:

> (1) For the purposes of sections 41 and 42 of this Act, development shall be taken to be begun on the earliest date on which any specified operation comprised in the development begins to be carried out.
> (2) In subsection (1) of this section "specified operation" means any of the following, that is to say —
> (a) any work of construction in the course of the erection of a building;
> D (b) the digging of a trench which is to contain the foundations, or part of the foundations, of a building;
> (c) the laying of any underground main or pipe to the foundations, or part of the foundations, of a building or to any such trench as is mentioned in the last preceding paragraph;
> (d) any operation in the course of laying out or constructing a road or part of a road;
> (e) any change in the use of any land, where that change constitutes material development.

The inspector came to the following conclusion:

> I consider that the marking out of the line and width of the first 250 ft length of the estate road with pegs, placed accurately by measurements taken E from the site boundaries, so that the road could be constructed with machines guided by these pegs, did amount to an "operation" in the course of laying out part of a road as defined in paragraph (d) of section 43(2). Therefore, it was a specified operation. Laying out does not appear to be defined in the 1971 Act, but its meaning to me is the marking out on the ground of the road in accordance with an approved plan; the *Concise Oxford Dictionary*, 6th ed, supports this interpretation. Laying out was a necessary preliminary to the construction of the road and for the first 250 ft the construction followed in April shortly after the expiry date. As well as completing the laying out of the first 250 ft, pegs were also placed to mark considerable stretches of the remaining estate roads before March 28 1979.

Counsel for the appellants submitted: (1) That the words "laying F out" involved an activity closely allied to construction itself and that "pegging out" the road was not such an activity. It was submitted that road material is laid out when it is placed on the road and that the expression in section 43 involves doing something that was a permanent contribution to the road.

(2) Alternatively, if the builders were engaged in laying out the road, then it is submitted that they had not begun to carry out any operation in the course thereof because the word "operation" itself in the Town and Country Planning Act referred to an activity which changes the physical character of land to some permanent extent.

(3) That as the minister did not apply either of these tests, he applied the wrong test in law.

The meaning of "laying out" G

The experienced builders who gave evidence to the inspector did not speak of laying out. They said that they were "setting out" the road. However, the phrase "laying out" has been used in a number of statutes. The Towns Improvement Clauses Act 1847, section 63, reads: "It shall not be lawful to make or lay out any new street until the same be", etc. The Metropolis Management Amendment Act 1862 provided that no existing road or way, being of less than 40 ft, should after 1862 be formed or laid out for building as a street for the purposes of carriage traffic unless it were widened to the full width of 40 ft. There the words "laid out" contemplate a stage before building has begun.

Section 157 of the Public Health Act 1875 provided: "Every urban H authority may make by-laws with respect to the following matters; (that is to say) (1) with respect to the level, width and construction of new streets and the provisions for the sewerage thereof". The section then goes on to say, "and they may further provide for the observance of such by-laws by enacting therein such provisions as they think necessary as to the giving of notices, as to the deposit of plans and sections by persons intending to lay out streets or to construct buildings", etc. The reference to "lay out streets or to construct buildings" might at first sight invoke the argument that to lay out a street was an activity similar to that of constructing a building. However, that argument could not prevail in my opinion when one appreciates that the first matter listed in the subsection J which I have reproduced above refers to the level, width and *construction* of new streets. Construction is clearly something different from laying out.

A number of expressions are to be found in the statutes relating to the provision or creation of roads and streets. We find "making" (section 146 of the Public Health Act 1875), we find "laying out", we find "constructing" and we find "building". (For example, the Metropolis Amendment Act above, and the Towns Improvement Clauses Act, section 63.)

These various expressions are used with precision. A street may be made without being laid out. It may simply grow. A street may be laid out without necessarily starting on the drawing-board or being K pegged out on the land. For example, by a deliberate construction of a line of houses facing each other and thus delineating a space between them which is intended to be used as a street. A road or street may be constructed without being laid out as where a pre-existing road which has grown is treated with some form of top surface. In some cases these terms may be interchangeable, but the authorities in my opinion clearly support the conclusion that laying out is the process of planning. This process may first be carried out on the drawing-board. It is then actually set out on the ground. In my opinion section 43 of the Town and Country Planning Act 1971 refers to the process of planning or delineating the road upon the ground. L

These different stages and the meaning of the words can clearly be seen in the judgments in *Robinson v Barton-Eccles Local Board* (1883) 21 Ch D 621. The Master of the Rolls at p 632 said:

> Now there are two ways in which a street may come into existence where there was no street before. A person may take a grass field or a country lane (for in my opinion, it makes no difference whether or not there was a public highway and lane, or a footpath existing before which is thrown into the street and is utilised, or whether there was nothing but a mere plot of grassland out of which a new roadway is made), he may take it and build continuous lines of houses so as to form what is commonly known as a street. When I say continuous lines, I do not mean that there are to be no breaks or intervals, but there must be a certain degree of continuity. A new street may arise in another M way, and that is, where it is not from the first laid out as a street in a formal manner, but may be considered to grow up, so to say, of itself. This often happens where there is an existing highway, and people build houses along the sides of that highway, so that, without any intention of laying out a street, the street grows. When does it become a street? This question cannot be answered until you know the locality. It must be a question in each particular case when the road becomes a street. At some time or other it becomes a street, and as soon as it does so it is a new street, and not the less a new street because some of the houses were built before it was a street.

At p 634 he said: "He is building his line of houses on the other side of the new street; he is laying out a new street by building a continuous line of houses". While in that case the building of the

A houses was the method of laying out the street, it does not follow that a street cannot be laid out by using some other method to delineate it upon the ground.

At p 636 Brett LJ said:

> New streets may be made under different circumstances. The whole land on both sides may belong to one owner. Then suppose he conceives a design of making a new street, and has a plan drawn. To my mind that does not begin the laying out the street within the meaning of the Act of Parliament. The Act of Parliament is not concerned with what people do upon paper, but with what they do in point of fact, and upon the land. When would such an owner begin to lay out and form a street? To my mind he would do so when he built his first house, having the intention to go on to make a street. He would then B have begun to lay out and to form a street, and it would from that moment begin to be a street.

At p 637 he said:

> One man at the end of the street may have begun to build, but the others have not built nor laid out their land as building land, nor advertised it as building land, nor done anything to show an intention to build. Then the street really forms itself as it were by gradual accretion without any common intent. In such a case you have to wait until you can see by the course of building that there is a common intent to build. The moment you see that, you come to the conclusion that a street has begun to be formed.

At the bottom of that page he said:

> The by-law does not say that it shall be laid out on a plan, and, as I said C before, the laying out on a plan is utterly immaterial. A plan laid out may show an intent to do a thing, but the laying out in this by-law does not mean laying out on paper. *It means laying out on the land.* It is laying out and forming it.

The cases establish that a street is not laid out for the purpose of the various Acts merely by being planned on the drawing-board. On the other hand, the fact that a permanent alteration to the land delineating the line of a street may, particularly when so intended, amount to laying out a street in no way indicates that some permanent demarcation is required before a street is laid out.

In the absense of authority to the contrary, I would myself regard the words as meaning that the plan of the road is being marked upon D the land. I have considered a number of authorities which have been concerned with whether or not an activity has amounted to laying out either a road or a street and I can see nothing in any of them to cast a doubt upon this meaning of the word (see *A-G v Dorin* [1912] 1 Ch 369 and the cases referred to in argument therein). Furthermore, the distinction that is to be found between laying out a street and the further distinction between those expressions and constructing a street (or road) compel me to reject the submission that "laying out" in section 43 is to be regarded as something analogous to constructing. Laying out is an earlier stage in the planned making of a road to that of construction. It does not follow, however, that some operation related to construction may not be carried out before the E laying out is embarked upon. For example, an old strip of roadway may be intended to form part of a new one or an old road may have to be remodelled to accommodate the weight of traffic, and I can envisage the situation where construction is undertaken in places before the intended road begins to be laid out on the ground. Thus it is necessary for section 43(2)(d) to allow for each of the two processes to be considered independently. I therefore cannot accept that the phrase "laying out" in section 43 requires that some part of the road shall have been formed by placing something in a permanent position on the land.

On the other hand, I do not think that "laying out" on a drawing-F board is enough. All the other specified operations in section 43 are activities carried out on the land itself and the phrase "laying out" must be regarded similarly.

As to the meaning of "operation"

The first step is to look at section 43 and attempt to discern its meaning without reference to decided cases. So often that which is relatively simple on the face of it is complicated by adopting the process of starting with the decided cases or indeed the last decided case instead of starting with the statute itself. In *Barrell v Fordree* [1932] AC 676 at p 682 Lord Warrington said: "In my opinion the safer and more correct course of dealing with a question of construction is to take the words themselves and arrive if possible at that meaning without in the first instance reference to cases". Doing G that, it seems to me that "operation" in section 43(1)(d) refers to any working activity on the land in the course of laying out a road whether or not that activity has resulted in a change in the character of the land or in anything that might be called development. As on the face of it I find no difficulty in understanding the meaning of the word "operation" in section 43(1)(d), I can see no reason for trying to introduce a difficulty.

At the beginning of the section the words "specified operation" are used as a label applicable to each of the examples listed. One of these (e) records "any change in the use of any land where that change constitutes material development". Although in *Parkes v Secretary of State for the Environment* [1979] 1 All ER 211 a H distinction was drawn between operations and use — the sorting of material was held to be a use and not an operation — yet for the purpose of section 43 any change in the *use* of land has been made a specified operation. It is not restricted to something which results in a physical change to the land. When I come to consider the meaning of the word "operation" in section 43 (2)(d) and find it intelligible in its natural meaning, why should I go outside the section to attract a more restricted meaning when the very section does itself employ the word more widely, so that it includes use which does not necessitate permanent change in the character of the land?

If the expression "laying out" is only applicable to a process which itself affects a change in the physical character of the land and J does not therefore include pegging, we have no need to consider whether the word "operation" should be given a restricted meaning. The stage of laying out would not have been reached and consequently the pegging could not be an operation in the course of it, no matter what the word "operation" means.

If, as I think, "laying out" means simply the process of delineating the shape and confines of the road upon the ground, the appellants have to argue that the word "operation" bears the restricted meaning in that some change in the physical condition of the land with some degree of permanence has to be effected by the operation. This argument would mean that something of a permanent nature has to be achieved within a process which is of K itself only temporary or preparatory. It seems to me to follow that the use of the word "operation" cannot of itself introduce the idea of permanent change.

Parkes' case did not decide that the carrying out of operations in or over or under the land does not begin until there is some change in the physical condition of the land. It contrasted use which did not result in some physical alteration to the land with "operations" which would. Lord Denning at p 213 said: "Looking at these various sections it seems to me that the first half, 'operations', comprises activities which result in some physical alteration to the land, which has some degree of permanence to the land itself, whereas the second half, 'use', comprises activities which are done in, alongside or on L the land but do not interfere with the actual physical characteristics of the land". In other words, as I understand it, the operations have to be carried out with the object of producing some physical alteration to the land. Lord Denning spoke of activities which result — not of activities which have resulted. At p 215 I said: "In my opinion the word means operations with the same kind of purpose as building operations, that is to say, to add to or to alter the condition or the state or the quality of the land, usually no doubt with the future or the continuing use of the land in mind". I therefore do not regard *Parkes'* case an as authority for the proposition that there cannot be an operation within section 43 unless some physical change has already been achieved. M

If an oil company constructs a drilling rig for testing purposes but the drill has not yet pierced the soil, are we to say that it is not carrying on engineering operations?

This conclusion seems to me to accord with the intention of the legislature. Sections 41 and 42 seek to ensure that land will not be held undeveloped for an indefinite period in the hands of speculators whose only intention is to sell the land at some future propitious date at the enhanced value that development permission attracts. Section 43 seeks some earnest of intention to develop. The specified operations are not necessarily very extensive. Very little need be done to satisfy the section. That which is done, however, must genuinely

A be done for the purpose of carrying out the development. Section 43 is a benevolent section which aims at avoiding hardship to a developer who is genuinely undertaking the development. The respondents in this case were clearly doing that. It would indeed be hard to interpret section 43 by importing a restriction on the natural meaning of the words by reference to the decision in *Parkes'* case.

I would just add that if circumstances indicated that the development was no longer desirable, then there is under section 45 a power to revoke any permission to develop.

In my opinion the minister applied the correct test and was entitled to find as he did. I therefore would dismiss this appeal.

B Agreeing that the appeal should be dismissed, WATKINS LJ said: The grounds of this appeal against a decision of Sir Douglas Frank QC, sitting as a deputy High Court judge, are that he erred in law in rejecting the submissions advanced on behalf of the appellants, the Malvern Hills District Council, that the word "operation" in section 43(2)(d) of the Town and Country Planning Act 1971 means an operation that changes the physical character of the land and has some degree of permanence and in deciding that whether or not there had been an operation in the course of laying out or constructing a road or part of a road for the purposes of section 43(2)(d) was purely a question of fact and degree for the first respondent, the Secretary of State for the Environment, in determining it.

C In support of the proposition that the word "operation" in section 43(2)(d) bears the meaning he contends for, counsel for the district council relies upon the judgment of Lord Denning MR in *Parkes v Secretary of State for the Environment* [1979] 1 All ER 211 in which at p 213H he stated with reference to sections 22, 45 and 51 of the Act of 1971:

> Looking at these various sections, it seems to me that the first half, "operations", comprises activities which result in some physical alteration to the land, which has some degree of permanence to the land itself, whereas the second half, "use", comprises activities which are done in, alongside or on the land but do not interfere with the actual physical characteristics of the
> D land.

By analogy counsel argues that the word "operation" as used in section 43(2)(d) must bear the same meaning as was attributed to "operations" by Lord Denning MR. Accordingly, since what was done by Robert Barnes & Co Ltd — the builders — did not, as is the fact, bring about any physical alteration to the land they had recently purchased, they had not begun to carry out an operation which could save the existing planning consent from lapsing by effluxion of time.

I am unable to accept that. The word "operation" in section 43(2)(d) must be regarded always, as is therein stated, as a "specified operation", the definition of which is contained within the subsection itself, which so far as material provides:

> E 43(2) In subsection (1) of this section "specified operation" means any of the following, that is to say . . .(d) any operation in the course of laying out or constructing a road or part of a road.

Accordingly, seeing that the definition of "specified operation" is so explicitly provided for, there is no warrant that I can see for importing into it the notion that the process of laying out a road has not begun unless what has been done has altered the land physically to some permanent extent. The meaning attributed to "operations" in *Parkes'* case is, I think, inapplicable to section 43(2)(d), the provisions in which are wholly different from those contained in
F those sections in which the word "operations" is used. It is, therefore, of no assistance in determining the vital question in this appeal which is, had the builders begun to lay out a road which was to form part of the permitted development before time ran out?

In *Spackman v Wiltshire County Council* (1977) 33 P&CR 430 Willis J said of the provisions of section 43, "these are in a sense penal provisions which take away from the subject, albeit in a fairly leisurely way, a vested entitlement, and should in my view be construed benevolently towards and not strictly against the G
subject".

I respectfully agree and would add that this is especially so where, as here, the bona fides of the developers of the land — the builders — are not assailed. The inspector found and the Secretary of State accepted that the builders intended to develop the appeal site in parallel to the nearby Huntingdon Estate which they were in the process of building and at first to construct a length of 250 ft of the estate road and to start on about seven houses in order to give continuity of work.

While the line of the road, pegged out to a substantial extent, may have been the subject of an order of variation by the district council before it was completed, the pegging out was, in my opinion, upon H
the whole of the accepted facts, an unequivocal manifestation of the intention of the builders to begin development within the permitted time and, weather permitting, to proceed with it from that time henceforward until development was complete.

I suspect that the understandable opposition of some, at least, of the villagers of Cradley and Westfields, who must have thought time was about to save them from an unwanted intrusion of new houses into their rural neighbourhood, has driven the district council to seize upon every legitimate tactic and argument to thwart the builders. By doing so they may be obliged not only to permit the development to go on but also to pay quite heavy damages to the builders. Doubtless they must view these twin prospects with dismay. J
Nevertheless, I feel bound to say that I see no alternative to their having to face up to them, since I am satisfied that the Secretary of State was fully entitled to accept the inspector's conclusion about the effect of pegging out the road and other relevant acts done by the builders.

We heard a number of submissions about what activity could be, in any given circumstance, properly described as the laying out of a road. There is apparently no legal authority upon the matter. We were not provided with any assistance from any other quarter to decide in what circumstance the process of laying out could be said to have begun.

The inspector's conclusions upon the matter, accepted by the K
Secretary of State, were as follows:

> On the other hand I consider that the marking out of the line and width of the first 250ft length of the estate road with pegs, placed accurately by measurements taken from the site boundaries, so that the road could be constructed with machines guided by these pegs, did amount to an "operation" in the course of laying out part of a road as defined in paragraph (d) of section 43(2). Therefore it was a specified operation. "Laying out" does not appear to be defined in the 1971 Act, but its meaning to me is the marking out on the ground of the road in accordance with an approved plan; the *Concise Oxford Dictionary*, 6th ed, supports this interpretation. Laying out was a necessary preliminary to the construction of the road and for the first 250ft the construction followed in April shortly after the expiry date. As L
> well as completing the laying out of the first 250ft, pegs were also placed to mark considerable stretches of the remaining estate roads before March 28 1979.

I can see no legal or other reason for doubting the soundness of that reasoning and none for questioning the decision which led from it.

There is neither error of law nor principle of construction detectable there. Accordingly, it seems to me that the determination of whether or not what had been done on the land in the circumstances of this case amounted to the beginning of a specified operation — the laying out of a road — was pre-eminently a matter of fact and degree from the Secretary of State. M

In his decision letter he appears to have made that determination, after having correctly applied the provisions of section 43(2)(d) to the facts found by the inspector. Accordingly, I deem it unnecessary for the matter to be referred back to him for further consideration. I would simply dismiss this appeal.

The appeal was dismissed with costs awarded to both respondents. Leave was given to appeal to the House of Lords.

LANDS TRIBUNAL
COMPENSATION

A GOSS v PADDINGTON CHURCHES HOUSING ASSOCIATION LTD

(REF/24/1981)

(W H REES FRICS)

December 14 1981

Estates Gazette January 30 1982

(1981) 261 EG 373

Disturbance claim by tenant under section 37 of 1973 Act for fitting and adapting new carpet on displacement — Floor area of new premises almost twice that of old — Tribunal rejects claim for cost
B **of fitting new carpet to whole area plus cost of adapting it to additional area, ie half cost of carpet and underlay — Cost of new carpet held not a natural and direct consequence of displacement and claimant got value for money — Approach in *Cole v Southwark LBC* followed — Correct basis either cost of lifting, removing and relaying old carpet or loss on forced sale, whichever is the less — Only in exceptional circumstances (not present here) can loss be determined by reference to cost of new carpet**

The claimant, N J B Goss, appeared in person; John Thompson (instructed by the solicitor for Paddington Churches Housing Association Ltd) for the compensating authority.

C Giving his decision, Mr Rees said: There is only one item in dispute in this reference by the claimant under section 37 of the Land Compensation Act 1973 as amended by Schedule 13, para 39, to the Housing Act 1974: it relates to a carpet. The basis of the amount which the claimant is entitled to receive is set out in section 38(1) and is as follows:

> The amount of a disturbance payment shall be equal to — (a) the reasonable expenses of the person entitled to the payment in removing from the land from which he is displaced.

Statement of facts

No evidence was called, but there is an agreed statement of facts.
D Slightly amended, this was as follows:

1. PCHA Ltd is a registered housing association within the meaning of the Housing Act 1974. It is a charity which buys houses in disrepair. Tenants are then decanted to other property in its stock while it carries out improvements to the houses with Government grants. "Spare" accommodation is let to persons in necessitous circumstances.
2. In 1974 PCHA purchased 25 St Stephen's Gardens, London W2. Mr Goss was the lawful tenant of the top-floor flat at the time of purchase.
3. In 1976 Mr Goss was permanently displaced from 25 St Stephen's Gardens by PCHA Ltd to enable it to improve the building. He was rehoused in Flat 1, 152 Westbourne Grove, London W11.
4. The following expenditure was incurred by Mr Goss when moving in:

E

Fitting of carpet	fitting cost	£55.95	
Cost of securing		4.50	
Tape		10.00	
Double Naplock		1.75	
Single Naplock		99	
			£ 73.19
Carpet			256.06
Underlay			9.60
			£338.85

5. The top flat, 25 St Stephen's Gardens, was carpeted throughout, except for the linoed bathroom and kitchen. F
6. Flat 1, 152 Westbourne Grove is a maisonette comprising the entire ground floor and basement, and is approximately twice the area of the top flat at 25 St Stephen's Gardens.
7. It is the general practice for tenants to buy new carpets on being decanted to a new flat.
8. Mr Goss is entitled to a disturbance payment under section 37(1)(d) of the Land Compensation Act 1973.
9. The following items of expenditure incurred by Mr Goss have been agreed as being reasonable expenses in removing from 25 St Stephen's Gardens:

Removal costs	£ 30.00	G
Cost of notifying changes of address: and communications with PCHA	3.00	
Reconnection of telephone	42.12	
Replacement cooker, secondhand	40.00	
	£115.12	

This is together with interest from the date of displacement (9.2.76) until the date of payment at the rate for the time being prescribed under section 32 of the Land Compensation Act 1961.

The issues in dispute were agreed to be as follows:

Mr Goss claims as part of the reasonable expenses in removing from the land the cost of: H

Carpet fitting	£ 73.19
The cost of adapting carpet to additional area, ie half the cost of the carpet and underlay (£256.06 + £9.60) ÷ 2	132.83
	£206.02

PCHA Ltd will allow only half the fitting cost: £36.60.

The claim under the sub-heading being admitted, the matter is really a question of "How should the expense be measured?"

I was not told the area of the flats at either no 25 or no 152 (see J para 6 above); in order to put the claimant's argument clearly I am going to take the floor area of no 25 as 40m² and that of no 152 as 79m². When he left no 25 the claimant did not take the carpeting with him: I was not told what happened to it. He took the opportunity afforded by the move to buy a new carpet and to lay it over the whole of the 79m² in no 152. He claimed that his reasonable expense in so doing was the cost of fitting the new carpet to the whole of the 79m² plus "the cost of adapting carpet to additional area", ie the cost of buying new carpet and underlay for 39(79 − 40)m². As already noted, the compensating authority offered him only half the cost of fitting.

The claimant referred me to *Glasgow Corporation v Anderson* K 1976 SLT 225 in which the provisions in section 35(1) of the Land Compensation (Scotland) Act 1973 were considered on an appeal by way of case stated by the Lands Tribunal for Scotland. Section 35 of the Scottish Act corresponds to section 38 of the Act which applies to England and Wales. He also referred me to three decisions of the Scottish tribunal determining amounts payable under section 35 and to two decisions of this tribunal concerning compensation under section 38. In addition he referred to 11 decisions of this tribunal, one of the Lands Tribunal for Scotland and one of the

Lands Tribunal for Northern Ireland: these 13 decisions all concerned claims made in respect of carpets under the general umbrella of "disturbance" where an interest in land had been acquired, ie the more usual disturbance cases. I do not propose to list these, as many are of no assistance. However, included in the decisions were: *Bryce v Motherwell District Council* [1980] RVR 282; *Succamore v London Borough of Newham*, a decision of this tribunal (R C Walmsley FRICS) (1977) 245 EG 404; *Nolan v Sheffield Metropolitan District Council*, a decision of this tribunal (E C Strathon FRICS) (1979) 251 EG 1179; *McGonigal v Department of the Environment for Northern Ireland* [1976] RVR 56; *Anderson v Corporation of the City of Glasgow*, a decision of the Lands Tribunal for Scotland (W Hall FRICS) [1974] RVR 398; *Cole v Southwark London Borough* (1979) 251 EG 477; *Appleby & Ireland Ltd v Hampshire County Council*, a decision of this tribunal (R C Walmsley FRICS) (1978) 247 EG 1183; 248 EG 54, 143, 235, 326.

Claimant's argument

The claimant based his argument mainly on the decision in *Bryce*. The relevant paragraph is on p 283 and is as follows:

> *Item 9. 9 Venetian Blinds. £120*
> Mr Bryce said that he had to buy new blinds because those in the house in Shotts were a different size. The new blinds cost £120 in 1976. The older blinds had been bought in 1974 for £108 and he had sold them for £20. In his opinion, the cost of adapting them for the windows of the Newmains house would have been as much as the cost of buying new ones. Mr Bryce produced a copy estimate of £120 for 1976 from the Venetian Blind Centre, Wishaw. Mr Tarelli resisted the claim for the cost of new blinds and said that if a loss were proved it should be based on the value of the older blinds less any price realised on the forced sale of the blinds.
>
> A schedule of agreed window sizes was produced and from this it was clear that the windows in the house at Shotts were about one foot less in height than the windows at Newmains. Five of the windows at Newmains were, however, the same width as four of the windows at Shotts and it may be that some adaptation by adding slats could have minimised the loss. The Newmains house has an extra bedroom with two windows and Mr Bryce has the benefit of the extra accommodation. We allow a contribution of £60 for the blinds.

The claimant said that the cost of buying and laying new carpet for the extra 39 m² in no 152 was directly comparable with the cost of adding the slats to the Venetian blinds.

Counsel for the respondent compensating authority pointed to the claim made in *Bryce* with regard to new curtains. The relevant paragraph is on p 284 and is as follows:

> *Items 10 and 17. Four pairs of new curtains at £32 and loss of use of curtains £180.*
> The claimant said that his wife changed curtains frequently. The curtains for Shotts were too short for Newmains. Accordingly the older curtains could not be reused. Mr Tarelli, who had personal experience of adapting curtains in three removals in five years, considered the problem of the curtains in some detail. He had examined the Shotts' curtains and measured them. His conclusion was that 16 pairs of curtains out of the 21 pairs brought from Shotts could be altered and taking into account the £2 claimed for altering two pairs (item 5 of the claim) an allowance of £16 might be reasonable for any alterations necessary. So far as the remaining five pairs were concerned, he had no information given to him about price or age. Doing the best he could he estimated the loss in value at £20. This gave a total of £36 for these items.
>
> We found the claimant's evidence on the curtains vague and were impressed by the detail which Mr Tarelli had gone into in investigating this part of the claim. We accept Mr Tarelli's allowance of £36.

The claimant further said that there was no pattern in the decisions of the tribunals to which he referred, but I think that the different approaches which have been adopted go to illustrate the words of Lord Cameron in *Glasgow Corporation v Anderson* at p 230:

> The precise extent of the permissible items or elements of expense, as well as the assessment of their reasonableness in amount, would necessarily depend on the particular circumstances of each case.

Authority's case

Counsel for the respondent authority pointed to the decision of the Lands Tribunal for Scotland in *Anderson*, in particular to the following [at p 400]:

> Section 35 deals with the reasonable expenses in removing from the house and it does seem that along with the principles set out in various cases about disturbance, there are three tests to be applied in deciding what the claimant is due. Firstly, what has been spent in removing? Secondly, does that expense flow directly from the displacement? Thirdly, was the expenditure reasonable?

He submitted that the proper basis was the cost of the adapting and fitting the old carpet if it were taken to the new premises: that was a reasonable expense and flowed directly from the displacement. If a claimant were compelled to leave carpets behind, he was entitled to compensation based on the loss on forced sale. The claimant put his claim here under the sub-heading of "adaptation"; however, this was not a case where the old carpet was adapted. Adding an almost equal area to the flat (ie 39m² added to 40m² = 79m²) was not adapting the old carpet: it was not the same as adding slats to a blind. *McGonigal* was a special case where there was a polished staircase in the premises acquired which did not have a staircarpet: it was unnecessary. It was not possible to polish the staircase in the new house and this was the reason why the cost of buying a new carpet was allowed there.

Decision

I reject Mr Goss's basis of claim. I do not consider that the cost of the new carpet in no 152 was a natural and direct consequence of the displacement and in that respect I follow *Nolan*. I also consider that the claimant got value for money as in *Succamore* and in *Cole*.

I asked the claimant what the position would be if I were to come to the conclusion that neither his approach nor that of the compensating authority was correct. He asked me, in that event, to make an interim decision: counsel for the compensating authority agreed.

The right approach here, I think, is to adopt that approved in *Cole*. The relevant part of that decision is set out on p 479 and is as follows:

> As to the remaining items, my understanding (on the vouchers submitted and from the claimant's answers in cross-examination) is that these relate to expenditure on 8 Valmar Road in the following ways:
>
> | | | |
> |---|---|---|
> | Items 43 to 47 and part 35 | Carpets (including laying) | £ 784.25 |
> | Item 46A | Curtains (including hanging) | 1,089.09 |
> | Items 96 and 97 | Fitted furniture | 139.95 |
> | | | £2,013.29 |
>
> Mr Berry's evidence for the council was:
> "I accept that this claim is valid under the heading of Disturbance Costs in so far as it relates to adaptation of curtains and carpets or loss on forced sale of curtains and carpets, whichever is the lesser amount. The bills, receipts and estimates provided are for new furnishings and it is argued do not constitute a proper basis for an assessment.
>
> It is unfortunate that the last time an inspection was made of 21 Kerfield Crescent by a member of the Borough Valuer's staff was in February 1975 and it is felt that it would be unfair to the claimant to assess the compensation payable for the adaptation of curtains and carpets on the basis of that inspection. It is felt that it would be more equitable to base the valuation on comparable settlements achieved within the area for similar property with furnishings in fairly good order. Appendix 1 lists three such comparables and on this basis for the two properties I consider that the appropriate award is £150."
>
> I agree that the claimant's expenditure on furnishings for his new house cannot be regarded as a "loss", for there is no reason to doubt that he got value for any money thus expended. The "loss" that he has sustained is in respect of the furnishings that were in the houses acquired.

The basis to be adopted in the instant case is therefore the cost of lifting the carpet which was in no 25, removing it to, and laying it in, no 152, or the loss on forced sale of the carpet in no 25, whichever is the less. Loss on forced sale would be the difference between the value of the carpet to the claimant *in situ* less the amount realised on sale. The same basis was approved in *Appleby & Ireland*. It may be that had the carpet in no 25 been taken up and laid in no 152, its value may have been depreciated: if that can be shown, that depreciation may be reflected in the amount payable (see item (i) in *McGonigal*), but only in exceptional circumstances (of which the cost of new carpet for the staircase, item (m) in *McGonigal* is a good example), in such a case as this, can the loss be determined by

A reference to the cost of purchasing a new item. There are no such exceptional circumstances here.

I have insufficient evidence to determine the amount of the reasonable expense in respect of the carpet here on the basis indicated: this is therefore an interim decision, but I hope that these considerations will enable the parties to come to an agreement, as was indicated to me at the hearing. There will be liberty to apply by January 31 1982; if no application has been received by then either in respect of the carpet or as to the costs of the reference, this decision is to be regarded as final.

B

ANDREW v COD BECK INTERNAL DRAINAGE BOARD

(REF/26/1981)

(WM HALL FRICS)

January 12 1982

Estates Gazette February 20 1982
(1982) 261 EG 683

Land drainage works to stell or watercourse near Thirsk — Fence erected after works not of standard of fence removed — Need of
C **strong fence to turn away sheep or lambs — Loss occurred five years before reference to tribunal — Tribunal disallows credits against costs — Advantage of having new material (pig netting) now, rather than having material by now some five years old, offset by disadvantage of having substandard fence for five years — Delay in referring dispute to tribunal not unilateral, as authority could also have initiated reference — Claim for £257.55 awarded in full**

C M Wilson, solicitor, of Latimer, Hinks, Marsham & Little, of Darlington, appeared for the claimant; J Jenyns, solicitor, of Arthur W Walker & Hiley, of Thirsk, for the compensating authority.

D Giving his decision, MR HALL said: This is a reference to determine the compensation under the Land Drainage Act 1976.

The Cod Beck Internal Drainage Board, in terms of their powers under the Land Drainage Act 1961, entered upon land belonging to Mr W A Andrew, Broads House Farm, Borrowby, near Thirsk, to improve and clear the watercourse known as Crosby Stell. It was intended to excavate and remove spoil from the east bank, but Mr Andrew proposed that, as the west bank was an incline, the work should be carried out from that side. Along the west bank Mr Andrew had erected, about two years previously, a stock-proof fence consisting of 10 railway sleeper straining posts, one at every change in direction of the stell, with heavy-duty tanalised stakes
E every 3 yds. Heavy-gauge pig wire was stapled to the posts with a top strand of barbed wire. The field on the west of the stell was used for grazing sheep and because of the danger of sheep or lambs falling into the stell a strong fence was needed to turn away the sheep or lambs from it.

Coxon Brothers were appointed the contractors and commenced work in or about October 1975. The specification for fencing was as follows:

> The contractor shall provide all materials necessary for the erection of fencing where grass land adjoins the watercourse and where fencing is in existence alongside the watercourse. The contractor shall erect the fencing in as straight a line as possible and keep to a minimum changes in direction
> F of the line of fencing and at each necessary change of direction a strainer post or posts shall be driven.
> The contractor shall supply tanalised or celcured larch posts 1.676 m long × 75 mm diameter. The posts shall be driven firmly into the ground so that the top of each post is 0.91 m above ground level. Posts shall be driven at intervals of 4 m and not nearer than 0.50 m from the top of the slope of the bank of the watercourse.
> Two strands of barbed wire shall be attached to the fence posts, the top strand shall not be more than 50 mm below the top of the post and the lower strand 380 mm-450 mm below the top strand of wire.

This specification is the normal one used by the board on schemes which are the subject of a Ministry of Agriculture grant, but it is agreed by the board that a fence of that type would not turn lambs. Neither the board nor the contractor discussed with Mr Andrew the G
type of fence to be erected.

The contractor removed the existing fence and laid aside the sleepers and the posts with the pig netting attached. Mr Coxon said in evidence that Mr Andrew agreed to dismantle the netting, but Mr Andrew disputed that.

A fence to the above ministry specification was erected; Mr Andrew objected to that fence after it was erected, but the contractor said that he could only carry out his instructions. Accordingly, Mr Andrew formally wrote to the board on January 26 1976 complaining that the fence erected was not of the standard of the fence removed. On May 3 1976, the board replied saying that "they are not prepared to carry out further work on the fencing which has H
been inspected by the ministry's regional engineer and a board member, and found to be erected to the required specification". Mr Andrew then took the matter up with the board's vice-chairman.

The board apparently offered to replace the pig wire, but this was rejected by Mr Andrew. Since then various negotiations have taken place. Correspondence, agreed as factual by the parties, was lodged and from that it was pointed out, on behalf of the board, that they offered the following amounts: in 1978 — £80, in 1980 — £120 and in July 1981 — £150.

In August 1977, Mr Andrew sent the board the following valuation prepared by R E A Newton BSc ARICS, of Northallerton Auctions Ltd, setting out the cost of replacing the fence so far as J
practicable to its original state:

1. Replace 10 railway sleeper strainer posts	£45
2. Five rolls of pig netting	80
3. Labour to erecting same	48
4. Stones to lift from dyke side where drinking spot abandoned	20
5. Time spent erecting temporary stock-proof fence	12
	205
Surveyor's fee	21.28
	£226.28

K

The reason for items (1) and (5) was explained by Mr Andrew as follows. He took the wire off the removal posts and fixed it temporarily to the posts which had been erected by the contractor, but the wire had deteriorated and the fence was not strong enough to be stockproof. The sleepers which had been left lying on the ground were, he said, washed away when in 1977 the land was flooded to a depth of about 3 or 4 ft by the Cod Beck overflowing. After the flood he was unable to trace the sleepers which he owned.

Section 17(5) of the Land Drainage Act 1976 states:

> Where injury is sustained by any person by reason of the exercise by a drainage authority of any of their powers under this section, the authority L
> shall be liable to make full compensation to the injured person; and in cases of dispute the amount of compensation shall be determined by the Lands Tribunal.

As the amount is in dispute, it has been referred to the tribunal.

The claim now presented to the tribunal has the same items as set out in the August 1977 claim, but the figures are adjusted to reflect current costs thus:

1. Sleepers as before	£45
2. Pig netting	74.75
3. Labour: four man days at £24.45	97.80
4. Removing stones	20
5. Erecting temporary netting	20
	£257.55

M

The respondent authority accepts the actual figures put forward if the tribunal finds that the items of claim were established. It was submitted, however, that if the tribunal did so find, there should be deducted from the compensation a sum of £55, being the board's opinion of the secondhand value at 1975 of 10 railway sleepers, the pig wire and 75 tanalised posts taken down and laid aside for the farmer to retain. In considering the items of claim, I was particularly asked to determine whether the £74.75 for pig netting should be

A allowed in view of the fact that the netting from the previous fence had been used. The board disputed that there were stones or roots in the spoil removed from the stell and therefore item 4 should not be allowed. The figure which the board asked the tribunal to determine was £90 calculated as follows:

Cost of sleepers	£45	
Labour erecting them	97.80	
	£142.80	say £145
Less credit for materials		£55
		£90

B

Along with the parties I inspected the Crosby Stell and the new fence between it and Mr Andrew's land.

This dispute appears to arise from a lack of communication at the beginning. The specification was drawn up on the assumption that the stell would be cleared from the east bank, where such a fence would have been acceptable. Mr Andrew proposed that it should be done from the west bank, which was accepted by the board. Mr Coxon had tendered for the specification fence and cannot be blamed for adhering to that. Mr Robertson, the engineer from the water authority who advised the board, said that the specified fence was used in all grant-aided schemes but if the occupier was not C satisfied negotiations should take place before commencement of the work. In this case, this was not done as Mr Andrew had not raised the matter until the fence was erected.

Having accepted that the work should be done from the west bank, it is not disputed by the board that they should pay full compensation.

I find that the claimant is entitled to have a fence with railway sleeper straining posts as he had before. The claimant, by his claim, implies that to have a fence of sufficient strength to carry heavy-gauge pig netting to turn sheep and lambs, it can be achieved by replacing the present straining posts with 10 sleepers and stapling new pig netting to the sleepers and to the posts erected by the D board. I have inspected the existing netting and consider it inadequate; in fact in one part new netting has been put up over the older netting. The existing sleepers have disappeared and I have no contrary evidence to refute Mr Andrew's explanation of what happened to them. I therefore accept the first three items of the claim totalling £217.55.

So far as the removal of stones is concerned, I was shown where the drinking space had been and I have no reason to doubt that it is likely that some stones may have been included in the spoil and had to be removed. I also accept that the attachment of the netting removed from the posts taken out may have been done to try to mitigate further loss. I therefore accept the £20 claim for each of E these items.

This leaves the question of credits against the cost. Mr Coxon said that normally when a fence is taken down the materials are either buried or removed to a dump or if they are good they are left for the farmer. I note that the specification item is "dismantle and clear existing fencing". There was no evidence to indicate that the farmer would have to pay for any materials left. I have accepted that the sleepers were swept away and that the pig netting removed from the previous posts is inadequate. There was no information as to what happened to the posts which were taken out. Mr Coxon placed a secondhand value of about £11 on them, being one half of the cost of new posts in 1975.

F Secondly, it was submitted for the respondents that the costs were being calculated at 1981 figures whereas the loss was in 1975. The claimant, therefore, it was said, was obtaining an advantage in getting new netting now rather than have netting which would now have been five or six years old. A discount should be made for this advantage.

Thirdly, over five years had elapsed before the claimant had referred the matter to the Lands Tribunal, which had escalated the claim and this should be taken into account.

In all the circumstances of this case, I find that no credit should be made against the accepted claim. The posts would probably have been left in any event. It may be necessary to insert posts to reduce the 4 m interval to firm the netting. The advantage of having new G netting is difficult to quantify and no evidence was given by the authority as to the assessment of this. In my opinion, any advantage has been offset by the disadvantage of Mr Andrew having a sub-standard fence for the period of five years. The delay in bringing the matter to the tribunal is not a unilateral one, as the board could also have initiated the reference. Mr Andrew has tried by other means to persuade the board to provide a fence to the standard he had and, having regard to my award, was justified in eventually having the dispute determined by the tribunal.

I therefore determine the compensation at £257.55 to which shall be added the surveyor's fee based on scale 5(a) of the Royal Institution of Chartered Surveyor's scale of charges and proper H legal costs incurred from October 1975 to the date of the notice of this reference.

The respondent authority will pay the claimant his costs of this reference in the agreed fixed sum of £180 plus value added tax.

PROVINCIAL BUILDING SOCIETY v HAMMERSMITH AND FULHAM LONDON BOROUGH

(REF/191/1980)

(W H REES FRICS) J

January 8 1982

Estates Gazette March 6 1982

(1982) 261 EG 895

Building society's claim as mortgagees in respect of freehold interest in London NW10 house — Effect of control order under s 73 of Housing Act 1964 — Whether compensation determinable by reference to vacant possession or subject to tenancies — Nearly whole of accommodation let to five tenants contrary to covenant against letting — County court order giving society possession not enforced — Control order made 10 months before notice to treat served — Society entitled to obtain, and sell with, vacant possession K before control order — Held that control order was a temporary matter and society re-entitled to vacant possession when order ceased to operate — Valuation at date of entry, when control order had ceased — Cessation of control order, not scheme under CPO, had increased value of interest, so that "Pointe Gourde" principle did not apply — Value to be with vacant possession

Michael Nield (instructed by Anthony J E Kidd) for the claimant; Nicholas Huskinson (instructed by the solicitor for Hammersmith and Fulham London Borough) for the acquiring authority.

Giving his decision, MR REES said: This reference was made by the claimant, a building society, to determine the compensation to be paid in respect of the freehold interest in a house in London L NW10. The claimant contended that the amount due to it, which is in the order of £10,000 (it could not be determined exactly at the date of the hearing before me, but it comprises capital, interest and costs), should be determined by reference to the value of the freehold interest with vacant possession, which was agreed between the parties to be £12,500. The acquiring authority contended that claimant's compensation should be determined by reference to the value of the freehold interest subject to tenancies, which was agreed between the parties to be £8,000.

Agreed facts

No evidence was called but there was a statement of agreed facts. M The relevant part of this (slightly amended) was as follows:

1. 35 Holberton Gardens, London NW10, is a two-storey terraced house.
2. The freehold owners are Basil Stephenson and Ida Stephenson.
3. The property is mortgaged to Provincial Building Society, Provincial House, Bradford, West Yorkshire.
4. A second mortgage on the property is held by Roshclose Ltd of 4 Woodstead Grove, Edgware, Middlesex.
5. The acquiring authority is the London Borough of Hammersmith and Fulham ("the council").
6. A direction order under section 19 of the Housing Act 1961 was made on February 6 1978 in respect of the property requiring the reduction of overcrowding in a house in multiple occupation.

7. On February 15 1978 the council made a control order under the provisions of the Housing Act 1964 as amended. No appeal was made against it.
8. The Hammersmith (35 Holberton Gardens NW10) Compulsory Purchase Order was made by the council on March 1 1978. This order was confirmed without modification by the Secretary of State for the Environment on September 15 1978.
9. A notice to treat and notice of entry were served on December 12 1978. This notice of entry stated that the council would take possession of the property on January 9 1979.

Further, it was not disputed that the subject house was vacant when the mortgage was taken (January 21 1976), that there was a covenant against subletting and that an order was made in the Willesden County Court on September 12 1977 that Mr and Mrs Stephenson were to give up possession of the subject house to the claimant on October 10 1977. That order was not enforced and the claimant has never taken possession. There is also no dispute that, contrary to the covenant against letting, some time after January 21 1976 and without the claimant's consent Mr and Mrs Stephenson let nearly the whole of the accommodation to five tenants, but in view of the agreement as to values I do not need to record details of the tenancies. The tenants were still in occupation in December 1978 and January 1979, but it appeared that Mr and Mrs Stephenson cannot be traced. The acquiring authority took possession on January 8 1979. It was agreed before me that in January 1978 (ie before the control order was made) the claimant was entitled to obtain vacant possession against the tenants and to sell the property with vacant possession. If the acquiring authority had made the compulsory purchase order at that time, it was conceded that the basis of compensation should be the vacant possession value.

The relevant parts of the statutory provisions to which I was referred were as follows:

Housing Act 1957

Section 98. Where a local authority are authorised to purchase compulsorily any house to be used for housing purposes under this Part of this Act and have acquired the right to enter on and take possession of the house by virtue of having served a notice under paragraph 3 of the Second Schedule to the Acquisition of Land (Authorisation Procedure) Act, 1946, the local authority may, instead of exercising that right by taking actual possession of the house, proceed by serving notice on any person then in occupation of the house or any part thereof authorising him to continue in occupation upon terms specified in the notice, or on such other terms as may be agreed; and accordingly where the authority proceed in the manner authorised by this section, —

(a) the like consequences shall then ensue, with respect to the determination of the rights and liabilities of any person arising out of any interest of his in the house or any part thereof, as would have ensued if the authority had taken actual possession on the date of the notice, and the authority may deal with the premises in all respects as if they had done so; and
(b) . . . .

Housing Act 1964

Section 73. (1) A local authority may make an order under this section (in this Part of this Act referred to as a "control order") as respects a house in their district which is occupied by persons who do not form a single household —

(a) if an order under section 12 of the Act of 1961 is in force as respects the house, or a notice has been served or direction given as respects the house under section 14, section 15 or section 19 of that Act, or
(b) if it appears to the local authority that the state or condition of the house is such as to call for the taking of any such action as may be taken under any of those sections,

and if it appears to the local authority that the living conditions in the house are such that it is necessary to make the control order in order to protect the safety, welfare or health of persons living in the house.
(2) . . . .
(3) As soon as practicable after making a control order the local authority shall —

(a) post a copy of the order, together with a notice as described in subsection (4) of this section, in some position in the house where it is accessible to those living in the house, and
(b) serve a copy of the order, together with such a notice, on every person who, to the knowledge of the local authority, —
    (i) was, immediately before the coming into force of the control order, a person managing, or having control of, the house within the meaning of this Part of this Act, or
    (ii) is an owner or lessee or mortgagee of the house.

(4) and (5) . . . .

Section 74 (1) While a control order is in force the local authority shall, subject to the provisions of the next following section relating to persons who are occupying parts of the house, have the right to possession of the premises and the right to do (and authorise others to do) in relation to the premises anything which any person having an estate or interest in the premises would, but for the making of the control order, be entitled to do without incurring any liability to any such person except as expressly provided by this Part of this Act.
(2) to (5) . . . .

Section 75 (1) This section applies to any person —

(a) who, at the time when a control order comes into force, is occupying any part of the house, and
(b) who at that time is not a person who has an estate or interest in the whole of the house.

(2) The last foregoing section shall not affect the rights or liabilities of a person to whom this section applies under any lease, licence or agreement, whether in writing or not, under which that person is occupying any part of the house at the time when the control order comes into force; and —

(a) any such lease, licence or agreement shall, while the control order is in force, have effect as if the local authority were substituted in the lease, licence or agreement for any party to the lease, licence or agreement who has an estate or interest in the house and who is not a person to whom this section applies, and
(b) any such lease shall continue to have effect as near as may be as a lease notwithstanding that the rights of the local authority, as substituted for the lessor, do not amount to an estate in law in the premises.

(3) to (6) . . . .

Section 82 (1) At any time after the making of a control order, but not later than the expiration of a period of six weeks from the date on which a copy of the relevant scheme is served in accordance with section 79 (1) of this Act, any person having an estate or interest in the house, or, subject to the following provisions of this section, any other person, may appeal to the county court against the control order.
(2) to (5) . . . .

Section 86 (1) A control order shall cease to have effect at the expiration of a period of five years beginning with the date on which it came into force.
(2) to (10) . . . .

Section 87 (1) Schedule 4 to this Act (which sets out the consequences of a control order ceasing to have effect) shall have effect for the purposes of this Part of this Act.
(2) and (3) . . . .

Schedule 4. 1. — (1) On and after the date on which the control order ceases to have effect any lease, licence or agreement in which the local authority were substituted for any other party by virtue of section 75 of this Act shall have effect as if for the local authority there were substituted in the lease, licence or agreement the original party or his successor in title.
(2) and (3) . . . .
(2) to (4) . . . .

Housing Act 1969

Section 63 (1) . . . .
(2) . . . .
(3) Where the compulsory purchase order is confirmed by the Minister and the local authority —

(a) enter into a contract to purchase the house, or
(b) in pursuance of a notice served under section 11 of the Compulsory Purchase Act 1965, either enter and take possession of the house or serve a notice under section 98 of the Act of 1957 authorising a person in occupation of the house or part of the house to continue in occupation,

the control order shall cease to have effect on the date when the contract is made or the notice under section 11 is served.

Hereinafter the above are referred to by their section or schedule number only.

I was referred to the following: *Rugby Joint Water Board v Shaw-Fox* [1973] AC 202; *Bradford Property Trust Ltd v Hertfordshire County Council* (1973) 229 EG 1226; *Midland Bank Trust Co Ltd (Executors) v Lewisham London Borough* (1975) 235 EG 59; *Metcalfe v Basildon Development Corporation* (1974) 230 EG

A 1140; *Re Morgan and London & North Western Railway Co* [1896] 2 QB 469; *Banham v Hackney London Borough* (1971) 217 EG 202; *Buckingham Street Investment Ltd v Greater London Council* (1976) 237 EG 503; *West Midland Baptist (Trust) Association Incorporated v Birmingham Corporation* (1967) 19 P & CR 9; *Pointe Gourde Quarrying & Transport Co v Sub-Intendent of Crown Lands* [1947] AC 565.

Briefly, the argument for the claimant was that nothing happened to deprive the claimant of the vacant possession value: the authority maintained that it did. The control order is at the heart of the matter.

B

Submission for the claimant

Counsel for the claimant submitted that the authority took over from the owners, but he agreed that the owners were not entitled to vacant possession. Under section 75 (2), when the control order took effect (February 15 1978) the authority stepped into the owners' shoes and so took over their rights and liabilities. Under section 63(3) the control order ceased with effect from December 12 1978 (the date of the notice to treat). Under Schedule 4 "the original party" were the owners, so the owners were substituted for the authority with effect from December 12 1978 and the owners C could not get vacant possession but, the control order having come and gone, the position was the same, ie the claimant could once again obtain vacant possession.

Submission for the authority

First, counsel for the respondent authority submitted that the value of the claimant's security was lowered when the control order was made and while it was in force. When on December 12 1978 (the date of the notice to treat) the control order ceased to have effect, the claimant did not revert to the advantageous position (in that it could obtain vacant possession) in which it was before the D control order was made. During the period from February 15 1978 (the date on which the control order was made) until December 12 1978 (when it ceased) the tenants in occupation were secure against all comers. Under Schedule 4 the owners were substituted for the authority and so the tenants continued to be protected against all persons: thus the claimant was still bound by the tenancies.

Second (if I were to hold against him on the *first*), on December 12 1978 (the date of the notice to treat) the claimant's interest was a charge on a property subject to a control order: that is the date on which the interest is determined (see *Rugby*), but he agreed that the *Bradford* and the *Midland Bank* decisions were against him. However, *Bradford* was, he submitted, neither clear nor consistent, and E having regard to *Midland Bank*, *Bradford* was no authority for the proposition that the interest should be determined at the date of entry. If the interest changed after the notice to treat the basis of compensation did not alter, but he submitted that the conclusion in *Midland Bank* was contrary to *Bradford* and *Rugby*.

Third (alternative to *second*, if I were to hold against him on the *second*), the *Pointe Gourde* principle applied: the compulsory purchase order had increased the value of the interest.

Reply for claimant

As to the *first*, the control order was a temporary matter only (see section 86): it came and went and left the claimant as it was in F January 1978. The claimant did not appeal against it and there were no provisions in the legislation relevant to control orders for those suffering to claim compensation.

As to the *second*, counsel for the authority would have to rely on there being a *scintilla temporis* between (1) the notice to treat, (2) the notice of entry, and (3) the operation of section 98: that was an artificiality. He relied on *Morgan* (at p 469) and *Rugby* (at p 216) and submitted that the three worked simultaneously and thus the property was freed from the control order.

As to the *third* submission, the termination of the control order was due not to the scheme but to section 69 and so the *Pointe Gourde* principle did not apply.

Decision G

It seems to me that the control order was a temporary matter (see section 86) and that when it ceased to operate the occupying tenants and the claimant were put back into the position in which they were in January 1978.

On the *second* submission I would follow the conclusion in *Midland Bank*, at p 272: "Our conclusion is that valuation as at the time of entry can only be a sensible exercise or principle if the interests are to be taken as they exist on that date". The date of entry here was January 8 1979: by that date the control order had ceased.

On the *third*, I would agree with counsel for the claimant that it was not the scheme which caused the interest to increase in value but the cessation of the control order. H

As the exact amount due to the claimant cannot be quantified, this is an interim decision. For the purpose of this reference the value of the claimant's interest (that of the building society) is to be determined by reference to the value of the freehold interest in the subject premises with vacant possession, which was agreed to be £12,500. There will be liberty to apply by February 28 1982 both in respect of the determination of the actual amount due and in respect of the costs of this reference. If no application has been received by that date this decision will then become final.

J

LEEK v BIRMINGHAM CITY COUNCIL

(REF/61/1980)

(V G WELLINGS QC)

January 29 1982

Estates Gazette March 20 1982

(1982) 261 EG 1101

Freehold Victorian two-storey buildings in Birmingham acquired by general vesting declaration — Subject to 14-year lease with 1½ years unexpired — Last used as shop with offices above but unoccupied for three years and out of repair at relevant date — Claimant's contention that compensation should be difference between value of property out of repair and its value in repair — Standing property value agreed at £3,145 and value in good repair at £15,000 — "Misconceived" submission for authority that breaches of repairing covenants had been waived by acceptance of rent — Claimant relies on section 9 of 1961 Act (disregard of depreciation due to prospect of compulsory acquisition) in support of case, but tribunal says no need to do so because losses directly attributable to CPO and general vesting declaration — Alternative claim under rule (6) not sustainable because loss calculable only by reference to value of land — Authority's valuer's method of adding back to standing property value an allowance in respect of a "speculative overbid" adopted by Tribunal — But allowance increased from 40 per cent to 80 per cent of difference in order to reflect appropriately the value of tenant's covenants having regard to fact that tenant was a subsidiary of an international company of repute (Philips Industries) K L

A M Putsman (solicitor of Maurice Putsman & Co, of Birmingham) appeared for the claimant, Mrs Rita Leek; R Barker (solicitor with City of Birmingham) for the acquiring authority.

Giving his decision, MR WELLINGS said: This is a reference for the determination of the amount of compensation to which the claimant is entitled for the compulsory purchase of her freehold interest M in properties known as 99-105 (odd numbers) Winson Green Road, Birmingham 18, West Midlands. The properties were acquired under the Birmingham (Rotton Park) Compulsory Purchase Order 1974, which order was confirmed on July 15 1977. On February 24 1978, the acquiring authority made a general vesting declaration vesting the reference properties in itself on April 10 1978. On that date the acquiring authority also took possession of the properties.

At all material times the freehold in the reference properties was subject to a lease of all of them. It was made on September 30 1965 between H Plotnek (Investments) Ltd of the first part, Collis Radio

A Ltd of the second part and certain guarantors of the third part. The lease was a full repairing lease and was for a term of 14 years from September 29 1965 at the yearly rent of £1,475 (without rent review) payable quarterly plus an insurance rent.

The reference properties were Victorian two-storey buildings which originally were dwelling-houses. Their last use, prior to vesting in the acquiring authority, was, however, as to the ground floors, as a television rental shop and otherwise as offices. These uses as offices of nos 103 and 105 began before 1964 and the use of the other properties as offices was authorised under express planning permissions. The official planning zoning since 1960 was for shopping purposes. The total site area was 407 sq yds or thereabouts.

B At the date of vesting the properties were out of repair and at that date and at all times since 1975 the properties had been unoccupied, but until that date the rent continued to be paid.

Immediately before the general vesting declaration took effect the lease was vested in a company known as East Coast Personal Finance Ltd, a subsidiary of Philips Industries Ltd, the well-known international company. The lease, too, was acquired by general vesting declaration and no claim for compensation in respect of it was made and accordingly none was paid.

It was common ground that the effect of the general vesting declaration, in so far as it vested the reversion in the acquiring
C authority, was to vest the claimant's right to sue in respect of the breaches of the repairing covenants on the part of the tenant in the lease in the acquiring authority as from April 10 1978 and to divest the claimant of her cause of action in respect thereof: see section 141(1) of the Law of Property Act 1925; *Re King, Robinson v Gray* [1963] Ch 459; *Woodfall on the Law of Landlord and Tenant* 28th ed, vol 1, para 1-1733.

The parties were in agreement that the value of the properties in their actual state of repair on the date of vesting was £8,000. The claimant's case, however, was that the properties should not be valued in their existing state of repair but that regard should be had to the right to enforce the repairing covenants on the part of the
D tenant, which right had been acquired by the acquiring authority. It was a valuable right and the properties should be valued in the condition in which they would have been if they had been put into good repair. The parties agreed that, if that were the right basis on which compensation should be assessed, the appropriate valuation would be £15,000.

In support of the claimant's case, Mr Putsman, her solicitor, relied upon section 9 of the Land Compensation Act 1961. That section provides:

> No account shall be taken of any depreciation of the value of the relevant interest which is attributable to the fact that (whether by way of designation, allocation or other particulars contained in the current development plan,
> E or by any other means) an indication has been given that the relevant land is, or is likely, to be acquired by an authority possessing compulsory purchase powers.

The indication on which he relied as having that effect was the indication contained in the compulsory purchase order and he referred to the words "any other means" as being material for that purpose. The claimant was entitled to be compensated for the loss of her right to sue her tenant in respect of dilapidations and the compensation should be the difference in value between the value of the property out of repair and its value if in repair. In the alternative to his main argument, he claimed compensation for that
F difference under rule (6) of section 5 of the Act of 1961. He referred to the decision of the High Court in *Hibernian Property Co v Liverpool Corporation* [1973] 2 All ER 1117.

Mr Putsman called no evidence. During the course of the hearing a schedule of dilapidations, dated July 12 1978, which had been prepared for the claimant by a firm of chartered surveyors, Leeson Son & Hackett, of Birmingham, was produced and agreed by the parties. During the hearing the parties also agreed that the cost of complying with the schedule was £10,000.

Mr Barker, for the acquiring authority, called Colin Massey ARICS, principal valuer in the acquiring authority's estates office. Mr Massey has had much experience in valuation over many years.

G He produced a valuation as follows:

1. *Standing property value*		
Reversion to estimated full rental at expiry of lease — say	£2,500 pa excl	
Deduct — management @ 7½%	188 pa	
	2,312 pa	
YP in perpetuity deferred 1½ years @ 14%	5.88	£13,595
Deduct (1) Estimated minimum capital expenditure required at expiry of lease to command rack rental — say £10,000 PV of £1 in 6 months @ 8% 0.95 (H)	£9,500	
(2) Risk profit etc — say 10% of £9,500	£ 950	10,450
		3,145
2. *Speculative "overbid"* Allowance to reflect purchasers' right to receive rent for the remainder of the lease term and to reflect the acquisition of the former freeholders' right to action against the lessee for breach of covenants, forfeiture of lease and damages for breach of contract. (J)		
Allow — Say 40% of differential between estimated value of property in good condition at £15,000 and valuation of standing property at £3,145		4,742
		£7,887
		Say £8,000

It will be apparent from that part of the valuation headed "Speculative 'overbid' ", that Mr Massey, like Mr Putsman,
K thought it right to give some credit to the claimant for the loss of her right to enforce the repairing covenants in the lease and he did that by allowing 40% of the differential in the estimated value of the properties in good condition at £15,000 and the standing property value at £3,145 as calculated in the first part of his valuation.

Mr Massey at the time when he prepared his valuation did not know that the lease was, as at April 10 1978, vested in a subsidiary of Philips Industries Ltd.

In cross-examination of Mr Massey, Mr Putsman accepted paragraph 1 of Mr Massey's valuation; that is to say, he accepted all the figures down to and including the figure of £3,145 shown for
L the standing property value. He put the rest of the valuation in issue.

Mr Barker relied upon the valuation of Mr Massey but also submitted that the breaches of the repairing covenants in the lease had been waived by the claimant prior to April 10 1978 because she had accepted rent with knowledge of the breaches. That submission was misconceived, because breaches of a repairing covenant are continuing breaches and the landlord, after ceasing to accept rent, is entitled to insist on performance of the covenants: see *Woodfall* (*op. cit*), paragraph 1-1914. Moreover, any waiver would debar the right to forfeit the lease and not the right to sue for damages.

On the other hand, I do not accept Mr Putsman's submissions. I
M do not think that it is right to say that the loss suffered by the claimant of her right to enforce the repairing covenants in the lease is attributable to an indication of the kind referred to in section 9 of the Act of 1961. There is no need to rely upon that section, because both the loss of the properties in their dilapidated state and the loss of the right to enforce the repairing covenants are losses which are directly attributable to the compulsory purchase order and the general vesting declaration. The claimant's case cannot be put under rule (6) because the claim in respect of the loss of the right to enforce the repairing covenant is necessarily "directly based on the value of land". That is how Mr Putsman puts the claim and that is how he is bound to put it, because the claimant's

A true loss in this context is necessarily related to the damage to her reversion, which cannot be calculated except by reference to the value of land.

In these circumstances, the only course open to me is to apply the basis which Mr Massey has applied in his valuation; that is to say, to accept that the standing property value was £3,145 as shown in paragraph 1 of his valuation and to add back into the valuation an allowance in respect of an overbid of the kind referred to in paragraph 2 of that valuation.

However, it appears to me that an allowance greater than 40 per cent of the difference between the value (£15,000) of the properties in good repair and the standing property value (£3,145) should be
B made. In my view that allowance should be 80 per cent of that difference in order to reflect appropriately the value of the tenant's covenants having regard to the fact that the tenant was a subsidiary of an international company of repute. Mr Massey was, as I have said, unaware of that fact at the time when he prepared his valuation.

The difference between £15,000 and £3,145 is £11,855: 80 per cent of that sum is £9,484. When the latter amount is added to £3,145, the total becomes £12,629. I award that sum by way of compensation to the claimant together with her surveyor's fees (if any) under scale 5(a) of the professional scales of the Royal Institution of Chartered Surveyors.

C The acquiring authority will pay to the claimant her costs, such costs if not agreed to be taxed by the Registrar of the Lands Tribunal on scale 3 of the County Court Scales of Costs.

LEESE v STOKE-ON-TRENT CITY COUNCIL

(REF/28/1981)

(W H REES FRICS)

February 18 1982

Estates Gazette March 27 1982
D (1982) 261 EG 1197

Shop in Tunstall, Stoke-on-Trent, used for sale of general household stores — Freehold interest valued by tribunal at £5,325 — Claimant also had 3-day-a-week stall in local market where he sold only some items from the shop, being restricted by licence to selling only certain types of goods — Claim for disturbance — Tribunal's finding of permanent loss of some part of claimant's goodwill from both outlets — No goodwill transferred to stall, which had different type of customers — Tribunal accepts basis of 3 per cent of adjusted average net profit apportioned between shop and stall — Amount of goodwill award higher than amount claimed owing to deduction of
E **lower rental value figure — Small deduction for scrap value made from claim for loss of fixtures and fittings — Costs of providing new storage accommodation at market allowed in full, being money spent in mitigating loss — Cost of removal of stock to stall also allowed in full**

I H Kinson (solicitor of Edward Hollinshead, of Tunstall, Stoke-on-Trent) appeared for the claimant, Herbert Leese; P N Wilson (solicitor with Stoke-on-Trent City Council) for the acquiring authority.

Giving his decision, MR REES said: This was a reference by the claimant to determine the compensation to be paid for freehold shop premises at 149 Furlong Road, Tunstall, Stoke-on-Trent. The
F claimant claimed the sum of £6,000 for the freehold and £2,275.68 for disturbance: the acquiring authority's valuer spoke to a valuation of £3,450 for the freehold and to disturbance "on basis of 'temporary loss' " of £413.30. He spoke to £575.50 on an alternative basis.

The claimant's solicitor called the claimant himself, J C Sutton FRICS and J E F Barker, a certified accountant. The acquiring authority's solicitor called H F Moore ARICS.

There is an agreed statement of facts. The relevant order is the City of Stoke-on-Trent Tunstall and Nos 71/73 (Furlong Road) Clearance Areas Compulsory Purchase Order 1973, which was made by the acquiring authority on October 29 1973 and confirmed on December 12 1974. Notice to treat was served on February 26 G 1975 and the acquiring authority took possession on April 13 1977, which is thus the relevant valuation date.

The premises were shown coloured grey on the map attached to the order and were used by the claimant for his business of general household stores. The subject premises were at the north-east end of a terrace of buildings probably originally constructed as houses. They were on the south side of Furlong Road between Clanway Street and an area known as Tunstall Greenway. From the evidence I find that the building was of brick under a slate roof; there had been an addition which was used for storage purposes and which was constructed just after the last war. It was not disputed that the original building was between 80 and 100 years old. Accommodation comprised: on the ground floor, a shop, two stores, an office H and an area described as "kitchen/store". On the first floor there were three stores. Agreed floor plans were put before me. The areas were agreed to be as shown in the valuations set out below.

Freehold interest

I consider first the valuation of the freehold interest. Mr Sutton's valuation was as follows:

Ground floor		
Shop Zone A 13 ft 9 in × 20 ft = 275 sq ft at £1.95 per sq ft		£536.25
Zone B 13 ft 9 in × 6 ft 6 in less chimney breast = 83 sq ft at 97½p per sq ft		80.925 J
Principal storage = 393 sq ft at 50p per sq ft		196.50
Secondary storage (allowing for stairs) = 90 sq ft at 25p per sq ft		22.50
First floor		
Net storage = 300 sq ft at 25p per sq ft (⅛ Zone A)		75.00
		£911.175
Less landlord's repairing liability	£75	
Insurance (£20,000 × 12½%)	12	87.0
		£824.175
		K
Net rental value, say		£825 pa
YP for 30 years at 12% and 3%		7.091
		£5,850
Residual value of site		
Site area 131 sq yds at £3.45 per sq ft	451	
PV of £1 in 30 years at 12%	0.033	nil
Capital value		£5,850
Say		£6,000

L

In support he relied on lettings of two premises in the open market, one of which he had afterwards sold, and to a transaction in land in Tunstall in 1974, a sale by auction. He also referred to general statistics.

Mr Moore's valuation was as follows:

Ground floor — Shop A space (20 ft): 275 sq ft at £1.15	£316.25
Remainder: 83 sq ft at 57½p	47.73
Store 1: 276 sq ft at 30p	82.80
Store 2: 72 sq ft at 25p	18.00
Office: 45 sq ft at 25p	11.25
Kitchen/store: 90 sq ft at 20p	18.00 M
First floor — stores 1 and 2: 300 sq ft at 15p	45.00
	539.03
Less repairs and insurance at 20%	108.00
	431
Years' purchase	8
	3,448
Say	£3,450

In support he relied on the analyses of two settlements reached in respect of properties acquired compulsorily by the respondent authority, one being in the same terrace as the subject property. In addition he referred to three sales by auction of premises in Burslem and Tunstall.

In so far as Mr Sutton relied on open market transactions, his evidence is to be preferred to that of Mr Moore, but Mr Moore's settlement in respect of a property in the same terrace was a closer comparison than any others referred to in terms of age, character and location.

Weighing all the evidence, I come to a net rental value of £750 for the subject premises, which I capitalised at Mr Sutton's 7.1 years' purchase. Thus I find for £5,325 as the value of the freehold interest in the subject premises.

Disturbance claim

The claim for disturbance was as follows:

Damage to goodwill	£1,705
Fixtures and fittings	348
Storage cupboards at market stall	174
Cost of removal of stock to stall	26
Accountants' fee for preparation of profits statements	22.68
Total	£2,275.68

There is no dispute that the claimant used the subject premises as the centre of his business, which was the sale, five-and-a-half days a week, of household goods, which included paints, gardening requisites, paraffin, fuel, crockery, electric items, cigarettes and tobacco. He also had a stall in Tunstall market three days a week from which he sold only some of the items which he sold from the shop; he was restricted by his licence to selling only certain types of goods.

Although the acquiring authority conceded that some disturbance was payable, they resisted the above claim, but I found their grounds for so doing confusing. In the same breath they said that the goodwill which adhered to the shop was transferred to the stall after the acquisition took place and thus there was no permanent loss of any part of the goodwill; they also said by inference that the claimant could have found other shop premises as alternative accommodation. They considered that only the following should be payable by way of disturbance. I set out Mr Moore's figures in summary form only:

1. As per claim — storage cupboards	£174
2. As per claim — cost of removal of stock	26
3. Loss of profit for three months: average net profit less rental value of shop and interest on capital (as below) £228 pa − $\frac{3}{12}$	60
4. Legal costs and agent's fees for purchase of alternative property	153.30
Total	£413.30

Loss from both outlets

The claimant said that no other shop premises were available in the area to which he could have transferred his business from the shop: Mr Moore, in answer to a question put to him by myself, said that he could not specify any premises which were available at the relevant time to which the claimant could have gone. I indicated to the claimant's solicitor that I considered that the claim for disturbance should be on the footing that part of a goodwill of the claimant's business (both that from the shop and that from the stall) was permanently lost due to the acquisition. Mr Moore put forward figures in the alternative "if it is decided that Mr H Leese could comply with the requirements of section 46 of the Land Compensation Act 1973, and compensation on a 'total extinction' basis is payable". Summarised, those figures were as follows:

1. Loss on forced sale of fixtures (37½% of £348 — the value as shown in the accounts)		£130.50
2. Loss on forced sale of about 20% of stock of £2,012 (as shown in the accounts) — 20%		80.00
3. Loss on forced sale of vehicle		—
4. Loss of goodwill:		
Net profit	£806	
Less rental value of shop and interest on capital	578	
	228	
YP	1.5	350.00
5. Auction sale fees for sale of fixtures		15.00
Total		£575.50

The question of section 46 did not arise at the hearing before me and, as indicated above, I am satisfied on the evidence that there was a permanent loss of some part of the claimant's goodwill from both outlets as a result of the acquisition.

I now consider the individual items of claim. Mr Sutton said that his experience covered the sale of premises including goodwill, and in the market he was satisfied that he could obtain 2½ years' purchase for the average profits derived from the shop, which were put at £806 per annum, a figure which was not seriously disputed. That gave a figure of about £2,000, but I do not put much weight on Mr Sutton's opinion, as no deduction had been made from the figure of £806 per annum for rental value or interest on capital.

Mr Barker valued the whole of the goodwill of the claimant's business by applying 3 years' purchase to the average net profit after deducting £825 per annum (Mr Sutton's estimate of the rental value of the subject premises) from the average net profits made over the three years prior to acquisition: that gave a total of £5,379. This he apportioned according to the takings at the shop and the stall to give £1,705 as the goodwill which was lost when the shop closed.

I am satisfied from the claimant's evidence that the type of customers who resorted to the shop were quite different from the customers who bought from the stall and that no goodwill transferred from the shop to the stall. In reply to a question put to him by myself, Mr Barker said that he valued the goodwill at the stall as the remainder, ie £5,379 less £1,705 = £3,674. The claimant said that after the acquisition, he could no longer take advantage of the same buying discounts as he had enjoyed before the acquisition, but that is not, I think, reflected in Mr Barker's figure of £1,705 and to some possibly small extent, that figure (£1,705) is low on that account.

Tribunal's findings

I prefer Mr Barker's figure of £1,705 to Mr Moore's £350 and would accept it but for the fact I have found the net rental value to be £750 per annum (as against £850 per annum, the figure used by Mr Barker). The necessary adjustment gives £1,775.

Mr Barker's evidence was that if fixtures and fittings were sold, very little was ever got for them and thus he claimed their full value. I am not convinced that their scrap value was nil, but I think Mr Moore puts the loss on forced sale including auction fees at far too low a figure, his total being £145.50. I allow a scrap value of £25.68 to give a figure under this heading of £322.32.

When the subject premises were taken from him, the claimant lost valuable storage space which he used to store goods at the shop which were sold at the stall. In order to provide storage accommodation at the market to make up for this loss, he had to spend £174 to have new cupboards made and fixed there. The amount of £174 was not disputed. Mr Moore's view was that that item should be allowed only if disturbance was to be on a "temporary loss" basis, but I am persuaded that the claimant spent this money in mitigating his loss and that it should be allowed. It was, quite rightly in my judgment, not suggested that the claimant got value for money, as the cupboards were installed in premises at the market which the claimant holds only on licence.

With regard to the item of £26, Mr Moore's argument was the same as for the item of £174, but I conclude that this was a proper item of claim and, the amount not being contested, I allow both in full.

I also allow in full the fee (£22.68) for preparation of the accounts, which work was certainly necessary to prepare the claim for disturbance; that item was not seriously contested by the acquiring authority.

A The result is that the amount to be allowed for disturbance is £2,320. Thus the total compensation is £7,645 to which will be added a surveyor's fee in accordance with scale 5(a) of the scale of charges of the Royal Institution of Chartered Surveyors and the usual legal costs.

I received submissions in writing with regard to costs. The acquiring authority will pay the claimant his costs of this reference, such costs, if not agreed, to be taxed by the Registrar of the Lands Tribunal on the High Court Scale of Costs.

B

COMMISSIONERS OF INLAND REVENUE v ALLIED BREWERIES (UK) LTD

(REF/253/1980)

(J H EMLYN JONES FRICS)

February 22 1982

Estates Gazette April 10 1982
(1982) 262 EG 153

Current use value of Birmingham public house — Buildings C demolished and site now used as vehicle park following sale to adjoining occupiers — Sale price included element of realised development value — Parties' valuers agree that most likely purchaser would have been a brewer, that the value to the brewer could best be arrived at by examination of actual trading figures and by estimating reasonably maintainable trade, and that state of structure and standard of maintenance have some bearing on value — Both valuers put in valuations on the "Kennedy" basis, but taxpayers' valuer only as "supporting" valuation as he considers the method less satisfactory — Evidence of market transactions and DLT and compensation settlements as comparables — Effect of location and quality of accommodation on value — Capital value D per weekly converted barrel — Calculation of brewer's profit — Methods of calculating tied rent — "Wet" and "dry" rents — Actual tied rent best starting point — Increase over "tone" to represent true values in 1976 — Tribunal increases basic brewer's profit by 25 per cent and actual tied rent by 15 per cent — Deduction for repairs and insurance of strict percentage (both valuers adopting 10 per cent) described by tribunal as only a rough and ready approach — As premises were in good structural condition, actual cost figure preferred — 10 YP applied as appropriate multiplier — Award of £36,000 against taxpayers' £47,500 and Inland Revenue's £31,500

Alan Fletcher (instructed by the Solicitor of Inland Revenue) E appeared for the applicant commissioners; W Glover QC and R L Fookes (instructed by Edge & Ellison, Hatwell Pritchett & Co, of Birmingham) for the respondent taxpayers.

Giving his decision, MR EMLYN JONES said: This reference concerns the valuation of a public house in Birmingham for the purpose of development land tax. The house, known as the Malt Shovel, was situated in Palmer Street, Birmingham, and formerly belonged to the brewers Ansells, who subsequently became a wholly-owned subsidiary of the taxpayer company, Allied Breweries (UK) Ltd. Ansells first occupied the Malt Shovel in March 1906 under a 50-year lease. They purchased the freehold interest in the premises subject to the lease in 1944 for £3,200. At that time, there was in F force an old beer-on licence, which was surrendered when a full licence was granted in 1964.

On January 21 1977 the freehold interest in the Malt Shovel was sold by the taxpayer to W Cannings Ltd for £57,500. The contract of sale was dated December 14 1976. Cannings are the occupiers of adjoining commercial premises, and the site of the Malt Shovel has now been brought into use as a park for heavy road vehicles. It is common ground that the price realised on the sale included an element of realised development value and that the transaction gave rise to an assessment of development land tax under the Development Land Tax Act 1976. Section 1(2) of the Act provides as follows:

Subject to the provisions of this Act, a person shall be chargeable to G development land tax on the realised development value, determined in accordance with this Act, which accrues to him on the disposal by him on or after the appointed day of an interest in land in the United Kingdom, and shall be so chargeable whether or not he is resident (for purposes of income tax or otherwise) in the United Kingdom.

The realised development value is the amount by which the net proceeds of the disposal exceeds the relevant base value of the interest (section 4(1)). The Act provides that the relevant base value shall be the highest of three alternative figures (section 5(1)).

It is agreed that in the present case the appropriate figure is that calculated in accordance with paragraph (b) of the subsection, that is to say 110 per cent of the current use value of the interest at the H time of the disposal. Current use value is defined by section 7(2) of the Act as the market value of the interest (as defined in section 7(1)) calculated on the assumptions that (a) planning permission would be granted in any class of development specified in Schedule 8 to the Town and Country Planning Act 1971, and (b) no material development of the land is or would be lawful other than a project of material development already begun or for which planning permission can be assumed in accordance with (a) above.

Applying these provisions to the facts of the present case it is agreed that the interest in the land and buildings which is to be valued is the freehold, that the relevant date of valuation is December 14 1976, that the required value is the current use value J and that that value is the value of the subject premises as a public house on the assumption that it could be continued to be so used and that it could never be lawfully used for any other purpose.

The Malt Shovel comprised a small, pre-1914 (probably built pre-1900) brick-built house with blue slate roof, being part two-storey, part single-storey and occupying a site of 572 sq yds or thereabouts. The accommodation was as follows:

Ground floor:	Front bar lounge, Rear bar with servery between the two rooms, Outsales; with sitting-room, kitchen, pantry and scullery built on at the rear.
First floor:	3 bedrooms and bathroom.
Basement:	2 beer cellars. K
Exterior:	Enclosed yard with ladies and gents toilets, coal-house and private wc.

The question for determination, therefore, is one of valuation. For the commissioners Mr Fletcher called E L Harris ARICS, who is the regional licensed property valuer, West Midland Region, for the Inland Revenue. For the taxpayer Mr Glover called J E Hill FRICS, senior partner in the firm of Edwards, Bigwood & Bewlay in Birmingham. Both expert witnesses have many years' experience in the valuation of licensed property and Mr Hill, in addition to serving in the office of the licensed property valuer has subsequently acted in private practice for a number of brewery companies for whom he has been concerned in valuations on sale and L purchase of licensed premises in the open market.

Both valuers are in agreement that the most likely purchaser of the Malt Shovel at the relevant date would have been a brewer and that its value to a brewer can best be arrived at by an examination of the actual trading figures and by making an estimate therefrom of the reasonably maintainable trade. They also agree that the state of the structure and the standard of maintenance have some bearing on the value.

Mr Harris attaches importance to the location of the premises; Mr Hill does not consider this to be a very important factor. Mr Hill's primary valuation is based on an estimated maintainable trade of eight weekly "equivalent or converted" barrels, which he M takes at £6,000 per barrel, giving a capital value for the Malt Shovel, valued as a public house, of £48,000. A weekly converted barrel represents the sum of the annual through-put of draught and bottled beer expressed in barrels and one-third of the annual through-put of wines and spirits expressed in gallons, the whole being divided by 52.

The valuation put forward by Mr Harris adopts what he described as a "Kennedy" basis, following as it does certain remarks made by Kennedy J in *Re Ashby's Cobham Brewery Co* [1906] 2 KB 754. Adopting this approach, the valuer arrives first at an annual value, which he then capitalises. The method can be

A summarised as follows: first it is necessary to arrive at a figure which represents the annual profit which a brewer might be expected to realise by the sale of intoxicating liquor to the premises. For this purpose it is necessary to make an estimate of the annual maintainable trade, taking separate figures for draught beer, for bottled beer and a further estimate of sales of wines and spirits. Draught and bottled beers are expressed in barrels of 36 gallons and quantities of wines and spirits are expressed in gallons, each gallon representing half-a-dozen bottles.

From these figures of estimated maintainable trade Mr Harris applies a multiplier to the different categories representing the brewer's profit, which he takes at £6.50 per barrel for bottled beer, B £3.90 per barrel for draught beer and 90p per gallon for wines and spirits. These were the figures which were generally applied for rating purposes for the preparation of the 1973 valuation list. Accordingly, Mr Harris increased his figures by 10 per cent to represent what he calls an increase over "tone". In his opinion this increase reflects the increase in the level of brewer's profit which was to be found at the relevant date in December 1976.

The second stage of a valuation on this basis is to add to the annual brewer's profit a figure to represent the tied rent, that is to say the other part of the income which a brewer purchaser could expect to receive from his ownership of the house. The Malt Shovel was let to a tenant from November 7 1974 at a tied rent of £1,560 C per annum determinable on the termination of the landlords' interest, so that the actual sale to Cannings was a sale with vacant possession. In valuing the tied rent element, Mr Harris takes the actual tied rent of £1,560 and as a check devalues it in terms of "units". The unit of tied rent is used by Mr Harris as an index which takes into account the retail profit which a tenant might be expected to realise and consists of a barrel of draught beer or half a barrel of bottled beer or half a gallon of wines and spirits, thus reflecting the greater margin of profit on the retail sale of bottled beer, wines and spirits.

From the resultant total of updated brewer's profit and tied rent Mr Harris deducts 10 per cent for repairs and insurance, thus D arriving at a figure which, in his estimate, represents the net income which a brewer might expect to receive. In order to capitalise this sum, Mr Harris applies a multiplier of 10 years' purchase and that produces in round terms a total of £31,500 which, in his opinion, represents the current use value. In his valuation, Mr Harris takes the actual trade figures for 1975 and 1976 and applies a yearly average. This produces 8.4 weekly converted barrels, and his valuation therefore represents £3,750 per weekly converted barrel.

Mr Hill also puts in a "Kennedy" valuation, which broadly follows the same pattern as that adopted by Mr Harris. In Mr Hill's opinion, this method of valuation is less satisfactory and he describes his calculation as a supporting valuation. He bases his E valuation on an estimate of the trade, based on his own experience, rather than on a two-year average as adopted by Mr Harris; but he splits his draught beer into three categories—mild, bitter and keg, the proportion being taken from the evidence of actual deliveries. He applies similar multipliers and there is very little difference between the two valuers on the basic brewer's profit, Mr Hill's figure being £1,726 compared with £1,744 adopted by Mr Harris. In Mr Hill's opinion, however, the increase to be applied to reflect the difference in the level of profit between that adopted in the 1973 valuation list and the appropriate level to be applied in 1976 is 50 per cent and not 10 per cent as suggested by Mr Harris.

Mr Hill calculated his tied rent by way of a formula taking 416 F annual converted barrels at £6.50 per barrel. From the resultant total he also makes a deduction of 10 per cent for repairs and insurance and applies a multiplier of 10 years' purchase, which produces in round terms a valuation of £47,500, equal to £5,955 per weekly converted barrel.

In putting forward valuations both the valuers relied on their experience over many years in dealing with licensed property matters. They also adduced a considerable amount of additional evidence upon which they relied to support their opinions. Mr Hill produced a list of six open market sales of public houses which took place in the period between November 1974 and November 1977. Two of these properties were in North Wales, where Allied Breweries (UK) Ltd had been the purchasers. In three other cases G brewers had sold to private individuals, in one case to the sitting tenant, and the sixth transaction was the further sale by that tenant by auction after an interval of about four months.

In analysing these sale prices Mr Hill took an average of the actual trade for the previous three years except in the case of the Welsh houses, where he took the first full year's trade following the purchase. These figures he then converted into equivalent weekly barrels and after making adjustments, where appropriate, to exclude the value of fixtures and fittings devalued the net sale price to produce a price per weekly converted barrel. These varied between £4,007 and £6,146.

Mr Hill also produced details of two settlements which he or his H firm had reached with licensed property valuers acting for the Commissioners of Inland Revenue on current use value for the purpose of calculating development land tax. Adopting a similar form of analysis, he arrived at a value per weekly converted barrel of about £6,500 in each case. One of the properties was in the centre of Walsall and the other in the centre of Leicester.

He also produced evidence of the level of liquor prices retail and wholesale over the periods from 1972 to 1976 and extracts from reports and accounts of Allied Breweries (UK) Ltd and other brewers over a period from 1972 to 1976. He relied on this information to support his opinion about the increase in the level of brewers' profit over the levels adopted for the 1973 valuation list in J order to reflect the values current at the relevant date in December 1976. He also referred to a report of the Price Commission published in April 1978 headed "Allied Breweries (UK) Ltd, Brewing and Wholesaling of Beer and Sales in Managed Houses". Finally, Mr Hill produced calculations in support of his estimate of tied rent as adopted in his supporting valuation.

Mr Harris also produced schedules of comparable transactions. The first list included details of 10 sales of licensed houses, being all the transactions he could find from his records of sales within the West Midland conurbation in urban and industrial areas during the period starting nine months before and ending 20 months after the relevant date. He did not consider that country houses were of any K value as comparables. His second list produced details of 14 transactions where agreement had been reached on compensation following compulsory acquisition of public houses. All save two of these negotiations had been carried out with Mr Hill's own firm.

Mr Harris devalued all these transactions in the same way. He took an average of the previous three years' trade (in two of the compensation cases he took a four-year average) calculating the brewer's profit in accordance with the 1973 rating basis. He then calculated the weekly converted barrels and added an increment over the "tone" of the 1973 valuation list. This varied between 7½ per cent for 1975-76, to 10 per cent for 1976-77, 15 per cent for 1977-78, 20 per cent for 1978 and in one case an increase of 25 per L cent on a compulsory acquisition in October 1978. He then took a figure for tied rent calculated on the basis of tied rent units, deducted 10 per cent of the gross income thus arrived at for repairs and insurance and divided the resultant net income into the sale price of the agreed compensation. The resultant figure gave him a figure of years' purchase which varied between 6.23 and 15.74 in the market sales and 7.68 and 10.33 on the compensation figures.

Finally, he analysed the sale price in terms of value per weekly converted barrel and this exercise produced figures which varied in the case of sales between £2,486 and £5,970 (half of them being between £3,000 and £4,000) and in the case of the compensation figures between £3,104 and £3,929. M

I have visited the site of the Malt Shovel and the surrounding area. As I have indicated, the buildings have now been demolished. The locality is primarily a commercial and industrial area containing a few pockets of residential property. A number of other licensed premises in the immediate locality have closed down in the past few years, but at the relevant date there were still seven public houses within a quarter-mile radius. I have also visited about a dozen of the other "comparable" houses which were referred to me, which represent nearly all of those which were still standing, including the two in North Wales. It is common ground that the most likely purchaser of the subject property at the relevant date

A would have been a brewer, and it is therefore right to consider first the evidence provided by the purchase by brewers of other licensed houses. In saying this, I do not ignore the other market transactions which, for reasons that I shall need to explain, must also be taken into account. As to the evidence of the agreed compensation on compulsory acquisition I find that this evidence is also relevant as is the evidence of the settlement of development land tax values in other cases.

In general, I accept the views of Mr Harris regarding the effect of location on value. It seems to me that in certain circumstances value can be affected by factors other than the estimated maintainable trade. Certain houses from their position possess an advertising B value, a factor that is not present in the case of the Malt Shovel, which in a locational sense was a back-street house. Similarly, the quality of living accommodation and situation can also have an influence either because it increases the value to a private purchaser who will be in competition with a brewer or because it will make it easier for a brewer to find a suitable tenant or indeed to receive a higher tied rent. This is particularly so in the case of houses where the trade is fairly small and where the value of the living accommodation must represent an important feature.

Four of Mr Hill's comparables were valued at a trade of less than 5½ weekly converted barrels. The two Welsh houses were nearer to the subject property in terms of trade, the Cross Keys, Buckley, C having 10.74 weekly converted barrels and the New Inn at Rhos 7.81 weekly converted barrels. The trade in both cases, however, was that which was actually done in the first full year following purchase and I heard no evidence about the subsequent trading pattern. In both cases, physical improvements have subsequently been carried out and I therefore find that I cannot put much reliance on the capital value per weekly converted barrel which emerges from Mr Hill's analysis. Any increase in trade over that achieved during the first year of trading under the brewer's ownership at a time when physical alterations were being carried out would of course lead to a reduction in the value per barrel. For this reason, therefore, I find it necessary to consider also sales to private D individuals.

Turning to Mr Harris' market transactions, I find that they are also to be taken with some reservation. Some of the sales were to sitting tenants, others were of houses in poor structural condition and some are no longer operating as licensed houses, having been delicensed or demolished.

As to the evidence of the settlements, the two cases mentioned by Mr Hill seem to me to be cases where the current use values agreed clearly reflected to some extent the advertisement or prestige value of town-centre houses. The relevant date in both cases was in any case considerably later than that with which I am concerned, one being February 1978 and the other June 1979. These agreements E do nothing in my opinion to support a value of £6,000 per weekly converted barrel for the subject premises.

Looking at the evidence as a whole and analysing it in terms of capital value per weekly converted barrel, it seems to me that the value to be attributed to the Malt Shovel at the relevant date should be something in the region of £4,250 to £4,500.

From this broad brush starting point I find it useful to look at the value arrived at following the alternative approach as adopted by Mr Hill and used by Mr Harris as his primary valuation. At first sight there is little between the two valuers in the calculation of brewer's profit, £1,744 by Mr Harris, £1,726 by Mr Hill. The position is complicated, however, because of different methods F which are adopted in the calculation of the tied rent. In the past, the general practice has been for the brewer to obtain his return from the tied tenant partly through the tied rent and partly through charging for his deliveries of beer at a higher price. In these circumstances, the tied rent was known as a "wet" rent. From about 1976 new tenants were offered tenancies on the basis of a "dry" rent, which was calculated at a higher level. Where this method was adopted the charge made to tied tenants for deliveries of beer was the same as that made to other customers. In his valuation Mr Hill had assessed his brewer's profits on the basis of wet rents but had calculated his tied rent as a dry rent. The actual tied rent was £1,560 per annum fixed in 1972, and I therefore consider that that forms the best starting point for a calculation of the appropriate tied rent G element in the valuation.

As to the increase over "tone," it is common ground that the figure of brewer's profit must be increased to represent the true values in December 1976. The starting point of both valuers is the values which were actually adopted for the 1973 valuation lists. These may well have been taken by licensed property valuers to reflect a projection forward from the date when the evidence was collected to April 1 1973, when the new lists came into force. I heard no evidence, however, from either valuer which would enable me to say at what precise date they were to be taken as accurate. In the light of the evidence extracted from the brewer's accounts and from the Price Commission report, I am satisfied, H however, that the 10 per cent addition applied by Mr Harris is quite insufficient.

In my opinion, Mr Harris' figure of £1,744 should be increased by 25 per cent to £2,180 and the actual tied rent of £1,560 should be increased by 15 per cent to £1,794 (which represents £2.57 per unit on Mr Harris' approach; and £4.31 per converted barrel on Mr Hill's approach). This produces a total of £3,974. As to the deduction for repairs and insurance it is of course a convenient practice to adopt a strict percentage, but this is only a rough and ready approach. The actual costs of repairs will depend on the physical condition of the premises and will not necessarily vary in proportion to the capital value of the house. The evidence of Mr Hill, who knew J the house when it was still trading, is that the premises were in good structural condition. In these circumstances, I adopt a sum of £374 to be deducted from the gross income to produce a net income.

Finally, I come to the question of the appropriate multiplier to convert net annual value into capital value. Both valuers adopt 10 years' purchase, but it is true to say that Mr Harris derives this figure from his analysis of the market transactions and agreed compensation incorporating the smaller increases which he applies to reflect increases over "tone" of the valuation list figures. As a matter of arithmetic the higher percentage increase which I have adopted, if applied to the analysis carried out by Mr Harris, would produce a lower rate of years' purchase. Nevertheless, I am not K persuaded that any other figure is appropriate when applied to the valuation of the subject property. I therefore adopt 10 years' purchase, which produces a capital value of £36,000. I observe that this represents £4,285 per converted barrel on Mr Harris' estimate and £4,500 per converted barrel on Mr Hill's estimate. This seems to be in line with the view I had provisionally formed looking at the evidence of the comparables devalued in terms of capital value per weekly converted barrel.

I therefore determine that the current use value of the Malt Shovel public house calculated in accordance with the statutory provisions governing the assessment of development land tax at the relevant date is £36,000 and I so award. L

I heard submissions as to costs from both sides. The applicants will pay the respondents' costs, such costs failing agreement to be taxed by the Registrar of the Lands Tribunal on the High Court scale.

PEAKTOP PROPERTIES HAMPSTEAD LTD v CAMDEN LONDON BOROUGH COUNCIL

M

(REF/1/1981)

(V G WELLINGS QC)

February 17 1982

Estates Gazette April 17 1982

(1982) 262 EG 239

Liability for compensation for refusal of planning permission for extension to two blocks of flats in one curtilage at Hampstead, London — Preliminary point of law on claim for £142,500 — Proposed extension the construction of additional storey on each block to form flats — Gross floorspace for residential purposes would

exceed by more than 10% the existing gross floorspace for residential purposes — Acquiring authority's submission that in present context dwellings were in different category from other premises, that subdivision of an existing dwelling into separate dwellings amounts to a material change of use requiring planning permission (s 22(3)(a) of 1971 Act) and that by analogy planning permission to enlarge a block of flats does not enable developer of enlarged area to create separate flats as opposed to mere enlargement of living space of one of original flats — Further submission that s 169(3)(c) defeated claim — Tribunal rejects these submissions — S 22(3)(a) had no application in circumstances of present case — No reason to distinguish between flats and other uses — S 169 (3)(c) does not require entire prospective use to be ignored but only that part which constitutes the excess — Acquiring authority liable to pay compensation

Malcolm Spence QC (instructed by Brecher & Co) appeared for the claimants; Jeremy Sullivan (instructed by the London Borough of Camden) for the acquiring authority.

Giving his decision, Mr Wellings said: The preliminary issue arises under section 169 of the Town and Country Planning Act 1971 and Schedule 8 thereto and is whether the council of the London Borough of Camden is liable in law to pay compensation to the claimants for a refusal of planning permission for an extension of two blocks of flats [Holmfield and Gilling Courts] belonging to the claimants at Belsize Grove, London NW3. The council's application for an order that points of law at issue between the parties be determined in a preliminary manner mentioned six questions of law. Only two of them were argued before me and those two questions were in effect subdivisions of the main question, namely, the question of liability.

Section 169 of the Act of 1971 provides as follows:

(1) The provisions of this section shall have effect where, on an application for planning permission to carry out development of any class specified in Part II of Schedule 8 to this Act, the Secretary of State, either on appeal or on the reference of the application to him for determination, refuses the permission or grants it subject to conditions.
(2) If, on a claim made to the local planning authority within the time and in the manner prescribed by regulations under this Act, it is shown that the value of the interest of any person in the land is less than it would have been if the permission had been granted, or had been granted unconditionally, as the case may be, the local planning authority shall pay to that person compensation of an amount equal to the difference.
(3) In determining, for the purposes of subsection (2) of this section, whether or to what extent the value of an interest in land is less than it would have been if the permission had been granted, or had been granted unconditionally:
(a) it shall be assumed that any subsequent application for the like planning permission would be determined in the same way; but
(b) if, in the case of a refusal of planning permission, the Secretary of State, on refusing that permission, undertook to grant planning permission for some other development of the land in the event of an application being made in that behalf, regard shall be had to that undertaking; and
(c) no account shall be taken of any prospective use which would contravene the condition set out in Schedule 18 to this Act . . .

The relevant paragraph of Schedule 8 is paragraph 3, which is as follows:

The enlargement, improvement or other alteration, as often as occasion may require, of any such building as is mentioned in paragraph 1(a) or (b) of this Schedule . . ., so long as the cubic content of the original building is not increased or exceeded:
(a) in the case of a dwellinghouse, by more than one-tenth or 1,750 cubic feet, whichever is the greater; and
(b) in any other case, by more than one-tenth.

It is sufficient to say that the reference to "any such building" as is mentioned in paragraph 1(a) or (b) of this schedule is a reference, *inter alia*, to any building which was in existence on the appointed day (July 1 1948), because both blocks of flats were in existence on that day.

In addition to the provisions of section 169 (3)(c), which I have already set out, section 278 (1) contains the following provision:

In any case where the value or depreciation in value of an interest in land falls to be determined on the assumption that planning permission would be granted for development of any class specified in Schedule 8 to this Act, it shall be further assumed, as regards development of any class specified in paragraph 1 or 3 of that Schedule, that such permission would be granted subject to the condition set out in Schedule 18 to this Act.

The relevant condition is contained in paragraph 1 of that schedule and is as follows:

Where the building to be rebuilt or altered is the original building, the amount of gross floor space in the building as rebuilt or altered which may be used for any purpose shall not exceed by more than 10 per cent the amount of gross floor space which was last used for that purpose in the original building.

The two blocks of flats are in effect comprised in one and the same curtilage. One of the blocks is known as Gilling Court, Belsize Grove. The claimants own the freehold interest in that block. The other block is Holmfield Court, Belsize Grove. The claimants hold that block under a lease dated September 28 1934 for the term of 999 years from December 25 1933. Each block contains 104 flats let to different tenants.

On December 20 1978 the claimants applied to the council for full planning permission for the following development: "Additional floor to existing buildings. Form flats similar to existing floors below".

The application was accompanied by drawings. It was refused on March 21 1979. The claimants then appealed to the Secretary of State, whose inspector, to whom the appeal had been delegated, dismissed the appeal by letter dated December 14 1979. The claimants' claim for compensation was made within the time and in the manner prescribed by the Town and Country Planning General Regulations 1976.

In the claimants' notice of reference they claim compensation in the sum of £142,500, "being the difference between the value of the interest in the said properties if the permission had been granted, namely, £542,500, and the value in the light of the planning decision, namely, £400,000, plus professional fees in connection with the submission of the planning application and the subsequent appeal proceedings." As I understand matters, there is no agreed valuation and the claim is in all respects disputed.

Since the two blocks are comprised in the same curtilage and since they are used as one unit for the purposes of the claimants' undertaking, the reference in paragraph 3 of Schedule 8 to the cubic content of the original building is to be construed as a reference to the aggregate cubic content of the two buildings (paragraph 11 of Schedule 8). For the purposes of the preliminary hearing, the parties agreed that it should be assumed that the aggregate enlargement of the two buildings proposed by the application for planning permission (after making proper allowance for the cubic content of certain temporary garages constructed during the 1960s) would not have increased the aggregate cubic content of the two buildings by more than one-tenth.

The parties also agreed that the gross floorspace (for use for residential purposes) shown on the drawings accompanying the application for planning permission exceeds by 11.49 per cent the existing amount of gross floorspace (for use for residential purposes) in the two buildings (after making proper allowance for the gross floorspace of the garages).

Malcolm Spence QC, counsel for the claimants, and Jeremy Sullivan, counsel for the council, were agreed upon the following propositions which had been put in the form of submissions by Mr Spence:

(1) A planning permission for an enlargement of a building of the kind contemplated by paragraph 3 of the Schedule 8 was necessarily a permission for carrying out the building operation involved and also for the use of the space created by it for the purposes for which the space was designed: see *Wilson v West Sussex County Council* [1963] 2 QB 764;

(2) In the present case the application was for planning permission for the building operation of constructing an additional storey on each block, and by virtue of the application itself and section 33(2) of the Act of 1971, for permission for use of the additional storey as individual flats, which was the purpose for which the additional storey was designed;

A (3) If the blocks had consisted, not of flats, but of suites of offices and the proposed additional storey in each case were designed and intended to be used as suites of offices, and planning permission for that development had been refused, the claimants would have been entitled to compensation, at all events, apart from the provisions of Schedule 18.

There Mr Spence and Mr Sullivan parted company. Mr Spence submitted that on the face of the statutory provisions the claimants were entitled to compensation notwithstanding that the amount of gross floorspace in the buildings as extended which may be used for any purpose (ie residential purpose) will exceed by more than 10 per cent the amount of gross floorspace which was last used for that B purpose (ie residential purpose) in the original building.

Mr Sullivan's first submission was that in this context dwellings were in a different category from other premises. This was because, while any building which was a factory, office block or warehouse could lawfully be subdivided so as to make an indefinite number of separate factories, suites of offices or warehouses without planning permission being required for that purpose (unless an intensification of use took place), the subdivision of an existing dwelling into separate dwellings as a matter of law amounts to a material change in the use of land requiring planning permission: see section 22(3) of the Act of 1971.

It followed by analogy that a planning permission to enlarge a C block of flats does not enable the developer to use the enlarged area so as to create separate flats as opposed to a mere enlargement of the living space of one of the original flats. If an existing dwelling could not, without planning permission, be subdivided into new dwellings, it would be strange if a block of flats could be enlarged so as to create several new flats. He found support for that submission in paragraph 1 of Schedule 18 and in the original provisions of Schedule 8 as contained in Schedule 3 to the Act of 1947, in which schedule the relevant provisions were described as "development included in existing use" whereas in Schedule 8 to the Act of 1971 they were described as "new development". The creation of new dwelling-houses within the extended area would, he said, not be D within the existing use.

Mr Sullivan's second submission was that, if his first submission failed, the claim for compensation was defeated by section 169(3)(c) of the Act of 1971 because the amount of gross floorspace intended for residential user in the building as extended exceeded the limit permitted by Schedule 18. The prospective use would contravene the condition set out in that schedule and the subsection directed that no account should be taken of it. Accordingly, as a matter of law, the compensation was nil.

In reply, Mr Spence said that there were no words of exception in the statutory provisions which justified Mr Sullivan's first submission. Section 22(3)(a) of the Act of 1971 was not concerned with E extensions but with subdivision of existing dwelling-houses. It did not constitute an analogy which was of assistance in the present case. As to Mr Sullivan's second submission, section 169(3)(c) was concerned not with entitlement to compensation but the amount of it. It was only that prospective use which would exceed the permitted percentage which would contravene the condition in Schedule 18 and not the entire residential use of the building as enlarged.

I am unable to accept either of Mr Sullivan's submissions. Section 22(3)(a) of the Act of 1971 has no application in the circumstances of the present case. In my opinion, there exists no reason, whether or not by reference to an analogy based on that subsection, to infer that the claimants are not entitled to compensation under section F 169 merely because their proposal is to create separate flats intended to be in separate occupations in the proposed new storeys. I see no reason to distinguish between flats and suites of offices or indeed other uses. The words of the statute are plain enough.

I agree with Mr Spence that section 169(3)(c) does not defeat the claim for compensation. The true view is that that subsection does not require the entire prospective use to be ignored but only that part which constitutes the excess. If that subsection were intended to defeat such a claim, rather than being in a subsection concerned with assessment of compensation, it would surely have been added at the end of either subsection (1) or (2) of that section. The comparison which the latter subsection requires to be made, therefore, is between the values of the interest of the claimants, first, on G the basis of the absence of planning permission for the proposed extensions and, secondly, on the assumption of the existence of planning permission which authorises the proposed extension but which also is subject to a condition, in effect, limiting the proposed increase in gross floorspace to 10 per cent of the original amount of gross floorspace: see section 278(1) of the Act of 1971. For these reasons I determine the preliminary issue, that is to say, the question of liability, in favour of the claimants.

The acquiring authority will pay the claimants' costs of and arising out of the preliminary issue, such costs if not agreed to be taxed by the Registrar of the Lands Tribunal on the High Court Scale. H

NUTTALL v LEEDS CITY COUNCIL

(REF/107/1981)

(C R MALLETT FRICS)

March 15 1982

Estates Gazette April 24 1982

(1982) 262 EG 349

Loss of goodwill on total extinguishment of newsagent, confectioner and tobacconist's business in Leeds — Turnover of around £2,000 per week at dispossession date — Tribunal prefers claimant's valuer's conventional valuation, based on adjusted average of last two years' net profit, to Inland Revenue valuer's "entirely different and novel" approach of estimating increased turnover figure from last three years' annual sales and using gross weekly takings and asking prices stated in advertisements — Very positive evidence required to show that a well-established pattern that had formed basis of agreement in the area for many years was unacceptable — Three fundamental defects in approach adopted for DV: (i) no means of accounting for profitability, (ii) any difference between market value and value to claimant is unsolved, and (iii) it depends on advertised offers of business where there is no evidence as to accuracy of information or as to profitability — Tribunal refers to dicta in *English Exporters (London) v Eldonwall* on hearsay evidence — Award of £30,439 for goodwill, being amount claimed J K

J M Norris, solicitor (of Lupton Fawcett Bell Cumming & Co, of Leeds) appeared for the claimant; S G Vowles, solicitor to Leeds City Council, for the acquiring authority.

Giving his decision, MR MALLETT said: This is a reference by the claimant, Mr James Nuttall, to determine the compensation payable for the extinguishment of his business as a newsagent, confectioner and tobacconist at 94 Domestic Street, Leeds, on the compulsory acquisition of these premises by the Leeds City Council. L

J N Norris, solicitor, appeared on behalf of the claimant and called the claimant himself and Kenneth Harrap ARICS, a partner in the firm of Weatherall Hollis & Gale. S G Vowles, solicitor for the Leeds City Council, appeared on their behalf and called Donald Smith Evans ARICS, senior valuer in the Inland Revenue District Valuer's Office, Leeds.

The acquiring authority acting under Part III of the Housing Act 1957 and the City of Leeds (Holbeck) (Cross Ingram Road) Clearance Area Number 2 (Part) Compulsory Purchase Order 1976 (the blue area). This order was made on September 22 1976 and confirmed with modifications on May 15 1978. The area was defined as M a clearance area by resolution of the council dated July 21 1976. The subject property is included in Part 1 of the schedule to the order as a property unfit for human habitation.

Under a general vesting declaration the freehold of the subject property became vested in the council on September 11 1978. The claimant vacated on August 11 1979 and this therefore becomes the valuation date.

The claimant held a seven-year lease on the premises which expired on February 1 1978 and thereafter, because of the compulsory purchase order, he agreed to take the property on a monthly tenancy at a rent of £37.50 per calendar month. This tenancy, subject to the provisions of the Landlord and Tenant Act 1954, subsisted at the date of dispossession.

The property was an end-terrace, back-to-back type dwelling-house and shop known as 94 Domestic Street, Leeds 11, in which the claimant lived and conducted his business as newsagent, confectioner and tobacconist.

Domestic Street was originally the main shopping street in this residential part of the city. The subject premises were situated in a densely populated area of back-to-back housing. Slum clearance schemes have affected the area over a number of years, the first such scheme in the immediate locality being in 1964.

No suitable alternative premises were available to which the business could be transferred at the date of dispossession and it is agreed that compensation is to be calculated on the total extinguishment of the business. As the property was classified unfit the claimant has no compensatable interest in the land and compensation falls to be assessed under sections 37 (1) (a), 38 (1) (a) (b) and 38 (2) of the Land Compensation Act 1973.

In the immediate locality there are two further clearance areas and associated compulsory purchase orders. One is the City of Leeds (Holbeck) (Cross Ingram Road) (Clearance Area Number 1) Compulsory Purchase Order 1974, which was made on July 31 1974 and confirmed on August 20 1975 (the green area). It is agreed that for the purpose of calculating compensation this forms part of the "scheme". The other is the City of Leeds (Holbeck) (Crescent Terrace) (Clearance Area Number 1 and 2) Compulsory Purchase Order 1975, which was made on November 19 1975 and confirmed with modification on August 12 1976 (the yellow area). This order contains a number of unfit houses and shops in an area on the opposite side of Domestic Street to the subject premises. The general vesting declaration date was November 12 1976.

The two surveyors have very helpfully agreed most of the facts and settled many of the minor differences on matters of valuation. They have agreed the compensation on the loss on forced sale of fixtures and fittings, the loss on forced sale of stock and the domestic disturbance in the sum total of £964.

The issue on valuation arises on the value of the compensation payable in respect of the loss of goodwill on a business that has been totally extinguished. Mr Harrap for the claimant values this at £30,439 and Mr Evans for the acquiring authority values it at £16,157.

The second issue arises on a dispute as to what areas are to be included in the "scheme" for the purpose of calculating compensation. The claimant contends that all three areas, that is the green, blue and yellow areas, come within the scheme, whereas the authority contend that the yellow area is outside the scheme. The surveyors agree that whether or not the yellow area is within the scheme makes little or no difference to the compensation in this case, but as the issue arises in a number of other cases that are still outstanding the parties asked that this matter be decided.

From the evidence I find that Mr Nuttall and his wife acquired the business in March 1963, together with the residue of the lease of the premises which expired in 1970. It was an established business but run down and it dealt with newsagency, cigarettes, sweets and toys. The weekly takings were in the region of £250 and Mr Nuttall paid £2,100 for the business. He had no previous experience of retail business, but he and his wife worked very hard for long hours and built up a business which at the time of dispossession had a turnover in the region of £2,000 per week.

Valuations

Mr Harrap approached the problem of valuing the compensation for extinguishment of the business in a conventional manner that has been used in the agreed settlements of the compensation in the area for a number of years and which he himself had adopted in agreeing compensation on five other retail businesses in Domestic Street and three elsewhere.

He first looked at the trading accounts of the business between 1966 and 1978. These indicated a steady and continuing increase in turnover and profit. Nevertheless, he took the average of the last two years' net profit and after adjusting this for interest on capital, depreciation, annual profit rental, domestic items (which were included in the accounts), and for wife's wages (which were not) he arrived at a figure of £8,419 to which he added 11¼ per cent to reflect the adverse effect of the "scheme" in those two years, making a total of £9,366. To this figure he applied a years' purchase of 3.25, giving a figure of £30,439.

He sought to support his years' purchase of 3.25 by reference to the profitability of the business and in particular to the agreed net profit and compensation of which he had personal experience in eight other retail businesses, five of which were in Domestic Street, two in other parts of Leeds and one in Wakefield in respect of a sweet and tobacco business. The agreed net profit in these cases varied between £1,122 and £9,845 and the years' purchase between two and three.

Notwithstanding a rising trend of inflation these agreements had been reached by taking the average net profit over two or three years. For this reason, therefore, Mr Harrap took an average of the last two years' net profit at the subject premises. By comparison, having regard to the nature of the business and the fact that it was well established and profitable, he considered that a years' purchase of 3.25 was appropriate.

Mr Evans approached the problem in an entirely different and novel way. He looked at the annual sales for the three years 1976 to 1978 and from these estimated an increased figure of turnover at the date of dispossession which represented £1,735 per week. He added 15 per cent for the effect of the "scheme", increasing the weekly takings to a figure of £1,995. He then multiplied this figure by 8.5 and deducted the agreed value of the fixtures and fittings to arrive at the value of the goodwill at £16,157.

Mr Evans justified his multiplier of 8.5 by reference to the asking prices for newsagent, sweet and tobacconist businesses with living accommodation advertised in 1979 and 1980 in respect of four premises in other parts of Leeds and 16 throughout a large area between Coventry in the south and Durham in the north. He had visited the four premises in Leeds but had no personal experience of any of the businesses advertised.

He extracted from the advertisements the stated gross weekly takings and the asking price for the business excluding stock but including fixtures and fittings. The gross weekly takings varied between £530 and £2,500 and the asking price expressed as a multiple of the gross weekly takings varied between 5.33 and 10. From this information, he concluded that a multiple of 8.5 would be appropriate for Mr Nuttall's business.

In support of his figure he referred to two recent settlements of goodwill compensation for total extinguishment of newsagent, sweet and tobacconist businesses in Leeds where the value of the goodwill, fixtures and fittings amounted to 7 and 9.3 times the gross weekly takings.

Mr Evans produced an alternative valuation more in line with Mr Harrap's conventional approach whereby he arrived at an adjusted estimated net profit in 1979 of £10,699, to which he applied a years' purchase of 1.5 to arrive at goodwill compensation of £16,048. His multiplier of 1.5 was derived from taking the value of the goodwill in his primary valuation and dividing by the estimated profit. He also compared the two settlements of goodwill compensation already referred to where the years' purchase amounted to 1.5 and 1.79, but where the ratio of profit to turnover was substantially less than that in Mr Nuttall's business.

The matter of what should be included in the "scheme" for the purpose of compensation was not fully argued before me and I was not referred to any authority. The two valuers agreed that whether or not the yellow area was included in the "scheme" made little or no difference to their valuation. Neither valuer submitted any alternative valuation. Any decision on this point will not affect the amount of compensation payable. Therefore it is unnecessary for me to decide the issue in this case.

On the matter of valuation, I think Mr Evans' approach to the problem suffers three fundamental defects. There is no means of accounting for profitability, which can vary considerably from business to business, even in the same trade, and where it has been shown that in Mr Nuttall's business there was a relatively high profit in relation to turnover. The method may, with proper evidence, form a guide as to the asking prices or even the market values of

A businesses but leaves unsolved any difference there may be between market value and the value to the claimant. Above all, the valuation depends on evidence of advertised offers on businesses none of which are within the personal knowledge of Mr Evans and where there is no evidence as to the accuracy of the information or as to the profitability of the businesses being offered. Mr Harrap did not accept this evidence, but no objection was made on the grounds that it was hearsay.

Hearsay evidence

I refer to *English Exporters (London) Ltd v Eldonwall Ltd* [1973] Ch 415 and the decision of Megarry J at pp 420-423 and the conclusion

B on the matter of hearsay evidence at p 423 as follows:

> Putting matters shortly, and leaving on one side the matters that I have mentioned, such as the Civil Evidence Act 1968 and anything made admissible by questions in cross-examination, in my judgment a valuer giving expert evidence in chief (or in re-examination): (a) may express the opinions that he has formed as to values even though substantial contributions to the formation of those opinions have been made by matters of which he has no first-hand knowledge; (b) may give evidence as to the details of any transactions within his personal knowledge, in order to establish them as matters of fact; and (c) may express his opinion as to the significance of any transactions which are or will be proved by admissible evidence (whether or not given by him) in relation to the valuation with which he is concerned; but (d) may not give hearsay evidence stating the details of any transactions not
> C within his personal knowledge in order to establish them as matters of fact. To those propositions I would add that for counsel to put in a list of comparables ought to amount to a warranty by him of his intention to tender admissible evidence of all that is shown on the list.
>
> I have spent some little time in dealing with this matter of evidence as it appears to be the subject of no direct modern authority, and experience suggests that it is a matter upon which there is considerable misunderstanding. When a list of comparables is being prepared for the trial, as is usual and convenient, it is all too common to include in the list transactions upon which there will be no admissible evidence but only hearsay of a greater or lesser degree of reliability. If the parties exchange lists of comparables at an early date, often much time and money can be saved by the experts on each side agreeing such of the transactions in each list as, after any necessary
> D inquiry, they feel they can accept as being reliably summarised; and in this way the additional expense of proving a favourable comparable not within an expert's own knowledge can be avoided. But if the other side will not accept the facts, then either the transaction must be proved by admissible evidence or it must be omitted as a comparable.

In this case it is a matter as to the weight that I should attach to the evidence.

I prefer Mr Harrap's valuation because it is founded on the facts of profits of the claimant's business and comparison can be made with the profits and settled compensation in respect of a number of similar businesses in the locality that are within Mr Harrap's personal knowledge. The method of valuation follows a well-

E established pattern that has formed the basis of agreement on goodwill compensation in the area for many years. It would require very positive evidence to show that such a well-established practice was unacceptable.

Therefore I find that the compensation payable to Mr Nuttall for the loss of goodwill on the total extinguishment of his business is the sum of £30,439 to which must be added the agreed amount of £964 in respect of the loss of forced sale of fixtures and fittings, the loss on forced sale of stock and domestic disturbance, giving a total figure of £31,403 plus legal and surveyor's fees in accordance with the appropriate scale.

I have opened the sealed offer of compensation lodged by the

F acquiring authority and find it to be less than the amount of my award. The acquiring authority will pay to the claimant his costs of this reference, such costs if not agreed to be taxed by the Registrar of the Lands Tribunal on Scale 3 of the County Court Scales of Costs.

SIMPSON v STOKE-ON-TRENT CITY COUNCIL G

(REF/116/1980)

(W H REES FRICS)

March 22 1982

Estates Gazette May 15 1982

(1982) 262 EG 673

House in Stoke-on-Trent clearance area — Whether bank interest on bridging loan payable as disturbance — During compensation negotiations lasting nearly two years claimant bought another house with aid of bridging loan — Wife and family moved to new premises but claimant remained in subject property — Advised not H **to move because of possibility of deterioration of condition and damage by vandals and to get "proper compensation" — Authority's practice to offer advances of 90% of value to claimants — Alternatively 75% made available where possession continued, but claimant not informed of this — Held that it was not reasonable for claimant to take out bridging loan and he was not bound to do so — He could have referred compensation dispute to tribunal and taken advance payment — To award sum claimed (£1,068.07) would place "unfair burden" on acquiring authority — No special circumstances in present case — Member points out that soon after reference tribunal can be asked to view subject property — Also draws attention to president's undertak-** J **ing on expedited procedures to obviate delays in small compensation cases**

W Wood (instructed by Pickering & Butters, of Stafford) appeared for the claimant; Miss Ingram, solicitor, for the acquiring authority.

Giving his decision, MR REES said: This reference was made by the claimant to determine the compensation to be paid in respect of his house in Longton, Stoke-on-Trent. Compensation for his freehold interest had been agreed in the sum of £4,875 plus all other items of disturbance (save one) in the sum of £285.75. The disputed item (bank interest) was agreed as to amount at £1,068.07 K for the whole of the period concerned. The claimant maintained that it was payable under the heading "disturbance": the acquiring authority disagreed.

Counsel for the claimant called the claimant himself, whose evidence was not challenged and related mainly to the sequence of events and the reasons why he pursued the course he did. He also called C J G Peel FRICS, who practises in Stoke and who had advised the claimant from the outset. He spoke mainly as to the course which the negotiations took. He is a partner in Louis Taylor & Sons, chartered surveyors, of Stoke-on-Trent. I refer to them hereinafter as "Taylors". The acquiring authority's solicitor called H F Moore ARICS, an estates officer employed by them, whose L evidence also related to the course of negotiations.

I shall refer hereinafter to the subject house as "no 29" and to the claimant's present house as "no 1": its full address is 1 and 1a Meir Road, Normacot, Stoke-on-Trent.

From the evidence and from statements which were read by counsel for the claimant I set out the relevant events in date order. There was very little dispute on these matters.

Date	
January 20 1977:	The acquiring authority made the relevant order: The City of Stoke-on-Trent Longton Numbers 145/148 (Blantyre Street) Clearance Area Compulsory Purchase Order 1977.
February 27 1978:	The order was confirmed. M
April 24 1978:	Notice to treat and notice of entry were served. The claimant decided to buy another house.
May 29 1978:	A formal claim was submitted.
June 1978:	Taylors wrote to the acquiring authority saying that the claimant had informed them that he was buying no 1.
December 21 1978:	The acquiring authority made an offer of £3,400 for the freehold of the subject house. Thereafter there were telephone calls and conversations between Taylors and the acquiring authority.
January 3 1979:	Taylors asked for £5,500 compensation for the freehold interest.

April 1979:	The claimant contracted to buy no 1.
May 1979:	A Mr Dutton, an officer of the acquiring authority, visited the claimant and was informed by him that he (the claimant) had purchased no 1 and had obtained a bridging loan. The acquiring authority say that they did not know of the loan until September 2 1980.
May 22 1979:	The claimant completed the purchase of no 1 and his wife and family moved in. The claimant, however, stayed on in the subject house for the reasons referred to below. The claimant obtained a loan of £3,400 from his bank, initially for a period of three months but that was extended.
May 23 1980:	The reference was made to this tribunal.
June 17 1980:	The acquiring authority verbally offered £4,000 for the freehold.
July 9 1980:	That offer was confirmed in writing, Taylors replying (in writing) that £5,500 was required.
August 28 1980:	A meeting took place between Taylors and the acquiring authority and £4,875 was reached in negotiations between them for the freehold (it seems there was reluctance on both sides) plus £285.75 for items of disturbance other than the disputed item.
November 14 1980:	The acquiring authority confirmed the figure of £4,875 in a letter to Taylors.
December 1 1980:	The matter was reported to the city council.
January 30 1981:	The purchase of no 29 was completed and the bridging loan repaid.

There was no dispute that it was the practice of the acquiring authority to offer advances of 90% of valuation (under section 52 of Land Compensation Act 1973) to claimants whose properties were acquired compulsorily. They also operated a local scheme under which a similar advance of 75% of valuation could be made, the claimants remaining in possession. Under cross-examination Mr Moore said that the "90% letter" had been sent to the claimant but he had assumed that Taylors would have informed the claimant of the arrangement regarding the possibility of a 75% advance.

The claimant said that he purchased no 1 because he had received notice of entry while he lived in no 29. His wife and family moved into no 1 and he had to pay rates on both properties, although he had made no claim for double rates. He knew of the possibility of obtaining an advance from the acquiring authority of 75% of £3,400 only a few weeks before the purchase of no 29 was completed. He had known of the possibility of obtaining an advance of 90% if he gave possession of no 29 but that he did not wish to do. He stayed in no 29 because he was conversant with the value of houses in the area and was not prepared to accept so low a figure as £3,400. He was advised not to move: he thought he would get a proper price if he remained in possession. He had maintained no 29 well and did not wish to risk the acquiring authority saying that no 29 was in poor condition. He also feared that no 29 might be damaged by vandals.

Mr Peel said that on some occasions he advised such claimants as Mr Simpson to remain in possession of their houses because of possible damage by vandals, because the house itself and those round about gave the appearance of being derelict if they were left vacant and to get proper compensation. It was particularly important that they should do so where there was a wide gap in the negotiations: that was so here, the difference being £5,500 on the one hand and £3,400 on the other. He knew that the claimant was buying a new house but he did not know that the claimant had obtained a bridging loan from his bank. In cross-examination he agreed that it was possible to agree a schedule of condition and photographs although he had reservations as to that course. In his experience it was not always completely satisfactory; he had known difficulties arise in agreeing the wording of schedules of condition.

Comments on the negotiations, their length and the form which they took were made both by counsel for the claimant and by the solicitor for the authority. The solicitor for the authority submitted that section 52 provided safeguards and submitted that the interest on the bridging loan was not a natural and reasonable consequence of the compulsory acquisition.

Counsel for the claimant submitted that the claimant was fully justified in taking the course he had: the claimant had stayed in no 29 at some cost to himself. He submitted that if the acquiring authority relied on the possibility of an advance of 75% of valuation being made, the onus was on them to tell the claimant of it. In reply to a question put to him by myself asking why the claimant had not made a reference to this tribunal in 1978 or 1979, he said that claimants wanted time to negotiate and that they were reluctant to come to the tribunal because of the possibility of incurring costs.

While I fully understand that in most cases it is desirable that time should be allowed for negotiations, a period of nearly two years (from, say, June 1978 ending two months after the notice of entry, May 1980) cannot be said to be a reasonable period for negotiations in a case of this sort. The claimant decided to buy another house in April 1978 when he received notice of entry. It was in fact not until the first quarter of 1979 that he found and decided to buy no 1. I fully understand his reasons for staying in no 29 as he did, but it was not reasonable that he should have taken a bridging loan from his bank in May 1979 and he was not bound to do so. Having decided to negotiate for no 1, he could at that stage have made a reference to this tribunal to have the compensation for no 29 determined and he could have taken an advance payment of 90%. If I were to award him the sum claimed it would place an unfair burden on the acquiring authority. That matter is referred to in *Horn v Sunderland Corporation* [1941] 2 KB 26. At p 49, Scott LJ said:

> The statutory compensation cannot, and must not, exceed the owner's total loss, for, if it does, it will put an unfair burden on the public authority or other promoters who on public grounds have been given the power of compulsory acquisition, and it will transgress the principle of equivalence which is at the root of statutory compensation, the principle that the owner shall be paid neither less nor more than his loss.

I asked counsel for the claimant if this item had been considered by this tribunal on another occasion and he referred me to: *Roberts v Greater London Council* (J Stuart Daniel QC) 229 EG 975 and *Clarke v London Borough of Lewisham* (JD Russell-Davis FRICS) Ref/283/1975. I have since had an opportunity of considering them. In *Roberts,* interest on a deposit was allowed because of "what Mr Roberts told me and from some figures he showed me and from some letters produced by the council." In *Clarke* £1,155 was claimed but only £105 (conceded by the acquiring authority) was confirmed by the tribunal. It appears that the relevant period was for two months from the date of acceptance of an offer in the amount agreed "to allow for conveyancing, which in the circumstances is thought to be reasonable". There follows the following paragraph:

> Had the claimants proved title as they were asked to do, they could have received about that time an advance payment of 90% of the valuation plus the home loss payment. Furthermore, it is significant that no action had been taken by the claimants to refer the dispute to the Lands Tribunal, which it was open to them reasonably to do by September 1972.

There appears to be special circumstances in each of these instances, but I find none in the present one. The only "evidence" which is related to my finding for the claimant, but in a sum lower than the £1,068.07 claimed here (as in *Clarke*), is the following contained in a document submitted by the acquiring authority:

> These payments (advance payments of 90 or 75%) would have been cheaper to the city council than reimbursement of the bank interest as the council's loss of interest on money deposited at a long-term rate of about 11½%, eg on a loan of 75% of £3,400 for the period 22nd May 1979 to 30th January 1981 the sum would be £494.00.

This seems to me to be irrelevant: the basis of compensation for such an item as this, if any is payable, is the loss to the claimant, not the cost to the authority. There was no evidence to the effect that the claimant could not have avoided borrowing from his bank, whatever course he had pursued.

There are two other matters upon which I should comment. First, in such circumstances as this requests are sometimes made soon after the reference has been made to this tribunal for the member to view the subject property before the hearing. That could have been done here. Second, the following is an undertaking given by the president of this tribunal which relates to small compensation cases:

A

Statement on Delays

The Committee on the Use of Valuers in the Public Service in their Report (Cmnd 5518) stated that delay had been advanced as a major criticism of the present procedures of the Lands Tribunal, it being said that the average time between a case being listed and being heard is about 7 months, but can be considerably longer. Concern over delay has been voiced elsewhere both in and out of parliament. The primary reason for delay has been the settling of cases after a date has been appointed for hearing, leaving it too late for the Member concerned to hear another case in its place. Another reason has been postponements at the request of parties.

I have with the help of the Members and the Registrar instituted new procedures to obviate these delays. Already an appreciable number of cases have been set down for hearing within 3 weeks of readiness. I am par-
B ticularly concerned with small compensation cases and it is my intention that, as a general rule, subject to the availability of courts, they shall be heard within one month of readiness. Further, I can say that, subject to there being no substantial increase in the Tribunal's work load and to exceptional circumstances, in future there should be no unreasonable delay in any type of case.

Applications for postponements, whether or not by agreement, will be allowed only in the most exceptional circumstances, for every postponement delays the hearing of another case. However, it is and always has been open to parties to fix with the Tribunal a date convenient to all. Where that is the wish of a party, the Registrar should be so informed within 7 days of receipt of notice that the case has been entered in the list for hearing; after the case has been appointed it will be too late.

C So, for the reasons set out above, I come to the conclusion that no additional compensation is payable in respect of the interest on the bridging loan.

I received submissions in writing with regard to costs. The claimant will pay the acquiring authority its costs of this reference, such costs if not agreed to be taxed by the Registrar of the Lands Tribunal on scale 2 of the County Court Scales of Costs.

D B THOMAS & SON LTD v GREATER LONDON COUNCIL

D

(REF/213/1980)

(W H REES FRICS)

April 1 1982

Estates Gazette June 5 1982 and June 12 1982

(1982) 262 EG 991 and 1086

Disturbance claim by clothing manufacturers in Poplar, East London—An "exceptional case" in which surveyor's fee for preparing and negotiating claim awarded on scale 5(a) increased by £100 — Claimant company purchased site from acquiring authority and built new factory on it — Losses sustained in transferring
E **business in very complicated though well-planned move — Abortive payments to architects and builders with regard to development of new site disallowed as not being direct and natural consequence of acquisition — Not right to allow claimant to indulge in speculative development at authority's expense — "Too great a burden", following dictum of Scott L J in *Horn v Sunderland* — Other items disallowed included: bank fees and interest on bridging finance and similar costs on ICFC loan (following *Service Welding v Tyne and Wear CC*); surveyor's fee for advising on rating assessment of new premises; and cost of automatic porter (as it would pass with new premises on sale) — Among items allowed were: surveyor's fee for negotiating and purchasing new site at amount above scale 17(a) (as**
F **purchase "by no means a straightforward matter"); installation of burglar alarm system (since it would not pass on sale of new premises); provision of special ceiling and door to comply with Factory Acts ("more closely allied to plant than to building"); director's time in planning and effecting removal and in formulating compensation claim ("Tobin" costs); casual labour to supplement removal specialists; and loss of discounts for prompt payment — Award of £19,740 against £48,911 claimed and £12,001 offered**

Paul Morgan (instructed by Forbes & Son) appeared for the claimant; John Male (instructed by the solicitor for the Greater London Council) for the acquiring authority.

Giving his decision, MR W H REES said: The claimant who made G
this reference is a family concern of clothing manufacturers established for 100 years in the East End of London. It owned and occupied its own freehold premises ("the old premises") which were acquired compulsorily and it purchased from the acquiring authority a site nearby ("the new site") on which it built a new factory which it now occupies ("the new factory"). ["The old premises" were a house, rooms over roadway, factory, office, roadway and premises, part of 14 Andrew Street, London E14, and house and premises at 12 Mills Grove, London E14. "The new factory" is at 164 Chrisp Street, London E14.]

Compensation for the freehold of the old premises was agreed in
the sum of £21,600 and that matter forms no part of this reference, H
which is to determine the amount payable in respect of disturbance. Some items were agreed, the total being £4,069.77 (see Appendix A hereto*). For the remaining items, at the beginning of the hearing a total of £44,841.60 was claimed, the acquiring authority offering £7,931.84 in reply.

The detailed claim in respect of the disputed items is set out in Appendix B to this decision. It is substantially the same as the document submitted by the claimant pursuant to rule 42 of the Lands Tribunal Rules 1975, but it incorporates certain amendments which were made during the hearing before me: references to other documents have been omitted together with some words which were more
in the nature of submissions and explanations. J

In this decision, individual items are referred to by the numbers and letters used in Appendix B.

[For the convenience of readers Appendix B has been split up so that each item in it appears immediately above that section of the tribunal's decision which refers to it.—Ed]

As originally submitted there was also an item:

"(5). Cost of security at old factory, 14 Andrew Street, following demolition of surrounding buildings."

but this item was withdrawn at the hearing before me. Agreed items have been omitted from Appendix B, but where the amount only was disputed the sum put forward in reply by the acquiring authority is
shown. In addition to the items in the rule 42 document the acquiring K
authority offered the sum of £820, which item I have numbered "(22)": it was described as "other losses". The claim below (21) in respect of surveyor's and accountant's fees was also disputed as to amount: I have numbered that "(23)".

The items in the claim were, where appropriate, supported by receipted bills and letters, a total of about 100 documents. Some items of claim were said by the respondent authority not to be supported by full documentation, but in a claim of this complexity it may not be possible to support every detail by documents: some have to rest on estimates. In no case where the claimant said that sums had been paid out by it did the acquiring authority dispute that the sums
concerned had in fact been paid. L

The relevant order is the Greater London Council (Byron Street, Tower Hamlets) Compulsory Purchase Order 1969, which was made under the Housing Act 1957 on October 17 1969, the council's resolution being passed on July 23 1968. The order was confirmed on June 30 1971. Notice to treat was dated December 1 1971 and the parties agreed that the date of valuation for business disturbance was August 7 1978 when possession of part of the old premises was taken. They had a floor area of just under 6,000 sq ft. The new factory is in Chrisp Street and the new building has a floor area of 6,345 sq ft, the site area being 10,500 sq ft. The claimant paid £16,000 for the
freehold of the new site on March 21 1972. M

Counsel for the claimant called J B Thomas, Managing Director of the claimant company, and H R Wightwick FRICS, whose professional practice in the City of London was founded on his experience in advising on compulsory purchase matters. Counsel for the respondent authority called R L Howell ARICS, an assistant divisional valuer in its valuation and estates department; A H Sumpter, a member of the Institute of Public Finance Accountants, who is employed by the acquiring authority and whose work is in connection with trade disturbance claims; A J Dolan, a civil

*Reproduced at the end of the decision.

A engineer, who was employed formerly by the acquiring authority but who is now retired. Both Mr Thomas' and Mr Wightwick's evidence gave the history of the whole matter in detail and they spoke as to the individual items of claim. Mr Howell spoke of the development of the new site and commented on some of the individual items of claim, namely (2), (3), (4), (5) and (6). Mr Sumpter mainly referred to (7), (8), (20), (21) and (22). Mr Dolan's evidence related to items (8), (9), (11) and (15).

For the sake of completeness and by way of clarification the sequence of events over the period of about 10 years prior to the hearing before me is summarised in date order in Appendix C hereto, the
B dates being taken from an agreed schedule and including only those which are relevant to the disputed items.

Appendix C

July 23 1970	GLC questionnaire for completion - No action till compulsory purchase order confirmed and notice to treat served.
April 19 1971	Date the initial particulars of the council's site at Rifle Street/Chrisp Street were passed to the claimant's surveyor.
April 27 1971	GLC offer site 10,500 sq ft - (old site 4,450 sq ft net). Price asked £1.75 per sq ft (£76,230 acre). GLC unable to divide site.
C March 1 1972	Architect Mr Brian Hayward ("Hayward"), who had previously been instructed, produced plans for two factories, 7,200 and 6,500 sq ft, on Chrisp Street site, one for occupation and use by D B Thomas & Son Ltd ("DBT & Son Ltd").
September 4 1972	Revised drawings from Hayward.
Early 1970 to autumn 1972	Many discussions with bankers (Barclays) re compulsory purchase order and possible finance, but bank non-committal until figures available.
October 17 1972	Hayward first account presented - £1,423.73. Tenders opened - £133,972, £95,800, £94,000, £79,976.
November 29 1972	GLC confirm site area at 10,500 sq ft. GLC asked to confirm amended memorandum of terms.
D February 12 1973	Contractors tender price for revised drawings received £79,976. Planning permission received.
February 12 1973	Date of possession of 12 Mills Grove.
June 11 1973	Werren & Co confirmed able to find merchant bankers to provide finance for scheme subject to details, at MLR + 5½%.
April 10 1974	Contracts exchanged re Chrisp Street site.
Winter 1972 to summer 1974	DBT & Son Ltd unable to obtain finance. Barclays not helpful and only interested in loan for 5 years. Merchant banks want all business (etc), with MLR and 5½% and 70% mortgage. Williams & Glyn's, North Lombard, Barclays District Head Office not helpful.
E Early August 1974	Meeting with Quill and Rosen to discuss assistance with land, building and finance.
August 12 1974	Rosen asked to employ E J Cook & Co, on behalf DBT & Son Ltd (not Quill). Eventual account £2,500 + VAT.
August 19 1974	Date of planning application submitted by M D Rosen on behalf of DBT & Son Ltd for a new two-storey factory building of 13,580 sq ft, gross.
November 7 1974	Plans produced.
November 19 1974	Date of planning application submitted by M D Rosen on behalf of DBT & Son Ltd for a new two-storey factory building of 13,500 sq ft.
December 18 1974	Tenders for building due, but did not arrive.
F February 5 1975	Date of GLC's letter to Forbes & Son (DBT & Son Ltd's solicitors) ". . . contract . . . will be in jeopardy if a factory is not built on the land by 10th April 1975 . . ."
February 20 1975	Quill's solicitor reports delay in dealing with land due to six weeks' absence in crucial matter.
April 10 1975	Expiry of contractual period for completion of new factory.
Late summer 1974 - summer 1975	Consideration of solving finance problem by selling half land to Quill. Quill's company eventually pulled out without explanation. Mr Rosen told E J Cook (Steel Constructions) that DBT & Son Ltd had given instructions for all steelwork planning, both DBT & Son Ltd and Quills to go ahead. Account comes to DBT & Son Ltd. Account paid by DBT & Son Ltd.
July 1975	DBT & Son Ltd decided to investigate possibility of one single-storey factory on site, with Mr Rosen, now familiar with requirements site etc. G
August 15 1975	Mr Rosen told to request tenders by September 1 1975.
August 22 1975	Date of planning application submitted by Mr Rosen on behalf of DBT & Son Ltd for a new factory of 6,200 sq ft gross.
September 1 1975	No tenders received.
September 5 1975	DBT & Son Ltd phoned all builders and found Mr Rosen had not yet sent plans to builders for tenders.
Summer 1975 - autumn 1975	DBT & Son Ltd decided only possible to consider financing single-storey factory on site. DBT & Son Ltd employed Mr Rosen as he had much information. H Result considered totally unsatisfactory by DBT & Son Ltd, who disputed his accounts, but after counsel's opinion, which was accepted, settled out of court.
Mid-September 1975	DBT & Son Ltd in contact with builders Parpak, who could do a complete service.
September 16 1975	Parpak submit first quote and specification £50,000 + £1,750 for clearing site.
October 2 1975	Meeting with Parpak.
October 7 1975	Other builders investigated.
October 17 1975	Parpak submit revised plans and quote £50,000 then £50,460.
November 20 1975	Verbal instructions given by DBT & Son Ltd to Parpak to proceed at £50,460. Finance not settled. J
December 1 1975	Parpak expected to start work.
December 3 1975	Meeting with Parpak agreeing to start December 4 1975.
December 10 1975	Consultants asked to quote for layout of machinery in new factory.
January 1 1976	Deemed date of possession of Chrisp Street site for calculation of interest.
January 16 1976	Parpak requested first payment.
January 20 1976	Parpak still do not have written signed agreement with DBT & Son Ltd.
February 3 1976	Parpak ceased work as no contract had been signed.
February 4 1976	First payment £3,283.00 made by DBT & Son Ltd to Parpak.
December 1975- - February 1976	Department of Industry Land Scheme to assist. Can only consider application if before start of work. K
March 1976	Information received by DBT & Son Ltd that Parpak were thought to be heading for liquidation. A E Symes Ltd asked to quote for work on existing scheme.
March 9 1976	Consultant quoted 16 days at £25 per day = £400.
April 5 1976	Parpak requested second payment £7,069. Not paid.
April 15 1976	Meeting with Parpak asking DBT & Son Ltd to pay Sanders & Foster direct for steel frame.
April 23 1976	A E Symes report they could take over at £57,100.00. They go into liquidation before any action could be taken.
July 2 1976	Orwell say they would take over - price £48,650.55. L
August 5 1976	Orwell asked for new quote; DBT & Son Ltd still trying to get assistance from Department of Industry.
August 6 1976	Sanders & Foster would want DBT & Son Ltd to pay their account.
August 12 1976	Orwell quote £55,740.
August 31 1976	Parpak in liquidation.
September 29 1976	Orwell could start October 3.
October 17/19 1976	Contract made out DBT & Son Ltd/Orwell but not signed until October 19 1977.
November 22 1976	Date of GLC letter to Forbes & Son ". . . I must now take steps to rescind the contract".
April 26 1977	GLC concerned re starting. Have other interested buyers.
September 13 1977	GLC suggest rescinding contract of sale alleging breach of clause as to date by which building should have been completed. M
December 12 1977	ICFC offer accepted and fee £800 paid.
December 12 1977	Meeting at National Westminster Bank Ltd.
February 1978	Rosen account. Counsel employed and DBT & Son Ltd settled out of court.
February 17 1978	Date of completion of the sale of council's freehold interest in the Chrisp Street site to DBT & Son Ltd, also date of supplemental agreement.
February 22 1978	Forbes fee re site paid, £653.67.
Week ending August 6 1978	Date of removal of the claimant to new premises.
August 7 1978	Date of possession 14 and 14a Andrew Street.

A August 10 1978	Date of completion of the purchase of the subject lands by the council.
August 2 1979	Date disturbance claim submitted.

I was referred to the following:

(i) *Harvey v Crawley Development Corporation* [1957] 1 QB 485
(ii) *Service Welding Ltd v Tyne and Wear County Council* (1979) 250 EG 1291
(iii) *Pearl Wallpaper Holdings Ltd v Cherwell District Council* (1977) 244 EG 474
(iv) *G E Widden & Co Ltd v Kensington and Chelsea London Borough Council* (1970) 213 EG 1442
B (v) *Bailey v Derby Corporation* [1965] 1 All ER 443
(vi) *Bryce v Motherwell District Council* [1980] R & VR 282
(vii) *Smith v Birmingham City Council* [1974] 232 EG 593
(viii) *Easton v Newcastle upon Tyne Metropolitan Borough Council* (1979) 250 EG 165
(ix) *Horn v Sunderland Corporation* [1941] 2 KB 26
(x) *Dodd Properties (Kent) Ltd v Canterbury City Council* [1980] 1 WLR 433
(xi) *Bloom (Kosher) & Sons Ltd v London Borough of Tower Hamlets* (1978) 247 EG 1091
(xii) *Smith v Strathclyde Regional Council* (1981) 257 EG 391, 501

The roman figures in brackets are included after the case name where it is used hereinafter. I was also referred to *Halsbury* (3rd ed) *on*
C *Damages*, paragraphs 1127-1131, 1139, 1140, 1174 and 1193 and to *McGregor on Damages* (14th ed), paragraphs 233, 234, 236-238 and 241.

I now consider the disputed items in turn.

Accrued interest

	Claimed £	*Offered* £
(1) *Accrued statutory interest on the price of no 12 Mills Grove, E14* The interest, over the period 12.2.73 (date of entry) to 15.8.78. (date of completion) is agreed in the amount of £853.24 net of tax. But because payment was not D received until 29.11.79 a further amount of £101.60 net is due.	954.84	853.24

Towards the close of the hearing before me counsel for the claimant withdrew this item: in his view it does not fall under the head of "disturbance". However, he asked me to record that the item was originally put forward; he said that as a result of an oversight on the part of the two solicitors concerned, payment of interest in respect of the price of part of the old premises was not received by the claimant until about 15 months after the date of completion of the sale.

E **Abortive payments in respect of the new site**

	Claimed £	*Offered* £
(2) *Abortive expenditure on alternative accommodation*		
(a) These amount to £1,423.73, and payment to Mr Hayward was made on November 17 1972. Adopting an average rate of interest of 14%, the interest due over the period of 5 years and 254 days to August 7 1978 is £1,135.30. The total claimed is therefore	2,559.03	Nil
(b) *Fees of Mr Michael Rosen* These comprise two accounts from Mr Rosen totalling £3,328, together with photocopies of four cheques (two for £800 and two for £500), ie a total of £2,600, this being the reduced fee F agreed as a result of the threatened litigation. Payment was made in stages, on October 22 1974, December 9 1974, June 6 1975 and February 9 1978. The interest is related to the amount of £2,100 – ie ignoring the February 1978 payment – and adopting a commencement date of March 6 1975, and adopting an average rate of interest of 15½%, the interest due over the period of 3 years and 154 days is £1,113.83. The total claimed is therefore	3,713.83	Nil
(c) *Fees of Messrs E J Cook* Payment of the amount of £2,500 was made in two stages – on July 24 1975 and January 1 1976. Adopting a commencement date of October 1 1975 and an average rate of interest of 15½%, the interest due over the period of 2 years and 311 days is £1,105.17. The amount therefore claimed is	3,605.17	Nil
(d) *Legal costs re: proceedings by Rosen and Cook* Since payment was not made until August 15 1978, no question of an interest payment arises. The amount claimed is thus the amount paid	386.00	Nil
(e) *Abortive payment to the Parpak (Building Group) Ltd* A letter from Parpak dated February 5 1976, acknowledges the payment of £3,283. That part of the payment relating to site clearance can be treated as part of the cost of the new building and the value of this work is put at £783. Thus the net amount claimed becomes £2,500 and interest at the average rate of 15½% over the period of 2 years and 183 days from February 5 1976 is calculated as being £969.28. The total claim figure therefore is	3,469.28	Nil

The new site was offered to the claimant by the respondent authority in April 1971: Mr J B Thomas considered that it was too large, but the respondent authority refused to divide it and no other sites, suitable in his view, were available. Mr Thomas instructed an architect, a Mr Hayward, to prepare plans for two buildings on the site, the intention being that the claimant should occupy one of them and that it should dispose of the other. This Mr Hayward did as well as obtaining competitive tenders. Mr Thomas discussed possible financing with banks and others but he could not obtain terms which he regarded as satisfactory to provide a viable scheme, so nothing came of it. One of the difficulties he encountered then and later was that he required only a fairly small sum of money: bankers and institutions were only interested in lending large sums.

The amount claimed under (2)(a) is the amount of Mr Hayward's fee paid by the claimant. In the first half of 1974, before contracts on the new site had been exchanged, Mr Thomas had discussions with a Mr Quill who owned an industrial site adjoining the new site. Mr Quill had plans for development of his site and there were discussions between Mr Thomas and Mr Quill on the basis that Mr Thomas, having purchased the new site, would sell the front land to Mr Quill, who would develop it together with his land and, at the same time, would build a factory on the back land for Mr Thomas. Mr Quill had an architect, a Mr Rosen, whom Mr Thomas instructed to prepare plans for this development. Mr Rosen was in touch with E J Cook Ltd ("Cooks") who were engineers concerned with the steelwork for the building proposed for Mr Quill's land and Mr Rosen instructed them to advise Mr Thomas. Mr Thomas lost confidence in Mr Rosen, and Mr Quill's company went into liquidation; the property market was down at that time. Thus that scheme foundered. Proceedings were taken by Mr Rosen and by Cooks against the claimant for recovery of fees: these proceedings were defended by the claimant and a settlement was made out of court. The sums claimed under (2)(b), (c) and (d) are the amounts of fees paid to Mr Rosen and to Cooks and the amount paid by the claimant to its solicitors respectively.

In the middle of 1975 Mr Thomas was in correspondence with a firm of builders, Parpak Ltd ("Parpak") with regard to the development of the new site, the building to be a single-storey one. By the end of that year he knew that the cost was likely to be in the order of £50,000. At a meeting between Mr Thomas and Parpak in December 1975, instructions were given to Parpak to start work on the site, Parpak also having prepared plans. No written contract was entered into and at this period Mr Thomas was investigating the possibility of obtaining a grant from the Department of Industry. £3,283 (see(2)(e)) was paid by the claimant to Parpak in February 1976 for plans and some site work, but further payment was refused after Mr Thomas had formed the opinion that Parpak were in financial difficulties. In August 1976 Parpak went into liquidation. Another builder said that he could take over the work, but he, too, went into liquidation before work could be started.

At least one of those who had been employed by Parpak, with others, formed a new company, Orwell Ltd ("Orwell"), and in July 1976 they had conversations with the claimant: eventually Orwell constructed the new factory on the new site and it was agreed that the value of the freehold interest in the new factory and site exceeds the total of its costs including all abortive amounts under (2).

From the evidence, I find that some of the work done on the site by Parpak and the plans which they prepared were of some use to Orwell and to some extent therefore the claimant may have obtained value for money. I am very doubtful whether any of the amounts paid under (a) to (d) went towards the ultimate development of the new site. At various dates between 1972 and 1977 the claimant was seeking finance to develop the new site, but it was not until the end of 1977 that it obtained finance on what it considered to be satisfactory terms from ICFC Ltd.

Counsel for the claimant submitted that the amounts claimed under (a) to (e) represented payments reasonably incurred by the claimant as a direct result of the compulsory acquisition for which it did not get value for money: they were not too remote to rank as compensatable items. The new site represented the only possible alternative premises and the claimant had a duty to mitigate its loss. He relied particularly on *Harvey* (i), *Bailey* (v), *Bryce* (vi) and *Dodd Properties* (x) and he referred to *McGregor*.

Counsel for the respondent authority submitted that the payment of these fees was not the direct and natural consequence of the acquisition and he submitted that the claimant's conduct was not reasonable. He, too, relied on *Harvey* (i), *Bailey* (v) and *Bryce* (vi). Mr Howell's view was that the claimant should have built a single building on the new site in 1972, but I come to the conclusion that he and counsel for the respondent authority are to some extent being wise after the event: in 1972 the claimant could not obtain finance on what it considered to be proper terms.

I find this a difficult matter but I come to the conclusion that it was the claimant's inability to get finance which caused at least one scheme to fail: that factor was a new cause analogous to the claimant's illness in *Bryce* (vi). Moreover, the claimant did not take advice as to whether the scheme, plans for which were drawn by Mr Hayward, was viable. To some extent the claimant adopted an expensive course in having detailed plans prepared: it might have sufficed to have had simple sketch plans drawn, at least at some early stages of the various schemes.

Particularly in respect of (a) and (b), the claimant was indulging in speculative development on the new site and in my judgment it would not be right to allow it to do so at the expense of the acquiring authority (see *Bibby (J) & Sons Ltd v Merseyside County Council* (1979) 251 EG 757, CA affirming (1977) 243 EG 467, Lands Tribunal, which was referred to briefly before me without being cited in detail). I accept the arguments put forward by counsel for the respondent authority. In addition, were I to allow any of these sums it would, I think, put too great a burden on the respondent authority and in that connection I have regard to the judgment of Scott LJ in *Horn* (ix). At p 49 he says:

> But here we come to the other side of the picture. The statutory compensation cannot, and must not, exceed the owner's total loss, for if it does, it will put an unfair burden on the public authority or other promoters who on public grounds have been given the power of compulsory acquisition, and it will transgress the principle of equivalence which is at the root of statutory compensation, the principle that the owner shall be paid neither less nor more than his loss.

In case this matter should go further, it is right that I should determine the amount which I would have allowed had I decided it right to admit these items as part of the claim. For the reasons indicated above, I consider that the necessary work for each scheme could have been carried out for less. I would therefore have allowed 60% of item (a), 40% of item (b), 50% of (e) and £250 for (d). I would not have allowed any sum in respect of (c), as the scheme based on Mr Rosen's plans did not, in my judgment, go sufficiently far to warrant fees in respect of structural steelwork being incurred.

Costs of new factory prior to occupation

(3) *Financial costs of existing new building from which no benefit was gained prior to full occupation*

A permanent loan in respect of the construction of the new building was obtained from Industrial and Commercial Finance Corporation Ltd (ICFC Ltd), and this is repayable over 14 years at a fixed rate of interest of 14%. However, the loan formalities, apart from involving an arrangement fee and a special audit, were not finalised until February 17 1978, when the first tranche was received.

By contrast, the building contract with Orwell Contractors Ltd took effect from October 17 1977, work commenced on October 21; the first stage payment of £8,900 was paid to the contractors on December 17 1977. As a consequence, it was necessary to obtain a bridging loan from the National Westminster Bank (for a period of 63 days) and further bridging finance was necessitated by delays on the part of ICFC Ltd in making payment of additional tranches.

(a) *Bank arrangement fees*

These are covered by four letters from the National Westminster Bank, and the amounts are:

	£		
	35.00		
	150.00		
	100.00		
	140.00		
		425.00	140.00

The last arrangement fee relates to the provision of finance in relation to the removal.

(b) *Bank bridging finance*

The first part of this calculation is straightforward, ie 63 days' interest at an average rate of 10% on £8,900, ie £153.62. But although the computation is a very difficult one, it is estimated that there was a running overdraft to the order of about £9,000, which endured for a further period of about 3 months (say 92 days). Adopting an average rate of 12%, produces an additional interest charge of £272.22, so that the total claimed under this heading becomes 425.84 150.00

(c) *ICFC loan arrangement costs*

ICFC negotiating fee	£800.00		
Audit fee - EK Frederick & Co	120.00		
		920.00	Nil

(d) *Interest paid to ICFC prior to August 8 1978*

February 17 to April 30 1978	£700.00		
May 1 to July 31 1978	1,465.20		
August 1 to 7 1978	121.10		
		2,286.30	Nil

(e) *Interest on purchase price of Chrisp Street site*

The price was £16,000 and the interest due to the GLC on the balance of the purchase money less deposit (£14,400) over the period January 1 1976 to February 17 1978. This sum is £2,631.05 and, in addition, there is interest on the deposit (£1,600) over the same period, which is calculated at £292.33; interest on the total purchase price (of £16,000) from February 18 to August 7 1978 - 171 days - at a rate of 12½% amounts to a further £936.99 (gross).

The total claimed under this heading is therefore 3,860.37 Nil

The first three sums claimed [in (a)], totalling £285, were fees paid to the claimant's bank to arrange bridging finance: the last payment was admitted by the respondent authority, since it relates to the provision of finance for the removal. (b) represents the interest on the bridging finance and, as in (a), part (£150) was admitted by the respondent authority, since it was in connection with finance for the removal. (c) and (d) are similar to the amounts in (a) and (b) but relate to the loan ultimately made by ICFC Ltd to the claimant. (e) is fully explained in the claim.

Counsel for the claimant submitted that the outstanding

items were not part of the price paid to develop the new site; alternatively, the claimant did not get value for money: The payments were made to proceed with the move and the claimant was mitigating its loss. The payments were in effect the same as double overheads, particularly rent paid for new premises not used.

Counsel for the respondent authority relied particularly on the decision in *Service Welding* (ii) and I see no difference between the bank interest which was paid in that instance and the disputed sums paid here. I therefore reject the disputed items. The total amount payable under (3) will therefore be £290, the amounts admitted by the respondent authority.

Costs of purchasing new site

(4)	*Costs of purchasing alternative site at no 164 Chrisp Street*		
(a)	*Legal costs*	614.35	474.35
(b)	*Surveyor's fees* The fee charged is considerably in excess of scale.	835.95	235.00
(c)	*Surveyor's fees for dealing with the rating assessment of new factory*	140.60	Nil
(5)	*Cost of security at old factory, no 14, Andrew Street, following demolition of surrounding buildings*	Withdrawn	

(a) Legal costs of £614.35 are claimed and £474.35 is conceded by the respondent authority. Part of the sum claimed relates to the abortive scheme discussed with Mr Quill. I reject that for the same reasons as I have rejected (2)(b). Part of the fees related to the claimant's not having conformed with the terms of the contract to buy the new site; it brought that on its own head. Part related to work connected with the loan for the new factory and so the claimant obtained value for money. I allow only the amount admitted, £474.35.

(b) Of the £835.95 claimed, it was agreed that £225 related to work done by Mr Wightwick before the date of the notice to treat. I therefore exclude that amount, following *Bloom (Kosher)* (xi) and *Widden* (iv) rather than *Smith v Strathclyde* (xii), which leaves £610.95 against which the respondent authority put £235, which is the amount based on scale 17(b) of the scales of charges of the Royal Institution of Chartered Surveyors. From the evidence I find that scale 17(a) is the more appropriate scale: the amount on that scale is £285, which would be the appropriate amount to allow here but for the fact that the purchase of the new site was by no means a straightforward matter. It was necessary for Mr Wightwick to advise on and negotiate various matters including a particular clause in the contract relating to recision. I therefore allow the sum of £350 for 4(b) plus VAT if appropriate and £5 out-of-pocket expenses.

(c) The claimant employed Mr Wightwick to advise on the rating assessment of its new factory and paid him a fee of £140.60 for that advice. Counsel for the respondent authority submitted that that was not a direct consequence of the compulsory acquisition and that the claimant got value for money. Counsel for the claimant submitted that this payment was similar to the surveyor's fee allowed in *Harvey* (i). I do not agree. I accept the submissions made by counsel for the respondent authority and delete this item of claim.

Costs of adaptation to new factory

(6) *Cost of adaptation to new factory*
Because the new factory premises occupy a very much more prominent position than the old, a burglar alarm system was considered essential, although that was not the case with the old premises until notice to treat was served. Since therefore, there is no statutory obligation upon a factory owner to install a burglar alarm system, it must follow that the expense cannot be treated as part of the cost of the new building, for which the company have obtained value for money. Similarly, whereas at 14 Andrew Street both the offices and factory had one access, the design of the new building is such that whereas the works and loading access front on to Cording Street, the offices, which are at first-floor level at the Chrisp Street end of the building, have had to be provided with their own separate pedestrian access (at ground-floor level) and because the office staff are not in a position to keep a constant check on who enters the factory through the door on the Chrisp Street frontage, it has been necessary for this to be provided with an automatic "porter".

		£		
(a)	*New burglar alarm system*			
	Cost of installation	1,621.00		
	Post Office connection charge	30.00		
	Additional keys	18.72		
			1,669.72	Nil
(b)	*Cost of automatic porter to door on Chrisp Street frontage*		92.00	Nil
(c)	*Special works to floor and boiler room of new factory*			
	Cost of sealing floor		95.00	95.00
	Cost of "Asbestolux" ceiling and louvred door to boiler room		175.00	Nil
(d)	—			

(a) Until notice to treat was served, the claimant had a caretaker and other employees living in accommodation adjoining the old premises and it had no need for a burglar alarm system. After the notice to treat, those conditions no longer obtained and so it installed a burglar alarm. The cost of so doing formed part of (5), which was withdrawn. The amount claimed is the cost of installing a new burglar alarm system at the new premises. Counsel for the respondent authority relied on *Smith v Birmingham* (vii) and submitted that the claimant got value for money and I would have accepted that argument if the alarm system at the new premises formed part of them, but it does not. The system is rented and there was no evidence to show that a purchaser of the new premises from the claimant would give any more because of the alarm system which would not pass with the new premises on sale. It seems to me to be somewhat similar to the costs of installation of the telephone, an item which was not disputed. I therefore allow £1,669.72 in full.

On the other hand, 6(b) would pass with the new premises on sale and in that respect I regard the £92 claimed as part of the cost of the new building. I therefore disallow it.

(c) The outstanding item here is the cost of providing an "Asbestolux" ceiling and louvred door to the boiler room to comply with the requirements of the Factory Acts. Counsel for the respondent authority submitted that the claimant got value for money in respect of this item but counsel for the claimant pointed out that this item related to a boiler used to provide steam for the claimant's machines and that this was more closely allied to plant than to the building. I agree and allow £175 in full.

Director's time in connection with the removal to the new factory

(7) *Claim for cost of director's time in relation to the provision of the removal to the new factory*
In the case of D B Thomas & Son Ltd, there are four working directors, Mr J B Thomas (chairman & managing), his sister, Miss V G Thomas, and his two sons, Mr N C Thomas and Mr M P Thomas. However, the last two directors were only appointed in June 1977, so it follows that much of the early work was dealt with solely by Mr J B Thomas.
It is considered that the appropriate rates per hour should be £10 for Mr J B Thomas, £5 for Mr N C Thomas and £3 for Mr M P Thomas. Subject therefore to these introductory comments, the compensation claimed can be categorised as follows:

A

(a) *Obtaining finance and organising erection of building*

It is estimated that Mr J B Thomas spent 41 hours in seeking, and ultimately obtaining, financial support for the new factory building, and prior to the negotiation of the loan from ICFC, already referred to, inquiries were made of, and discussions took place with, the following:

Barclays Bank Ltd (Poplar)
Barclays Bank District Head Office
Hambros Bank Ltd
Williams & Glyn's Bank (Romford)
B
North Lombard Finance Ltd (Ilford)
National Westminster Bank (Stratford)
Spicers (Finance Brokers) (Weston-super-Mare)
Warren & Co (Finance Brokers) (Bristol)

As a consequence, it is estimated that Mr Thomas travelled some 420 miles.

The assessment of time spent in organising the building of the new factory is a very much more difficult proposition, for it is realised that in adopting a reasonable approach, regard must be had to the level of abortive architect's and other professional fees claimed earlier.

C

Nevertheless, a further 20 hours would seem to be fair in the circumstances.

The total number of hours under this subheading thus becomes 61, and if the mileage travelled is costed at 10p per mile, the claim figure becomes — 652.00 — Nil

(b) *Planning the removal*

Initially, the company considered employing a firm of management production consultants to plan and organise the move from the old factory to the new. But when it was found that the consultants would charge a fee of a minimum of £1,500 and possibly as high as £2,500 and,

D

moreover, the consultants would require Mr J B Thomas to work very closely with them, it was decided that it would be cheaper and a more practical expedient if Mr J B Thomas dealt with the exercise himself.

To this end, he purchased a magnetic board, which was cut so as to provide a large-scale model of the new factory floor. Then each item of machinery and plant was carefully measured and a scale template cut from magnetic tiles, a coloured tape being used for the purposes of differentiation. The cost of the materials for this was £34.25.

E

It is estimated that the time spent in constructing the new factory "model", together with time spent in exploratory discussions with consultants, is in the order of 25 hours: but the ultimate arrangement of the machinery and plant took far longer, because consultations with other directors and senior staff extended over a considerable time.

Although no detailed records were kept, it is estimated that the time involved must have been about 100 hours of Mr J B Thomas' time, other time being "thrown in".

The total under this heading is thus 125 hours at £10, plus the cost of some items purchased, and

F

it is a lesser amount than would have been incurred had specialist consultants been employed; the careful planning served to reduce very considerably the time spent on the actual removal. — 1,284.25 — 34.25

(c) *Effecting the removal*

As a result of very careful planning, the move from the old factory to the new was completed in a maximum of 7 days, and, indeed, much of the major work was completed in the remarkably short space of 48 hours (the weekend of August 5/6 1978).

G

But in order to minimise disruption to production, Messrs N C and M P Thomas worked the whole of each of the preceding weekends from June 24 1978 onwards, in preparation for the removal, providing a work "bank" for operators to work on while various departments were being moved, checking new positions for machinery and completion of installations etc.

It is estimated that each of the junior directors worked a total of 85 hours, which at the rate of £5 per hour for Mr N C Thomas equals £425, and at £3 per hour for Mr M P Thomas, £255.

H

And although Mr J B Thomas was just as much involved as both of his sons, an additional claim of 10 hours at £10 per hour is made for him. — 780.00 — Nil

(d) *Assembling information to enable disturbance claim to be prepared*

The work in this respect was again handled by Mr J B Thomas, and involved three meetings with Humphrey Wightwick & Associates, totalling 8 hours, and a further period of about 40 hours for the assembly of copy invoices, labour costs and other computations.

48 hours at £10 per hour — 480.00 — Nil

J

(a) This item is closely connected with items (2) and (3) and almost all of the amount claimed (£652) falls with them. Nevertheless, part of Mr J B Thomas' time must have been spent in obtaining finance for the move and so I allow £50 for this.

K

(b) It will be seen from the details of claim that Mr J B Thomas considered employing consultants to advise him on the removal. He decided not to do so and he himself did the work. The basis of the claim is an estimated 125 hours spent at an estimated cost of £10 per hour. Counsel for the respondent said that the matter should be approached by considering the loss in profits and he referred to *Pearl Wallpaper* (iii) and submitted that (7)(b) is closely connected with (20), (21) and (22). A document which resulted in a figure of £1,640 was submitted. This sum was arrived at by taking the gross profit for the year ended November 30 1977, adding the wages, deducting the discount allowed (see (21) below) and taking one fiftieth (one week) of the result: that figure has to be looked at together with (22), which is a spot figure, half of £1,640 put forward by Mr Sumpter and it covers in whole or in part other items claimed, as well as (7)(b). Both the hours and the rates in the claim were criticised by the respondent authority.

L

Certainly an approach using loss of profits is possible, rather than pricing a number of hours, but the former tends to be notional rather than actual and I have no doubt that it was necessary for Mr J B Thomas to spend a great amount of time on planning the move in order to cut down the loss of profit on removal ((20) below). Much of the time spent was late at night and early in the morning, some at weekends. On the evidence I do not see that either the number of hours estimated nor the price per hour claimed is grossly unreasonable in all the circumstances, but balancing the two approaches I allow £1,100 for this item. The same considerations apply in respect of (7)(c) and I allow £700 for this item.

M

With regard to (7)(d), a copy of the memorandum of the terms agreed for the sale and purchase of the freehold of the old premises was put before me. It contains the following paragraph:

> The purchase money agreed on is the sum of £21,600 which is to include compensation for all items of claim against the Council whether specifically mentioned or otherwise except claims in respect of loss of and disturbance to business and costs of removal which claims remain to be agreed with the council's valuer and estates surveyor.

The matter of expenses incurred in preparing a claim for compensation was considered in *LCC v Tobin* [1959] 1

WLR 354 CA affirming *Tobin v LCC* (1957) 8 P & CR 453, Lands Tribunal. The tribunal held as follows:

> We are satisfied therefore that the costs incurred by a claimant in formulating his claim are proper subjects of compensation.

It follows that if a claimant himself suffers loss as a result of having to prepare a claim, that loss should form part of his compensation under the umbrella of "rule (6)" and I do not think that the clause as set out above in the memorandum of terms should be construed narrowly. I therefore allow this item in principle, but there may be some overlap between this item and Mr Wightwick's fees (23). I therefore allow the sum of £400. Neither accountant's nor legal fees were claimed as part of these "Tobin" costs.

Labour and transport costs

(8) *Removal - own labour and transport costs*

Although, as mentioned above, the move took in effect 7 days, for practical reasons this period was spread over some 2 weeks. In other words, the company were operating in two factories for a short period, because, for instance, the cutting room was moved the weekend before the pressing department.

As has been mentioned under an earlier heading, work banks had to be built up to cover lapses in production by both the cutting and machine rooms; and this involved the working of overtime by employees as well as directors.

In addition, casual labour had to be brought in to assist the specialist contractors dealing with the removal and reinstallation of the cutting and machine rooms, and boiler/presses, and in respect of the removal of work in progress. Finally, certain operatives had to be laid off for short periods, due to lack of work, so the wages paid to them were abortive.

A statement of staff overtime, casual labour costs, and abortive wages totals £867.07.

Transport for items not dealt with by specialist contractors, viz stock, work in progress and certain racking, dexion etc was provided by "Taxitruck" in the sum of £325. It may be mentioned that an alternative quotation for this work was obtained from Ryans Removal Service; and although this was a lower sum, of £310, the second contractor could give no guarantee that the work would be complete by August 7. The higher quotation was therefore accepted.

The total under this heading is	1,192.07	325.00

As will be seen from Appendix B, the claimant employed a firm, "Taxitruck," to move certain items: the respondent authority allowed the cost incurred but disputed the sum of £867.07 which was paid by the claimant for casual labour, staff overtime and abortive wages. The wording in Appendix B is, I think, somewhat misleading: what happened was that the specialist contractors dealt with certain items and the casual labour moved others. The respondent authority considered that that work could have been done by the claimant's employees or should have been done by "Taxitruck". Many of the claimant's employees are women: heavy items had to be humped from the old premises to the new. I am convinced that the claimant had to employ casual labour in addition to the specialists to do this work and were justified in so doing and I allow the amount claimed in full.

Removal of band knife

(9) *Removals - cutting room*

The cost of removing the band knife and cutting tables, together with reinstallation of same in the new factory, is covered by accounts from Aublet Homer (Engineering) Ltd, in the respective sums of £1,600 and £825, ie a total of	2,425.00	1,990.00
(10) —	Agreed	—

The sum of £1,600 was not disputed, but Mr Dolan considered that the amount of £825 was excessive. He estimated from his experience that that part of the work could have been done for £390. It was also said that there have been some improvements to the drive mechanism to this machine on removal, but I think that the claimant pursued a reasonable course in having the work done as he did and I prefer to allow the actual amount spent by the claimant rather than to rely on Mr Dolan's estimate. The claimant might perhaps have got some small benefit from the new drive mechanism, but the value of that would be a very small proportion of £825 and I prefer not to regard it as an improvement.

Removal of boilers and presses

(11) *Removals - boilers, presses and fusing*

This was at a cost of	3,630.00	1,125.00
(12) —	Agreed	—
(13) —	Agreed	—
(14) —	Agreed	—

Mr Dolan's evidence was to the effect that there had been some improvements made on removal of these items to the new premises and again he considered that the actual amount paid (£3,630) was excessive. The claimant obtained two quotations. As with (9) I prefer to have more regard to the actual sum paid by the claimant, but on the evidence I consider that there were some improvements. I therefore allow only £3,000 for this item.

New name sign

(15) *Name sign on new factory*

The company had a sign board above the only entrance to 14 Andrew Street. But apart from the fact that reuse of this signboard at the new factory would have been quite inappropriate; indeed it is doubtful whether planning permission would have been forthcoming - as mentioned earlier, the new factory has both front and rear access. As a consequence, two new name signs have been necessary.

The company chose the cheapest expedient of plastic lettering superimposed straight on to the brickwork at a cost of	145.00	50.00
(16) — to (19) —	Agreed	—

On the evidence, I come to the conclusion that the claimant acted reasonably in making its choice and that it was not practical to take any other course, as was suggested by the respondent authority. Four quotations were obtained by the claimant. I allow the amount claimed in full.

Temporary loss of profit

(20) *Claim for temporary loss of profit*

As has been mentioned, considerable care was taken in planning the move, in order to ensure the minimum possible disruption to production; nevertheless the removal period was spread over 2 weeks, the company had to operate for a time in two factories, and some departments and operatives were not as productive as might otherwise have been the case.

Moreover, in the period immediately following the removal, the reinstalled boiler and presses suffered some fairly serious teething problems, which resulted in the loss of at least one fairly substantial order.

The loss of production is computed by reference to three weeks loss of the average gross profit over the years ending November 30 1977 and November 30 1978.

An average figure was adopted, partly because the gross profit margin in 1977 (17.4% of sales) was considerably lower than the average margin in 1975 and 1976 and partly because, although the accounts for the year ended November 30 1978 show a substantial upturn both in sales and gross profit and to a lesser extent in the percentage margin, it is thought that the latter figures may be distorted by the fact that the company traded in modern, and hence more efficient, premises for about one-third of the 1978 financial year.

Hence

	£			
Gross profit for 1977	28,061.00			
Gross profit for 1978	42,136.00			
Average	35,098.00	$\times \frac{3}{52} =$	2,025.00	1,640.00

The use of 1978 figures of profit by the claimant as set out in Appendix B was criticised by the respondent authority who also considered that three weeks was excessive, although orders were lost (including one large one) owing to the move. As explained in connection with (7)(b) above, the respondent authority took one week's loss of gross profit plus wages less discounts to give £1,640, but that figure was put partly against other items in the claim. Having regard to those criticisms, other factors and to double counting I allow £1,600.

Lost discounts

(21)	*Claim for additional loss of profit due to lost discounts*	6,000.00	Nil

Mr J B Thomas' evidence, which was not challenged, was to the effect that the claimant company was in the habit of buying articles offered to it, paying immediate cash for which it was allowed discounts. In respect of other items it purchased, it paid bills promptly and so also obtained discounts for such payments. The fact that it had not received compensation for disturbance caused difficulty with cash flow and, said Mr Thomas, because of this, it was unable to follow its previous practice and so had lost a total of about £6,750 in discounts between June 1978 and December 1981. Mr Wightwick found the average annual loss to be £1,500, which he capitalised at four years' purchase to give £6,000 against which the respondent authority offered no particular amount but reflected it in the £1,640 referred to under (20).

On Mr Thomas' evidence, I find that this was much more in the nature of a temporary loss than a permanent loss, although he said there may be some permanent loss of goodwill; those from whom they previously purchased goods for spot cash would no longer come to them. That permanent loss of goodwill must be fairly small, but I find that there is a temporary loss of profit over and above (20) for these discounts. The amount of the loss cannot be as great as £6,750; indeed Mr Wightwick's evidence, which appeared to me to be on a permanent-loss basis, amounted only to £6,000.

I asked the parties to consider the amount of this loss on a discounted cash flow footing. They agreed the rate of interest to be 16% with quarterly rests; on that basis the loss comes to a little under £5,000, but since no addition back for these discounts was made by Mr Wightwick in computing the loss under (20) above, the amount to be awarded here must be less than £5,000. The respondent authority considered that the claimant got into difficulties with its cash flow because of the payment it had made in respect of fees under (2) above and there may be some substance in this.

The period used to ascertain the total of £6,750, three-and-a-half years, could have been shortened had the claimant referred this matter to the tribunal before it did. However, it necessarily took some time to prepare the claim for compensation; the respondent authority suggested that the claimant could have borrowed money to enable it to continue its practice of taking discounts, but from Mr Thomas' evidence I am not convinced that it would have been profitable to do this. It might have been possible to obtain further advance payments of compensation from the respondent authority. Shortage of cash could have been caused by expansion of the business. Taking all factors into account I allow the sum of £2,000 for this item.

Other losses

I have already referred to this item. It represents half of the figure computed by the respondent authority to meet (20) and other items of claim including (7). By offering it, the respondent authority recognised that the claimant suffered loss both before and after the move and, like their figure of £1,640, it was a spot figure put forward to meet a number of items of claim. In general I prefer to start with the arithmetical approach on the actual or estimated figures used by the claimant in its various items of claim. However, taking the £820 and £1,640 together on a very broad view it seems to me that the respondent authority's total amount of £11,951.61 for disturbance (leaving aside (23) below) is too small a figure to meet the losses which the claimant sustained by transferring its business from the old premises to the new in what was necessarily a very complicated though well-planned move.

(23)
In addition, there are claimed the surveyor's and accountant's fees for preparing and negotiating the claim, surveyor's out-of-pocket expenses, the proper legal costs of completing the matter, and statutory interest in accordance with section 32 of the Land Compensation Act 1961 from August 8 1978 to the date (or dates) that payment of the disturbance compensation is made.

Mr Wightwick submitted an account to the claimant, which it paid, in the sum of £2,030 plus VAT at 15%, £304.50, plus out-of-pocket and travelling expenses, £63.94, total £2,398.44. That was expressed to be on a *quantum meruit* basis. I think that a better description might perhaps be "on a time basis", 158½ hours. Against that, the respondent authority offered a fee in accordance with scale 5(a) of the scales of charges of the Royal Institution of Chartered Surveyors. Mr Howell's evidence was that this was not a complicated case, but I do not accept that view in the light of the evidence given by Mr Thomas and Mr Wightwick. The scale 5(a) fee, based on a figure of about £33,000 which the respondent authority consider should be the total amount of the compensation, is in the order of £370. Based on £72,000, the total figure contended for by the claimant, the fee amounts to about £670. Of the amount of £2,030 in Mr Wightwick's account, £300 relates to work after the date of the notice of reference to this tribunal and must therefore be excluded for that reason. I was urged by counsel for the claimant to find that the fee should be equal to the amount paid by the claimant to Mr Wightwick. Counsel for the acquiring authority, however, submitted that this was not an unusual claim and that scale 5(a) fees were appropriate. This matter was considered by the tribunal in *Widden* (iv) and in *Easton* (viii). In the latter decision the Member said [at p 166]:

> There has been a recent revision of scale 5, which is referred to as Ryde's Scale, 1978 edition; it is effective from June 10 1978. Evidence was submitted of the publication of the 1978 scale with the practice note which appeared in the *Chartered Surveyor*, July 1978. The note stated that after lengthy discussions a revised scale 5 had been agreed between the Royal Institution of Chartered Surveyors and the Valuation Office, Inland Revenue, local authority associations and relevant government departments. It has been revised twice by simple addition in 1929 and 1948 and an increase to the minimum fee in 1971 completed the amount of amendment made in the whole history of the scale. The note continues that in the light of complaints received from members of the institution and changes in the nature of compulsory purchase, coupled with complex changes in land law, a committee was appointed to consider all aspects of the scale. The new scale aims to provide a more appropriate fee in respect of lower-value compensation and to take account of claims relating to leasehold properties and those which include an element of disturbance. The note states that no scale of charges can ensure fair payment in all circumstances. Scale 5 is intended to provide a reasonable basis appropriate in the great majority of cases, for the reimbursement of claimants' valuers' fees by acquiring and compensating authorities. It is nevertheless recognised that there may be exceptional cases in which the reimbursement of fees on a basis which departs from the scale would be justified.
>
> In my judgment the revised scale provides an appropriate measurement of fees in the instant case; the revised scale is designed to take account of a claim relating to leasehold property and which includes an element of "disturbance"; it provides, as it is intended to provide, a reasonable basis appropriate to the instant case, and I do not accept the claimant's contention that the instant case is an exceptional case in which a departure from the scale would be justified. I find that the claimants' surveyor's fees payable by the acquiring authority are to be on scale 5.

I do not think that it would be right for me to disregard the agreement between the institution, the valuation office, local authority associations and relevant government departments. Nevertheless, as I have said above, I find this to be an exceptional case and I direct that the acquiring authority shall pay, in addition to the compensation, a surveyor's fee in accordance with scale 5(a) on the amount of my award (less the total of the fees in (4) above) increased by £100 plus VAT if appropriate and out-of-pocket expenses amounting to £53.74 less any amounts already paid.

No evidence was put before me as to any accountants' fees. The claim for the legal costs of the conveyance and interest on the amount payable are not matters for this tribunal: they are covered in the relevant statutory provisions.

The total amount for disturbance is thus:

Agreed items	£4,069.77	(summarised in Appendix A hereto)
Items other than (4)	14,841.79	(summarised in Appendix D hereto)
(4)	829.35	
	£19,740.91	

to which is to be added a surveyor's fee in accordance with the finding in (23) above.

Having opened the sealed offer of compensation lodged by the acquiring authority I find that the amount thereof is less than the amount of my award. I heard submissions as to costs. In view of the time spent on items 2 and 3, I order that the acquiring authority will pay the claimant 5/6th of its costs of this reference, such costs, if not agreed, to be taxed by the Registrar of the Lands Tribunal on the High Court Scale.

Appendix A

Summary of agreed items of disturbance

Reference in claim	*Item*	*Amount £s*
(6) (d)	Stepladder	47.26
(10)	Removal of sewing machines	1,879.98
(12)	Electrical reconnections	997.25
(13)	Telephone reconnections	110.00
(14)	Removal of time-clock	16.00
(16)	Removal of gas appliances	64.50
(17)	Notification of change of address	167.63)
		236.00)
(18)	Stationery	188.15
(19)	Fixtures and fittings	363.00
	Total	£4,069.77

[Appendix B and C appear in the text of the decision.]

Appendix D

Summary of amounts of disturbance determined by the tribunal

Reference in Appendix A	*Item*		*Amount £s*
(3)(a)	Bank arrangement fee		140.00
(b)	Bank bridging finance		150.00
(6)(a)	Cost of adaptation		1,669.72
(c)			270.00
(7)(a)	Director's time		50.00
(b)			1,100.00
(c)			700.00
(d)			400.00
(8)	Costs of own labour on removal		1,192.07
(9)	Removals - cutting room		2,425.00
(11)	Removal of boilers etc		3,000.00
(15)	Name on new factory		145.00
(16)	Temporary loss of profit		1,600.00
(21)	Loss of discounts		2,000.00
			£14,841.79
(4)(a)	Cost of purchasing alternative site	£ 474.35	
(b)	Surveyor's fees	355.00	829.35
		TOTAL	£15,671.14

FLANAGAN v STOKE-ON-TRENT CITY COUNCIL

(REF/68/1980)

(W H REES FRICS)

April 6 1982

Estates Gazette June 19 1982

(1982) 262 EG 1201

General store of "corner shop" type, with living accommodation, in Stoke-on-Trent — Claim for injurious affection under s 10 of 1965 Act — Situation "drastically altered" following road construction works which converted frontage road into dual carriageway further away and cut off side road with brick wall, leaving premises at end of cul-de-sac — Business closed when trade declined to point where unprofitable to continue — Physical damage to building also incurred — Findings that "very serious injury" done to premises and that claimant "deprived of the easy access to his premises which he before enjoyed" — Authority's submission that remedy did not lie under s 10 but might lie in negligence rejected — Right to compensation upheld on authorities cited — Relevant date for assessment agreed by parties to be either date of completion of works or date of hearing — Member finds for former date but notes that he had no argument as to whether any other date appropriate — Entirely illogical, in his view, for loss to be assessed at date later than date of claim, which was some 15 months before completion of works — Tribunal rejects authority's valuer's contention that freehold interest not depreciated by works — Nothing wrong in principle with claimant's valuer's approach to valuation, but his view that value of freehold interest enhanced by goodwill of trade carried on "entirely" rejected — Depreciation reduced to 20% from "far too high" 40% — Award of £3,220 against £8,220 claimed and a nil offer — Alternative award of £5,300 at hearing date — Three-quarters costs awarded

Christopher Beaumont (instructed by Heath & Cantlay, of Stoke-on-Trent) appeared for the claimant; Mark Lowe (instructed by P N Wilson, solicitor to Stoke-on-Trent City Council) for the compensating authority.

Giving his decision, Mr Rees said: The claimant in this matter owns the freehold of a shop with living accommodation and garage in Stoke-on-Trent, [280 Shelton New Road, Cliff Vale, Stoke-on-Trent]. Its main frontage was to Shelton New Road and it had a side passage to give vehicular access to the garage and to the rear of other premises nearby. It was almost on the corner of Valley Road. For many years the claimant had carried on the trade of general stores including greengrocery: it was indeed a "corner shop", a general store. Shelton New Road was a main road: Valley Road carried less traffic than did Shelton New Road but more than some of the side roads nearby from the houses in which the claimant drew his customers.

Beginning at the end of 1974 and ending in November 1977 the respondent authority carried out major works to both roads. Valley Road became part of a "D" ring road and was formed into a dual carriageway. Shelton New Road also became a dual carriageway and was carried on a fly-over above the ring road. At the same time it was realigned and taken somewhat further away from the subject premises than the former road. The subject premises are now at the end of a cul-de-sac. A brick wall 3 m high was built parallel to the west boundary of the site of the subject premises between them and the ring road in order to minimise traffic noise and a pedestrian access was formed along the side of Shelton New Road over the bridge to houses on the west. The situation of the subject premises was thus drastically altered.

The claimant closed his business in August 1977. No part of the subject premises was acquired in connection with these road-works and the claimant now claims compensation for injurious affection amounting to £10,500 under section 10 of the Compulsory Purchase Act 1965. The respondent authority contest the claimant's right to compensation, but, if they are wrong in that, they say that the damage is nil. The parties are also at issue as to the date on which compensation, if any, should be assessed. The claimant said that it should be assessed at the date of the hearing before me; the respondent authority say that the correct date is November 1977, the date when the works were completed.

Counsel for the claimant called the claimant himself and P E C Blood, a chartered surveyor. Counsel for the respondent authority called M V Astbury, a chartered surveyor, the principal valuer in the City Estates Department, and his instructing solicitor, P N Wilson, who is employed by the respondent authority.

I consider first whether the claimant is entitled to compensation at all. Counsel for the respondent authority agreed that the claimant had a right not to have his property damaged by the construction of the works, but in that connection, first, he submitted that the claimant's remedy did not lie under section 10 but might lie in an action for negligence; such an action was started on January 25 1977 by the claimant against the compensating authority, and the contractors and subcontractors who carried out the works were later joined as defendants. Second, he submitted that the claimant still had both vehicular and pedestrian access to his premises and so no legal right had been interfered with. His claim therefore failed to pass the second of the four well-known tests in *Metropolitan Board of Works v McCarthy* (1874) LR 7 HL 243. Alternatively, he submitted that it was not the works which had caused damage to the claimant's property but the order closing Shelton New Road which left the subject premises at the end of the cul-de-sac; that order was the City of Stoke-on-Trent "D" Road (Classified Road) (Side Roads) Order 1971. In that connection he relied on *Jolliffe v Exeter Corporation* [1967] 1 WLR 993.

References to authorities

Counsel for the claimant submitted that compensation under section 10 included all loss suffered by the claimant in respect of his freehold interest and that included damage caused by any negligence of the contractors. As to the argument concerning the second rule in *McCarthy*, he referred me to the following: *Caledonian Railway Co v Walker's Trustees* (1882) 7 App Cas 259 HL; *Chamberlain v West End of London & Crystal Palace Railway Cos* (1863) 2 B & S 617; *R v Saint Luke's, Chelsea* (1871) LR 7 QB 148; and *Metropolitan Board of Works v Howard* (1889) 5 TLR 732.

I was also referred to *Compulsory Acquisition and Compensation* by Corfield and Carnwath at p 324 and to an article by W A Leach FRICS in ESTATES GAZETTE for October 17 1970. In connection with the amount of compensation, but not in connection with the claimant's right to it, I was referred to *Argyle Motors (Birkenhead) Ltd v Birkenhead Corporation* (1973) 27 P & CR 122.

Again not in connection with the claimant's right to compensation but with reference to the date at which it should be assessed I was referred to *Dodd Properties (Kent) Ltd* v *Canterbury City Council* [1980] 1 All ER 928. I include both here, so keeping all the references together.

It seems to me that if damage is done to premises by the carrying out of work it matters not for the purposes of section 10 whether that damage arises through negligence or not and I therefore reject the submission made by counsel for the respondent authority on that point.

Counsel for the claimant referred me to Lord Selborne LC at p 273 of *Caledonian Railway:*

> My Lords, the only facts in this case which I think material to the question of principle to be determined are that, before the construction of the appellants' works under the authority given by their Act of 1873, the property of the respondents had a frontage to Canal Street, in Glasgow, and had by that street a direct, straight, and practically level access (at the distance of about ninety yards), for all sorts of traffic, to Eglinton Street, one of the main thoroughfares of that city; and that, by the works of the appellants, that direct access to Eglinton Street has been altogether cut off and taken away, a more distant and circuitous access, crossing the railway by a bridge with a rather steep gradient, being substituted for it. Another direct access to Eglinton Street, from the back of the appellants' premises, has been rendered less convenient, but upon that part of the case I do not think it necessary to dwell.

The claim was upheld.

He also referred me to the whole of the report of *Howard:*

> This was an action upon an inquisition finding £1,031 to be due from the defendants to the plaintiff in respect of the injurious affection of the plaintiff's property by certain street improvements made by the defendants. The plaintiff is the tenant of a fully licensed public house, called the Eight Bells, in Bridge Street, Fulham, about 250 ft distant from old Putney Bridge. Bridge Street was the main street on the Middlesex side of the Thames leading to old Putney Bridge. The Metropolitan Board of Works recently built a new bridge across the Thames at Putney, a short distance higher up the river, and made a new thoroughfare on the Middlesex side leading to this new bridge. The old bridge was then closed, and Bridge Street led only down to the water, and in consequence the traffic, which formerly went along Bridge Street past the plaintiff's public house, is now diverted at a point before the plaintiff's house is reached, passes along the new thoroughfare, and so over the new bridge. The plaintiff claimed compensation under the Lands Clauses Act 1845, in respect of his property being "injuriously affected" by the works carried out by the defendants, and on the inquiry before the jury the plaintiff produced evidence that in consequence of the diversion of the traffic the trade of the public house had greatly diminished. The jury awarded the plaintiff £1,031 compensation. Upon the hearing of the action the defendants contended that the inquisition was bad, as it found solely, or to a great extent, money due to the plaintiff for loss of profits of the trade, which could not be the subject of compensation, the only subject for compensation being the depreciation in the value of the premises. Mr Justice Denman, who tried the case, held that he was not justified in treating the inquisition as a nullity, and gave judgment for the plaintiff. The Court of Appeal affirmed the decision of the learned Judge.
>
> **Their Lordships now gave judgment, dismissing the appeal.**
>
> LORD HERSCHELL did not think it could be doubted that an interference of this character with the access to the house of the respondent by means of thus dealing with the road or highway on which it was situated was an injurious affecting of his premises which would give him a right to compensation if those premises had been rendered less valuable than they were before. When he said that, he meant that the house itself must have been affected in value, having regard to all the purposes for which such a house might be used. It would not be sufficient that there had been an injury to the respondent in respect of the particular use which he was making of the house, which did not affect its value in the market for all purposes. It was, therefore, clear that if the respondent could establish before the jury that his house was, in the sense which he had described, diminished in value he would make out a claim to compensation, and it was for the jury to assess the amount. The only question that that House had jurisdiction in this particular case to entertain was whether there was jurisdiction in the sheriff and jury to entertain the claim to compensation and to award some damages in respect of it. If there were such jurisdiction, and if any evidence was before the jury which warranted the award of any damages, then, in such an action as this, the plaintiff must recover, however excessive the amount of damages, however erroneous the law laid down to the jury, however wrong the principle which they adopted. If a house were rendered less accessible to customers, that might diminish the value of the house for all purposes, although the evidence which was given with reference to the actual loss of trade and the decreased number of years' purchase that it would fetch on sale ought not to have been admitted. In his opinion the judgment of the Court below was right and ought to be affirmed, and the appeal dismissed with costs.

In *St Luke's*, he relied on the following passage from the judgment of Kelly J at p 153:

> I cannot but observe in a case like this, that, whenever it appears that the case is one in which it is plain that very serious injury may have been done to the premises of the party claiming compensation, I think that we must put a liberal construction upon the Acts of Parliament before us in determining the points raised. Unless it is perfectly clear that the language of the different Acts is not sufficiently ample or extensive to embrace the case in question, we ought to hold that a party whose property is injuriously affected, and to a very great extent, by the operations of a public body, shall be entitled in a court of law to compensation.

I have no doubt from the evidence that "very serious injury" was done to the claimant's premises in this instance. He also relied on the following passage from the judgment of Erle CJ in *Chamberlain* at p 636:

> Still, if any person have sustained a particular damage therefrom beyond that of his fellow citizens, he may maintain an action in respect of that particular damnification. I think that the fact found by the umpire, that the plaintiff's houses have been injuriously affected, is a finding that he has suffered "particular damnification;" and, he finds specially how that injurious affecting would be occasioned; — viz, that by the obstruction to the thoroughfare the number of persons passing by the plaintiff's houses would be diminished, and consequently the prospect of customers to the occupiers of the houses in respect of any branches of industry carried on in them would be injured. Therefore I am of opinion that a particular damage to the plaintiff

A by the obstruction of the highway is made out. And I am supported in this by the analogy of several cases. *Moore v The Great Southern and Western Railway Co* 10 Irish Com Law Rep 46 is in principle very closely allied to the present, for there the Company did not take any part of the plaintiff's land, but, in order to lower the road, made a deep cutting along the boundary of his land, so that he was deprived of the easy access to his house and premises which he before enjoyed; and it was held that this was a permanent injury for which he was entitled to compensation. Here the Railway Company have substituted a new highway for the old one; the old highway is blocked up, and the plaintiff's houses are as inaccessible as was the plaintiff's house in *Moore v Great Southern and Western Railway Co.* The same principle was acted upon in *Reg v Eastern Counties Railway Co,* 2 QB 347, where the Company, in order to carry their railway over a highway which before pass-
B ed the prosecutor's land on a level, lowered the road so that his land was left on a high bank.

On the evidence there is no doubt in my mind that the claimant "was deprived of the easy access to his house and premises which he before enjoyed" and that seems to me to summarise exactly the position in this instance.

With regard to *Jolliffe,* there is a vital difference on the facts between that case and this instance. In *Jolliffe* the works did not affect the claimant's land, but in this instance the works done undoubtedly do. Counsel for the claimant referred me to the judgment of Lord Denning MR at p 998:

Mr Jolliffe is, therefore, entitled to compensation for injurious affection due
C to the execution of the works. What are the "works"? The works are the works authorised by the "special Act." The "special Act" is defined by the Act of 1946 as the "enactment under which the purchase is authorised and the compulsory purchase order." In this case that is section 214(1) of the Highways Act, 1959, and the compulsory purchase order. The works authorised by those provisions are the works on the pink land, and not the works in Coombe Street itself. The result is this: if the roadworks executed on the two parcels of pink land had injuriously affected Mr Jolliffe's land, he would have been entitled to compensation. But it is apparent that the works on those two parcels (taken by themselves) did not affect Mr Jolliffe's land at all. They were near Coombe Street, but not across it. So they did not affect the use of Coombe Street. The thing which affected Mr Jolliffe's land was the stopping up of the highway, for which there is no compensation.

D On these authorities, I come to the conclusion that the claimant has a right to compensation in this instance.

Relevant date

As to the date at which the damage should be assessed, both counsel referred me to *Dodd Properties* and having considered the arguments I come to the conclusion that the earlier of the two dates, November 1977, is the more appropriate. The parties agreed that the relevant date was either November 1977 or the date of the hearing before me. In this connection I have regard to the provisions of section 63 of the Land Compensation Act 1973, which is as follows:

E 63(1) Compensation under section 68 of the Lands Clauses Consolidation Act 1845 or section 10 of the Compulsory Purchase Act 1965 (compensation for injurious affection where no land taken) shall carry interest, at the rate for the time being prescribed under section 32 of the Land Compensation Act 1961, from the date of the claim until payment.

That section seems to me to contemplate an earlier rather than a later date, possibly the date of the claim. In this instance the parties agreed that the date of the claim was August 10 1976. It would, it seems to me, be entirely illogical for the loss to be assessed at a date later than the date of the claim (in this instance the loss is greater than at any earlier date) and for compensation to carry interest in accordance with the section set out above from the date of the claim. I also have regard to the decision in *Dodd Properties* at p 939, where Donaldson
F LJ says:

Counsel for the defendants submits, and I for my part would readily accept, that the general rule is that damages fall to be assessed as at the date when the cause of action arose. The rule is so stated by Lord Wilberforce in *Miliangos v George Frank (Textiles) Ltd* [1976] AC 443 at 468. And I am inclined to think that in normal circumstances this would be applicable where the relevant measure of damage was diminution in the capital value of the property. But it is only a general or basic rule and is subject to many exceptions. Thus damages for personal injury, excluding consequential loss to which other principles apply, are assessed in the light of the value of money at the date of the hearing. The issue here is whether the assessment of damages based on the cost of repair or reinstatement is another exception, as counsel for the plaintiffs contends. I think that it is.

G I therefore decide that the correct date for the assessment of loss is November 1977 when the works were completed, but I note that I had no argument as to whether any date other than the two agreed was appropriate. However, should this matter go further I have also determined the amount of compensation at the second agreed date, the date of the hearing before me.

Before considering the evidence of the experts it is convenient to summarise that given by the claimant, who said that he was born in 1912 and purchased the subject premises and the business carried on therein in 1945. At that time, few goods were sold so he continued both to follow other employment and to run the business, but trade increased and he devoted all his time to the shop. He increased the
H range of goods sold and made deliveries to nearby houses. He also had custom from those employed in nearby factories, particularly in the early mornings, and from many school children. A bus stop close by attracted some passing trade. It was now 200 yds away.

The increase in trade continued, but after about five years it levelled off and gave him a good living until a year or two after the works were started in 1974. Thereafter trade declined to the point where it became unprofitable to continue and so he closed the business in August 1977 but continued to live in the subject premises. His takings, which had levelled off at £300 a week, had dropped to £70 a week when he closed. He produced a schedule of his accounts. Under cross-examination he agreed that the maximum annual profit
J was £1,000.

Sales dropped in 1972 when tenants of local authority houses, which were affected by the works or had to be demolished in consequence of them, left the area. Customers would not cross either of the two new roads once the works had started. In 1977 he could physically have continued to run the business: his health was good. No one would run a shop in the subject premises now: there was too little profit to warrant it. He produced sets of photographs showing the position around the subject premises as it was during the construction of the works. He now has no choice but to live in the subject premises: no one would wish to buy them; the environment is too poor. The works have a depressing effect; he suffers from traffic
K noise but not from vibration. Before the works were begun the subject premises were in good condition. The new roads were opened in November 1977. Cracks appeared in the brickwork and a window was damaged when piling was carried out. There were other items of damage to the building. He suffered much disruption and loss of custom during the construction of the works, but, had he remained trading, when they were completed the customers would not have come back. By then they had gone elsewhere.

Mr Blood's valuations were as follows:

			Loss Injurious affection	Remaining values
1. Value of shop and living accommodation assuming no injurious affection.				£9,000 (13,000)
2. Losses resulting from the roadworks.				
(a) Extinguishment of business element in land and buildings (ie goodwill £4,000)			£4,000 (4,000)	5,000* (9,000*)
(b) Depreciation in value of residential part (made up as follows):				
Shop and dwelling	£9,000	(£13,000)		
Less goodwill	4,000	(4,000)		
Residential value	5,000	(9,000)		
Depreciation at 40%	2,000	(3,600)	2,000 (3,600)	3,000 (5,400)
(c) Cost of reconverting shop to living accommodation			320* (500*)	2,680 (5,000)
(d) Further diminution in value due to structural damage and resultant constraints on marketability (made up as follows):				
Remaining value	£2,680	(£5,000)		
Less forced sale value	800	(2,000)		
ie approx cost of essential repairs agreed at £1,900* (3,100*)	1,880	(3,000)	say 1,900 (3,000)	800 (2,000)

L

M

3. Total claim under section 10, Compulsory Purchase Act 1965	£8,200 (£10,500)

Agreed figures are marked with asterisks: the figures are as at November 1977. Those in brackets are as at the date of the hearing before me. Mr Blood also submitted a "goodwill valuation" in the sum of £4,000, but for the reasons given below I do not need to set that out in detail.

Mr Astbury spoke formally to the agreed figures of £5,000 and £9,000 included in Mr Blood's valuation above. Those values ignored the road construction and structural damage and made no addition for business. In his view if the road construction were taken into account but structural damage was ignored and no addition was made for any business element, the values were the same. In his view, therefore, there was no depreciation in the value of the freehold interest in the subject premises due to the road construction.

By way of support he pointed out that the number of shops in the area had been reduced by 75% due to properties being acquired compulsorily in connection with the road-works but that the number of houses had been reduced by less than 20% and whereas the subject property previously fronted directly on to a busy main road it was now situated at the end of a cul-de-sac, which in his view improved the value of the residential accommodation. He disagreed with some of the figures put forward by Mr Blood, in particular the 40% for depreciation. He agreed that there might be some reduction in the "trading value" of the subject premises but that was offset by the increase in the value of the residential portion.

"Very much for the worse"

I viewed the subject premises internally and, perhaps more important, I saw the surroundings as they now are for myself. On the evidence, supported by my view, I cannot accept that the subject premises have not been depreciated by the construction of the road-works, which have entirely altered, very much for the worse, the immediate surroundings of the subject property. I see nothing wrong in principle with Mr Blood's approach. However, I reject entirely his view that the value of the freehold interest in the subject premises is enhanced by the goodwill of the trade carried on: that view was rightly criticised by Mr Astbury. Loss of profitability can in some circumstances affect the value of an interest in land: Mr Astbury instanced a public house. The matter was summarised in the following paragraph which appears on p 126 of *Argyle Motors* included in the judgment of Lord Wilberforce:

> I must, therefore, agree with both courts below that if the right to compensation in the present case depends upon section 68 the appellants cannot succeed in obtaining compensation for business losses as such. I make it clear, as did the Court of Appeal, that if they can prove that a loss of profitability affects the value of their interest in the land they can recover compensation for this loss of value.

I therefore delete the figure of £4,000, 2(a) in Mr Blood's valuation. On the evidence I also come to the view that 40% depreciation in 2(b) is far too high and I take 20% to give a loss of £1,000. That of course affects Mr Blood's values in 2(d), but it seems to me to be reasonable to equate diminution in value here with the cost of the essential repairs which were agreed at £1,900 and I therefore take that figure: the forced sale value may be considerably above £800. Thus the total compensation becomes £1,000 plus the two agreed figures of £320 and £1,900, £3,220. If the correct date be the date of the hearing before me, the compensation would amount to £5,300 (£1,800 plus £500 plus £3,000).

In arriving at these figures I take into account all damage (including physical damage) done to the subject premises including damage arising from any negligence. There was no evidence before me that there was any negligence and certainly none to enable me to distinguish between damage caused by negligence and damage not so caused.

I heard submissions with regard to costs. The compensating authority will pay the claimant three-quarters of his costs of this reference, such costs, if not agreed, to be taxed by the Registrar of the Lands Tribunal on the High Court Scale.

LANDS TRIBUNAL
LEASEHOLD REFORM

A **LLOYD-JONES v CHURCH COMMISSIONERS FOR ENGLAND**

(LR/29/1980)

(W H REES FRICS)

December 4 1981

Estates Gazette February 6 1982
(1981) 261 EG 471

Freehold price of house with RV over £1,000 in Maida Vale, London — Second case under section 9(1A) before tribunal — 12 years of lease unexpired — Certificate of fair rent for £4,250
B pa — Agreed value of unencumbered freehold £187,500 — Tribunal follows "Indian case" and *Lambe v Secretary of State for War* in upholding lessors' valuation approach — "Basis to be adopted is not a nominal amount above the sum which any person other than the lessee would pay but what the lessee would be willing to pay in a friendly negotiation with the freeholder" — Both parties' valuers calculate value of freehold interest and add proportion of share of "marriage" value — Lessors' valuation substantiated by 57 settlements on Grosvenor and Cadogan London Estates — 10 per cent deduction from vacant possession value overwhelmingly backed by settlement evidence — Tribunal criticises lessee's valuation approach to calculation of freehold excluding marriage value as
C "out of touch with reality" — Valuation assumed worst position which could arise, that lessee would remain at a fair rent — Purchaser making such assumption would be easily outbid by competitor taking more optimistic view — Tribunal agrees with lessors' valuer's assumption that parties were of equal bargaining strengths and would divide marriage value equally — Review of fair rent evidence — Summary of negotiating practice of Grosvenor and Cadogan Estates — "Massive" support from settlement figures, which were not from analyses but from actual calculations — No "Delaforce effect" one way or the other — Lessors' valuation of £113,000 (against lessee's £45,725) accepted "entirely"

D Michael Barnes QC (instructed by Farrer & Co) appeared for the lessee; Nigel Hague QC (instructed by Radcliffe & Co) for the lessors.

Giving his decision, MR REES said: This reference was made by the freeholders (with the consent of the lessee) of a house at 18 Warwick Avenue in that part of Maida Vale, London W9, known as "Little Venice", to determine the price to be paid for the freehold under section 9(1A) of the Leasehold Reform Act 1967 as amended. It is only the second such reference to be heard by the tribunal and raises the question of the meaning of the words ". . . the price payable for a house and premises . . . shall be the amount which at the relevant time the house and premises, if sold in
E the open market by a willing seller, might be expected to realise . . ." as well as matters of valuation. The lessor sought a sum of £113,000: the lessee (before me) £45,725.

Counsel for the lessor called M St J Hopper FRICS [partner, Gerald Eve & Co], whose valuation is set out below, and H A C Edwards FRICS [partner, Chestertons], who gave evidence as to the fair rent of the subject house (Rent Act 1977). Counsel for the lessee called W F H Gibbon FRICS [partner, Gibbon & Rogers], whose evidence also related to the fair rent, and F K Hewitt FRICS [partner, Hampton & Sons], whose valuation is set out below.

Agreed facts F

From an agreed statement of facts I note that the lessee's notice to purchase the freehold was dated October 18 1978, when the lease had about 12 years unexpired at a rent of £45 per annum. The lease was a full repairing and insuring one: the rating assessment was gross value £1,750, rateable value £1,430. It was agreed that the value of the notional unencumbered freehold interest as at October 18 1978 with vacant possession but disregarding the value of tenant's improvements was £187,500. The value of the leasehold interest as at the same date with vacant possession but disregarding both the tenant's right to claim the freehold and the tenant's improvements was agreed at £45,000. It was also agreed that following an application to the rent officer by the lessee for a certifi- G
cate of fair rent for the subject house, certificate of fair rent no 64018/1 dated July 29 1980 and signed by E J B Brown FRO, rent officer, certifies that £4,250 per annum would be a fair rent for the subject house under a regulated tenancy on the terms specified by the certificate.

The subject house is on the corner of Blomfield Road and Warwick Avenue where the latter crosses the Grand Union Canal. The house stands on a plot of 50 ft by 127 ft and comprises a four-storey semi-detached dwelling-house of brick under a slate roof. There is a small garden at the rear. Agreed floor plans were put before me. Accommodation comprises: on the lower ground floor, dining-room, scullery with boiler room, tank room with wine store and H
three vaults; on the ground floor there are a sitting-room, kitchen and a wc; on the first floor there are two double bedrooms, a bathroom and a wc; on the second (top) floor there are two single bedrooms, a double bedroom and a bathroom.

The basis of valuation is set out in section 118 of the Housing Act 1974, which inserted a new subsection (1A) in section 9 of the Leasehold Reform Act 1967. That new subsection reads as follows:

Notwithstanding, the foregoing subsection, the price payable for a house and premises, the rateable value of which is above £1,000 in Greater London and £500 elsewhere, on a conveyance under section 8 above, *shall be the amount which* at the relevant time *the house and premises, if sold in the open market by a willing seller, might be expected to realise* on the following J
assumptions:

(a) on the assumption that the vendor was selling for an estate in fee simple, subject to the tenancy, but on the assumption that this Part of this Act conferred no right to acquire the freehold;
(b) on the assumption that at the end of the tenancy the tenant has the right to remain in possession of the house and premises under the provisions of Part I of the Landlord and Tenant Act 1954;
(c) on the assumption that the tenant has no liability to carry out any repairs, maintenance or redecorations under the terms of the tenancy or Part I of the Landlord and Tenant Act 1954.

(The italics are mine)

I omit (d) to (f): it was agreed that they did not affect the valuations.
There is no section 19 scheme ("Retention of management powers K
for general benefit of neighbourhood"). Both parties considered that the bid of the lessee was to be taken into account and that under (b) above the lessee would not necessarily exercise his right at the end of the current lease to stay in the house paying a fair rent.

Mr Hopper assumed that the lessee was a willing but not an anxious buyer and that the lessor was a willing seller. In a friendly negotiation the first would start at a low figure and the latter at a higher one and they would move to consensus. This approach was supported in argument by counsel for the lessor. Conversely Mr

A Hewitt, supported by the argument of counsel for the lessee, assumed that the lessor was, in the words of the statute, "a willing seller" but that the lessee, knowing that to be so, would need to bid only a nominal amount more than any other purchaser to obtain the freehold. He would be able to break off the negotiations if he wished to do so; he was not obliged to purchase, whereas the lessor was obliged to sell, albeit not at any price. In negotiation the lessee would take advantage of this. Mr Hopper assumed that there would be direct negotiations between the lessor and the lessee, but Mr Hewitt did not approach the valuation in that way. He considered what other purchasers would pay and added a nominal amount for the lessee's direct bid; that amount was never quantified.

B In this connection I was referred to the following:

Cited authorities

Norfolk v Masters, Fellows and Scholars of Trinity College Cambridge (1976) (a decision of this tribunal, my own) 32 P & CR 147 [238 EG 421].

Stokes v Cambridge Corporation (1961) (a decision of this tribunal, J R Laird FRICS and John Watson FRICS) 180 EG 839.

Siggs v Royal Arsenal Co-operative Society Ltd (1971) (a decision of this tribunal — R C Walmsley FRICS) 220 EG 39.

Re S J C Construction Co Ltd's Application (1974) 230 EG 906 and *S J C Construction Co Ltd v Sutton London Borough Council* (1975) 234 EG 363 (CA). C

Inland Revenue Commissioners v Clay and Buchanan [1914] 3 KB 466.

F R Evans (Leeds) Ltd v English Electric Co Ltd (1977) 36 P & CR 185 [245 EG 657].

Wrotham Park Estate Co Ltd v Parkside Homes Ltd [1974] 1 WLR 798.

Kemp v Josephine Trust Ltd (1970) (a decision of this tribunal, R C Walmsley FRICS) 22 P & CR 804 at p 814 [217 EG 351].

Raja Vyricherla Narayana Gajapatiraju v Revenue Divisional Officer, Vizagapatam [1939] AC 302 ("The Indian Case").

D *Lambe v Secretary of State for War* [1955] 2 QB 612.

London Investment & Mortgage Co Ltd v Middlesex County Council (1952) (a decision of this tribunal — James L Milne FRICS) 2 P & CR 381 at p 384 [159 EG 364].

Hearts of Oak Benefit Society v Lewisham Borough Council (1978) (a decision of this tribunal — my own) 249 EG 967.

Counsel for the lessee relied particularly on *Clay* (pp 471, 472, 475 and 476), *Norfolk* (pp 151, 154, 156 and 157) and *Evans* (pp 187, 188, 190-192) and stressed that in the last there was a willing lessor and willing lessee whereas in this instance, in his submission, the lessor was willing but not the lessee.

Counsel for the lessor relied particularly on the following passage in "The Indian Case" at p 315: E

> Proceeding, therefore, with the imaginary auction at which are present two classes of buyers — namely, the "poramboke buyers" and the "potentiality buyers", — the former will disappear from the bidding as soon as the "poramboke" value has been reached, and the bidding will thereafter be confined to the "potentiality buyers". But at what figure will this bidding stop? As already pointed out it cannot be imagined as going on until the ultimate purchaser has been driven by the competition up to a fantastic price. For he is ex hypothesi a willing purchaser and not one who is by circumstances forced to buy. Nor can the bidding be imagined to stop at the first advance on the "poramboke" value. For the vendor is a willing vendor and not one compelled by circumstances to sell his potentiality for anything that he can get. The arbitrator will, therefore, continue the imaginary F bidding until a bid is reached which, in the arbitrator's estimate, represents the true value to the vendor of the potentiality.

He also relied on *Lambe*. At p 618 the basis is referred to by Parker LJ:

> Before referring to these three bases of valuation, I would refer, so far as it is material, to the general rules laid down by section 2 of the Acquisition of Land (Assessment of Compensation) Act 1919. By rule (2), the value of the land, which includes by definition any interest therein, "*shall*, subject as hereinafter provided, *be taken to be the amount which the land if sold in 'the open market by a willing seller might be expected to realise'* "

(The italics are mine)

G At p 621 Parker LJ continues:

> Prior to the introduction of rule (3) by the Act of 1919 it was always the law — and it was forcibly restated by Lord Romer in the Indian case — that the amount which the authority, in a friendly negotiation, would be willing to pay in acquiring a freehold interest and as though no compulsory powers of acquisition had been obtained, was a matter to be taken into consideration.

The words in section 2 of the 1919 Act are almost identical with those in subsection (1A) inserted into section 9 by the Act of 1974. At p 623 Parker LJ concludes:

> In the present case, basis 3 is specifically stated to be on the basis of the value which the acquiring authority, in a friendly negotiation, would be willing to H pay in acquiring a freehold interest for its purposes, and as though no compulsory powers of acquisition had been obtained. I would uphold the decision of the tribunal and dismiss the appeal.

So I come to the view that the approach adopted by counsel for the lessor and by Mr Hopper in his valuation is the correct one, that the basis to be adopted is not a nominal amount above the sum which any person other than the lessee would pay but what the lessee would be willing to pay in a friendly negotiation with the freeholder.

Parties valuation and analysis

Mr Hopper's valuation was as shown opposite. J

There are similarities between the two, more on the surface than in reality. Both calculate the value of the freehold interest subject to the lease excluding marriage value (referred to before me as "stage A"): each then adds a proportion of the share of that value ("stage B"). There the similarity ends. The difference of £11 (£388-£377) is of no significance, even in Mr Hewitt's figure of £45,725. Mr Hopper's calculation of the value of the reversion in stage A was substantiated by reference to 57 enfranchisement prices settled under section 9(1A) in respect of houses on the Grosvenor and Cadogan London Estates. Full details of the calculations used were put before me: they were on the same lines as Mr Hopper's valuation above. In almost all cases the percentages used were 10 per K cent for the deduction and 6 per cent for the deferment. On the other hand, Mr Hewitt's use of 10 per cent for deferment was unsupported by reference to any transaction.

I asked Mr Hewitt if he had analysed any of the 57 settlements using the approach he had adopted; he said he had not attempted this. Mr Hopper, when asked the same question, said that he had not done so using reversions to fair rents because of the difficulty of assessing them for this sort of property in London and because tenants did not exercise their rights to stay in such houses paying fair rents. Nor had he endeavoured to analyse the settlements using terms agreed for surrenders and grants of new leases, because that would be very complicated. L

Mr Hewitt criticised the use of settlement figures generally. Although a valuation had to be done at the date of the tenant's notice to enfranchise, in a rising market, by the time negotiations took place, values had increased, in some cases quite considerably: in cross-examination Mr Hopper said that that period was often two years. This made lessees somewhat indifferent as to the prices they paid. Similarly, under the terms of the Leasehold Reform Act, whether under section 9(1) or section 9(1A), lessees got such good bargains that very frequently they did not challenge the prices asked. Counsel for the lessee submitted that the 10 per cent deduction made by Mr Hopper was not backed by market evidence. It is, however, overwhelmingly backed by settlement evidence. Mr M Hewitt has no backing for his approach at all.

In cross-examination Mr Hopper agreed that his stage A "allows a deal with the occupier at some future date on some terms or other". However, he denied that the tenant's bid was reflected in his stage A. As to whether one deducted 10 per cent from the vacant possession value or 78 per cent as was reflected in Mr Hewitt's figures (see below) was a valuation matter: I agree.

Counsel for the lessee said that the settlement evidence only showed that the lessors adopted a consistent method, no more. As to settlements made under the Leasehold Reform Act, generally speaking a lessee would be in a weaker financial position than a

A

(A) *Value of lessors' interest excluding prospects of "marriage" of interests*			
Term			
Rent receivable	£45		
YP 12 years at 5½ per cent	8.62	£388	
Reversion			
Estimated value of unencumbered freehold interest	187,500*		
Less for risk of tenant claiming a tenancy under Part I of Landlord and Tenant Act 1954 10 per cent	18,750		
	168,750		
Defer 12 years at 6 per cent	0.497	83,869	£84,257
(B) *Lessors' share of "marriage" value*			
Estimated value of unencumbered freehold interest		187,500*	
Less (i) Value of lessors' interest excluding prospects of "marriage"	84,257*		
(ii) Value of lessee's interest excluding prospects of "marriage"	45,000*	129,257	
		58,243	
Gain on marriage 50 per cent to the lessors			29,121
			113,378
		Say	113,000

B

C

* All values exclusive of the value, if any, of lessee's improvements.

Mr Hewitt's valuation was:

D

Value of lessors' interest excluding "marriage" value				
Term	Ground rent	£45.00 pa		
	YP 12 years at 6 per cent	8.38	£377	
Reversion	Fair rent (net of outgoings)	4,115		
	YP in perpetuity def 12 years at 10 per cent	3.18	13,085	£13,462
Lessors' share of "marriage" value				
Value of freehold in possession			187,500	
Value of lessors' interest exclusive of "marriage" value		13,462		
Value of leasehold interest		45,000	58,462	
			£129,038	
Gain on "marriage" of interests				
Lessors' share at 25 per cent				32,259
				45,721
			But say	£45,725

E

lessor: he personally would have to bear the risks of costs in coming to this tribunal. In the case of an estate, costs would be spread over a
F number of properties. Those matters should be kept in mind when one looked at settlements.

Counsel for the lessor, however, submitted that it is important for a lessee that he should buy the freehold because (1) the value of his leasehold interest is depreciating as the lease gets shorter (2) capital values are increasing all the time (3) if he delays, the price will only be greater later on.

Mr Hopper and Mr Hewitt agreed that it was almost unknown for a lessee to continue in occupation of such a house as this in London at a fair rent and that was confirmed by Mr Edwards. Lessees usually came to terms for a surrender and grant of a new lease, if they did not buy the freeholds.

Tribunal's views on valuation G

I regard Mr Hewitt's approach in stage A as being so out of touch with reality that I reject it and I accept Mr Hopper's. Mr Hewitt's figure of £13,462 represents the lowest bid ("the lowest bid") which a purchaser other than the lessee would make, ie the bid from one who entirely disregarded the possibility that the lessor and lessee would come to terms for the transfer of the freehold between them. It assumed the worst position which could arise, ie that the lessee would remain in the house at a fair rent. In reality such a purchaser would easily be outbid by a competitor, who would take a more optimistic view. Counsel for the lessee submitted that stage A should not reflect *any* possible bid by the lessee and in that connection he relied on *Norfolk* at p 157, otherwise there would be double H counting.

Counsel for the lessor submitted that in the real market, if the freehold subject to the lease were to be offered for sale and the lessee were not interested in purchasing, the possibility that he or his successor in title might later on wish to buy would be a factor affecting the bid of potential purchasers: he referred to that possibility as the "hope value" in contrast to the *actual realisation* of the share of the marriage value (stage B).

I observe that Mr Hewitt could have got to the figure of £13,085 by taking the capital value of £41,150 (£4,115 at 10 years' purchase) and deferring it at 10 per cent: £41,150 is 22 per cent of the agreed vacant possession value, ie it represents a deduction of 78 J per cent from that figure. The evidence pointed strongly to 78 being far too high a percentage and that is another reason why I reject Mr Hewitt's figure.

In cross-examination Mr Hopper said that he considered the value of the freehold interest in the subject house if it were let by the freeholder at a fair rent would be not less than £93,000. Such houses as the subject house were not normally let at fair rents, such cases were very rare. The actual amounts of the fair rents of such houses (ie of houses let by freeholders at fair rents) were only relevant to some extent: purchasers of such freeholds would look more at the vacant possession values than at the fair rents. Mr Hewitt agreed. The difference between the valuers at stage A can K therefore be regarded as that between a 10 per cent and a 78 per cent reduction.

Criticising Mr Hewitt's valuation, Mr Hopper said that, in his view, the primary approach to a valuation of a freehold interest in a house which is let at a fair rent was to take a percentage of the vacant possession value, not to capitalise the net income.

I accept Mr Hopper's figure supported as it is by reference to the figures of the 57 settlements. Apart from that support, the gain on the marriage of interests of £129,038 as determined by Mr Hewitt looks to be an extraordinarily large percentage of the agreed figure of £187,500, freehold with vacant possession. Mr Hopper's corresponding figure is £58,243 and on the face of it is more reasonable. L

The difference at stage B is that between 25 and 50 per cent, although the percentages are of amounts calculated on different bases and represent different concepts. Mr Hewitt's 25 per cent represented the amount which a potential purchaser other than the lessee would pay above the lowest bid for the possibility that the lessee or a successor in title would be interested in buying the freehold. The actual lessee would be bidding, but he would only need to bid just above £45,725 in negotiation with the lessor to secure the freehold now. I have already rejected that approach.

Mr Hopper's addition as stage B represents the actual amount which the lessee in friendly negotiations with the lessor would be willing to bid above the amount which any other purchaser would M pay: that amount would take his bid well above that of any other potential purchaser. Mr Hopper assumed that the parties were of equal bargaining strengths. I agree: neither can unlock the marriage value without the other. In friendly negotiations they would agree to divide it equally as they had done in the 57 settlements.

Determination of fair rent

There are other matters to which I should refer. First, the likely fair rent. According to the certificate of fair rent the amount of £4,250 per annum related to a term of three years, the tenants being responsible for minor repairs and internal decorations and the landlord for all other repairs and decorations. There was no disagreement between the parties that the amount of the deduction for the landlord's outgoings should be £750 per annum. According to the certificate of fair rent, that would leave a net income of £3,500 per annum. Mr Hewitt himself regarded that as too low, since he used a figure of £4,115 per annum in his valuation: that was based on Mr Gibbon's estimate.

Counsel for the lessor invited me to disregard the certificate because it did not result from a *bona fide* application. In this instance it was in the lessee's interest to obtain a low figure. Mr Gibbon said that he regarded the figure of £3,000 per annum contained in the application for the fair rent certificate as being too low. The grounds of the application were not completed: these are:

(a) conversion of tenancy from rent control to rent regulation;
(b) erection/conversion/improvement of premises (otherwise than (a) above);
(c) proposal to let premises on a regulated tenancy, or on a tenancy to which Part VI of the Rent Act 1977 will apply.

Neither (a), (b) nor (c) applied; counsel for the lessor said that there was no evidence that the lessee, on whose behalf the application was made, intended to convert a tenancy or to improve the subject premises or to let them. Mr Edwards' evidence was to the effect that the fair rent on the same terms as the certificate would be £8,000 per annum, which would give a net income of £7,250 pa.

I was referred to *Tormes Property Co Ltd v Landau* [1971] 1 QB 261, but for the reasons given above I do not need to determine the fair rent. I would find it difficult to do so; there is little pertinent evidence. Details of a considerable number of properties let at fair rents were put before me, but practically none of these related to *houses* close to the subject house. Some related to flats and maisonettes nearby: others were in different parts of London. General figures relating to applications and determinations of fair rent were put before me relating to England, Wales and Scotland. Included in the evidence was much which related to the relationship, if any, between capital values and fair rents. There was evidence of rental value in the open market, in the rent at which a nearby house was let, but if one were to take that as a basis for determining a fair rent of the subject premises the problem would be to determine what the discount should be to eliminate scarcity. It seems to me that that is a matter of opinion for the rent officer subject to appeal to the rent assessment committee: that is another reason for rejecting Mr Hewitt's approach at stage A.

Negotiations by Grosvenor and Cadogan Estates

Second, should it be thought that the figures relied on by Mr Hopper, which were obtained from the 57 settlements, were the results of analyses made by him of the prices reached in negotiation, I summarise how these negotiations were approached and how the percentages of 10, 6 and 50 in Mr Hopper's valuation set out above were supported.

I take first the Grosvenor Estate. When an approach is received by the lessor from a lessee, Mr Hopper's firm is instructed to advise. They then seek the assistance of another firm who advise them (1) on the value of the house concerned, freehold in possession at the relevant date and (2) on the value of the leasehold interest at the same date, assuming that the lessee did not have any right to enfranchise under the Leasehold Reform Act 1967, as amended. Mr Hopper's firm then reports to the estate setting out a full calculation on the lines of Mr Hopper's valuation set out above. Sometimes the figures are altered after consultation with the other firm and with the estate's own surveyors. The calculation is then put to the lessee in full or in outline. If negotiations follow, the figures in the calculation may be altered somewhat. Bargains are struck. The price might be the same as in the original calculation or it would be as shown in the amended one.

With regard to the Cadogan Estate, no outside firm is used to advise but otherwise the same practice is followed. So the figures relating to the 57 settlements supporting Mr Hopper's percentages were not derived from his analyses but from the actual figures shown in a calculation although the lessee did not always agree each figure in it. That support was described by counsel for the lessor as "massive" and I agree.

Third, Mr Hopper said that, in his opinion, with such large estates as the Grosvenor and Cadogan the "*Delaforce*" effect (see *Delaforce v Evans* (1970) 215 EG 315, a decision of this tribunal (R C Walmsley FRICS)), was often a negative amount, since lessors did not wish to press matters to a hearing before this tribunal for fear of the publicity and the force of the authority which would emerge should the finding be against them: rather than risk that, they would settle at a lower figure. In Mr Hewitt's view that carried little weight: settlements were rapidly and widely known; they "got out". There was an association of tenants on the Grosvenor Estate. I come to the conclusion that there is no *Delaforce* effect here one way or the other.

I accept Mr Hopper's valuation entirely and determine that the price which the lessee shall pay to the lessor for the freehold interest in the subject house is £113,000.

I was told that there were certain matters outstanding between the parties which could affect the provisions to be contained in the conveyance (see sections 10 and 21(2)(a) of the Leasehold Reform Act 1967). It was agreed, therefore, that this should be an interim decision with liberty to apply.

I heard submissions as to costs. The lessor will pay the lessee's costs for the first day of the hearing. The lessee will pay the lessor's costs of the reference (except for the first day of the hearing), all such costs, if not agreed, to be taxed by the Registrar of the Lands Tribunal on the High Court Scale of Costs.

LANDS TRIBUNAL
RATING

A JERMY v JONES (VO)

(LVC/24/1981)

(C R MALLETT FRICS)

October 22 1981

Estates Gazette January 16 1982
(1981) 261 EG 161

Premises on three floors designed and built as shop with living accommodation over but used solely for residential purposes — In terrace of shop units — Matter fundamental to rating law —
B Whether to be valued as "house" or, as VO contends, "shop, maisonette and premises (used wholly for domestic purposes)" — Only first floor fully used — Ground-floor shop used only occasionally — Third floor and remainder of ground floor unused or virtually unused — Residential-only restriction in mortgage deed held to be disregarded — Premises can be used for shop purposes without physical change and there is demand for shop use — Tribunal decides these factors can be taken into account following *Midland Bank v Lanham (VO)* — VO's view upheld

Both parties appeared in person.

Giving his decision, MR MALLETT said: This is an appeal against a decision of the Greater London (North Eastern) Valuation Panel
C confirming the assessment in the list of gross value £760, rateable value £605, in respect of a shop, maisonette and premises (used wholly for domestic purposes) at 173 Clarence Road, E5.

The appeal raises a matter which is fundamental to the law of rating. The appellant ratepayer contends that the appeal premises should be valued as a house having regard to the actual use and the fact that the mortgage deed on the premises restricts the use to residential only. The respondent valuation officer contends that the premises should be valued as a shop, maisonette and premises (used wholly for domestic purposes).

It is agreed between the parties that in the event of the appellant's contentions being found to be correct the assessment should be
D gross value £360, rateable value £274, and in the event of the respondent's contention being found to be correct the assessment should be gross value £680, rateable value £538.

The appeal property is one of a terrace of shops, with living accommodation over, built in the latter part of the 19th century and is constructed mainly of brick with a slated roof. There are 19 similar units in the same terrace and on the opposite side of the road six similar units. Sixteen of the units are currently used as retail shops, doctors' surgeries or bookmakers' shops. Five of the units are being used as non-retail workshops. Three of the units are empty. The appeal premises are the only unit used solely for residential purposes.

E The appeal premises consist of a shop and two rooms on the ground floor, three rooms and wc on the first floor and two rooms on the second floor.

Mr and Mrs Jermy have been sitting tenants of the appeal premises for many years and for some nine years prior to 1977 they sublet the shop on a weekly basis and continued to occupy the remainder of the premises as living accommodation. In 1977 they purchased the freehold of the premises and obtained a mortgage from the London Borough of Hackney. A clause in the mortgage deed reads:

that the premises shall be used exclusively as a residence for himself and F
for his wife and children PROVIDED ALWAYS that the Council at any time by notice in writing to the borrower waive or modify this requirement.

In effect, Mr and Mrs Jermy use only the first floor of the premises as a residence, the second floor being undecorated and unused and the ground-floor shop, although containing some isolated items of domestic furniture, is only occasionally in use and the two rear rooms on the ground floor are undecorated and virtually unused. The shop window and doorway is still in existence and there is a separate door leading to the upper floors. The division between the ground-floor shop and the passageway to the upper floors is by means of a flimsy pegboard partition to three-quarters
of the ceiling height. G

Mrs D L Jermy, who conducted her own case and gave evidence, contended that because of the condition contained in the mortgage deed she could only occupy the premises for domestic purposes and it should be assessed accordingly. She admitted that were it not for the conditions contained in the mortgage deed the premises could be used as a shop and they had had several inquiries from prospective shop-keepers seeking to occupy the shop. In her eyes the local authority which imposes a restriction on her occupation is the same authority which demands rates on a more valuable use, albeit on the basis of the domestic rate and not the commercial rate.

A D Jones ARICS, the valuation officer, submitted that the
conditions contained in the mortgage deed are not a matter that H
could be taken into account in arriving at the rating assessment and he drew my attention to the words of Lord Pearce in the House of Lords decision in *Dawkins v Ash Bros & Heaton Ltd* [1969] RA 205 at p 213:

> so one must assume a hypothetical letting (which in many cases would never in fact occur) in order to do the best one can to form some estimate of what value should be attributed to a hereditament on the universal standard, viz a letting "from year to year".
> But that only excludes the human realities to a limited and necessary extent, since it is only the human realities that give any value at all to hereditaments. They are excluded in so far as they are accidental to the letting of a
> hereditament. They are acknowledged in so far as they are essential to the J
> hereditament itself. It is, for instance, essential to the hereditament itself that it is close to the sea and that humans will pay more highly for a house close to the sea. One can therefore take that into account in the hypothetical letting. It is, however, accidental to the house that its owner was shrewd or that the rich man happened to want it and that therefore the rent being paid is extremely high. In the same way I think it would be accidental to the hereditament that its owner intended to pull it down in the near future. For the hereditament might have had a different owner who would not pull it down. So the actual owner's intentions are thus immaterial since it is the hypothetical owner who is being considered.

Mr Jones contended that the appeal premises were contained
within a shopping parade where clearly there was a demand for K
shops. The appeal premises could be used as a shop and indeed were designed and built to be so used and the appellant admitted that they could be so used.

He drew my attention to decisions of the Lands Tribunal dealing with similar circumstances in the case of *Spencer v Hampshire* (1968) 206 EG 1274, *Prince v Baker* (1973) 228 EG 2297 and *Tappenden v Stratford* LVC/107/1980. He suggested that in valuing for rating purposes it was possible to have regard to other alternative uses as well as the existing use. As there was clearly a demand for use as a shop with living accommodation, and as the premises could be so occupied without any structural alterations, this factor could be taken into account.

He admitted that the change from residential use to shop and residential use would require planning permission, but it was reasonable to suppose that this could readily be obtained and prospective tenants seeking to occupy the ground floor as a shop had not been deterred from making a bid.

He pointed out that although the local valuation court had confirmed an assessment of gross value £760 he had since formed the view that the passageway on the ground floor forming access to the upper floors should be excluded from the shop value and for this reason he had reached his valuation. The figure of gross value £680 was agreed between the parties.

Tribunal's decision

I viewed the appeal premises and viewed externally the shopping parades in Clarence Road.

The rating statutes require that all premises must be valued for rating purposes on the assumption that they are vacant and to let on a standard form of tenancy. The object of the legislation is no doubt to ensure that all premises are valued for rating purposes to a similar standard. This applies whether in fact premises are let or not, or whether let on different terms or conditions.

In the present case the local authority were presumably only entitled to grant mortgages on residential premises and this accounts for the restriction in the mortgage deed. Another mortgagee might apply a different restriction or none at all. Therefore these are matters which are not essential to the hereditament for the consideration of the hypothetical tenancy for rating purposes, and the actual restriction imposed by the mortgage deed must be disregarded.

Nevertheless, the existing use of the appeal premises is wholly for residential purposes and *prima facie* that is the use to be valued. Clearly, the appeal premises can also be used for shop purposes without any physical change. Clearly, there is a demand for such use. Can these additional factors be taken into account? I think, in the circumstances of the case, they can. See *Midland Bank v Lanham (VO)* (1978) 246 EG 1017 at p 1120 where the tribunal (J H Emlyn Jones FRICS) said:

> I thus accept the submission of counsel for the ratepayers and counsel for the valuation officer's alternative submission so far as it goes — that in valuing for rating purposes it is necessary to take the hereditament as it stands *rebus sic stantibus*; that the value as thus restricted must relate to the hereditament in its existing physical state; and that the use of the hereditament must be taken to be within the same mode or category as the existing use. So much, I think, is established by authority, as I have interpreted it; and certainly now confirmed by statute law in section 20 of the General Rate Act 1967. Finally, all alternative uses to which the hereditament in its existing state could be put in the real world, and which would be in the minds of competing bidders in the market, are to be taken as being within the same mode or category, where the existence of such competition can be established by evidence . . .
> The second observation is with regard to planning law in England and Wales. As the tribunal has said in earlier cases, it does not seem to be part of the role of the tribunal to determine issues of planning. Planning legislation can be regarded as a modern innovation when compared with the ancient institution of the poor rate, and quite clearly it cannot be taken to have amended any of the principles on which valuations for rating are to be made. Its significance in my opinion is that the existence of planning restrictions and the necessity of obtaining planning permission for certain changes of use are all factors which affect the minds of potential tenants in the real world, and to that extent they must have an influence on value. But the interpretation of these factors remains a matter of evidence and of expert valuation evidence.

Therefore the appellant's submission fails, but in view of the valuation officer's revised valuation I direct that the assessment in the list be amended to gross value £680, rateable value £538. To that extent the appeal succeeds.

I heard submissions as to costs, but in the special circumstances of this case direct the appellant to pay a contribution to the valuation officer's costs in the sum of £30.

COOK (VO) v ROSS POULTRY LTD

(LVC/850-1/1979)

(J H EMLYN JONES FRICS)

March 23 1982

Estates Gazette May 29 1982
(1982) 262 EG 887

Two parcels of land near Woodhall Spa occupied by firm engaged in breeding and keeping pheasants and partridges for sale — Whether agricultural hereditaments — 12-month cycle of operation — Various parts of larger parcel occupied by birds at different periods of a few months each and then left empty or used for sheep grazing for remainder of year — Smaller parcel occupied by birds for 4½ months, then left empty and not grazed — Tribunal rejects submissions that land used as pasture ground only and that birds which run and scratch about are grazing — Argument that pheasants and partridges are poultry also rejected — Held that none of the land is "agricultural land" within s 26(3) of 1967 Act — Finding that pheasants and partridges in instant case not kept for use in farming of land nor are they "livestock" for purposes of 1971 Act — Therefore buildings on land not "agricultural buildings" — Further held that rearing of pheasants and partridges not an agricultural operation and that buildings not used solely in connection with agricultural operations on land — VO's proposals for valuation list entries at agreed values successful

Alan Fletcher (instructed by the Solicitor of Inland Revenue) appeared for the appellant valuation officer, E Cook ARICS; G P E Curry (solicitor to Imperial Foods Ltd) for the respondents.

Giving his decision, MR EMLYN JONES said: These two appeals were heard together. They are concerned with two separate parcels of land near Woodhall Spa in Lincolnshire, both occupied at the material date as part of a single undertaking operated by the respondents Ross Poultry Ltd. The undertaking was known as the Sterling Game Farm and was engaged in the breeding and keeping of pheasants and partridges for sale. The question for determination is whether the premises qualify for rating purposes as agricultural hereditaments.

At Whitehall Farm there is a dwelling-house occupied by the farm manager; a large Dutch barn used for storage of straw, crates, tractors, grass cutters and other implements used in the running of the farm; three old cow stalls used as a store for feedingstuffs and disinfectants; an "L"-shaped building containing the hatcheries and various stores and an office. With the buildings are about 23 acres of land divided up into areas containing wintering pens and small and large breeding pens reserved for the pheasants and a plot of about one acre occupied by the partridges.

The other parcel of land at Kirkby Lane extends to an area of 1.1 acres, which is largely covered by rearing sheds and enclosed pens used in connection with the rearing of the pheasants.

The house at Whitehall Farm was included in the valuation list for the East Lindsey district rating area with a description "House (AG) and Garage" and an assessment of gross value £142, rateable value £92. On March 28 1977 a proposal for the alteration of the valuation list was made by the duly authorised valuation officer for the area. He proposed that the entire premises should be included in the valuation list with the description "Pheasantry, House, Garage and Premises" with an assessment of gross value £800, rateable value £638. On the same day, he made a similar proposal for an entry in the valuation list in respect of the Kirkby Lane premises to be described as "Pheasantry and Premises" with an assessment of gross value £750, rateable value £597.

The land and buildings had previously enjoyed exemption from rates as agricultural hereditaments. The valuation officer's proposals were based on his contention that the premises no longer qualified for this exemption and should therefore be rated at their full value in the valuation list.

The matter was referred to the Lincolnshire Valuation Court, who dismissed the valuation officer's appeals on the grounds that the premises qualified as agricultural hereditaments. The court therefore confirmed the assessment of Whitehall Farm, treating only the house as an agricultural dwelling-house and garage, and directed

A that there should be a nil entry in respect of the Kirkby Lane premises.

The valuation officer now appeals to the Lands Tribunal. It is agreed that if he is successful in his appeals the appropriate entries to be made in the valuation list are as follows: LVC/850/1979, Pheasantry and Premises, Kirkby Lane, Kirkby on Bain, gross value £500, rateable value £388. LVC/851/1979, Pheasantry, House, Garage and Premises, Whitehall Farm, Horncastle Road, Roughton, gross value £680, rateable value £538. If in either case it is held that the open land is agricultural land but that the buildings are not agricultural buildings, then the assessments are to be respectively gross value £495, rateable value £384, and gross value £420, rateable
B value £322. If the appeals are dismissed then it is accepted that the decisions of the local valuation court shall stand.

It will be helpful if at this stage I summarise the activities which were carried on by the respondents at the relevant date on the appeal premises. There is no dispute about the facts and what I have to say here is taken from an agreed statement.

The operations follow an annual cycle. In the month of October parent stock are placed in the wintering pens at Whitehall Farm, consisting in round terms of 5,000 hen pheasants and 1,000 cocks and in addition 500 pairs of partridges. I should explain that a cock pheasant will run with up to six hens but that partridges are monogamous. The birds remain in the winter pens until the middle of February,
C when the partridges are moved to an area occupied by what are described as verandahs. These are large wooden frame cages on legs. At one end is a covered trough for food and water and at the other a covered nesting area. The cages are each approximately 8 ft long by 2 ft wide and are constructed in groups of four. At the same time the pheasants are split up into smaller group and put into breeding pens.

Towards the end of April the hen birds begin to lay and in a season will produce between 30 and 40 hatching eggs. Eggs are removed and stored prior to hatching in the buildings, but must be turned regularly during storage and set within 10 days and placed within incubating units in the hatcheries. The hatcheries each contain 12 incubating
D units each consisting of a setter, where the eggs remain for the first three weeks of incubation, and a hatcher, in which the eggs are placed for the last three days. A high level of humidity is maintained so that in due course the chicks are able to break through the tough membrane lining the shell of the egg. The hatching cycle lasts for about 24 days.

Between 10% and 20% of the day-old chicks are taken to Kirkby Lane and the remainder are sold either direct or through other game farmers, to landowners and farmers for the purpose of being released on to the land, subsequently shot and sold mainly for human consumption. In a good season about 65,000 day-old chicks would be produced.

E The first day-old chicks transferred to Kirkby Lane normally arrive during the second half of May and would continue to arrive until the beginning of September. There they would be placed in brooder pens, each holding about 250 chicks. There are 42 pens in all, situated on either side of a central passage running through each of the two buildings which had formerly been used as poultry houses. Each pen has a water fount, two food-hoppers, a gas heater and electric light; and access to a run through an access way known as a pophole. The runs are wired grass enclosures covered with a fine nylon mesh. At an early stage the birds normally sleep in the sheds, but with the arrival of warmer weather sleep outside in the runs. In hot sunny weather, however, the birds still seek shade inside the rearing sheds.

F After six or seven weeks at Kirkby Lane most of the birds are sold as poults to farmers and estate owners, perhaps as many as 16,500 in one season, and the remainder are then transferred as breeding birds to Whitehall Farm in October, when the cycle begins again.

At the end of their breeding life 500 breeding birds are sold either to farmers or estate owners or to game farmers and some 4,500 are slaughtered and sold to a firm of poulterers and game dealers who operate from Smithfield Market and sell them for human consumption.

These operations can be summed up as follows: At Whitehall Farm the winter pens are occupied by birds from October until the middle of February. For the rest of the year they then remain empty until May, when a neighbouring farmer brings sheep to graze G
on the land until September. The breeding pens are occupied from mid-February until the end of July, when the day-old chicks are moved out and the partridges sold. They then remain empty until November, when sheep come in to graze until the end of December. The breeding pens are then empty until occupied again in mid-February. The hatcheries and associated buildings are prepared for use in February and March and continue to be used until the beginning of August. They then remain empty until February.

At Kirkby Lane the runs and the buildings are occupied and used from the middle of May until the end of September, when the buildings are cleaned and disinfected and remain empty for the rest of the 12-month cycle. H

It is agreed that the dwelling-house at Whitehall Farm is to be rated. The question for determination is whether the remaining buildings in both cases are agricultural buildings and whether the land is agricultural land.

Mr Curry called evidence from Dr David Lloyd BSc PhD ARAgS and from Mr Kenneth John Foster, a chick production director of Ross Poultry (GB) Ltd. Dr Lloyd is senior lecturer in agricultural management at the University of Reading, where he has been for more than 25 years. In that capacity he has assisted with the administration of the university's extensive farming estates in Berkshire which have included a game fowl enterprise.

In Dr Lloyd's opinion the operations carried on in 1977 at the J
Sterling Game Farm were agricultural operations producing game which were ultimately intended for human consumption. The methods employed involved the use of pasture where the birds could run, take exercise and eat the grass. For this purpose, it was necessary that the grass length should be controlled and for this reason it was necessary to introduce the sheep who grazed upon the land. In Dr Lloyd's opinion, the raising of pheasants and partridges for the table would be an uneconomic proposition without the income to be obtained from shooting. Dr Lloyd did not claim that the game birds were poultry but emphasised the many similarities between the two types of bird. For example, game birds such as pheasants and partridges, said Dr Lloyd, are agricultural livestock. Their genetic K
origins are similar to those of poultry; they all belong to the genetic family *phasianidae*. Equally, methods of husbandry and the state of domestication of the birds are similar.

In concluding his evidence, Dr Lloyd said that the process carried on at the Sterling Game Farm, involving specialised breeding, incubation and rearing, formed an essential and integral part of a longer and continuous food production system which is clearly agricultural in nature.

Mr Foster described in some detail the operations actually carried on. He referred to the importance of allowing the birds space in which to run about. The intensive rearing of game birds had proved impossible. The birds did not prosper under these conditions L
and were prone to cannibalism and disease. In Mr Foster's opinion, the growth of the cover and the space for exercise was more important than the nutritional value of the grass.

This, then, is the evidence and these are the facts which I have found. I turn next to the law and first of all consider the provisions for agricultural derating contained in the General Rate Act 1967. Section 26, so far as it is relevant, provides as follows:

> (1) No agricultural land or agricultural buildings shall be liable to be rated or be included in any valuation list or in any rate.
> (2) ......
> (3) In this section the expression "agricultural land"—
> (a) means any land used as arable, meadow or pasture ground only, land used for a plantation or a wood or for the growth of saleable underwood, land exceeding one quarter of an acre used for the purposes of poultry farming, cottage gardens exceeding one quarter of an acre, market gardens, nursery grounds, orchards or allotments, including allotment gardens within the meaning of the Allotments Act 1922, but does not include land occupied together with a house as a park, gardens (other than as aforesaid), pleasure grounds, or land kept or preserved mainly or exclusively for purposes of sport or recreation, or land used as a racecourse; and for the purposes of this paragraph the expression "cottage garden" means a garden attached to a house occupied as a dwelling by a person of the labouring classes; M
> (b) ......

(4) In this section, the expression "agricultural buildings"—
(a) means buildings (other than dwellings) occupied together with agricultural land or being or forming part of a market garden, and in either case used solely in connection with agricultural operations thereon;
(b) ......

It will be seen that in considering these provisions it is first necessary to identify the agricultural land and there can be no question under this section of finding any agricultural buildings — apart from market gardens — which are not occupied together with agricultural land. For the appellant valuation officer, Mr Fletcher submits that the land in the present appeals does not satisfy any part of the statutory definition and, in particular, that it is not land used as pasture ground only or land used for the purposes of poultry farming. Mr Curry for the respondents submits that the land is used as pasture ground only. Sheep graze and birds run on the land and take some of their feed from the grass and the land is used for no other purpose. He also submits that pheasants are poultry and, in the alternative, that even if they are not properly to be described as poultry nevertheless the operations carried on have all the characteristics of poultry farming. The land therefore satisfies in that sense the statutory definition.

In my opinion the land at Kirkby Lane does not qualify as land used as pasture ground only. Nor do I consider that the birds which run and scratch about in the pens can be said to be grazing. The land at Whitehall Farm when the sheep are grazing can properly be described as pasture ground, and as a description that phrase may well be considered apt during the rest of the year, when the sheep are not there. It is not enough, however, that land should be described as pasture ground in order to qualify as agricultural land. It must also be used as pasture ground only.

When the birds are running on the land it is true that they consume a certain amount of the grass; but this forms a comparatively small proportion of their total nutrition. According to Dr Lloyd, this would not amount to more than between 15% and 20% of their total food intake. It seems to me that the main contribution made by the land was as exercise ground, and it also provided food; but the food consisted not only of the grass but also of grubs, insects and worms. Mr Fletcher referred me to *Westropp v Elligott* (1884) 9 AC 815. This was an action brought for the recovery of possession of land under the Land Law (Ireland) Act 1881. Although the case was brought under a very different statute, I derive some help from three passages in the speeches. At the bottom of p 819 the Earl of Selbourne said:

I entertain no doubt that "pasture" in this place means feeding cattle or other live stock upon the land. This is both the etymological sense of the word and its sense according to ordinary agricultural use.

On p 830 Lord Watson said:

I have never seen any reason to doubt that words "depasturage" . . ., "pasturing" . . ., and "pasture" . . . in the Act of 1881, were all intended by the legislature to signify one and the same thing, viz., the grazing of cattle, or other livestock, upon grass lands.

On p 838 Lord Fitzgerald said:

Then what is the meaning of the term "for the purpose of pasture"? We were supplied with a good many definitions of the word "pasture", and some went to the extent that it meant "land under grass", or even the grass itself without more, but I am prepared to adopt the ordinary and popular meaning that "for the purpose of pasture" means for the feeding of cattle, whether it be for grazing or fattening, or for dairy purposes.

Adopting this approach, I do not think it can be said that the land at Whitehall Farm is used as pasture ground only, even though it is used for grazing sheep for six months of the year. For the rest of the year it is used for the exercising of game birds. Adopting the ordinary use of language, the game birds did not graze upon the land, nor can the land be said to be used as pasture ground. This same conclusion must be reached *a fortiori* in the case of the land at Kirkby Lane, which in no sense can be described as pasture ground.

The next leg of Mr Curry's argument is that the land, exceeding as it does one quarter of an acre, is used for the purposes of poultry farming. Mr Curry submitted that the pheasants and partridges were poultry for the purposes of the subsection. I cannot accept this submission. In the first place, the evidence of Dr Lloyd clearly recognises that pheasants and partridges are not poultry in the emphasis which he gave to pointing out the similarities between the category of poultry on the one hand and the category of game birds on the other. But the matter is in any case settled by authority.

In *Hardwick Game Farm v Suffolk Agricultural Poultry Producers' Association* [1969] 2 AC 31 the plaintiffs bought quantities of compounded meals for feeding to pheasants and partridges and their chicks. The meal unfortunately contained some toxic material which led to the death of many of the chicks. The action was brought *inter alia* for a breach of implied warranty under section 2(2) of the Fertilisers and Feeding Stuffs Act 1926. This subsection referred to food "for cattle or poultry". It was held in the House of Lords that the word "poultry" in the subsection must be given its ordinary meaning and, accordingly, did not include pheasants and partridges, since those would not normally be so described. On p 85 Lord Reid said:

The second argument requires more consideration. In the first place they say that pheasants are not poultry and therefore the subsection does not apply. I do not think that these pheasants were poultry. They were reared for the purpose of being released to serve as targets for sportsmen, and pheasants which have never been in captivity are clearly not poultry. It may well be that, if it should prove profitable to rear and keep pheasants in captivity until killed for human consumption, such pheasants should be regarded as poultry. But the mere fact that these pheasants like other game will come to the table after they have been shot seems to me to be immaterial. It would not in my view be in accordance with the ordinary use of language to say that they were poultry until released and then became game. They were game throughout and the farm where they were reared was properly called a game farm.

Lord Morris of Borth-y-Gest, Lord Guest, Lord Pearce and Lord Wilberforce all expressed similar opinions. The meaning of the word "poultry" in the instant case has to be considered in the context of the General Rate Act 1967 rather than of the Fertilisers and Feeding Stuffs Act of 1926. But in both statutes the word is being considered in an agricultural context, and I can see no sufficient reason why, as a matter of construction, an interpretation should apply when considering rating relief for agricultural premises different from that which is applied when considering the sale of agricultural foodstuffs.

In my judgment, therefore, the pheasants and partridges reared at the Sterling Game Farm are not poultry as the word is applied in the ordinary use of language, nor are they to be taken as poultry for the purposes of the General Rate Act 1967.

Mr Curry says, however, that even if the birds are not poultry in the strict sense, nevertheless the land can properly be described as land used for poultry farming because the operations carried on there possess all the characteristics of poultry farming, that is to say, the breeding, rearing and keeping of birds for the table. Such a conclusion seems to me to involve an exercise in mental gymnastics in which I am not prepared to follow Mr Curry.

In my opinion, the words of the statute are clear. Land used for poultry farming means land used for the farming of poultry and if it had been the intention of Parliament to include also land used for the farming of other birds like poultry, it would have been easy to say so.

I therefore find that none of the land which is the subject of these appeals falls within the definition of agricultural land in section 26(3). It follows that under the provisions of that section the buildings cannot be said to be agricultural buildings.

I should, however, deal with the situation which would arise if I were wrong and if the land at Whitehall Farm or at Kirkby Lane or at both places was properly to be taken as agricultural land. In such an event, further consideration would have to be given to the question whether the buildings were agricultural buildings. In those circumstances they would be occupied together with agricultural land. It would then become necessary to consider whether the buildings were used solely in connection with agricultural operations on that land. This is a question to which I shall shortly need to return.

Before doing so, I prefer to deal with the extended definition of agricultural land and buildings contained in the Rating Act 1971. Whereas in the 1967 Act the role of the buildings can be interpreted as ancillary to the land, in the 1971 Act the roles are reversed and agricultural land is defined so that it includes land occupied and used solely in connection with the use of agricultural buildings.

The relevant provisions are contained in sections 1 and 2 of the Act, which provide:

!. (1) In section 26 of the General Rate Act 1967 (in this Part of this Act referred to as "the principal section") –

(a) the expression "agricultural buildings" shall include any building which is an agricultural building by virtue of section 2, 3, or 4 of this Act; and

(b) the expression "agricultural land" shall include land occupied with, and used solely in connection with the use of, one or more such buildings.

(2) In this Part of this Act and in the principal section "building" includes a separate part of a building; and in determining for the purposes of this Part and that section whether a building used in any way is solely so used no account shall be taken of any time during which it is used in any other way, if that time does not amount to a substantial part of the time during which the building is used.

(3) In this Part of this Act "livestock" includes any mammal or bird kept for the production of food or wool or for the purpose of its use in the farming of land.

...

2. (1) Subject to subsections (2) to (4) of this section, each of the following is an agricultural building by virtue of this section –

(a) any building used for the keeping or breeding of livestock; and

(b) any building (other than a dwelling) which is occupied together with one or more buildings falling within paragraph (a) above and is used in connection with the operations carried on in that building or those buildings.

(There is no dispute in these appeals about the application of subsections (2), (3) and (4) of section 2.)

In a consideration of these extended definitions it is necessary to ask first whether the buildings at Whitehall Farm or Kirkby Lane are buildings "used for the keeping or breeding of livestock". That immediately raises the question — are pheasants and partridges livestock for the purposes of rating legislation, as defined in section 1(3)? This necessarily depends on whether they are birds kept for the production of food or for the purpose of use in the farming of land.

The case for the ratepayers is that these birds all finish their careers on the table, that they are raised on the land, so that even if the land does not qualify as agricultural land under the restricted definition in section 26(3) of the 1967 Act they nevertheless form part of a food production system carried on in an agricultural context. The reply of Mr Fletcher for the valuation officer is that they are not kept for the production of food but for sporting purposes, and the fact that the birds are eventually eaten is an incidental by-product. As to their use in the farming of land, Mr Fletcher's submission is that these words refer to farm animals such as carthorses or sheepdogs which would not otherwise qualify as livestock but quite clearly form an essential part of the farmer's workforce.

I have no difficulty in accepting Mr Fletcher's submissions on this last point. The pheasants and partridges in the instant cases are not kept for their use in the farming of land. The use to which the birds are to be put does not, in my judgment, contribute in any relevant sense to the farming of land. The other proposition I have found more difficult. Mr Curry's submission that the birds are kept for the production of food I find superficially attractive at first sight. A closer consideration, however, leads me to the conclusion that the pheasants and partridges in these appeals are not livestock in the statutory sense. It is difficult to see how the birds can be said to be kept when nearly all of them are in fact released into the wild.

Any doubts which might be entertained are effectively removed by a consideration of the case of the *Earl of Normanton v Giles* [1980] 1 WLR 28. This was a case where a gamekeeper reared pheasants for shoots, catching birds for laying, clipping their wings, putting them into pens and, after the incubation of the eggs, putting the poults into release pens which they subsequently left, becoming wild and available for shooting. Most of the birds killed were sold to gamedealers and butchers, the proceeds being used to maintain the shoots. All the birds killed and retrieved were consumed as food. After the termination of his employment the gamekeeper was required to give up possession of his cottage but refused to do so on the grounds that he was a protected occupier by virtue of section 2(1) of the Rent (Agriculture) Act 1976, which gave protection in certain circumstances to persons employed in agriculture. In section 1(1) of the Act agriculture included "livestock keeping and breeding", and in subsection (2) it was provided:

For the purposes of the definition in subsection 1(a) above — . . . "livestock" includes any animal which is kept for the production of food, wool, skins or fur, or for the purpose of its use in the carrying on of any agricultural activity, and for the purposes of this definition "animal" includes bird but does not include fish.

The case proceeded from the county court to the Court of Appeal and to the House of Lords, and at all stages the gamekeeper was unsuccessful in his action. On p 33 of the report Lord Wilberforce, after quoting from the passage in the speech of Lord Reid in the *Hardwick Game Farm* case to which I have already referred, went on to say:

Some similar observations appear in the speeches of Lord Guest and myself. I think that these reinforce the common impression that the rearing of pheasants for sport is *sui generis* and is not to be equated with the rearing of other livestock.

A little later Lord Wilberforce referred to a passage in the judgment of Scarman LJ, as he then was, in the Court of Appeal in *Lord Glendyne v Rapley* [1978] 1WLR 601, where he said at pp 603-604:

. . . not every rural or country activity is intended to be included in the definition of agriculture. Fishing, for example, is clearly excluded. The definition is really directed towards including all operations involved in farming land for commercial purposes of which the one relevant to this appeal is the production of food. The finding that these pheasants were kept for sport, though 80% of those killed and retrieved were in fact sold, is, in our judgment, conclusive. M Rapley's employment was to promote not agriculture but a field sport. This is a country activity but not an agricultural one.

Lord Wilberforce continued:

Apart from the reference to the "production of food", which to my mind is not the critical issue in this case, I agree with this passage.

Viscount Dilhorne considered the dictionary definition of livestock as: "Domestic animals generally; animals of any kind kept or dealt in for use or profit". At p 35 he went on:

I do not think that pheasants kept and reared for shooting, and which when sufficiently grown are released from captivity, can properly be regarded as "domestic animals" or as being kept for use or profit. They cease to be kept when they are let to go free. I do not therefore consider that the pheasants looked after by Mr Giles were within the definition of livestock in the dictionary and consequently the rearing and keeping of them does not come within the definition in the dictionary of agriculture.

On p 39 Lord Russell of Killowen said:

In my opinion "agriculture" per se is not a word appropriate to the preservation and breeding of pheasants for sporting purposes. Further the fact that in the Act "agriculture" includes "livestock keeping and breeding" cannot in my opinion per se be regarded in the context as embracing within the word "livestock" pheasants bred and kept for that purpose. I consider that the provisions of section 1(2) reinforce that last conclusion. These pheasants are in no sense birds "kept for the production of food". In so far as both home-bred pheasants and others are "preserved" in the woods on the estate this is done so that they may by way of sport be in due season shot at. The fact that those which are successfully shot at are for the most part sold and eaten is not the reason for their keeping, they are not "kept for" that. Accordingly in my opinion the provisions of subsection (2) do not extend "livestock keeping and breeding" in subsection (1)(a) to cover this case. If anything the stress in subsection (2) on "kept for the production of food" tends to confirm the view that the language of subsection (1)(a) does not embrace a case such as the present.

In *Normanton v Giles*, as in *Hardwick Game Farms*, it is true to say that the statutes under consideration are not those governing exemption from rates. Nevertheless, the inquiry carried on in both cases is concerned with the question of pheasants and partridges in an agricultural context, and I can see no reason why dicta expressed by the learned Law Lords in *Normanton v Giles* should not apply in the instant case. I therefore find that the pheasants and partridges at Whitehall Farm and Kirkby Lane are not livestock for the purposes of the Rating Act 1971 and that the buildings accordingly are not agricultural buildings. It also follows that the land cannot qualify as agricultural land under the 1971 Act in circumstances where there are no agricultural buildings.

Finally, I must return to the question of the buildings considered in the light of the definition in section 26(4) of the 1967 Act, if I am

A wrong in finding that the land is not agricultural land, as defined in section 26(3). In those circumstances, the buildings would clearly be occupied with agricultural land. For the reasons which I have given in rejecting the submissions made on behalf of the ratepayers on the interpretation of the provisions of the 1971 Act, I am of the opinion that the rearing of pheasants and partridges cannot be said to be an agricultural operation and in so far as that is the primary use to which the land is put the buildings cannot be said to be used solely in connection with agricultural operations on the land.

For reasons which I have already given, however, I do not consider that the land is agricultural land. It follows that the valuation officer's appeals succeed and the decisions of the local valuation court
B are reversed. The appropriate orders will be issued directing that entries be made in the valuation list in accordance with the terms provisionally agreed, namely: C

LVC/850/1979	Pheasantry and Premises, Kirkby Lane, Kirkby on Bain	gross value £500 rateable value £388
LVC/851/1979	Pheasantry, House, Garage and Premises, Whitehall Farm, Horncastle Road, Roughton.	gross value £680 rateable value £538

The respondents will pay the costs of the appellant valuation officer, such costs failing agreement to be taxed by the Registrar of the Lands Tribunal on the High Court Scale of Costs. D